Richard Sharpe.   Dec. 1993

Molecular Biology
of the Male
Reproductive System

Molecular Biology of the Male Reproductive System

Edited by

David de Kretser

Institute of Reproduction and Development
Monash University
Monash Medical Centre
Melbourne, Victoria
Australia

Academic Press, Inc.

A Division of Harcourt Brace & Company

San Diego   New York   Boston   London   Sydney   Tokyo   Toronto

Academic Press, Inc.
1250 Sixth Avenue, San Diego, California 92101-4311

United Kingdom Edition published by
Academic Press Limited
24–28 Oval Road, London NW1 7DX

Library of Congress Cataloging-in-Publication Data

Molecular biology of the male reproductive system / edited by D.M. de Kretser.
        p.   cm.
    Includes bibliographical references and index.
    ISBN 0-12-209030-6
    1. Testis--Molecular aspects.   I. De Kretser, D. M. (David M.)
QP255,M633 1993
612.6'1--dc20   93-17236
                                                          CIP

PRINTED IN THE UNITED STATES OF AMERICA
93   94   95   96   97   98   BB   9   8   7   6   5   4   3   2   1

Contents

1

Genetic Control of Testis Determination

V. R. Harley

2

Cell Biology of Testicular Development

Lauri J. Pelliniemi, Kim Fröjdman, & Jorma Paranko

3

Nuclear Morphogenesis during Spermiogenesis

Marvin L. Meistrich

4

Hormonal Control of Spermatogenesis

Gerhard F. Weinbauer & Eberhard Nieschlag

5

Patterns of Expression and Potential Functions of Proto-oncogenes during Mammalian Spermatogenesis

Martin A. Winer & Debra J. Wolgemuth

6

Gene Expression during Spermatogenesis

E. M. Eddy, Jeffrey E. Welch, & Deborah A. O'Brien

7

Molecular Basis of Signaling in Spermatozoa

Daniel M. Hardy & David L. Garbers

8

Paracrine Mechanisms in Testicular Control

B. Jégou & R. M. Sharpe

9

Molecular Biology of Iron Transport in the Testis

Steven R. Sylvester & Michael D. Griswold

10

Molecular Biology of Testicular Steroid Secretion

Peter F. Hall

11

Hormonal Control Mechanisms of Leydig Cells

Ilpo Huhtaniemi

12

Growth Factors in the Control of Testicular Function

David M. Robertson, Gail P. Risbridger, Mark Hedger,
& Robert I. McLachlan

13

Vascular Controls in Testicular Physiology

Anders Bergh & Jan-Erik Damber

Contributors

Numbers in parentheses indicate the pages on which the authors' contributions begin.

Anders Bergh (439), Department of Urology and Andrology, University of Umeå, S-901 87 Umeå, Sweden

Jan-Erik Damber (439), Department of Urology and Andrology, University of Umeå, S-901 87 Umeå, Sweden

E. M. Eddy (181), Gamete Biology Section, Laboratory of Reproductive and Developmental Toxicology, National Institute of Environmental Health Sciences, National Institutes of Health, Research Triangle Park, North Carolina 27709

Kim Fröjdman (21), Laboratory of Electron Microscopy, University of Turku, SF-20520 Turku, Finland

David L. Garbers (233), Howard Hughes Medical Institute, and Department of Pharmacology, Southwestern Medical Center, University of Texas, 5323 Harry Hines Boulevard, Dallas, Texas 75231

Michael D. Griswold (311), Department of Biochemistry and Biophysics, Washington State University, Pullman, Washington 99164

Peter F. Hall (328), Department of Endocrinology, Prince of Wales Hospital, Randwick, NSW 2031, Australia

Daniel M. Hardy (233), Howard Hughes Medical Institute, and Department of Pharmacology, Southwestern Medical Center, University of Texas, Dallas, Texas 75231

V. R. Harley (1), Department of Genetics, University of Cambridge, Cambridge CB2 3EH, England

Mark Hedger (411), Institute of Reproduction and Development, Monash University Medical Center, Clayton, 3168 Victoria, Australia

Ilpo Huhtaniemi (383), Department of Physiology, University of Turku, 20520 Turku, Finland

B. Jégou (271), Groupe d'Etude de la Reproduction chez le Mâle (G.E.R.M.), Université de Rennes I, 35042 Rennes Cedex, Bretagne, France

Robert I. McLachlan (411), Prince Henry's Institute of Medical Research, Monash Medical Center, Clayton, 3168 Victoria, Australia

Marvin L. Meistrich (67), Department of Experimental Radiotherapy, The University of Texas, M.D. Anderson Cancer Center, Houston, Texas 77030

Eberhard Nieschlag (99), Institute of Reproductive Medicine of the University, WHO Collaborating Centre for Research in Human Reproduction, Westfällischen Wilhelms University, D-4400 Münster, Germany

Deborah A. O'Brien (181), Gamete Biology Section, Laboratory of Reproductive and Developmental Toxicology, National Institute of Environmental Health Sciences, National Institutes of Health, Research Triangle Park, North Carolina 27709, and Laboratories for Reproductive Biology, and Departments of Pediatrics and Cell Biology and Anatomy, University of North Carolina at Chapel Hill, Chapel Hill, North Carolina 27599

Jorma Paranko (21), Department of Anatomy, University of Turku, SF-20520 Turku, Finland

Lauri J. Pelliniemi (21), Laboratory of Electron Microscopy, University of Turku, SF-20520 Turku, Finland

Gail P. Risbridger (411), Institute of Reproduction and Development, Monash University Medical Center, Clayton, 3168 Victoria, Australia

David M. Robertson (411), Prince Henry's Institute of Medical Research, Monash Medical Center, Clayton, 3168 Victoria, Australia

R. M. Sharpe (271), MRC Reproductive Biology Unit, Centre for Reproductive Biology, Edinburgh EH3 9EW, Scotland

Steven R. Sylvester (311), Department of Biochemistry and Biophysics, Washington State University, Pullman, Washington 99164

Gerhard F. Weinbauer (99), Institute of Reproductive Medicine, WHO Collaborating Centre for Research in Human Reproduction, Westfällischen Wilhelms University, D-4400 Münster, Germany

Jeffrey E. Welch (181), Gamete Biology Section, Laboratory of Reproductive and Developmental Toxicology, National Institute of Environmental Health Sciences, National Institues of Health, Research Triangle Park, North Carolina 27709

Martin A. Winer (143), The Center for Reproductive Sciences, Columbia University College of Physicians and Surgeons, New York, New York 10032

Debra J. Wolgemuth (143), The Center for Reproductive Sciences, and Departments of Genetics and Development, and Obstetrics and Gynecology, Columbia University College of Physicians and Surgeons, New York, New York 10032

Preface

The last decade has seen an explosion in our knowledge concerning the cellular and molecular mechanisms involved in the control of male reproduction. This volume brings the reader up to date with some of the major advances in these fields. The chapters have been written by acknowledged leaders in the field and provide the reader with the key elements in each of the areas, thereby providing a sound base for a further acquisition of knowledge by those who seek more detailed information. The volume covers the developments in the control of testis determination and the subsequent cell biology of testicular development. The process of spermatogenesis is dealt with from several angles, providing the reader with a basic understanding of the hormonal control of spermatogenesis and our rapidly expanding knowledge of the role of growth factors and proto-oncogenes in the control of this process. Although the details of all aspects of the cell biology of spermiogenesis are beyond the scope of this volume, the description of the factors controlling nuclear morphogenesis during spermiogenesis gives the reader an indication of the complexity of the process.

No modern description of testicular function would be complete without a discussion of our emerging knowledge of the paracrine interrelationships between the components of the testis. In particular, the function of the Sertoli cells and the potential control of their function by germ cells is a subject of one of the chapters of this volume. An expansion of one of these areas can be found in the chapter concerning the control of iron transport in the testis. Details of the mechanisms by which molecular signaling occurs in spermatozoa demonstrate the evolution of parallel systems through a range of species.

The dependence of spermatogenesis on testosterone secretion by the testis makes it necessary for any volume on testicular function to consider the molecular biology of the control of testosterone secretion. A chapter on this and one on the control systems in Leydig cells, particularly as they evolve during development, complete a picture of the hormonal control of spermatogenetic function. No organ can function without an adequate vascular supply, and the chapter concerning the control of the vasculature within the testis and its intriguing and novel features rounds off the volume.

1

Genetic Control of Testis Determination

V. R. HARLEY

I. Introduction

Jost (1947; Jost et al., 1973) showed that castrated rabbit embryos of either chromosomal sex develop as females, indicating that the presence of testes is necessary for the development of male characteristics. During embryogenesis, the gonad arises as an indifferent tissue, the fate of which follows either the testicular or the ovarian pathway. In mammals, this developmental switch is controlled chromosomally; females have two X chromosomes and males have an X and a Y chromosome. Individuals with a single X chromosome and no Y chromosome (that is, XO) are female (Ford et al., 1959). The presence of a Y chromosome results in male development, regardless of the number of X chromosomes the individual possesses (Jacobs and Strong, 1959; Table 1). Thus, the Y chromosome must encode a dominant inducer of testis formation. The Y-linked gene(s) controlling this process has been named the testis determining factor (TDF; Tdy in mouse). TDF is activated at some time during embryogenesis to commit the undifferentiated genital ridge to the testicular pathway. As a consequence of the action of TDF, subsequent hormonal production induces male sexual differentiation.

The gonad is composed of cells derived from four lineages: supporting cells, steroid-producing cells, connective tissue cells, and germ cells. The first signs of male development can be recognized in the genital ridges of mice at 12.5 days post coitum (dpc) when Sertoli cells organize into "testis cords." Female gonads show no apparent change until 13.5 dpc when the germ cells

TABLE 1　Chromosomal Basis of
Sex Determination[a]

Genotype	Phenotype
Female	
45, X	Turner's syndrome
46, XX	Normal
47, XXX	
48, XXXX	
49, XXXXX	
Male	
45, X/46, XY	
46, XY	Normal
47, XXY	Klinefelter's syndrome
48, XXYY	
49, XXXYY	

[a] The presence of the Y chromosome confers male-
ness, regardless of the number of X chromosomes.

first enter meiosis (McLaren, 1984). *Tdy* is thought to act solely in the
supporting cell lineage to divert the indifferent gonad away from the ovarian
pathway and into the testicular pathway, by triggering supporting cells to
differentiate into Sertoli cells rather than into the follicle cells of the ovary
(Palmer and Burgoyne, 1991a). Sertoli cells, without further *Tdy* involvement, then commit the fate of the steroid-producing cells to produce Leydig
cells and induce mitotic arrest in the germ cells (Burgoyne *et al.*, 1988). As in
other developmental processes, *Tdy* is presumed to act with other regulatory
molecules in the genital ridge to give rise to testes and subsequent male
development.

Among humans, rare individuals arise who carry two X chromosomes
but are phenotypically male (XX males) or carry a Y chromosome but are
phenotypically female (XY females). The molecular analysis of the genomes
of these so-called sex-reversed patients led to the isolation of the male
sex-determining region Y gene, *SRY*. Prior to this discovery, candidate Y-lo-
cated genes had been isolated that subsequently failed to meet established
criteria.

This chapter chronicles the identification of a new candidate for the testis
determining gene: *SRY*. Evidence, both circumstantial and direct, that *SRY* is
TDF is outlined. Emphasis is placed on biochemical properties of *SRY*. Detailed information on the biology of mouse *Sry* can be found elsewhere
(Capel and Lovell-Badge, 1992). The cell and tissue biological processes
during sexual differentiation of the gonad are discussed in subsequent
chapters.

II. *SRY*, a New Candidate for the Testis Determining Factor

A. Mapping the Human Y Chromosome

The approach that proved successful for the isolation of *TDF* was deletion mapping of the Y chromosome. Almost 60 years ago, Koller and Darlington (1934) observed that X and Y rat chromosomes were dissimilar but could pair along part of their lengths. These researchers proposed that the Y chromosome was composed of a shared region and a Y specific region. The shared or "pseudoautosomal" region (PAR) is terminal on the short arms of the X and Y chromosomes (Pearson and Bobrow, 1970) and would be required for correct pairing during male meiosis. The Y-specific region would encode *TDF*. Recombination in the PAR region would maintain homology between the X and Y shared regions but recombination should not occur in the Y-specific region (Ferguson-Smith, 1966; Fig. 1). Although cytogenetic

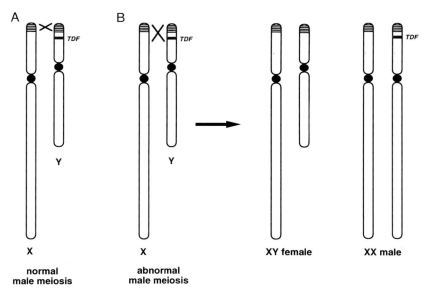

FIGURE 1   Human X and Y chromosomes. The solid black circle is the centromere, the shaded region is the pseudoautosomal region (PAR), which is the same on both chromosomes, and the solid black line represents the testis determining factor, *TDF*. Normally, during male meiosis a reciprocal recombination occurs between the X and Y chromosomes that is wholly within the PAR. Occasionally, abnormal crossing over occurs and includes the testis determining factor outside the PAR. This gives rise to two abnormal chromosomes: a Y lacking *TDF* and an X containing *TDF*. If these chromosomes are transmitted to the next generation, they give rise to an XY female and an XX male, respectively.

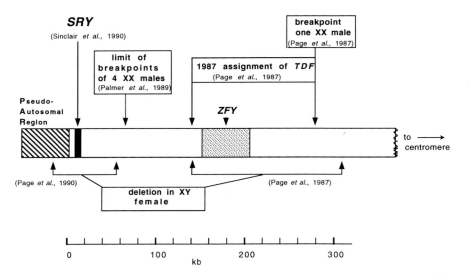

FIGURE 2 Localization of the sex determining region gene *SRY* on the Y chromosome. The zinc finger gene *ZFY*, an earlier candidate for TDF, is shown as is the limit of the breakpoints observed in four 46XX males (Palmer *et al.*, 1989). The breakpoints in one XX male and one XY female analyzed by Page *et al.* (1987) are indicated also. Note that this second deletion, which includes *SRY*, subsequently was detected in this 46XY female (Page *et al.*, 1990).

analysis of individuals with Y chromosome deletions suggested that *TDF* was located on the short arm (Goodfellow *et al.*, 1985), precise localization was achieved by studying XX males.

Analysis of the genomes of some XX males showed the presence of Y-derived sequences (Guellen *et al.*, 1984) that presumably arose by aberrant recombination between the X and Y chromosomes (Ferguson-Smith, 1966). Different XX males inherited different terminal fragments, which allowed the construction of a deletion map of the Y chromosome (Affara *et al.*, 1986; Muller *et al.*, 1986; Vergnaud *et al.*, 1986). These maps began to define the minimum region required for male sex determination (Page *et al.*, 1987; Palmer *et al.*, 1989).

B. *ZFY*

Page *et al.* (1987) further narrowed the region in which *TDF* must lie as between 140 and 280 kb from the PAR boundary, because an XX male had inherited only 280 kb of Y-derived sequence and part of this region was deleted in an XY female patient with a Y;22 translocation (Fig. 2). On cloning this region and screening by Southern analysis for Y-specific sequences among eutherian mammals, a gene dubbed ZFY was identified. Since ZFY could encode a zinc finger protein and therefore might be a transcription factor, and since mouse Y homologs (*Zfy-1* and -2) mapped to a small region known to contain *Tdy* (McLaren *et al.*, 1988; Roberts *et al.*, 1988; Mardon *et al.*, 1989; Nagamine *et al.*, 1989), ZFY seemed a good candidate for *TDF*.

However, inconsistencies arose. First it was shown by Sinclair *et al.* (1988) that the genes homologous to ZFY were not on the marsupial Y chromosome. Further, Koopman *et al.* (1989) showed that, although *Zfy-1* was expressed in the fetal testis of normal mice, expression was absent in homozygous WW mice, which have normal testes but lack germ cells. Finally, Palmer *et al.* (1989) found four XX males/intersexes who carried less than 60 kb of Y-specific DNA derived from the region adjacent to the PAR boundary but lacked ZFY in their genomes. Collectively, these data firmly rule out ZFY as *TDF*. The variation in sexual phenotype was suggested to be caused by the proximity of the Y breakpoints to *SRY* (Palmer *et al.*, 1989).

C. Discovery of *SRY*

Palmer *et al.* (1989) found that different sized DNA fragments were detected from the four ZFY-negative XX patients when hybridized to a probe 35 kb from the PAR boundary. This result implied that *TDF* must lie within 35 kb of the PAR boundary. Of the 50 DNA probes prepared from this region, only one probe recognized a Y-specific fragment on blots that was conserved among all eutherian DNA tested (Sinclair *et al.*, 1990). This probe is derived from a fragment located 5 kb proximal to the PAR boundary and defines the gene *SRY* (sex-determining region Y-linked gene). More recently, Page *et al.* (1990) reported that the ZFY-negative X,t(Y;22) female described previously (Page *et al.*, 1987) also lacked the region containing *SRY* (Fig. 2). *Sry*, the mouse homolog, mapped to the Sxr' region, the smallest part of the mouse Y known to be testis determining. Further, *Sry* is deleted from a mutant Y chromosome that has lost its sex determining activity (Gubbay *et al.*, 1990,1992).

DNA sequence comparisons of the conserved regions from human, rabbit, and mouse indicated the presence of an open reading frame that shares considerable homology over a 79-amino-acid region to a recently identified DNA binding motif dubbed "HMG box" (Fig. 3). HMG box-containing proteins form a diverse family of DNA binding proteins with a variety of DNA binding properties: some appear to be chromatin-associated, a second group are involved in rRNA gene transcription, and a third group, which probably includes SRY, play a role as tissue specific Pol II transcription factors.

The open reading frame for the human *SRY* gene appears to be within a single exon and could encode a 204-amino-acid protein (Fig. 3). The structure of the mouse transcript is not yet clear and is flanked, in the genome, by a

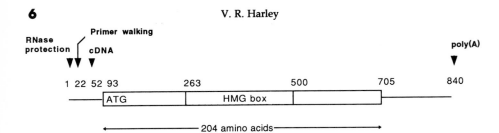

FIGURE 3 *SRY* gene structure. Ambiguity remains about the precise 5′ end of the *SRY* transcript. RNase protection has defined the furthest 5′ end of the transcript at base 1 whereas primer walking suggests its location to be base 22. The cDNA clone isolated is colinear with the genomic sequence, starts 5′ at base 22, and extends to the polyadenylation (poly(A) site at base 840. The known coding region starts at base 93 (defined by an initiating methionine codon, ATG), ends at base 705, and encodes 204 amino acids. In the middle of the open reading frame is the conserved HMG box extending from base 263 to base 500.

large inverted repeat extending at least 15.5 kb on either side (Gubbay *et al.*, 1992). The amino acid identity between the human, mouse, and rabbit SRY sequences outside the HMG box region is less conserved (Sinclair *et al.*, 1990), suggesting that DNA binding activity might be the only biochemical function of this protein. The HMG box of SRY from several species has been sequenced and reveals conservation (about 50% overall amino acid identity; Fig. 4) between eutherian and metatherian mammals, consistent with a common Y-linked sex determining mechanism (Foster *et al.*, 1992).

D. Direct Evidence that *SRY* Is *TDF*

Since the isolation of *SRY* in 1990, several pieces of evidence have been gathered that provide proof that *SRY* can be equated with *TDF*.

1. *Sry* expression during testis differentiation

Northern blot analysis detects *SRY* transcripts of ~1.2 kb in the adult testis and in no other adult tissue in humans and mice (Gubbay *et al.*, 1990; Sinclair *et al.*, 1990 Fig. 3). *Sry* expression is germ cell dependent in the adult mouse. Adult XXSxr and XXSxr′ mice lack germ cells and fail to express *Sry* in their testes (Koopman *et al.*, 1990). Adult expression of *Sry* is thought to occur directly in postmeiotic germ cells (possibly in round spermatids) rather than in a somatic lineage that depends on feedback from the germ cells (Koopman and Gubbay, 1991). The function, if any, of SRY in adult testis remains unclear; one possibility is involvement in the maturation of spermatids into spermatozoa.

Sry transcripts are detectable in mouse embryos by the highly sensitive reverse transcriptase polymerase chain reaction (RT-PCR) method and by *in situ* hybridization. *Sry* expression is apparent in genital ridge cells at 10.5 dpc, which corresponds to the first appearance of the urogenital ridge, and also at

```
           58  L   G   I                            T                    M       R
Human          D R V K R P M N A F   I V W S R D Q R R K   M A L E N P R M R N   S E I S K Q L G Y Q
Mouse          G H V K R P M N A F   M V W S R G E R H K   L A Q Q N P S M Q N   T E I S K Q L G C R
Rabbit         E R V K R P M N A F   M V W S Q H Q R R Q   V A L E N P K M R N   S D I S K Q L G H Q
Wallaby        D R V K R P M N A F   M I W S R Q R R K     V A L E N P K M H N   S E I S K H L G F T
M.mouse        S R V K R P M N A F   M V W S Q T Q R R K   V A L Q N P K M H N   S E I S K Q L G V T
Sheep                        A F     I V W S R E R R R K   V A L E N P K L Q N   S E I S K Q L G Y E
               * * * * * * * *       * *         *         *     * *     *   *     * * *     * *


                              I       S                                         W       138
Human          W K M L T E A E K W   P F F Q E A Q K L Q   A M H R E K Y P N Y   K Y R P R R K A K M
Mouse          W K S L T E A E K R   P F F Q E A Q R L K   I L H R E K Y P N Y   K Y Q P H R R A K V
Rabbit         W K M L S E A E K W   P F F Q E A Q R L Q   A M H K E K Y P D Y   K Y R P R R K V K I
Wallaby        W F M L P D N E K Q   P F I D E A E R L R   A K H R E E F P D Y   K Y Q P R R K K F
M.mouse        W E I L S D S E K I   P F I D E A K R L R   D K H K - Q V S D Y   K Y Q P R R
Sheep          W K R L T D A E K R   P F F E E A Q R L L   A I H R D K Y P
               *       *     * *     * *     * *     *         *             * *   * *     *   *
```

FIGURE 4 SRY HMG box from six mammals and point mutations in XY females. Amino acids constituting the HMG box region of SRY from human and rabbit (Sinclair *et al.*, 1990), mouse (Gubbay *et al.*, 1990), wallaby and marsupial mouse (Foster *et al.*, 1992), and sheep (c. Cotinot, unpublished results). Asterisks below the sequences denote overall identity. Letters above the sequences denote point mutations in human SRY that cause sex reversal (see Section II,D).

11.5 dpc. By 12.5 dpc *Sry* is still detectable, although at lower levels (Koopman *et al.*, 1990; Fig. 5). Although the chromosomal sex is established at fertilization, no sex specific differences are apparent until some point between 11.5 and 12.5 dpc, when the male gonad takes on a striped appearance thought to arise when cells of the supporting cell lineage differentiate into Sertoli cells and align into testis cords. The temporal expression profile of *Sry* meets the expectation that *Tdy* should be expressed just prior to the overt differentiation of the genital ridge. At later stages of testis growth, *Sry* is not detected (Gubbay *et al.*, 1990; Koopman *et al.*, 1990).

The type of cell in the genital ridge that expresses *Sry* has not been established unequivocally. Clearly germ cells are not required for testis determination. Rat fetuses whose germ cells have been destroyed by busulfan treatment still form normal testes (Merchant, 1975). Mice homozygous for the W (white spotting) mutation lack germ cells (McLaren, 1985) yet somatically normal testis are able to form. Genital ridges from 11.5 dpc W^{e}W^e mouse embryos were shown to have *Sry* expression levels comparable to those of genital ridges from normal mice (Koopman *et al.*, 1990). Therefore, *Sry* is expressed in somatic cells of the genital ridge. Analysis of chimeric mice constructed by aggregation of male and female embryos suggests that Sertoli cell precursor cells are the site of *Sry* expression; the sex of the chimera is dependent on the number of Sertoli cells (Singh *et al.*, 1987), almost all of which bear a Y chromosome (Burgoyne *et al.*, 1988).

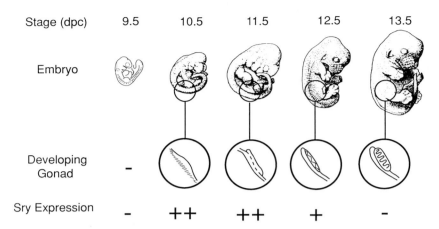

Stage (dpc)	9.5	10.5	11.5	12.5	13.5
Embryo					
Developing Gonad	−				
Sry Expression	−	++	++	+	−

FIGURE 5 *Sry* expression during gonadal development in the mouse embryo. The time course of *Sry* expression in relation to the schedule of gonad differentiation in the mouse. The primordial gonad, or genital ridge, is first seen at 10.5 days post coitum (dpc) as a swelling of epithelium; by 11.5 dpc, this swelling has differentiated into separate genital ridge and mesonephric components. Males and females are morphologically identical at this stage. Alignment of Sertoli cell precursors into cords (seen as stripes) indicates differentiation of the testis in male embryos by 12.5 dpc. The transcriptional activity of *Sry* is shown beneath each stage of testis development. Reproduced with permission of W. B. Saunders, Orlando, Florida.

2. Transgenic mouse studies

To establish that *Sry* is the only gene on the Y chromosome necessary for testis determination, constructing mice transgenic for *Sry* was necessary. Fertilized mouse eggs were injected with a 14-kb genomic fragment carrying *Sry* (Koopman *et al.*, 1991). The rationale for injecting such a large fragment was (1) to allow for the possibility of undiscovered exons and (2) to provide the regulatory sequences for correct spatial and temporal expression. After 14 days *in utero,* embryos were sexed by gonad morphology and their chromosomal sex was verified by the presence of sex chromatin in amnion cells and by Southern analysis. In 25% of transgenic XX embryos, testicular development was found. The failure of the 14-kb *Sry* fragment to cause sex reversal in most cases can be explained by (1) disruption of *Sry* expression due to position effects or (2) mosaicism, since 4 of the 7 transgenic XX female embryos had fewer than one copy of *Sry* per cell.

Of the embryos allowed to come to term, one transgenic mouse, m33.13, is phenotypically male but chromosomally female. This mouse displays normal copulatory behavior and the only external anomaly is that his testes are small. Histological examination of his testis revealed normal tubules and the presence of all somatic cell types—Sertoli, Leydig, and peritubular myoid cells. However, spermatogenesis is absent. This phenotype is identical to that of XXSxr or XXsxr' male mice, for whom the presence of two X chromosomes is not compatible with spermatogenesis. DNA sequence analysis of the 14-kb region fails to identify any genes other than *Sry*, leading to the conclusion that the *Sry* gene alone is able to initiate formation of testes and subsequent male development (Koopman *et al.*, 1991).

Not all XX transgenic mice show sex reversal. Founder mouse m32.10 had multiple copies of the *Sry* transgene and was a fertile female. This mouse was bred and *Sry* expression was examined in the progeny. *Sry* was expressed in these embryos and some of the XX progeny developed as males, suggesting that the biochemical activity of *Sry* in this transgenic line is near its threshold level (N. Vivian, P. Koopman, and R. Lovell-Badge, unpublished results).

3. *SRY* mutations in XY females

If *SRY* is required for testis formation, mutations in *SRY* would be predicted to be associated with male to female sex reversal. XY females are a rare class of individuals with an XY karyotype but no testes. Such patients have mutations in the sex determination pathway or in early stages of testis differentiation. On detailed histological investigation of gonadal material, patients can be classified into two groups: (1) pure or complete gonadal dysgenesis, in which patients have dysgenic or "streak" ovaries identical to those in individuals with Turner syndrome (XO females) and (2) mixed or partial gonadal dysgenesis, in which some testicular material is present in the gonads.

The coding region of *SRY* has been amplified by PCR from 41 XY females and analyzed by DNA sequencing (Berta *et al.*, 1990; Jager *et al.*, 1990; Harley *et al.*, 1992; Hawkins *et al.*, 1992a,b; McElreavey *et al.*, 1992). In 23 cases, sequencing was preceded by the single-strand conformation polymorphism (SSCP) assay, which detects mutations in large stretches of DNA. Six patients had mutations in *SRY* and, interestingly, no sequence variation was detected in the *SRY* HMG box from over 100 normal males. Four of these mutations were *de novo* whereas two mutations are shared by XY female and male members of the same pedigree. All mutations cluster in the HMG box (79 of the 204 amino acids of SRY). The simplest explanation for these findings is that non-box mutations are subclinical.

Hawkins *et al.* (1992b) have partitioned, after careful histological examination, five patients with complete gonadal dysgenesis and found that three had *SRY* mutations. Therefore, *SRY* mutations appear to be more likely to occur in patients with pure rather than partial gonadal dysgenesis. Clearly, mutations in *SRY* in XY females provide strong genetic evidence that *SRY* is necessary for testis formation. One would predict from the site of these mutations that the DNA binding motif is likely to function *in vivo*; biochemical evidence confirming this prediction is considered in Section III.

E. *SRY* in True Hermaphrodites

Hermaphroditism can be caused by XX/XY chimerism as a result of double fertilization of an ovum and its polar body. Most hermaphrodites with a normal XX female karyotype have no Y-derived sequences present in their genome. Individuals with both testicular and ovarian development are termed true hermaphrodites. The form of gonadal differentiation is dependent on the proportion of testis determining and ovary determining cells. Many XX/XY chimeras show normal male differentiation, suggesting that testis determination is predominant; such patterns of sex differentiation also are observed in artificial XX/XY mouse chimeras (Palmer and Burgoyne, 1991a).

In one case (Berkovitz *et al.*, 1992) of XX true hermaphroditism, Yp-specific sequences including *SRY*, *RPS4Y*, and *ZFY* were exchanged and associated with part of a deletion of the X. The interchanged X was inactivated preferentially so the partial sex reversal could have been caused by variable spread of inactivation into the Y segment. In three other cases (Palmer *et al.*, 1989), only a small part of the Y including *SRY* had been transferred, so random X inactivation affecting the *SRY* locus may have been responsible for the presence of both testicular and ovarian development.

III. Biochemical Properties of SRY

A. SRY Is an "HMG Box" Protein with Sequence- and Structure-Specific DNA Binding Activity

The high mobility group protein HMG1 is a nonhistone nuclear protein with an acidic region and two homologous regions of about 80 amino acids that were recognized as novel DNA binding domains by analysis of the RNA Pol I transcription factor hUBF and named the "HMG box" (Jantzen *et al.*, 1990). Several HMG box proteins have been identified since, including the fungal mating type genes *Mat-Mc* of *Schizosaccharomyces pombe* (Kelly *et al.*, 1988) and *MT a-1* of *Neurospora crassa* (Staben and Yanofsky, 1990) and the mitochondrial transcription factor mtTF1 (Parisi and Clayton, 1991). The average amino acid sequence homology between individual HMG boxes is roughly 25% (40% similarity); proteins with one, two, or four HMG boxes have been reported (Fig. 6). Some HMG box factors show little DNA sequence specificity: UBF binds to a GC-rich stretch in the promoter of rRNA genes (Jantzen *et al.*, 1990) whereas a second class of factors, including HMG1 and NHP6, shows DNA binding that is relatively sequence independent but shows some preferential binding to single stranded DNA (Butler *et al.*, 1985; Kolodrubetz, 1990).

A third class of proteins contains only a single HMG box that shows significant DNA sequence specificity and high affinity in its interaction with DNA. The presence of a single HMG box in the sequence of SRY suggests

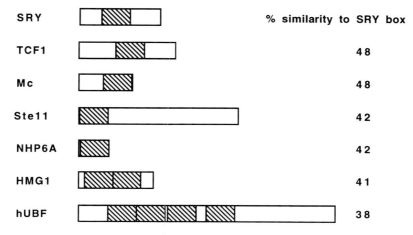

FIGURE 6 Domain structure of SRY and other HMG box proteins. Assigned HMG boxes are shaded. The degree of similarity (%) between SRY and homologs is calculated based on the amino acid groupings K,R; E,D; Q,N; S,T; F,Y,W,H; and L,I,V,A. References for sequences are as follows: SRY, Sinclair *et al.* (1990); TCF1, van de Wettering *et al.* (1991); Mc, Kelly *et al.* (1988); Ste 11, Sugimoto *et al.* (1990); NHP6A, Kolodrubetz (1990); HMG1, Wen *et al.* (1990); hUBF, Jantzen *et al.* (1990).

that it may belong to this group. For several members of this group, the *in vivo* target DNA sequences have been identified (Table 2). For example, TCF-1 originally was cloned based on its affinity to the AACAAAG motif in the CD3ε enhancer (van de Wettering *et al.*, 1991). A highly related transcription factor (93% amino acid identity in the HMG box) was cloned and termed TCF-1α or LEF-1, with binding specificity very similar, if not identical, to that of TCF-1 (Giese *et al.*, 1991; Travis *et al.*, 1991; Waterman *et al.*, 1991). Remarkably, ste11, a yeast HMG box factor with only 25% amino acid identity to TCF-1, also binds the sequence AACAAAG present in the mating type genes *matP, matM,* and *mei2* (Sugimoto *et al.*, 1990). Analogous to ste11, SRY, with only 26% amino acid identity to TCF-1, recognizes the TCF motifs AACAAAG and TTCAAAG in gel retardation assays (Nasrin *et al.*, 1991; Harley *et al.*, 1992). Binding is sequence specific, since point mutations in certain positions in the site abolish binding. Unlike HMG-1, SRY shows no affinity for single-stranded DNA (Harley *et al.*, 1992).

Using an iterative selection assay (Pollock and Treisman, 1990), we have derived a consensus DNA binding site of A/TAACAAT for SRY from a random oligonucleotide pool (V. Harley, R. Lovell-Badge, and P. Goodfellow unpublished results). This sequence is similar, although not identical, to the TCF-1/ste11 target sequence. We find that the 7th position of the original heptamer sequence does not contribute to binding in a sequence-dependent manner. Our consensus sequence is consistent with footprinting data for SRY that suggest that the −1 position and not the 7th position contacts the

TABLE 2　HMG Box Proteins and Their DNA Sequence Specificities

Protein	Target DNA sequence	Reference
Ste11	AACAAAGAA TR boxes of *matP*, *matM*, and *mei2*	Sugimoto *et al.* (1990)
TCF-1	AACAAAG Enhancer of *CD3ε*	van de Wettering *et al.* (1991)
TCF-1α (LEF1)	CAAAG$^{G}_{A}$ Enhancers of *TCR-α, -β, -δ*	Waterman *et al.* (1991); Travis *et al.* (1991)
IRE-ABP	TTCAAAGG Insulin responsive element of *GAPDH*	Nasrin *et al.* (1991)
SRY	$^{AA}_{CT}$CAAA$^{G}_{T}$	Nasrin *et al.* (1991); Harley *et al.* (1992)
ROX-1	AACAAAG *ANB1* promoter	Lowry *et al.* (1990)

protein (van de Wettering and Clevers, 1992). The DNA consensus that we derived for SRY matches that for a recently reported autosomal *SRY*-like gene termed *SOX-5* (Denny *et al.*, 1992b), further illustrating that members of this subfamily of HMG box proteins tend to bind the same DNA sequences. New HMG box proteins are being discovered at a rapid rate; autosomal *SRY*-like *SOX* genes have been described in the mouse (Gubbay *et al.*, 1990), rat (Nasrin *et al.*, 1991), human (Denny *et al.*, 1992a), *Drosophila* (Denny *et al.*, 1992a), and bird (Griffiths, 1991). At present, how a vast array of HMG proteins, both closely and distantly related to SRY in amino acid sequence, interacts *in vivo* with their cognate motifs without interference is unclear, unless their temporal and/or spatial expression is regulated strictly.

Some HMG proteins also appear to have structural requirements for DNA that may be as important as sequence-specific requirements. Four-way junctions or "cruciform" DNA can be constructed from synthetic oligonu-cleotides; these structures are thought to resemble the node or crossover region in a loop of DNA. Bianchi *et al.* (1989) identified a protein from rat liver nuclei that bound to cruciform DNA; they identified this protein as HMG1. Ferrari *et al.* (1992) showed that SRY, like HMG-1, can recognize cruciform DNA. In addition, HMG1 failed to bind linear DNA, so the se-quence-specific binding to linear DNA appears to be a unique property of the SRY/TCF-1/ste11 class of HMG box factors. For this class, cruciform struc-tures may resemble bent or looped DNA intermediates that are presumed to exist in the cell during transcription. Giese *et al.* (1992) and Ferrari *et al.* (1992) showed that SRY can bend DNA almost back on itself. Such proteins may activate transcription by bringing regulatory molecules adjacent to SRY on the DNA into close proximity, a phenomenon observed in the bacterial integration host factor (IHF) protein (Giese *et al.*, 1992).

B. DNA Binding Activity of Recombinant SRY from XY Females

When the *SRY* genes of human, mouse, rabbit, tammar wallaby, marsupial mouse, and sheep are aligned, all point mutations of *SRY* found in human XY females fall on amino acid residues that are completely conserved among species (Fig. 4). This observation prompted us to test whether these mutations altered the *in vitro* DNA binding activity of SRY. Recombinant SRY proteins encoded by XY females were produced in bacteria and tested for DNA binding activity in an electrophoretic gel mobility shift assay. In all cases, DNA binding activity is abolished or reduced (Nasrin *et al.*, 1991; Harley *et al.*, 1992). This result provided strong evidence that the observed mutations were responsible for sex reversal in these patients.

In one familial case (190M), DNA binding activity is detectable, which is consistent with the partial penetrance of the mutation. XY sex reversal with variable penetrance is seen in mice when the POS Y chromosome is crossed into a B6 inbred background. The POS Sry protein, which differs by one amino acid from B6 Sry, is thought to be late acting so its action is preempted by the process of ovary determination (Palmer and Burgoyne, 1991b). In the second familial case, DNA binding activity is abolished completely; the simplest explanation of this result is that our *in vitro* assay is more stringent than normal *in vivo* conditions (Harley *et al.*, 1992). More recently, a third familial case carrying the mutation F109S has been analyzed and found to produce normal DNA binding activity (Jager *et al.*, 1993). These results imply that the F109S mutation is either a neutral sequence variant or produces an SRY with slightly altered *in vivo* activity. The resulting sexual phenotype depends on the genetic background or environmental factors.

Clearly SRY has a sequence-specific DNA binding activity that is necessary for male sex determination. What the consequence is of SRY binding of DNA remains unknown—it is likely to exert its regulatory effect during transcription of downstream genes in the male determining pathway.

C. Possible SRY Function *in Vivo*

Singh and Dixon (1990) have used inhibition of transcription in known genes and relief of inhibition with exogenously added HMG1 or -2 to demonstrate an absolute requirement for these HMGs for class II transcription. Other studies *in vitro* of several HMG box proteins (HMG1, LEF-1, ste11) clearly have shown stimulation of transcription (Watt and Molloy, 1988; Sugimoto *et al.*, 1990; Giese *et al.*, 1991). On the other hand, the rox-1 protein of *Saccharomyces cerevisiae* has an HMG box and regulates gene expression negatively (Lowry *et al.*, 1990). For SRY, no overt activation domains (Pro-rich, Gln-rich, acidic) are apparent (Johnson and McKnight, 1989), suggesting that this protein exerts its action *in vivo* solely through its HMG box.

LEF-1 was shown to substitute for IHF in a DNA bending-dependent *in vitro* recombination assay (Giese *et al.*, 1992). Thus, SRY may act purely as a structural element to allow the assembly of multiprotein complexes. Since SRY is expressed in a strict cell-type and temporal manner, its DNA bending action may bring ubiquitous factors into proximity to produce an SRY-mediated transcription complex. Alternatively, such bending may derepress downstream genes. A third hypothesis is that SRY competes with a second factor (possibly an HMG box factor) that shares the same binding site. Direct contact of SRY with protein factors cannot be ruled out; on binding of SRY to DNA, a regulatory protein may contact SRY or adjacent DNA sequences to affect transcription. Such is the case for components of the SL-1 complex, which contact one or more of the HMG boxes of hUBF and mediate Pol I transcription (Learned *et al.*, 1986).

IV. Perspectives and Conclusions

Not all cases of sex reversal can be explained by alterations of SRY. Gain-of-function and loss-of-function mutations, as well as gene dosage of other genes in the sex determination pathway, may affect testis determination. A major task facing current researchers is finding other genes in the sex determination pathway. Some evidence from karyotype–phenotype correlation exists for the presence of X and autosomal loci involved in testis formation.

About 20% of XX males lack Y-chromosome-derived material (Palmer *et al.*, 1989). Most of these patients have ambiguous genitalia. The molecular defect in these patients could be explained by the constitutive activation of an X-linked gene, normally regulated by SRY, or perhaps by a mutation in an *SRY*-like gene, causing molecular mimicry.

Several XY female patients have been described with duplications in the Xp region onto the X (Bernstein *et al.*, 1980; Scherer *et al.*, 1989; Stern *et al.*, 1990), the Y (Ogata *et al.*, 1991), or the 16th chromosome (May *et al.*, 1991). Analysis of these patients, in conjunction with four cases of Xp duplications in XY males that do not cause sex reversal (Narahara *et al.*, 1979; Langkjaer, 1982; Bardini *et al.*, 1991) suggests that the region Xp21.3-p22.11 is critical for sex determination (Fig. 7). If the presence of two active copies of the Xp distal region correlates with impaired testis formation, then a gene normally subject to X inactivation may exist that, when present in two active copies, disturbs the testis determination process (Ogata *et al.*, 1992). This model is

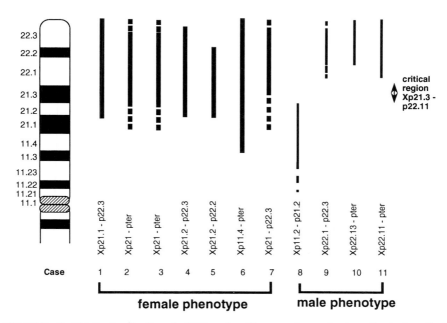

FIGURE 7 Partial Xp duplications in XY females. Case numbers refer to patients described by Ogata *et al.* (1991; case 1), Bernstein *et al.* (1980; case 2, 3), Scherer *et al.* (1989; case 4, 5), May *et al.* (1991; case 6), Stern *et al.* (1990; case 7), Neilsen and Langkjaer (1982; case 8), Narahara *et al.* (1979; case 9, 10), and Bardini *et al.* (1991; case 11).

consistent with patients with only one active copy of the gene, for example, 47 XXY and 48 XXXY males who masculinize as do normal 46 XY males. Possibly some XY females have submicroscopic duplications of Xp sequences that lead to sex reversal. Genetic mapping of the Xp21.3-p22.11 interval (about 10 Mb) and characterization of the duplications in XY females may lead to the cloning of the Xp testis determining gene.

Eicher and Washburn (1986) suggested that several autosomal genes may control a pathway for testis formation. The *Tda* autosomal locus affecting testis formation in the mouse has been assigned to chromosome 4, part of which is homologous to human chromosome 9p. To date, four cases of ambiguous genitalia in association with a *de novo* translocation involving 9p have been reported. All four showed a deletion of 9p with concurrent duplication of 2p (Hoo *et al.*, 1989), 3p (Fryns *et al.*, 1986), 7q (Crocker *et al.*, 1988), or 13q (Lotterand and Juillard, 1976). The shortest overlap with respect to the deletion is 9p24. Hoo *et al.* (1989) postulated that, since most 9p deletion cases do not show sex reversal, a recessive gene might be present on 9p24 and the lack of the gene product might cause delayed or incomplete testis formation.

The next stage in the pathway to testis formation could be the locus regulated by the *TDF* product. Analysis of the genomes of patients with sex reversal, as done for the discovery of *SRY*, may lead to the discovery of other testis determining genes. Alternatively, if an appropriate tissue culture system were established, biochemical studies of SRY action *in vitro* may lead the way by identifying genes affected by SRY action. At this stage, however, we do not know the action of SRY, in which cell type it is expressed, or whether the protein is an activator or a repressor of transcription.

Acknowledgments

I thank Peter Goodfellow, Robin Lovell-Badge, and Tom Ogata for critical reading of the manuscript. For results ahead of publication, I thank Corinne Cotinot and Robin Lovell-Badge. Figure 7 was compiled by T. Ogata. I am grateful to June Hunt for secretarial assistance.

References

Affara, N. A., Ferguson-Smith, M. A., Tolmie, J., Kwok, K., Mitchell, M., Jamieson, D., Cooke, A., and Florentin, L. (1986). Variable transfer of Y specific sequences in XX males. *Nucl. Acids Res.* **14**, 5375–5387.

Bardoni, B., Zuffardi, O., Guioli, S., Ballabio, A., Simi, P., Cavalli, P., Grimaldi, M. G., Fraccaro, M., and Camerino, G. (1991). A deletion map of the human Yq11 region: Implications for the evolution of the Y chromosome 3 and tentative mapping of a locus involved in spermatogenesis. *Genomics* **11**, 443–451.

Berkovitz, G. D., Fechner, P. Y., Marcantonio, S. M., Bland, G., Stetten, G., Goodfellow, P. N., Smith, K. D., and Midgeon, C. J. (1992). The role of the sex determining region of the Y chromosome (SRY) in the etiology of 46, XX true hermaphroditism. *Hum. Genet.* **88**, 411–422.

Bernstein, R., Koo, G. C., and Wachtel, S. S. (1980). Abnormality of the X chromosome in human XY female siblings with dysgenic gonads. *Science* **207**, 768–769.

Berta, P., Hawkins, J. R., Sinclair, A. H., Taylor, A., Griffiths, B. L., Goodfellow, P. N., and Fellous, M. (1990). Genetic evidence equating *SRY* and the testis determining factor. *Nature (London)* **348**, 448–450.

Bianchi, M. E., Beltrame, M., and Paonessa, G. (1989). Specific recognition of cruciform DNA by nuclear protein, HMG 1. *Science* **243**, 1056–1059.

Bianchi, M. E., Faliola, L., Ferrari, S., and Lilley, D. M. J. (1992). The DNA binding site of HMG1 protein is composed of two similar segments (HMG boxes), both of which have counterparts in other eukaryotic proteins. *EMBO J.* **11**, 1056–1063.

Burgoyne, P. S., Buehr, M., Koopman, P., Rossant, J., and McLaren, A. (1988). Cell-autonomous action of the testis-determining gene: Sertoli cells are exclusively XY in XX-XY chimeric mouse testes. *Development* **102**, 443–450.

Butler, A. P., Mardian, J. K. W., and Olins, D. E. (1985). Nonhistone chromosomal protein HMG 1 interactions with DNA. *J. Biol. Chem.* **19**, 10613–10620.

Capel, B., and Lovell-Badge, R. The *Sry* gene and sex determination in mammals. Advances in Developmental Biology. Vol. 2 (P. Wasserman, Ed.) J. A. I. Press Inc. Greenwich, CT. In Press.

Crocker, M., Coghill, S. B., and Cortinho, R. (1988). An unbalanced autosomal translocation (7;9) associated with feminisation. *Clin. Genet.* **34**, 70–73.

Denny, P., Swift, S., Brand, N., Dabhad, N., Barton, P., and Ashworth, A. (1992a). A conserved family of genes related to the testis determining gene, SRY. *Nucl. Acids Res.* **20**, 2887.

Denny, P., Swift, S., Connor, F., and Ashworth, A. (1992b). A testis specific transcript related to SRY encodes a sequence specific DNA binding protein. *EMBO J.* 2705–3712.

Diffley, J. F. X., and Stillman, B. (1991). A close relative of the nuclear, chromosomal high mobility group protein HMG1 in yeast mitochondria. *Proc. Natl. Acad. Sci. U.S.A.* **88**, 7864–7868.

Eicher, E. M., and Washburn, L. L. (1986). Genetic control of primary sex determination in mice. *Ann. Rev. Genet.* **30**, 327–360.

Eicher, E. M., Washburn, L. L., Whitney, J. B., III, and Morrow, K. E. (1982). *Mus poschiavinus* Y chromosome in the C57BL/6J murine genome causes sex reversal. *Science* **217**, 535–537.

Ellis, N. A. (1991). The human Y chromosome. *Sem. Dev. Biol.* **2**, 231–240.

Ferguson-Smith, M. A. (1966). X-Y chromosomal interchange in the aetiology of true hermaphroditism and of XX Klinefelter's syndrome. *Lancet* **II**, 475.

Ferrari, S., Harley, V. R., Pontiggia, A., Lovell-Badge, R., Goodfellow, P. N., and Bianchi, M. E. (1992). SRY, like HMG1, recognizes a sharp angle in DNA. *EMBO J.* **11**, 4497–4506.

Ford, C. E., Jones, K. W., Polani, P., de Almedia, J. C., and Brigg, J. H. (1959). A sex chromosome anomaly in a case of gonadal sex dysgenesis (Turner's syndrome). *Lancet* **i**, 711–714.

Foster, J. W., Brennan, F. E., Hampikian, G. K., Goodfellow, P. N., Sinclair, A. H., Lovell-Badge, R., Selwood, L., Renfree, M. B., Cooper, D. W., and Graves, J. A. M. (1992). Evolution of sex determination and the Y chromosome: SRY-related sequences in marsupials. *Nature (London)* **359**, 531–533.

Fryns, J., Kleczowska, A., Caesar, P., and Van Den Bergh, H. (1986). Double autosomal chromosomal aberration (3p trisomy and 9p monosomy) and sex reversal. *Ann. Genet.* **29**, 49–52.

Giese, K., Amsterdam, A., and Grosschedl, R. (1991). DNA-binding properties of the HMG domain of the lymphoid-specific transcriptional regulator, LEF-1. *Genes Dev.* **5**, 2567–2578.

Giese, K., Cox, J., and Grosschedl, R. (1992). The HMG domain of lymphoid enhancer factor 1 bends DNA and facilitates assembly of functional nucleoprotein structures. *Cell* **69**, 1–20.

Goodfellow, P. N., Darling, S., and Wolfe, J. (1985). The human Y chromosome. *J. Med. Genet.* **22**, 332–344.

Griffiths, R. (1991). The isolation of conserved DNA sequences related to the human sex-determining region Y gene from the lesser black-backed gull *(Larus fuscus)*. *Proc. R. Soc. Lond. B* **244**, 123–128.

Gubbay, J., Collignon, J., Koopman, P., Capel, B., Economou, A., Munsterberg, A., Vivian, N., Goodfellow, P. N., and Lovell-Badge, R. (1990). A gene mapping to the sex-determining region of the mouse Y chromosome is a member of novel family of embryonically expressed genes. *Nature (London)* **346**, 245–250.

Gubbay, J., Vivian, N., Economou, A., Jackson, D. I., Goodfellow, P. N., and Lovell-Badge, R. (1992). Inverted repeat structure of the Sry locus in mice. *Proc. Natl. Acad. Sci. U.S.A.* **89**, 7953–7957.

Guellaen, G., Casanova, M., Bishop, C., Geldwerth, D., Audre, G., Fellous, M., and Weissenbach, J. (1984). Human XX males with Y single-copy DNA fragments. *Nature (London)* **307**, 172–173.

Harley, V. R., Jackson, D. I., Hextall, P., Hawkins, J. R., Berkovitz, G. D., Sockanathan, S., Lovell-Badge, R., and Goodfellow, P. N. (1992). DNA binding activity of recombinant SRY from normal males and XY females. *Science* **255**, 453–456.

Hawkins, J. R., Taylor, A., Berta, P., Levilliers, J., Van der Auwera, B., and Goodfellow, P. N. (1992a). Mutational analysis of SRY: Nonsense and missense mutations in XY sex reversal. *Hum. Genet.* **88**, 471–475.

Hawkins, J. R., Taylor, A., Goodfellow, P. N., Midgeon, C. J., Smith, K. D., and Berkovitz, G. D. (1992b). Evidence for increased prevalence of SRY mutations in XY females with complete rather than partial gonadal dysgenesis. *Am. J. Hum. Genet.* **51**, 979–984.

Hoo, J. J., Salafsky, I. S., Lin, C. C., and Pinsky, L. (1989). Possible location of a recessive testis forming gene on 9p24. *Am. J. Hum. Genet.* **45**, A78.

Jacobs, P. A., and Strong, J. A. (1959). A case of human intersexuality having a possible XXY sex-determining mechanism. *Nature (London)* **183**, 302–303.

Jager, R. J., Anvret, M., Hall, K., and Scherer, G. (1990). A human XY female with a frame shift mutation in the candidate testis-determining gene SRY. *Nature (London)* **348**, 452–454.

Jager, R. J., Harley, V. R., Pfeiffer, R. A., Goodfellow, P. N., and Scherer, G. (1993). Familial mutation in the testis determining gene, SRY shared by both sexes. *Hum. Genet.* **90**, 350–355.

Jantzen, H.-M., Admon, A., Bell, S. P., and Tjian, R. (1990). Nucleolar transcription factor hUBF contains a DNA-binding motif with homology to HMG proteins. *Nature (London)* **344**, 830–835.

Johnson, P. F., and McKnight, S. L. (1989). Eukaryotic transcriptional regulatory proteins. *Ann. Rev. Biochem.* **58**, 799–839.

Jost, A. (1947). Recherches sur la differentiation sexuelle de l'embryon de lapin. *Arch. d'Anat. Microsc. Morphol. Exp.* **36**, 271–315.

Jost, A., Vigier, B., Prepin, J., and Perchellet, J. (1973). Studies on sex differentiation in mammals. *Rec. Prog. Horm. Res.* **29**, 1–41.

Kelly, M., Burke, J., Smith, M., Klar, A., and Beach, D. (1988). Four mating-type genes control sexual differentiation in the fission yeast. *EMBO J.* **7**, 1537–1547.

Koller, P. C., and Darlington, C. D. (1934). The genetical and mechanical properties of the sex chromosomes. 1. *Rattus norvegicus. J. Genet.* **29**, 159–173.

Kolodrubetz, D. (1990). Consensus sequence or HMG-1 like DNA binding domains. *Nucl. Acids Res.* **18**, 5565.

Koopman, P., and Gubbay, J. (1991). The biology of *Sry*. Sex determination in man and mouse. *Sem. Dev. Biol.* **2**, 259–264.

Koopman, P., Gubbay, J., Collignon, J., and Lovell-Badge, R. (1989). *Zfy* gene expression is not compatible with a primary role in mouse sex determination. *Nature (London)* **342**, 940–942.

Koopman, P., Munsterberg, A., Capel, B., Vivian, N., and Lovell-Badge, R. (1990). Expression of a candidate sex-determining gene during mouse testis differentiation. *Nature (London)* **248**, 450–452.

Koopman, P., Gubbay, J., Vivian, N., Goodfellow, P. N., and Lovell-Badge, R. (1991). Male development of chromosomally female mice transgenic for *Sry*. *Nature (London)* **351**, 117–121.

Learned, R. M., Learned, T. K., Haltiner, M. M., and Tjian, R. T. (1986). Human rRNA transcription is modulated by the co-ordinate binding of two factors to an upstream control element. *Cell* **45**, 847–857.

Lotterand, M., and Juillard, E. (1976). A new case of trisomy for the distal part of 13q due to maternal translocation, (9;13)(p21;q21). *Hum. Genet.* **33**, 213–222.

Lowry, C. V., Cerdan, M. E., and Zitomer, R. S. (1990). A hypoxic consensus operator and a constitutive activation region regulate the ANB1 gene of Saccharomyces cerevisiae. *Mol. Cell. Biol.* **10**, 5921–5926.

McElreavey, K. D., Vilain, E., Boucekkine, C., Vidaud, M., Jaubert, F., Richaud, F., and Fellous, M. (1992). XY sex reversal associated with a nonsense mutation in SRY. *Genomics* **13**, 838–840.

McLaren, A. (1984). Male sexual differentiation in mice lacking H-Y antigen. *Nature (London)* **312**, 552–555.

McLaren, A. (1985). Relating of germ cell sex to gonadal differentiation. In "The Origin and Evolution of Sex." (H. O. Halvorsen and A. Monroy, eds.), pp. 289–300. Liss, New York.

McLaren, A. (1988). The developmental history of female germ cells in mammals. *Oxford Rev. Reprod. Biol.* **10**, 163–179.

Mardon, G., Mosher, R., Disteche, C. M., Nishioka, Y. U., McLaren, A., and Page, D. C. (1989). Duplication, deletion and polymorphism in the sex-determining region of the mouse Y chromosome. *Science* **243**, 78–80.

May, K. M., Grinzald, K. A., and Blackston, R. D. (1991). Sex reversal and multiple abnormalities due to segregation of t(X:16)(p11.4;p13.3). Proceedings of the 8th International Congress of Human Genetics, Washington, D.C.

Merchant, H. (1975). Rat gonadal and ovarian organogenesis with and without germ cells. An ultrastructural study. *Dev. Biol.* **44**, 1–21.

Muller, U., Donlon, T., Schmid, M., Fitch, N., Richer, C-L., Lalande, M., and Latt, S. A. (1986). Deletion mapping of the testis determining locus with DNA probes in 46,XX males and 46,XY and 46X,dic(YC) females. *Nucl. Acids Res.* **14**, 6489–6505.

Nagamine, C. M., Chan, K., Kozak, C. A., and Low, Y-F. (1989). Chromosome mapping and expression of a putative testis determining gene in the mouse. *Science* **243**, 80–83.

Narahara, K., Kodama, Y., Kimusa, S., and Kimoto, H. (1979). Probable tandem duplication of Xp in a 46Xp+ Y boy. *Jap. J. Human Gene* **24**, 105–110.

Nasrin, N., Buggs, C., Kong, X. F., Carnazza, J., Goebl, M., and Alexander-Bridges, M. (1991). DNA-binding properties of the product of the tesitis-determining gene and a related protein. *Nature (London)* **354**, 317–320.

Nielsen, K. B., and Langkjaer, F. (1982). Inherited partial X chromosome duplication in a mentally retarded male. *J. Med. Genet.* **19**, 222–236.

Ogata, T., Hawkins, J. R., Taylor, A., Matsuo, N., Hata, J-i *et al.* (1992). Sex reversal in a child with a 46,X,Ypt karyotype: support for the existence of a gene(s), located in distal xp, involved in testis formation. *J. Med. Genet.* **29**, 226–230.

Page, D. C., Brown, L. G., and de la Chapelle, A. (1987). Exchange of terminal portion of X-Y chromosomal short arm in human XX males. *Nature (London)* **328**, 437–440.

Page, D. C., Fisher, E. M. C., McGillivray, G., and Brown, L. G. (1990). Additional deletion in sex-determining region of human Y chromosome resolves paradox of X,t(Y;22) female. *Nature (London)* **348**, 279–235.

Palmer, S. J., and Burgoyne, P. S. (1991a). *In situ* analysis of fetal, prepuberal and adult XX/XY chimaeric mouse testes: Sertoli cells are predominantly, but not exclusively, XY. *Development* **112**, 265–268.

Palmer, S. J., and Burgoyne, P. S. (1991b). The *Mus musculus domesticus Tdy* allele acts later than the *Mus musculus domesticus Tdy* allele: A basis for XY sex reversal in C57BI/6-Y^pos mice. *Development* **113**, 709–714.

Palmer, M. S., Sinclair, A. H., Berta, P., Ellis, N., Goodfellow, P. N., Abbas, N. E., and Fellous, M. (1989). Genetic evidence that ZFY is not the testis-determining factor. *Nature (London)* **342**, 937–939.

Parisi, M. A., and Clayton, D. A. (1991). Similarity of human mitochondrial transcription factor 1 to high mobility group proteins. *Science* **252**, 965–969.

Pearson, P. L., and Bobrow, M. (1970). Definitive evidence form the short arm of the Y chromosome associating with the X during meiosis in the human male. *Nature (London)* **226**, 959–961.

Pollock, R., and Treisman, R. (1990). A sensitive method for the determination of protein-DNA binding specificities. *Nucl. Acids Res.* **18**, 6197–6204.

Roberts, C., Weith, A., Passage, E., Michot, J. L., Mattei, M. G., and Bishop, C. E. (1988). Molecular and cytogenetic evidence for the location of Tdy and Hya on the mouse Y chromosome short arm. *Proc. Natl. Acad. Sci. U.S.A.* **85**, 6446–6449.

Scherer, G., Schempp, W., Baccichetti, C., Lenzini, E., Bricarelli, F. D., Carbone, L. D. L., and Wolf, U. (1989). Duplication of an Xp segment that includes the ZFX locus causes sex inversion in man. *Hum. Genet.* **81**, 291–294.

Sinclair, A. H., Foster, J. W., Spencer, J. A., Page, D. C., Palmer, M., Goodfellow, P. N., and Graves, J. A. M. (1988). Sequences homologous to ZFY, a candidate human sex determining gene, are autosomal in marsupials. A gene from the human sex-determining region encodes a protein with homology to a conserved DNA-binding motif. *Nature (London)* **346**, 240–244.

Sinclair, A. H., Berta, P., Palmer, M. S., Hawkins, J. R., Griffiths, B. L., Smith, M. J., Foster, J. W., Frischauf, A.-M., Lovell-Badge, R., and Goodfellow, P. N. (1990). A gene from the human sex-determinating region encodes a protein with homology to a conserved DNA-binding motif. *Nature (London)* **346**, 240–244.

Singh, L., and Dixon, G. (1990). High mobility group proteins 1 and 2 function as general class II transcription factors. *Biochemistry* **29**, 6295–6302.

Singh, L., Matsukuma, S., and Jones, K. W. (1987). The use of Y-chromosome-specific repeated DNA sequences in the analysis of testis development in an XX-XY mouse. *Development* **101 Suppl,** 143–149.

Sugimoto, A., Iino, Y., Maeda, T., Watanabe, Y., and Masayuki, Y. (1990). *Schizosaccharomyces pombe ste11*+ encodes a transcription factor with an HMG motif that is a critical regulator of sexual development. *Genes Dev.* **5**, 1990–1999.

Staben, C., and Yanofsky, C. (1990). *Neurospora crassa a* mating-type region. *Proc. Natl. Acad. Sci. U.S.A.* **87**, 4917–4921.

Stern, H. J., Garrity, A. M., Saal, H. M., Wangsa, D., and Dusteche, C. M. (1990). Duplication of Xp21 and sex reversal: Insight into the mechanism of sex determination. *Am. J. Hum. Genet.* **47**, A41.

Travis, A., Amsterdam, A., Belanger, C., and Grosschedl, R. (1991). LEF-1, a gene encoding a lymphoid-specific with protein, and HMG domain, regulates T-cell receptor a enhancer function. *Genes Dev.* **5**, 880–894.

van de Wetering, M., and Clevers, H. (1992). Sequence-specific interaction of the HMG box proteins TCF-1 and SRY within the minor groove of a Watson–Crick double helix. *EMBO J.* **11**, 3039–3044.

van de Wetering, M., Oosterwegel, M., Dooijes, D., and Clevers, H. (1991). Identification and cloning of TCF-1, a T lymphocyte-specific transcription factor containing a sequence-specific HMG box. *EMBO J.* **10**, 123–132.

Vergnaud, G., Page, D. C., Simmler, M.-C., Brown, L., Rouyer, F., Noel, B., Botstein, D., de la Chapelle, A., and Weissenbach, J. (1986). A deletion map of the human Y chromosome based on DNA hybridisation. *Am. J. Hum. Genet.* **38**, 109–124.

Waterman, M. L., Fischer, W. H., and Jones, K. A. (1991). A thymus-specific member of the HMG protein family regulates the human T cell receptor alpha enhancer. *Genes Dev.* **5**, 565–669.

Watt, F., and Molloy, P. (1988). High mobility group proteins 1 and 2 stimulate binding of a specific transcription factor to the adenovirus major late promoter. *Nucl. Acids Res.* **16**, 1471–1486.

Wen, L., Huang, J.-K., Johnson, B. H., and Reeck, G. R. (1989). A human placental cDNA clone that encodes nonhistone chromosomal protein HMG-1. *Nucl. Acids Res.* **17**, 1197–1214.

2

Cell Biology of Testicular Development

LAURI J. PELLINIEMI, KIM FRÖJDMAN,
& JORMA PARANKO

I. Introduction

The mammalian gonad develops as a local derivative of the mesonephros. The beginning in both sexes is a placode formed by condensation of surface epithelial and subjacent mesenchymal tissue in the ventral or medial cortex along the longitudinal axis of the mesonephros. These epithelial and mesenchymal cells are the precursors of all somatic cell types of the testis, with the possible exception of the macrophages, which may derive from the blood. The germ cells protodifferentiate in extraembryonic mesenchyme and migrate into the developing gonad at an early phase before the histological differentiation of the gonad.

The embryonic, or prenatal, development of the testis can be divided into four different phases: (1) the pregonadal placode on mesonephros (Fig. 1), (2) the gonadal ridge (Figs. 2, 3), (3) the sexually indifferent gonad (Fig. 4), and (4) the testis (Figs. 5, 6). The number of cell types increases with development. After sexual differentiation of the gonad into a testis, the basic components of a functional male gonad are present by the time of the birth. However, the spermatogonia do not undergo meiosis until spermatogenesis begins at puberty.

The cell differentiation events that take place in the developing gonads differ significantly from those in nongenital organs because the gonadal rudiment has two different developmental pathways, according to the genetic sex of the individual. The testicular pathway of differentiation is regulated by a testis determining gene and factor system (McLaren, 1991a,b), which becomes active after a short initial period of sexually indifferent gonadal development. The cell biology of male gonadal differentiation includes organization of epithelial tissues and endocrine cells from primitive nonepithelial precursors as well as the homing of germ cells into the gonad. The decisive role of the testis determining gene system is guiding the growth of the epithelial gonadal cords into continuous loops rather than into separated islet-like follicles, which is the organization of the epithelial component in the ovary. The masculinization of other genital organs is regulated by anti-Müllerian hormone from Sertoli cells (Josso, 1991) and by androgenic steroids from the interstitial Leydig cells, which differentiate soon after the formation of the testicular cords (Huhtaniemi and Pelliniemi, 1992; Pelliniemi and Dym, 1993).

II. Pregonadal Placode

A. Surface Epithelial Cells

At first, the surface epithelial cells (Figs. 1 and 2) of the presumptive gonadal primordium are actually "hybrids" by tissue classification. The apical portions of the cells are epithelial with some microvilli and ordinary junctional complexes. The basal portions are mesenchymal with loose contacts, wide intercellular spaces, long cytoplasmic processes, and no basement membrane (Pelliniemi, 1975a; Fröjdman *et al.*, 1992a). The surface cells in the rat contain only desmin intermediate filaments (Fig. 3). Other filaments such as those made of cytokeratins and vimentin have not been detected to date (Fröjdman *et al.*, 1992a). The surface epithelial cells of the mesonephros outside the gonadal placode have a basement membrane.

B. Mesenchymal Cells

Under the surface layer are mesenchymal cells that contain both desmin and vimentin (Fig. 3). Small accumulations of extracellular matrix can be detected by electron microscopy and immunocytochemistry in the rat, but no organized basement membrane is detectable (Fröjdman *et al.*, 1992a). Cytokeratins are not detected in the cells of the actual pregonadal mesenchyme, but are present in other mesenchymal cells of the surrounding tissues. Desmin, on the other hand, is present in all mesenchymal cells of the pregonadal mesonephros (Fröjdman *et al.*, 1992a).

C. Primordial Germ Cells

The primordial germ cells (PGC) in several mammalian species are located first in the yolk sac endoderm as large spherical cells that contain

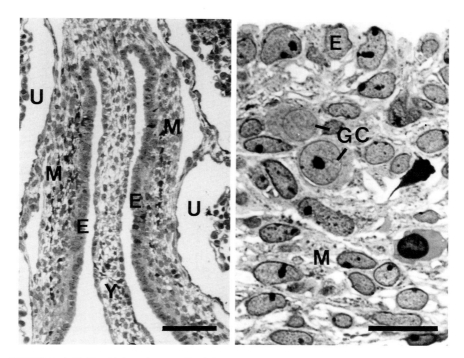

FIGURE 1, 2 (left) Light micrograph of the pregonadal placodes in the medial surfaces of the mesonephros in pig embryo at the age of 21 days post coitum (dpc). The surface epithelium (E) has thickened slightly and the mesenchyme (M) is still undifferentiated. Y, Mesentery; U, urinary space of mesonephric corpuscle. Bar, 80 μm. **(right)** Light micrograph of the cross section through the whole thickness of early gonadal ridge in pig embryo at the age of 22 dpc. Under the surface epithelium (E) are two primordial germ cells (GC) among the pleomorphic mesenchymal cells (M). The urinary space (U, Fig. 1) starts right below the bar. Bar, 10 μm.

alkaline phosphatase (Witschi, 1948; McKay *et al.*, 1953; Chiquoine, 1954; Spiegelman and Bennett, 1973). In the mouse, the earliest location of primordial germ cells is posterior to the primitive streak in the extraembryonic mesenchyme (Ginsburg *et al.*, 1990). These cells migrate via the hindgut mesentery toward the gonadal ridge using pseudopodia, and divide mitotically as they travel. *In vitro*, mouse migratory PGC adhere to laminin and fibronectin but not to type I collagen or glycosaminoglycans (de Felici and Dolci, 1989). These findings suggest that PGC use laminin, fibronectin, or both as a substrate for migration. The PGC are guided by an unknown, possibly chemotactic, mechanism because individual cells may take different routes.

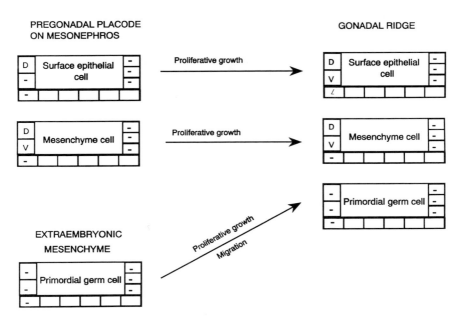

FIGURE 3 Changes in cells of the pregonadal placode on mesonephros during and after the differentiation into the gonadal ridge. The rectangles represent cells. The small squares on the left and right sides represent cytoplasmic intermediate filaments. The small squares under the rectangles represent the components of the basement membrane and extracellular matrix. The solid arrows show a confirmed development. Bold letters indicate full presence of a component; italic letters indicate incipient or incomplete presence of the same. An empty square means no data available; a dash (-) means tested but not found. D, Desmin; V, vimentin; L, laminin. The components have been identified by immunocytochemistry in the rat (Paranko *et al.*, 1983; Pelliniemi *et al.*, 1984; Paranko, 1987; Fröjdman *et al.*, 1989,1992a). (See explanation of dashes in Fig. 4.)

III. Gonadal Ridge and Indifferent Gonad

A. Surface Epithelial Cells

The thickness of the gonadal tissue increases by proliferative growth of the surface and mesenchymal cells that form the gonadal ridge (Figs. 2, 3). The epithelium-derived cells in the basal layer of the surface differentiate toward the mesenchymal type and the surface cells in an epithelial direction. In the rat, vimentin appears transiently in the epithelial cells and small fragments of laminin-containing basement membrane are seen along the base of some epithelial cells (Pelliniemi, 1975a,1976a; Magre and Jost, 1991; Fröjdman *et al.*, 1992a). In the rat, all somatic cells now have both vimentin and desmin in their cytoplasm. In the rabbit, the surface epithelial cells contain cytokeratin and vimentin (Wartenberg *et al.*, 1989).

With growth in the undifferentiated pig gonad, the surface epithelial cells form areas of real epithelium with a basement membrane, but regions also form where the surface epithelium is continuous with the undifferentiated "epithelioid" blastema tissue inside the gonad (Pelliniemi, 1976b).

B. Mesenchymal Cells

The mesenchymal cells multiply and form condensed tissue that also contributes to the thickening of the ridge (Figs. 2, 3). The cells grow in size and make closer contacts with each other, leading to narrowing of the interstitial space (Pelliniemi, 1975a). Desmin and vimentin but not cytokeratins are present in the mesenchymal cells of the rat gonadal ridge (Fröjdman *et al.*, 1992a).

C. Primordial Germ Cells

The migration leads the PGC into the gonadal ridge (Figs. 2, 3). The PGC make contacts by cell membrane apposition with proliferating epithelial cells and maintain their active pseudopodia, by which some PGC invade the epithelial tissues (Pelliniemi, 1975a). The adhesiveness of the PGC to laminin and fibronectin (de Felici and Dolci, 1989) may play a role in homing these cells to the gonadal ridge, where the synthesis of these compounds is just beginning (Fröjdman *et al.*, 1992a). Both the PGC and the somatic gonadal cells synthesize asparagine-linked glycopeptides and glycosaminoglycans, but the PGC also synthesize lactosaminoglycans, notably in different proportions in males and females (de Felici *et al.*, 1985). These components probably participate in creating specific contacts with epithelial cells and also may be important in avoiding adhesion to the mesenchymal cells. The germ cells continue to divide mitotically as they associate with the organizing epithelial cells inside the gonad (Vergouwen *et al.*, 1991). The PGC become spherical and the pseudopodia disappear as the cells settle in the forming gonadal cords (Figs. 4, 5). No cytokeratin, vimentin, or desmin has been detected in the PGC. In fact, no changes in their cytoskeletal composition have been observed although their surface components and adhesiveness change considerably (de Felici *et al.*, 1985; de Felici and Dolci, 1989).

GONADAL RIDGE   INDIFFERENT GONAD

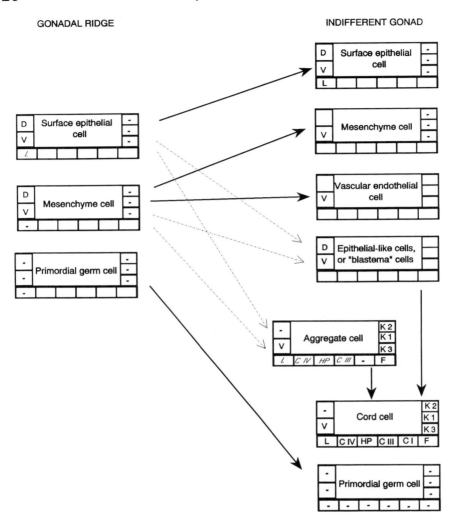

FIGURE 4 Changes in cells of the gonadal ridge during and after the differentiation into the indifferent gonad. The rectangles represent cells. The small squares on the left and right sides represent cytoplasmic intermediate filaments. The small squares under the cell rectangle represent the components of the basement membrane and extracellular matrix. The solid arrows show a confirmed development; the dashed arrows indicate unconfirmed possibilities. Bold letters indicate full presence of a component; italic letters indicate incipient or incomplete presence of the same. An empty square means no data available; a dash (-) means tested but not found. D, Desmin; V, vimentin; K1, K2, and K3, cytokeratins recognized by the PKK1, -2, and -3 antibodies, respectively (Holthöfer *et al.* 1983; Virtanen *et al.*, 1985); L, laminin; HP, heparan sulfate proteoglycan; C V, C IV, C III, C I collagen type V, IV, III, and I, respectively; F, fibronectin. The components have been identified by immunocytochemistry in the rat (Paranko *et al.*, 1983,1986; Pelliniemi *et al.*, 1984; Paranko, 1987; Fröjdman *et al.*, 1989,1992a).

D. Vascular Endothelial Cells

No comprehensive histological, histochemical, immunocytochemical, or ultrastructural studies are available of angiogenesis in the fetal mammalian testis, whereas details of the prepubertal maturation of the blood–testis barrier are well known (Setchell and Brooks, 1988).

Intratesticular vasculature differentiates in the gonadal mesenchyme with the growth of the epithelial components (Pelliniemi, 1975b,1976b; Magre and Jost, 1983; Paranko et al., 1986). The wall of early blood vessels consists of a simple epithelium with one cell or a few cells per circumference. These vessels remain in the interstitium and do not invade the cords at any phase. The differentiation of the vasculature may modulate the morphogenesis of the testis cords in some way, and vice versa, but such interactions have not yet been demonstrated (Pelliniemi, 1975b). However, a testis-specific distribution of blood vessels is obvious already in an early phase of testicular differentiation.

E. Gonadal Blastema and Somatic Cord Cells

The formation of the gonadal cords in the rat is a rapid process that takes place via transitory epithelial cell aggregates (Paranko et al., 1983; Fröjdman et al., 1992a). In slowly developing species such as cattle and pigs, the event proceeds via a large central accumulation of epithelial cells to form a transient tissue called the gonadal blastema (Ohno, 1967; Pelliniemi, 1976b). The ingrowing cells from the surface epithelium, and perhaps cells differentiating from the internal mesenchyme, form a large cluster of epithelial-like cells that usually are in close membrane apposition to each other (Ohno, 1967; Pelliniemi, 1976b). These blastema cells begin to organize into gonadal cords, first by alignment of a few cells like a piece of a cord wall, with specific intercellular junctions and patches of basement membrane. This process is analogous to the formation of epithelial aggregates as seen in some species. Blastema formation precedes the separate islets, or pieces of incomplete epithelial tissue, joining together to form continuous cords or sheets of epithelial tissue (Fig. 5). The male and female gonads organize similar gonadal cords in the beginning, although at different times (Pelliniemi and Lauteala, 1981). The cords organize in different species in different ways in three-dimensional space, but at the cellular and histological level the processes are similar. In the rabbit, in which the testicular cords differentiate via the formation of a gonadal blastema, the gonadal blastema cells contain both cytokeratin and vimentin (Wartenberg et al., 1989). The karyotypic male gonad is referred to as a testis as soon as it can be distinguished from the female gonad by any structural, immunocytochemical, biochemical, or other criterion. Therefore, the definition of testis changes over time with the progress of research.

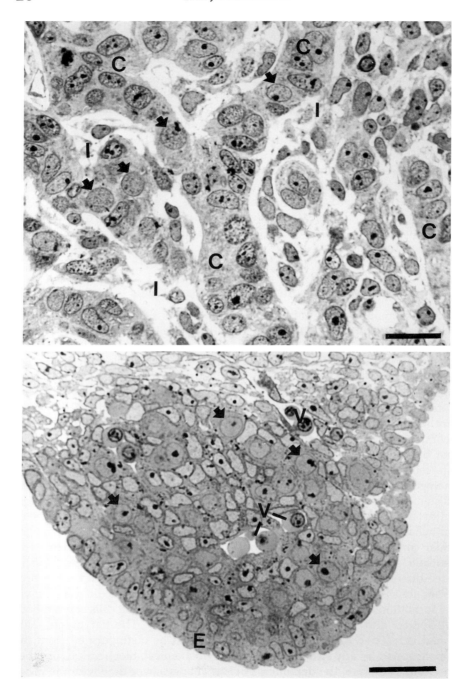

IV. Testis

A. Tunica Albuginea

1. Surface epithelial cells

The surface epithelial cells flatten out to form a continuous sheet (Fig. 6; see also Figs. 9, 16, 17) on a continuous basement membrane composed of laminin, collagen types IV and V, heparan sulfate proteoglycan, and associated fibronectin (Pelliniemi, 1975b; Paranko et al., 1983,1986; Fröjdman et al., 1992a). With epithelial specialization in the rat, cytokeratins appear and vimentin disappears in the flattened cells (Fröjdman et al., 1992a).

2. Fibroblasts

Some of the mesenchymal cells in the cortex differentiate into fibroblasts (see Figs. 9, 16, 17). In the rat, initially these cells synthesize type III collagen; later they produce type I collagen, which is a sign of connective tissue maturation (Paranko, 1987). The fibroblasts in the rat continue to have both vimentin and desmin in their cytoplasm (Fröjdman et al., 1992a). Smooth muscle myoblasts, the precursors of the smooth muscle cells, or myofibroblasts of the tunica albuginea appear in the rat among the fibroblasts (Virtanen et al., 1986; Paranko and Pelliniemi, 1992). In newborn rat (Leeson, 1975), elongated myoblasts contain bundles of microfilaments and are connected through short gap junctions. Similar cellular structures are already present in the myoblasts of 17-day-old fetal rat testis. Even before that time, at the age of 16 days, appearance of alkaline phosphatase and smooth muscle myosin with filamentous actin and desmin indicates smooth muscle cell differentiation (Paranko and Pelliniemi, 1992). A predominant growth of myoblasts from the posterior part of the testis (Leeson, 1975; Virtanen et al., 1986; Paranko and Pelliniemi, 1992) suggests an active role of the hilar tissue in the formation of the tunica albuginea.

The regulatory mechanisms of smooth muscle cell differentiation are not yet resolved, but temporal relationships with the differentiation of Leydig cells would correlate with androgen stimulation (Roosen-Runge and Anderson, 1959; Lording and de Kretser, 1972; Picon, 1976b; Feldman and Bloch, 1978; Pelliniemi, 1985; Kuopio et al., 1989; Paranko and Pelliniemi, 1992). Concomitant development of smooth muscle cells of the testicular blood

---

FIGURE 5 (above) Light micrograph of the early testicular cords (C) in the pig (28 dpc) are irregular in shape and size. In places, the demarcation of the cords from the interstitium (I) is still incomplete. Leydig cells have not yet differentiated. The arrows point at germ cells. Bar, 20 μm. **FIGURE 6 (below)** Light micrograph of a cross section of an early rat testis (13 dpc) shows germ cells (arrows) and somatic cells below the differentiating surface epithelium (E). V, blood vessels. Bar, 30 μm.

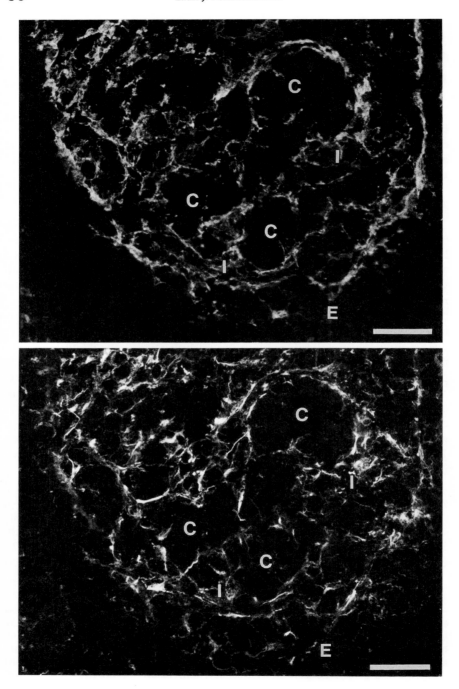

vessels, on the other hand, may be caused by circulating connective tissue growth factors (Florini *et al.*, 1991) that could induce myogenic differentiation. Leydig cell-derived sex steroids seem, however, to be more probable regulators because, for example, in the B6.Y DOM XY mouse ovotestis, tunica-like differentiation is visible only in the testicular but not at all in the ovarian parts (Suganuma *et al.*, 1991).

B. Testicular Cords

1. Sertoli cells

The somatic cells of the testicular cords usually are referred to as Sertoli cells after the male and female gonad can be distinguished by cell biological criteria (Figs. 5, 6, 7, 8; see also Fig. 14). The ultrastructure of the fetal Sertoli cells (Fig. 9; see also Figs. 12, 14) has been studied in several mammals at different phases of development (Pelliniemi *et al.*, 1993).

Species with a short prenatal period do not have a proper indifferent phase of the gonadal cords; features of Sertoli cells emerge at the same time as epithelial differentiation occurs (Fröjdman *et al.*, 1992a). In the rat testis (Fig. 6), the formation of the cord precursors begins by organization of somatic cells and germ cells into cell aggregates (Paranko *et al.*, 1983; Paranko, 1987; Fröjdman *et al.*, 1989,1992a,b; Fröjdman and Pelliniemi, 1990; Malmi *et al.*, 1990b). The differentiating aggregate cells have basolateral junctions and an immature and discontinuous laminin-containing basement membrane (Fig. 7). The aggregate cells, or Sertoli cell precursors, soon become organized into elongated gonadal cords with a continuous basement membrane.

In the Sertoli cells of the newly formed testicular cords in the rat (Fig. 9), the synthesis of desmin (Fig. 8) ceases concomitantly with aggregation and epithelial differentiation, whereas the synthesis of vimentin continues, but the filaments become reorganized (Fig. 10). As a new feature, cytokeratins appear (Fig. 11) in the cytoplasm of the organizing Sertoli cells (see Fig. 16). In addition to these changes, the synthesis of the basement membrane components begins. A continuous basement membrane is laid along the interstitium-facing surface of the testicular cords (Fröjdman *et al.*, 1992a).

FIGURE 7 (above) Light micrograph of a section corresponding to that seen in Fig. 6 shows the presence of laminin, detected by immunofluorescence, around the early testicular cell aggregates (C), in the interstitium (I), and below the surface epithelium (E) in rats (13 dpc). Bar, 30 μm. **FIGURE 8 (below)** Light micrograph of double immunoreaction in the same section (13 dpc) as seen in Fig. 7 shows desmin in the cells around the cell aggregates (C), which are devoid of desmin. Desmin is seen also in other cells of the early interstitium (I) and in the surface epithelium (E), although it is less prominent. Bar, 30 μm.

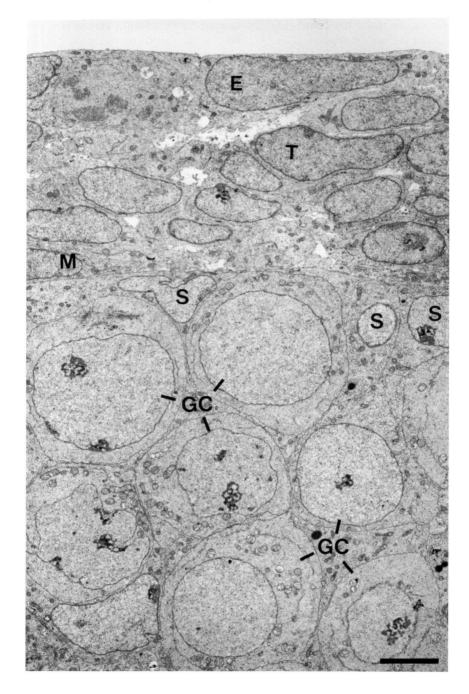

Fibronectin also disappears from the intercellular space inside the cords (Paranko *et al.*, 1983). Unlike other cytokeratin polypeptides, cytokeratins recognized by the PKK2 antibody (Fröjdman *et al.*, 1992a), including cytokeratin 19 (Fridmacher *et al.*, 1992), gradually disappear from the rat Sertoli cells after the age of 15 days.

Human fetal Sertoli cells form irregular cords separated by a basal lamina from the interstitium by the eighth week (Pelliniemi, 1970). The Sertoli cells are organized histologically as stratified cuboidal epithelium. The cytoplasm contains mitochondria with transverse lamellar cristae, granular but not agranular endoplasmic reticulum, and numerous free polysomes scattered evenly in the cytoplasm. Some coated vesicles and lipid droplets are observed among the intermediate filaments in the cytoplasm. Coated vesicles also have been described in Sertoli cells of several other mammals (Pelliniemi, 1975b; Merchant-Larios, 1979; Magre and Jost, 1991). In the basal part of the cells, these vesicles may take up nutrients from the interstitial space. The coated vesicles near the Golgi complex apparently are made for excretion. The cell membranes of neighboring cells are closely apposed; adherent junctions occasionally are seen between the cells. Gradually the irregular testicular cords become organized into loops, but no intracordal secretory activity or lumen formation occurs before puberty. Degenerative changes appear in some human Sertoli cells in the fourth month. The cells have shrunken dark cytoplasm and irregular nuclear infoldings. The degeneration may result from the lack of an interactive factor from the Leydig cells, which already have started their physiological involution.

The rat Sertoli cells are ultrastructurally similar to those in humans. They contain alkaline phosphatase, which disappears soon after their organization into the gonadal cords (Paranko and Pelliniemi, 1992). The Sertoli cells, unlike the rete cells, do not express neural cell adhesion molecules (N-CAM) (Møller *et al.*, 1991). Reactive groups of carbohydrates, especially those that bind *Ricinus communis* agglutinin I, decline gradually during the differentiation of the gonadal cords (Malmi *et al.*, 1990a,b; Fröjdman *et al.*, 1992b). In the late fetal stages, twisted or convoluted testicular cords constitute ~35% of the gonadal volume (Hargitt, 1926; Roosen-Runge and Anderson, 1959). The Sertoli cells acquire irregular shape and populate the periphery of the cords.

A major product of the fetal Sertoli cells is the anti-Müllerian hormone

---

FIGURE 9 Electron micrograph of the 14 dpc rat testis. Sertoli cells (S) are located along the periphery of the cords. The spermatogonia (GC) are relatively large and have spherical nuclei. Elongated precursors of the myoid cells (M) surround the cords in the interstitium. Fibroblasts of the early tunica albuginea (T) are seen below the surface epithelium (E). Bar, 5 μm.

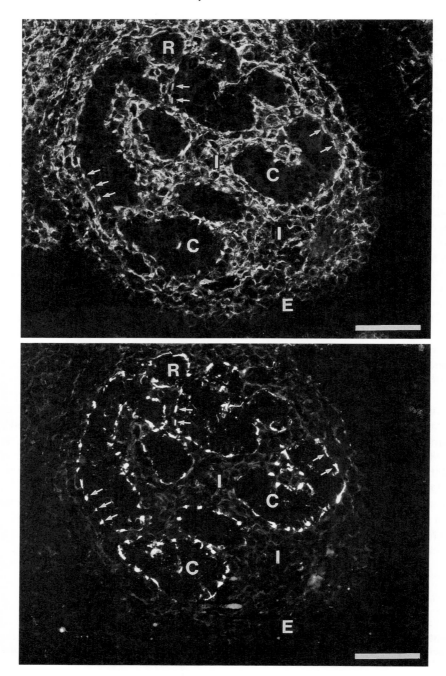

(AMH) or Müllerian inhibiting substance (see Fig. 17), which induces regression of the paramesonephric duct (Jost, 1947,1953; Josso *et al.*, 1977; Donahoe *et al.*, 1982; Josso, 1991). AMH, a 140-kDa glycoprotein (Picard and Josso, 1976,1984; Picard *et al.*, 1978; Josso *et al.*, 1981; Budzik *et al.*, 1983), has been localized in the endoplasmic reticulum of fetal (Tran and Josso, 1982) and neonatal (Hayashi *et al.*, 1984) bovine Sertoli cells by immunocytochemistry, and has been measured in biological fluids by radioimmunoassay (Vigier *et al.*, 1982a; Necklaws *et al.*, 1986; Burdzy *et al.*, 1991). Bovine and human cDNAs encoding AMH have been cloned (Cate *et al.*, 1986; Picard *et al.*, 1986). The conserved C-terminal region of AMH has portions homologous with TGF-β and related disulfide-linked proteins such as inhibin, bone morphogenetic proteins, the decapentaplegic gene complex, and the VG$_1$ gene in *Xenopus* (Mason *et al.*, 1985; Tannahill and Melton, 1989; Cate *et al.*, 1990; Massagué, 1990). AMH therefore is regarded as a member of the TGF-β family (Josso, 1991). Results of studies of the early mouse gonad suggest that a function of AMH, related to TGF-β, may be the inhibition of both chemotaxis and proliferation of the primordial germ cells (Godin and Wylie, 1991).

AMH from one species is biologically active in several other species (Tran *et al.*, 1977) but monoclonal antibodies against AMH of various species show only low cross-reactivity (Vigier *et al.*, 1982b; Necklaws *et al.*, 1986). The conserved C-terminal region of the molecule may contain the wide-range biological activity (Pepinsky *et al.*, 1988; Cate *et al.*, 1990). The possible mechanisms of AMH action on the regression of the paramesonephric duct, in testicular descent, and as a cytotoxic agent against cancers of the human female genital tract have been reviewed by Josso (1991) and Donahoe *et al.* (1984). Considerations of an intratesticular role of AMH are based on induction of testis cord-like structures in the fetal rat ovary by AMH (Vigier *et al.*, 1987), and on similar observations in transgenic mice that produce AMH (Behringer *et al.*, 1990). These findings suggest that AMH may cause the masculinization of the bovine freemartin ovary (Vigier *et al.*, 1984b,1987).

AMH is present in the newborn, prepubertal, and adult testis in amounts that decrease with age (Josso, 1991). Regulation of AMH synthesis in the fetal Sertoli cells is not known. Autosomal localization of the human gene for

---

FIGURE 10 (above) Light micrograph of the vimentin immunoreaction (arrows) is seen in the interstitium (I) and in the Sertoli cells located in the periphery of the testicular cords (C) in the rat (14 dpc). With the differentiation of the surface epithelium (E), the reaction for vimentin in epithelial cells decreases. R, rete cords. Bar, 60 μm. **FIGURE 11 (below)** Light micrograph of the double immunoreaction in the same section as seen in Fig. 10 shows cytokeratins colocalized with vimentin in the peripheral Sertoli cells (arrows) of the cords (C) and in the rete cords (R) of the rat (14 dpc). The interstitium (I) is devoid of cytokeratins and no reaction is yet found in the surface epithelium (E). Bar, 60 μm.

AMH in the short arm of chromosome 19 (Cate *et al.*, 1986; Cohen-Haguen-auer *et al.*, 1987) is in agreement with the expression of AMH in the postnatal ovary as well (Vigier *et al.*, 1984a; Picard *et al.*, 1986).

Expression of AMH coincides with the formation of fetal testicular cords (Meyers-Wallen *et al.*, 1991; Münsterberg and Lovell-Badge, 1991), but the Sertoli cells produce AMH even if the cord formation is inhibited (Magre and Jost, 1984; Jost *et al.*, 1988). Steroid hormones or gonadotropins are improb-able regulators of the fetal synthesis of AMH because follicle stimulating hormone (FSH) and testosterone fail to stimulate the production of AMH in fetal Sertoli cells *in vitro* (Vigier *et al.*, 1985; Voutilainen and Miller, 1987). Luteinizing hormone (LH) (Kuroda *et al.*, 1990), estrogens, progesterone, and gonadotropin releasing hormone (GnRH) also are not able to induce AMH production in cultured Sertoli cells (LaQuaglia *et al.*, 1986). The postnatal decrease in the synthesis of AMH may be caused by FSH (Kuroda *et al.*, 1990).

Inhibin is a dimeric glycoprotein hormone (see Fig. 17) produced in the adult testis by Sertoli cells (Vale *et al.*, 1988; Robertson *et al.*, 1989). Inhibin reduces FSH production in the pituitary gland. Subunits of this protein have been localized in Sertoli cells, Leydig cells, and germ cells (Albers *et al.*, 1989; Shaha *et al.*, 1989; Rabinovici *et al.*, 1991). The physiological role of gonadal inhibin in the male as well as female fetus is apparently the same as in the adult, as shown by *in vitro* bioassay technique (Torney *et al.*, 1990). Tran-scription of the α and β_A, but not β_B subunits of inhibin has been localized in the testicular cords of bovine fetal testis. The concentration of the α subunit increases with gestation and is considerably higher than that of the β subunit (Torney *et al.*, 1990). Another imbalance is that immunoactivity of the pro-tein increases with development clearly more than does bioactivity, appar-ently because of the synthesis of high concentrations of immunoactive α subunit precursor fragments (Torney *et al.*, 1990,1992). The possible roles of the observed differences in inhibin subunits produced by the fetal testicular cells and the regulation of their synthesis are not yet understood.

Sertoli cells in the fetus apparently are not capable of synthesizing androgens, although they contain agranular endoplasmic reticulum in guinea pigs and rabbits (Black and Christensen, 1969; Bjerregaard *et al.*, 1974). Also no reports exist of androgen receptors in the fetal Sertoli cells. However, the Sertoli cells still may regulate testicular androgen synthesis indirectly, as described in Section IV,C,5.

2. Spermatogonia

The germ cells are referred to as spermatogonia when they become embedded in the testicular cords (Figs. 5, 6, 9, 12; see also Fig. 16). The fetal spermatogonia first are located randomly among the Sertoli cells but move actively into the periphery of the growing cords, as shown in the rat (Franchi and Mandl, 1964), rabbit (Vihko *et al.*, 1987), and human (Gondos *et al.*,

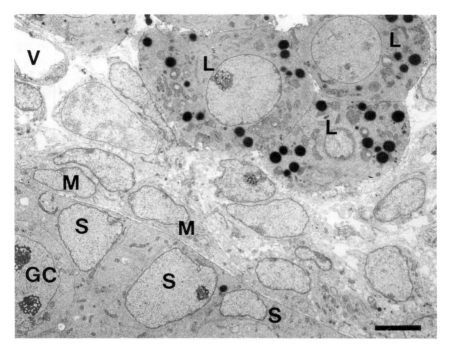

FIGURE 12   Electron micrograph of the fetal rat testis at the age of 19 days shows a few Sertoli cells (S) and a spermatogonim (GC) in a cord. In the interstitium, fetal Leydig cells (L) have formed aggregates. These cells have spherical nuclei and numerous lipid droplets. Between the myoid cells (M) and the Leydig cells there are fibroblasts, mesenchymal cells, and a blood vessel (V). Bar, 5 μm.

1971). The spermatogonia have a large spherical nucleus with one to three nucleoli. The few mitochondria are spherical, the Golgi complex is small, the granular endoplasmic reticulum is organized as a few cisternae, and numerous free polysomes are detectable (Pelliniemi, 1975b). The cell membrane is generally even with a few cytoplasmic processes and patches of adherent junctions with Sertoli cells are seen. The lack of intermediate filaments and their components continues, but alkaline phosphatase activity is high (Paranko et al., 1986; Fröjdman et al., 1992a; Paranko and Pelliniemi, 1992). The mitotic and meiotic activities as well as the relationships with Sertoli cells are presented in separate sections.

3. Rete cells

The intratesticular rete develops from the gonadal cords in the hilar area of the testis (Roosen-Runge, 1961). The rete cords also initially contain germ cells (Roosen-Runge, 1961). The early rete cord cells are ultrastructurally similar and contain distributions of vimentin and cytokeratins (Figs. 10, 11) similar to the early Sertoli cells in the rat (Fröjdman *et al.*, 1992a). Differences such as the previously described disappearance of PKK2 cytokeratins from the Sertoli but not from the rete cells in the rat (Fröjdman *et al.*, 1992a) have been shown also in the adult human (Amlani and Vogl, 1988) and ram (Tung *et al.*, 1987). The rete cells in humans contain vimentin and cytokeratin at the ages of 12 and 35 weeks, but only vimentin is found in 10-week-old fetuses (Dinges *et al.*, 1991).

The rete cells show a strong reaction for N-CAM during the embryonic and fetal stages (Møller *et al.*, 1991). This reaction may be associated with a low cellular motility and the formation of close cell associations. Certain epithelial rete cells of fetal male human and monkey express pregnancy-associated plasma protein-A (PAPP-A). The biological role for PAPP-A is still unknown, but in the adult it might have an immunoprotective role by coating the surface of the spermatozoa (Schindler *et al.*, 1986). The rete testis cells in fetal pig show membrane-bound carbonic anhydrase activity, but the role of this protein in the differentiation of the developing testis remains to be clarified (Rodriguez-Martinez *et al.*, 1990).

4. Interactions between the somatic and germ cells

Entry of primordial germ cells into the gonad and then into the testicular cords results in a decreased capacity for invasive locomotion (Donovan *et al.*, 1986) and affinity to fibronectin and laminin (de Felici and Dolci, 1989). This process is likely to include the observed changes in their surface properties, especially in the composition of the glycoproteins (de Felici *et al.*, 1985; Donovan *et al.*, 1986; de Felici and Dolci, 1989; Kanai *et al.*, 1989a,1991; Malmi *et al.*, 1990b; Escalante-Alcalde and Merchant-Larios, 1992; Fröjdman *et al.*, 1992b). Factors that would promote incorporation of germ cells into the fetal cords are, however, poorly understood. Unexplained also is the observation that fetal germ cells adhere *in vitro* to follicular and Sertoli cells from the adult mouse but not to the somatic cells from the fetal gonads (de Felici and Siracusa, 1985).

Specific cell junctions and cell contacts apparently are involved in the stabilization of the Sertoli–germ cell interactions. Intact fetal and newborn mammalian gonads have various types of junctions. Between the Sertoli cells, gap junctions of various sizes have been described in rat, mouse, guinea pig, and human (Gilula *et al.*, 1976; Nagano and Suzuki, 1978; Gondos, 1981; Pelletier and Friend, 1983; McGinley and Posalaky, 1986). Occluding junctions have been found in rat and mouse (Gilula *et al.*, 1976; Nagano and Suzuki, 1978) and desmosome-like junctions have been found in mouse (Nagano and Suzuki, 1978). Between Sertoli and germ cells are desmosome-like junctions in mouse (Nagano and Suzuki, 1978) and focal adherent contacts in several species. Gap junctions occur between the Sertoli cells, but not between Sertoli and germ cells. The number of gap junctions between Sertoli cells decreases with development but the number of tight junctions increases (Gilula *et al.*, 1976), leading to respective changes in the permeability of the blood–testis barrier and in the structural organization of the seminiferous epithelium. Inhibition of glycolysis by tunicamycin in fetal mouse testis has shown that intercellular junctions between Sertoli cells are more stable than those between Sertoli and germ cells (Kanai *et al.*, 1991).

Vinculin with actin has been found in the basal occluding junctions and in ectoplasmic specializations between Sertoli cells and spermatids of postnatal, prepubertal, and adult rat testis (Grove and Vogl, 1989; Pfeiffer and Vogl, 1991), but reports of these molecules in the fetal testis are not available. The possible role of vinculin (Ungar *et al.*, 1986) in the regulation of the cell contacts in the testis remains to be clarified. Plasminogen activators apparently regulate germ cell movements in the mature seminiferous epithelium (Vihko *et al.*, 1987) but their role in the fetal testis has not been studied.

Intercellular bridges between developing germ cells appear during the period of accelerated mitotic activity in the rabbit fetus (Gondos and Conner, 1973). Specific conformational and functional interactions between densely packed Sertoli and germ cells evidently are involved in bridge formation, because the germ cells located outside the cords lack intercellular bridges.

The fetal germ cells move actively from the inner parts of the cords to their periphery in the rat (Franchi and Mandl, 1964), rabbit (Gondos and Conner, 1973), and human (Gondos *et al.*, 1971). Germ cells even may migrate out from the testicular cords through the basement membrane (Gould and Haddad, 1978). Testosterone secreted by interstitial Leydig cells (Figs. 12, 13; see also Fig. 17) may serve as a chemoattractant for the germ cells in their movement toward the periphery of the cords (Gondos and Conner, 1973). Other interstitial factors also may affect the cells inside the seminiferous epithelium because the fetal blood–testis barrier is still immature (see Fig. 17).

C. Interstitium

1. Undifferentiated mesenchymal cells

Through the rest of the fetal period and into adulthood, the interstitium contains undifferentiated cells (Figs. 5, 6, 12; see also Figs. 16, 17) that apparently are able to differentiate into Leydig cells (Hardy *et al.*, 1989). These cells also may serve as precursors for other types of interstitial cells. They can be identified only by morphological characteristics: small irregular shape of the nucleus and of the whole cell. The cells in the rat contain vimentin and desmin (Fröjdman *et al.*, 1992a), in the rabbit, they contain only vimentin (Wartenberg *et al.*, 1989).

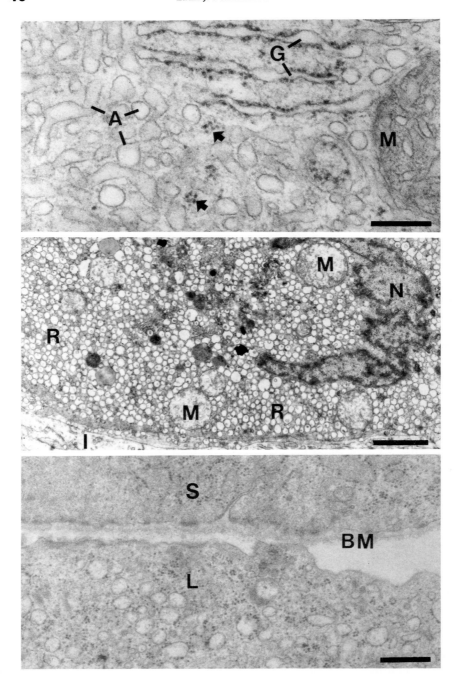

2. Fibroblasts

Some mesenchymal cells inside the testis differentiate into fibroblasts (Fig. 12; see also Figs. 16, 17) around the time of testicular cord formation in a way similar to that of cells forming the tunica albuginea. These cells begin to produce collagen type I in addition to type III, which has been produced already by the mesenchymal cells (Paranko, 1987). The fibroblasts settle between groups of cords and make up the septa of the testis.

3. Macrophages

Phagocytic cells in the fetal testis (see Figs. 16, 17) appear last of all cell types. In humans these cells differentiate between 16 and 20 weeks of age (Dechelotte et al., 1989); in the rat, differentiation takes place at the age of 19 days (Hutson, 1990). The most probable precursor cells for these phagocytic cells are local mesenchymal cells (see Fig. 16), although a hematogenous origin cannot be ruled out. The cells often are located in the vicinity of Leydig cells and acquire the ultrastructural characteristics of macrophages during the involution phase of interstitium (Dechelotte et al., 1989).

4. Myoid cells

The mesenchymal cells in direct contact with the testicular cords secrete type III collagen and become elongated (see Fig. 16), already at the indifferent gonadal phase (Paranko, 1987). In the rat, smooth muscle myosin is detected first at the age of 16 days in occasional early pericordal cells. The precursor cells contain desmin (Fig. 8) and vimentin (Fig. 10). With their differentiation into myoid cells, they contain actin in increasing amounts (Fröjdman et al., 1992a; Paranko and Pelliniemi, 1992). The cytoplasm of the myoid cells contains the microfilaments typical of smooth muscle cells, but the shape of the cells in the rat becomes squamous. The myoid cells organize into a continuous sheet, in which the cells are connected by adherent junctions. These cells also form a basement membrane on both sides, as do

---

FIGURE 13 (top) Electron micrograph of a Leydig cell in a 19-week-old human fetus. The cytoplasm contains agranular (A) and granular (G) endoplasmic reticulum, round profile of a mitochondrion (M), and occasional polyribosomes (arrows). Bar, 0.25 μm. **FIGURE 14 (middle)** After the age of 14 to 16 weeks, several Leydig cells in the human fetal testis degenerate. This electron micrograph of Leydig cells at the age of 23 weeks shows the remnants of the nucleus (N) and mitochondria (M) and the cytoplasm packed with disintegrating membranes of vesiculated endoplasmic reticulum (R). I, interstitium. Bar, 2 μm. **FIGURE 15 (bottom)** During the differentiation of the testicular cords, a uniform basement membrane (BM) develops on the basal surface of the Sertoli cells (S), seen in this electron micrograph. The testicular cords are not yet enveloped by the precursors of the myoid cells, which allows direct contacts with Leydig (L) and other interstitial cells in the pig (28 dpc). Bar, 0.5 μm.

ordinary smooth muscle cells. The histological organization of the myoid cells varies in different species.

The embryonic and fetal pericordal cells label strongly with the lectins of *Canavalia ensiformis, Triticum vulgaris, Bandeirea simplicifolia,* and *Ricinus communis;* moderately with succinylated *Triticum vulgaris;* and not at all with *Arachis hypogaea, Helix pomatia, Glycine maximum,* and *Ulex europeus* (Malmi *et al.,* 1990b; Fröjdman *et al.,* 1992b). Although the myoid cells are rich in carbohydrates, only minor changes in their carbohydrate composition currently are known to be present during the embryonic and fetal stages of differentiation.

The close relationship and matching pace of differentiation suggest regulatory interactions between the pericordal myoid and cord cells. The role of the myoid cells in the differentiation of the Sertoli cells has been studied *in vitro.* Positive effects of the myoid cells on the formation of the extracellular matrix, on Sertoli cell migration, and on differentiation (Tung and Fritz, 1980,1986a,b; Tung *et al.,* 1984) have been observed. This cooperativity seems to benefit both cell types and to promote myoid cell differentiation.

The presence of smooth muscle myosin in the developing pericordal cells, in contrast to the corresponding cells of the ovary, suggests that the initiation of myoid cell development is under androgenic control (Paranko and Pelliniemi, 1992). Testosterone also stimulates alkaline phosphatase, which is abundant in myoid cells (Anthony and Skinner, 1989). Postnatal myoid cells are responsive to androgens and participate interactively in the histological and biochemical differentiation of Sertoli cells (Hölund *et al.,* 1981; Skinner, 1987). A paracrine nonmitogenic factor, P-Mod-S (see Fig. 17), found in postnatal myoid cells (Skinner and Fritz, 1986; Skinner *et al.,* 1988) may be present already in the fetal period and may stimulate the differentiation of fetal Sertoli cells.

5. Leydig cells

The differentiation of the Leydig cells begins after the sexual differentiation of the testis (Figs. 12, 14) (Narbaitz and Adler, 1967; Pelliniemi and Niemi, 1969; Pelliniemi and Lauteala, 1981; Pelliniemi, 1985). The main function of these cells before birth and until puberty is the endocrine regulation by androgenic steroids of masculine differentiation of the extragonadal genital organs and their sex-related functions (Huhtaniemi and Pelliniemi, 1992). The development of the Leydig cells in the human fetal testis can be divided into three different developmental phases: the differentiation phase (fetal ages 8–14 weeks), the fetal maturity phase (14–18 weeks), and the involution phase (18–38 weeks) (Pelliniemi and Niemi, 1969).

The human fetal Leydig cell precursors appear among the undifferentiated mesenchymal cells in the 8th week of fetal life (Pelliniemi and Niemi, 1969). The cytoplasm grows by addition of mitochondria, granular endoplasmic reticulum, lipid droplets, and particularly agranular endoplasmic reticulum (Fig. 13). The cells become first elongated and finally polygonal. The triggering factor of Leydig cell differentiation is not known, but the differentiation probably is not regulated directly by the testis determining gene (McLaren, 1990b,1991b).

The functional maturation of human Leydig cells proceeds with morphological differentiation, as is indicated by the appearance of the histochemical reaction for the 3β-hydroxysteroid dehydrogenase (3β-HSD) enzyme at the age of 8 weeks (Niemi et al., 1967). Later, the Leydig cell number correlates with the plasma human chorionic gonadotropin (hCG) concentration (Zondek and Zondek, 1976); the cells also are stimulated in vitro by hCG (Huhtaniemi et al., 1977; Kellokumpu-Lehtinen and Pelliniemi, 1988).

With advancing differentiation in the human, several cells often become closely attached to each other and form clusters. By the 14th week, tightly packed Leydig cells occupy most of the interstitial space in humans. During the next week, they reach a maximum of 48×10^6 cells/pair of testes and 50% of the area of the testis section (Niemi et al., 1967; Pelliniemi and Niemi, 1969; Codesal et al., 1990). The Leydig cells are connected by gap junctions (Nagano and Suzuki, 1976), but the functional significance of these connections is not understood. Another surprising feature are patches of basement membrane around the Leydig cells in some species (Kuopio et al., 1989).

The number of human Leydig cells per unit volume begins to decrease at the age of ~ 16 weeks because of the fast growth of the other components of the testis, although the absolute number remains constant until $\sim$ the 24th week (Niemi et al., 1967; Codesal et al., 1990). During the rest of the prenatal period, the absolute number of Leydig cells decreases progressively to 18×10^6 cells/pair of testes just before birth (Codesal et al., 1990). The decrease in Leydig cell number results from degeneration and complete destruction of some cells (Fig. 15) and from a 50% reduction in the volume of the remaining cells (Pelliniemi and Niemi, 1969; Codesal et al., 1990). This process is simultaneous with the decrease in plasma hCG concentration (Reyes et al., 1989), but a possible role of hCG has not been shown. An obvious explanation for Leydig cell involution would be that the demand for androgens is lower during the second half of pregnancy, after the differentiation of the male genital organs has been completed.

In the rat, the cytodifferentiation of Leydig cells begins from undifferentiated mesenchymal cells between embryonic–fetal ages of 15–17 days and proceeds through ultrastructural changes similar to those described for humans (Roosen-Runge and Anderson, 1959; Narbaitz and Adler, 1967; Lording and de Kretser, 1972; Merchant-Larios, 1975; Magre and Jost, 1980). Evidence of functional maturation of the enzyme system in the form of 3β-HSD activity has been detected as early as 15 days of age (Niemi and Ikonen, 1961). The numerical profile of the rat Leydig cells has only one early maximum in the absolute number of Leydig cells per pair of testes (260×10^3) at the age of 19.5 days (Tapanainen et al., 1984).

INDIFFERENT GONAD

TESTIS

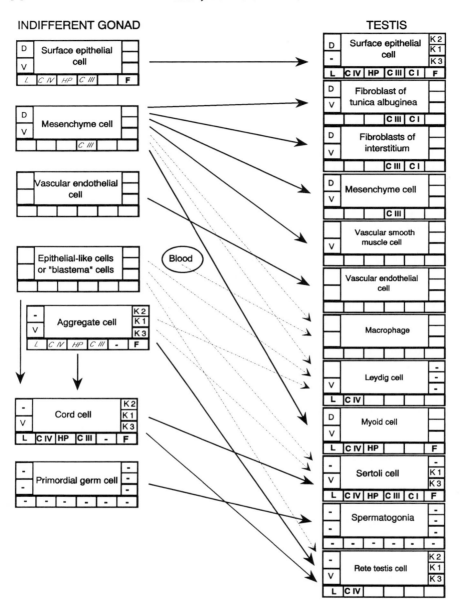

The fetal Leydig cells in the rat remain first as individual cells, but toward the end of pregnancy they form groups (Fig. 12) of closely attached cells (Kuopio *et al.*, 1989). The differentiating Leydig cells have small patches of basement membrane (see Fig. 17) that increase with development (Kuopio *et al.*, 1989). The presence of cell adhesion molecules, which would explain the aggregation of Leydig cells during the fetal period (Kuopio *et al.*, 1989), has not been demonstrated. Fetal Leydig cells seem to lack N-CAM (Møller *et al.*, 1991), but conflicting results still exist (Mayerhofer *et al.*, 1992). Additional studies are required to clarify this issue. Addition of L-azetidine 2-carboxylic acid, a proline competitor, into the medium of organ culture of differentiating testes inhibits the differentiation of Leydig cells and their steroidogenic function (Jost *et al.*, 1988). The effect is prevented by an excess of proline, which suggests that the extracellular matrix has a role in Leydig cell differentiation, perhaps in their association into glandular islets. The finding of basement membrane means that Leydig cells produce laminin, collagen type IV, and other basement membrane components (Fig. 16). Other important properties of the Leydig cell surface are specific nonimmunological binding of adult mouse Leydig cells to lymphocytes and macrophages (Born and Wekerle, 1981; Rivenson *et al.*, 1981). Macrophages are known to produce several regulatory factors, which may have local regulatory effects on Leydig cells (Hutson, 1990).

Androgen production by the fetal Leydig cells begins in the human at the fetal age of 8 weeks (Pelliniemi and Niemi, 1969), and reaches a maximum at the age of 12–14 weeks. After this time, steroidogenesis decreases in pace with the number of Leydig cells. In the rat, testosterone synthesis starts simultaneously with the differentiation of Leydig cells at fetal age 15.5–16.5 days (Niemi and Ikonen, 1961; Warren *et al.*, 1972,1975; Magre and Jost, 1980).

The LH/hCG receptors in the rat testis become measurable at the age of 15 days when the fetal testis becomes responsive to LH stimulation (Warren *et al.*, 1972,1984; Feldman and Bloch, 1978; Gangnerau *et al.*, 1982). Capac-

FIGURE 16 Cellular changes in the differentiation of the indifferent gonad into the testis. The rectangles represent cells. The small squares on the left and right sides represent cytoplasmic intermediate filaments. The small squares under the cell rectangle represent the components of the basement membrane and extracellular matrix. The solid arrows show a confirmed development; the dashed arrows indicate unconfirmed possibilities. Bold letters indicate full presence of a component; italic letters indicate incipient or incomplete presence of the same. An empty square means no data available; a dash (-) means tested but not found. D, Desmin; V, vimentin; K1, K2, and K3, cytokeratins recognized by the PKK1, -2, and -3 antibodies, respectively; L, laminin; C IV–V collagen type IV and V; HP, heparan sulfate proteoglycan; C III, C I, interstitial collagen type III and I, respectively; F, fibronectin. The components have been identified by immunocytochemistry in the rat (Paranko *et al.*, 1983,1986; Pelliniemi *et al.*, 1984; Paranko, 1987; Fröjdman *et al.*, 1989,1992a).

ity to synthesize testosterone and responsiveness to LH increase with development of the fetal testis (Feldman and Bloch, 1978; Warren *et al.*, 1984).

Fetal testis synthesizes androgens in organ culture also (Moon *et al.*, 1973; Picon, 1976a,b; Stewart and Raeside, 1976; Habert *et al.*, 1991). These results also support an autonomous start of testosterone production in the fetal testis without required external stimulatory factors. Subsequent regulation of testosterone production might be through ultrashort feedback by androgens (Pointis *et al.*, 1984).

6. Vascular endothelial cells

The vascular wall in most blood vessels in fetal testicular interstitium consists of one layer of endothelial cells (Fig. 12), all of which are connected to each other by adherent junctions (Pelliniemi, 1975b). The fetal blood vessels are not associated directly with the Leydig cells as are those in the adult (Roosen-Runge and Anderson, 1959). However, the permeability of the endothelial cells and their basement membrane allows transport of hormones and nutrients between the blood and testicular tissues (Figs. 12, 17).

7. Vascular smooth muscle cells

The development of vascular smooth muscle cells begins in the testicular artery in the tunica albuginea and in the largest intratesticular branches located in septulae testis. Some mesenchymal cells in the vicinity of early blood vessels elongate (Fig. 17). In the rat, smooth muscle myosin appears in their cytoplasm ~2 days after the formation of the testicular cords (Paranko and Pelliniemi, 1992). Alkaline phosphatase appears 1–2 days later and reaches maximal reactivity at prepubertal maturation (Kormano, 1967; Paranko and Pelliniemi, 1992). The smooth muscle cells that form the tunica media of the blood vessels also develop a normal basement membrane as they differentiate.

V. Basement Membranes and Extracellular Matrix

The morphological aspects of the differentiating basement membrane, as well as proteins and glycoproteins of the basement membranes of the testicular cells (Figs. 4, 7, 16), have been studied more extensively than the role and distribution of other interstitial material in the differentiating testis (Paranko *et al.*, 1983; Pelliniemi *et al.*, 1984a,b; Agelopoulou and Magre, 1987; Paranko, 1987; Gelly *et al.*, 1989; Fröjdman *et al.*, 1992a,b).

A. Gonadal Ridge

The interstitial tissue of the gonadal ridge in the rat contains type III collagen and fragments of basement membrane components, including laminin (Fig. 3) and fibronectin, (Paranko *et al.*, 1983; Pelliniemi *et al.*, 1984a,b;

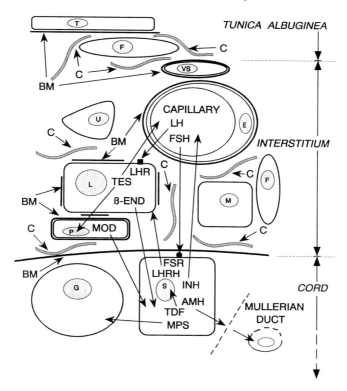

FIGURE 17 Regulatory relationships and substances known or postulated to be active in the fetal testis. The Sertoli cells (S) and spermatogonia (G) are inside the cord, and macrophages (M), peritubular cells (P), Leydig cells (L), fibroblasts (F), undifferentiated mesenchymal cells (U), vascular endothelial cells (E), and vascular smooth muscle cells (VS) are in the interstitium. The tunica albuginea contains surface epithelial cells (T) and fibroblasts. The ovals with one letter represent nuclei. The arrows indicate the origin and target of each substance. The target of the anti-Müllerian hormone (AMH) is outside the testis as indicated by the vertical dashed line and two consecutive arrows. The basement membranes (BM) are drawn to show possible penetration barriers; collagen fibers (C) represent matrix molecules. The macrophages and undifferentiated mesenchymal cells are not yet active nodes in the present regulatory network. END, Endorphin; FSR, receptor for FSH; INH, inhibin; LHRH, peptide factor; MOD, peritubular cell factor P-Mod-S; MPS, meiosis preventing substance; TDF, testis determining factor; TES, testosterone.

Magre and Jost, 1991; Fröjdman *et al.*, 1992a). The absence of type I collagen in the early embryonic rat gonads is consistent with the primitive nature of the tissue (Paranko, 1987).

B. Indifferent Gonad and Testis

The epithelial precursors of cord cells that contact the mesenchyme cells have several small microvilli on their basal surface. The microvilli disappear and leave a relatively wide extracellular space between the cord cells and the interstitial cells. Granular and fibrous extracellular material accumulates around the organizing epithelial cell aggregates but not inside the aggregates. The extracellular space around the cells and aggregates contains fibronectin, laminin, heparan sulfate proteoglycan, and collagen types III, IV, and V (Figs. 4, 7, 16). The appearance of epithelium-like cell groups excluding these matrix components (Paranko *et al.*, 1983; Pelliniemi *et al.*, 1984a,b; Paranko, 1987; Fröjdman *et al.*, 1989,1992a) and the changes in the cytoskeleton (Figs. 4, 10, 11, 16) (Fröjdman *et al.*, 1992a) are the first signs of the formation of the epithelial Sertoli cell precursors. The localization of laminin is useful in delineating testicular cords in fetal testis (Pelliniemi *et al.*, 1984a,b; Agelopoulou and Magre, 1987; Fröjdman *et al.*, 1989,1992a,b; Gelly *et al.*, 1989; Malmi *et al.*, 1990b). Deposition of type III collagen along the testicular cords with the formation of the basement membrane may reflect epithelio-mesenchymal interactions in the morphogenesis of the cords (Paranko, 1987). Another phenomenon is the rapid differentiation of the pericordal, or future myoid, cells with their cytoskeletal components concomitantly with the differentiation of the Sertoli cells (Fröjdman *et al.*, 1992a). The basement membrane probably is produced jointly, as shown by Sertoli and peritubular myoid cells under experimental conditions in the adult (Tung *et al.*, 1984; Tung and Fritz, 1986a,b,1987). Concomitantly with testicular cord differentiation, collagen type I (Figs. 16, 17) appears in the interstitial tissue; by the age of 17 days in the rat, this collagen is organized similarly to the type III collagen (Paranko, 1987), both of which support the cords and other testicular tissues.

The subtypes of the basement membrane components of the fetal Sertoli cells are not yet known. The different portions of basement membranes in adult human testis (Leu *et al.*, 1986) and the different time schedules in basement membrane formation in rat testis and ovary (Grund and Pelliniemi, 1987; Kuopio and Pelliniemi, 1989; Kuopio *et al.*, 1989) suggest that the Sertoli and Leydig cells regulate the composition and differentiation of their basement membranes (Kuopio and Pelliniemi, 1989; Kuopio *et al.*, 1989). The role of the different molecular types of laminin and laminin-like molecules in other cells of the urogenital tissues (Klein *et al.*, 1988,1990; Ekblom *et al.*, 1990; Engvall *et al.*, 1990) and the role of carbohydrates and glycosylations (Kanai *et al.*, 1989a,b,1991; Malmi *et al.*, 1990b; Fröjdman *et al.*, 1992b) in the ontogenesis of the basement membranes remain to be clarified.

VI. Regulatory Mechanisms

A. Initiation of Gonadal Differentiation

The ancestors of the somatic testicular cells become defined through the determination of the mesonephric region that forms the gonadal primordium. The regulation of these initial events may involve homeotic selector genes. One example is *HOX-5.2*, which is expressed in mouse embryos in temporally and spatially defined regions, one of which is the gonad (Dollé and Duboule, 1989). The regulatory requirements for the testis are the same as those for any organ: activation of mitoses, increase in cell size, and further structural and functional differentiation. The general growth regulatory substances (Fauser *et al.*, 1988) are likely to be active in gonads also, even at the early stages.

A dramatic change in gonadal differentiation is the epithelial transformation. The aggregation of the future cells into epithelial clusters (Paranko *et al.*, 1988; Fröjdman *et al.*, 1992a,b) seems to take place under the rules of general epithelial morphogenesis (Fristrom, 1988). The regulatory factors probably are not related to sex because the process is similar in both sexes (Pelliniemi, 1976b; Pelliniemi and Lauteala, 1981; Fröjdman *et al.*, 1989). Little actual information is available, but findings of specific differentiation initiation genes such as *Myo D* for muscle (Davis *et al.*, 1987) encourage the search for similar factors in the male gonad.

B. Regulation of Gonadal Sex Differentiation

The primary sexual differentiation of the testis is actually differentiation of the Sertoli cell (Pelliniemi and Dym, 1993; Pelliniemi *et al.*, 1993). Regulatory substances from the testicular Sertoli, Leydig, and other cells (Fig. 17) control the subsequent steps of sexual differentiation. Several theories of various differentiation factors have been proposed (Jost *et al.*, 1973; Pelliniemi, 1975b; Hall and Wachtel, 1980; Wachtel, 1983; Jost, 1988). The latest ones have a cell biological basis and range from the H−Y (histocompatibility−Y chromosomal) theory (Wachtel, 1983) via a zinc finger-containing protein factor (ZFY) (Page *et al.*, 1987) to the present candidate for the testis determining factor (TDF) (Palmer *et al.*, 1989), partially characterized in human and mouse (McLaren, 1990b,1991a,b).

The TDF (Fig. 17) produced by the Sertoli cell is most likely to be a protein that, in humans, contains a motif of less than 80 amino acids in its open reading frame. This amino acid sequence of TDF is homologous to the HMG box in human RNA polymerase I transcription factor and to the Mc protein of the *Schizosaccharomyces pombe* fission yeast (Palmer *et al.*, 1989; Sinclair *et al.*, 1990). The mouse testis cells, but not those in the ovary, contain a similar gene that is active during gonadal differentiation at the age of 11.5 days. This gene is missing in XY female mutants, but is present in gonads of male W/W mutants that lack germ cells. The gene induces a testis when transferred into XX female embryos (Gubbay *et al.*, 1990; Koopman *et al.*, 1990,1991). The involvement of Sertoli cells in the organization of testicular cords even without germ cells justifies the assumption (Fig. 17) that Sertoli cells produce TDF and also are its primary target (McLaren, 1991a).

Another regulatory factor produced by the fetal Sertoli cells is the previously described anti-Müllerian hormone, the transcription of which starts in the mouse at the age of 12.5 days (Münsterberg and Lovell-Badge, 1991). Expression of AMH does not depend on male germ cells and remains constant until several days after birth. The expression of TDF in the gonadal ridge is detectable from the age of 10.5–12.5 days, correlating well with the onset of testicular differentiation (Koopman *et al.*, 1990). These temporal and spatial correspondences are supportive of the idea that AMH could be regulated by TDF. Although many functions of AMH are clearly outside the testis, AMH might act in the fetal testis by eliminating any germ cells that enter meiosis despite inhibition by the Sertoli cell (McLaren, 1990a).

The approach of this laboratory has been to analyze the cell and tissue biological processes in gonadal sex differentiation and to identify the earliest developmental changes between the sexes (Pelliniemi and Lauteala, 1981; Pelliniemi *et al.*, 1984a; Paranko *et al.*, 1988; Fröjdman *et al.*, 1989,1992a). One characteristic of TDF is that its action begins earlier than the sex-specific differentiation processes in the female (Pelliniemi and Lauteala, 1981; Baker, 1989). From cell biological observations, we conclude that the primary male-specific action of TDF is organizing the indifferent epithelial aggregates into looping continuous cords, which otherwise would develop into separated follicles, a process that seems to be the default pathway in mammals (Pelliniemi and Lauteala, 1981; Pelliniemi *et al.*, 1984a,b; Fröjdman *et al.*, 1989, 1992a). Other sex-specific events, the mechanisms of which are unknown, include organization of the testicular stroma into male tunica albuginea and interstitial tissue including the Leydig cells. Male-specific structural cell surface or diffusible components such as H–Y antigen or other factors also may be involved (Ohno, 1980; Wachtel, 1983; Burgoyne, 1988). Transcription of the *AMH* gene is an early biochemical sex-specific event in Sertoli cells, but may not be related to TDF, which is transcribed 48 hr earlier in the mouse (Münsterberg and Lovell-Badge, 1991). Mutual interactions between germ cells and Sertoli cells probably also are needed for gonadal differentiation (McLaren, 1991a). These interactions include inhibition of meiosis in the fetal spermatogonia by Sertoli cells (Fig. 17). Clearly one product of one structural gene is unable to accomplish directly all the various tasks of TDF in testicular differentiation. A probable regulatory mechanism for TDF is regulating the transcription of other genes (McLaren, 1991a). This protein also might act by choosing the male or female differentiation program, as does a local homeotic selector gene (Dollé and Duboule, 1989), or by processing RNA through differential splicing of a hierarchy of regulatory genes, as shown in *Drosophila* (Baker, 1989).

C. Regulation of Sertoli Cell Proliferation

The growth of the fetal testis depends to a large extent on the proliferation of Sertoli cells. A clear stimulator of fetal Sertoli cells is FSH. The receptors for FSH are present in 17-day-old fetal rat Sertoli cells and their number increases by the age of 19 days (Warren *et al.*, 1984). These receptors probably are located on the Sertoli cell membrane (Fig. 17), as are those in early neonatal rats (Bortolussi *et al.*, 1990).

Organ culture experiments (Steinberger *et al.*, 1964; Byskov and Grinsted, 1981; Taketo and Koide, 1981; Agelopoulou *et al.*, 1984; D'Agostino *et al.*, 1984; Pelliniemi *et al.*, 1984a,b; Jost *et al.*, 1985; Grund *et al.*, 1986; Agelopoulou and Magre, 1987; Kanai *et al.*, 1992) suggest that auto- and paracrine factors probably regulate fetal Sertoli cell proliferation. Endogenous FSH present in the testis may stimulate Sertoli cell proliferation in short-term cultures (Orth, 1984). Leydig cells may play an active role in Sertoli cell regulation, as demonstrated in the late fetal and postnatal rat testis (Orth, 1986; Orth and Boehm, 1990). The factor to suppress Sertoli cell proliferation from Leydig cells is possibly β-endorphin, which might interact with GTP-binding proteins of the Sertoli cell (Fig. 17). The Leydig cells probably play an important role in the regulation of the final size of the Sertoli cell population of the testis (Orth, 1982).

D. Regulation of Germ Cells

Primordial germ cells migrate to the gonadal ridge and divide mitotically as they progress toward the ridge, where they proliferate only for a short period, for example, until the age of 16 days in the rat (Orth, 1982). The differentiation of fetal male germ cells leads, at the prospermatogenic stage, into a premeiotic arrest at the G_1 phase of mitosis (Wartenberg, 1980) which lasts until the onset of spermatogenesis at puberty. This arrest lasts from the average fetal age of 17 days in rat and 14 to 16 days in mouse (Kluin *et al.*, 1984; Vergouwen *et al.*, 1991) until the first few days of postnatal life (Sapsford, 1962; Beaumont and Mandl, 1963; Huckins and Clermont, 1968; Hilscher *et al.*, 1974). In the rabbit, the arrest begins in the first postnatal week (Gondos and Byskov, 1981) and ends 7–8 weeks after birth (Gondos *et al.*, 1973). Premeiotic proliferation in the fetal human testis slows gradually and ceases finally between fetal weeks 12 and 18 (Wartenberg, 1978).

Sertoli cells are necessary for the survival and maturation of the germ cells, as indicated by degeneration of the germ cells outside the cords (Byskov, 1986), for example, in the adrenal gland (Upadhyay and Zamboni, 1982). Germ cells, on the other hand, are not needed for the formation of testicular cords (Merchant, 1975).

Some of the germ cells, especially those in the central part of the cords, however, are destined to degenerate and are phagocytosed by Sertoli cells, as shown in the fetal and newborn rat (Franchi and Mandl, 1964; Roosen-Runge and Leik, 1968), rabbit (Gondos and Conner, 1973), guinea pig (Black, 1971), and human (Gondos and Hobel, 1971) testis. This ability is maintained even in dissociated fetal mouse Sertoli cells (Kanai *et al.*, 1991).

E. Inhibition of Spermatogonial Meiosis

Sertoli cells (Figs. 12, 17) are likely to be the source for a meiosis-preventing substance (MPS) (Jost, 1970; Byskov and Saxén, 1976). Fetal mouse testis with differentiated Sertoli cells prevents meiosis of oocytes in organ culture also (Byskov and Saxén, 1976). The mouse ovary can induce meiosis in male germ cells, suggesting that the ovary secretes a meiosis-inducing substance (MIS). Meiotic cells are not found in mouse fetal testis treated with cAMP (Taketo *et al.*, 1984). Manipulation of cellular interactions may disturb the factors controlling meiosis, as shown in dissociation–reaggregation experiments of chimeric fetal mouse gonads *in vitro* (Dolci and de Felici, 1990), in which MPS is present in the testicular tissue but MIS is not found in the ovary. The role of Sertoli cells in the regulation of meiosis is supported by the observations that mouse germ cells, dislocated in the interstitium or in the adrenal gland, enter meiosis but degenerate soon (Upadhyay and Zamboni, 1982; Byskov, 1986).

The biochemistry and mechanisms of action of MIS and MPS are not known. MIS seems to cross-react with mouse and bovine cells (Grinsted *et al.*, 1979). The MIS and MPS molecules may be related structurally (Byskov, 1986). MIS has steroid-like properties and might regulate meiosis by the plasma membrane-associated inositol–secondary messenger pathway (Byskov *et al.*, 1986).

F. Endocrine and Paracrine Regulatory Relationships

The differentiation of fetal Leydig cells and their steroid biosynthesis are under the control of pituitary LH (Fig. 17). The extragonadal regulation of testicular development by the hypothalamo-pituitary system has started already during the fetal period; the best known target is the Leydig cell.

Secretion of LH into the fetal circulation begins in the rat at the age of 17 days (Aubert *et al.*, 1985). This event takes place after the initiation of testicular testosterone synthesis and the appearance of LH/hCG receptors. The mechanism by which LH receptors are induced is not known. Truncated versions of LH receptor mRNA are expressed at least 10 days before the full-size message in the ovary (Sokka *et al.*, 1992), suggesting an autonomous induction of receptor gene action in the precursor cells. The gonadal cords (Byskov, 1986) or Sertoli cells (Kerr and Donachie, 1986) somehow seem to stimulate the growth of adult Leydig cells. A proline competitor suppresses the formation of fetal testis cords and Leydig cell differentiation *in vitro* (Jost *et al.*, 1988), probably by alterations in the collagen matrix. On the other hand, serum in culture of rat fetal gonads disturbs cord formation but not Leydig cell differentiation (Patsavoudi *et al.*, 1985). The possible stimulatory factor of Sertoli cells in Leydig cell development, as well as their measured ability to secrete AMH, were in this case independent of cord formation. A possible new action for Sertoli cells is modifying undulating fetal testosterone production via the GnRH-like peptides found in cultures of fetal rat gonadal tissue at various ages (Habert *et al.*, 1991).

Current knowledge of the endocrine, paracrine, and autocrine mechanisms of the different cell types of the testis in the prenatal period, as described earlier in this chapter (Fig. 17), is scarce in comparison with information about postnatal cells. The fetal testis is hormonally more active than the ovary. This difference depends in some cases on differences in time schedule. Several potential auto- and paracrine factors have been found, some of which may act already at the prenatal phases of development (Bellvé and Zheng, 1989). Regulatory mechanisms in the adult testis are not necessarily directly applicable to the processes in the fetal tissues, which often are different from those in the adult.

G. Other Genetic Factors Related to Testicular Differentiation

Proto-oncogenes are cellular homologs of viral oncogenes, and often are related to cellular proliferation or differentiation. A proto-oncogene c-*kit* is expressed in mouse primordial germ cells and spermatogonia but becomes inactive when their proliferation ceases (Manova and Bachvarova, 1991). This gene encodes a tyrosine kinase receptor that is a member of the PDGF/CSF-1 receptor family. The c-*kit* mRNA and protein that are encoded by the *white spotting* (W) locus are present in adult mice in differentiating type A spermatogonia and in Leydig cells (Manova *et al.*, 1990; Yoshinaga *et al.*, 1991). The ligand for the c-*kit* protein is a hematopoietic stem cell growth factor encoded by the *Steel (Sl)* locus in mice (Matsui *et al.*, 1990). Both the W and the *Sl* mutation result in lack of primordial germ cells, but the Sertoli cells and testicular cords differentiate normally. Although no information is available about the c-*kit* gene product or its ligand in Sertoli cells, the cessation of spermatogonial mitosis in embryonic testis (Manova and Bachvarova, 1991) may result from a Sertoli cell action that would prevent the mediation of the growth factor signal to the receptor on the spermatogonial plasma membrane.

The c-*abl* gene encodes a protein tyrosine kinase and also is expressed in haploid male germ cells of the adult mouse (Meijer *et al.*, 1987). mRNA encoded by this gene is most abundant in elongating spermatids, suggesting that the c-*abl* protein may be related to spermatogenesis. The c-*myc* gene apparently plays a general role in cell differentiation; its mRNA and protein are present in spermatogonia but not in spermatocytes, spermatids, Leydig cells, or Sertoli cells (Koji *et al.*, 1988). The Sertoli cells in 6-day-old postnatal mice have a high amount of c-*myc* mRNA, correlating with the proliferation phase of Sertoli cells (Stewart *et al.*, 1984). The c-*myc* proto-oncogene probably has cell-autonomous functions already in the fetal period, and may be involved in cell–cell interactions of developing testis.

The Wilms tumor suppression gene *WT1* encodes a putative transcription factor (Pritchard-Jones *et al.*, 1990; Pelletier *et al.*, 1991; Armstrong *et al.*, 1992). The gene is expressed in early human and mouse testis and other embryonic and fetal organs in which the epithelia arise from mesenchyme-to-epithelium transition. During differentiation the expression changes; in the postnatal human and mouse gonads, the gene is expressed in the Sertoli and the surface epithelial cells (Armstrong *et al.*, 1992). These findings suggest a regional regulatory role for the *WT1* gene, but the gene product is not likely to be involved in the sexual differentiation of the testis because the gene is present and expressed in both sexes.

References

Agelopoulou, R., and Magre, S. (1987). Expression of fibronectin and laminin in fetal male gonads *in vivo* and *in vitro* with and without testicular morphogenesis. *Cell Diff.* **21**, 31–36.

Agelopoulou, R., Magre, S., Patsavoudi, E., and Jost, A. (1984). Initial phases of the rat testis differentiation in vitro. *J. Embryol. Exp. Morphol.* **83**, 15–31.

Albers, N., Bettendorf, M., Hart, C. S., Kaplan, S. L., and Grumbach, M. M. (1989). Hormone ontogeny in the ovine fetus. XXIII. Pulsatile administration of follicle-stimulating hormone stimulates inhibin production and decreases testosterone synthesis in the ovine fetal gonad. *Endocrinology* **124**, 3089–3094.

Amlani, S., and Vogl, A. W. (1988). Changes in the distribution of microtubules and intermediate filaments in mammalian Sertoli cells during spermatogenesis. *Anat. Rec.* **220**, 143–160.

Anthony, C. T., and Skinner, M. K. (1989). Cytochemical and biochemical characterization of testicular of testicular peritubular myoid cells. *Biol. Reprod.* **40**, 811–823.

Armstrong, J. F., Pritchard-Jones, K., Bickmore, W. A., Hastie, N. D., and Bard, J. B. L. (1993). The expression of the Wilms' tumour gene, WT1, in the developing mammalian embryo. *Mech. Dev.* **40**, 85–97.

Aubert, M. L., Begeot, M., Winiger, B. P., Morel, G., Sizonenko, P. C., and Duboid, P. M. (1985). Ontogeny of hypothalamic luteinizing hormone-releasing hormone (GnRH) and pituitary GnRH receptors in fetal and neonatal rats. *Endocrinology* **116**, 1565–1576.

Baker, B. S. (1989). Sex in flies: The splice of life. *Nature (London)* **340**, 521–524.

Beaumont, H. M., and Mandl, A. M. (1963). A quantitative study of primordial germ cells in the male rat. *J. Embryol. Exp. Morphol.* **11**, 715–740.

Behringer, R. R., Cate, R. L., Froelick, G. J., Palmiter, R. D., and Brinster, R. L. (1990). Abnormal sexual development in transgenic mice chronically expressing Müllerian inhibiting substance. *Nature (London)* **345**, 167–170.

Bellvé, A. R., and Zheng, W. (1989). Growth factors as autocrine and paracrine modulators of male gonadal functions. *J. Reprod. Fertil.* **85**, 771–793.

Bjerregaard, P., Bro-Rasmussen, F., and Reumert, T. (1974). Ultrastructural development of fetal rabbit testis. *Z. Zellforsch.* **147**, 401–413.

Black, V. H. (1971). Gonocytes in fetal guinea pig testes: Phagocytosis of degenerating gonocytes by Sertoli cells. *Am. J. Anat.* **131**, 415–426.

Black, V. H., and Christensen, A. K. (1969). Differentiation of interstitial cells and Sertoli cells in fetal guinea pig testes. *Am. J. Anat.* **124**, 211–238.

Born, W., and Wekerle, H. (1981). Selective, immunologically nonspecific adherence of lymphoid and myeloid cells to Leydig cells. *Eur. J. Cell Biol.* **25**, 76–81.

Bortolussi, M., Zanchetta, R., Belvedere, P., and Colombo, L. (1990). Sertoli and Leydig cell numbers and gonadotropin receptors in rat testis from birth to puberty. *Cell Tissue Res.* **260**, 185–191.

Budzik, G. P., Powell, S. M., Kamagata, S., and Donahoe, P. K. (1983). Mullerian inhibiting substance fractionation by dye affinity chromatography. *Cell* **34**, 307–314.

Burdzy, K., Tung, P. S., and Fritz, I. B. (1991). High-resolution scanning electron micrographs of freeze-cracked cells in testes from normal and irradiated rats at different stages of gonadal development. *Microsc. Res. Tech.* **19**, 189–202.

Burgoyne, P. S. (1988). Role of mammalian Y chromosome in sex determination. *Phil. Trans. R. Soc. Lond. (Biol.)* **322**, 63–72.

Byskov, A. G. (1986). Differentiation of mammalian embryonic gonad. *Physiol. Rev.* **66**, 71–117.

Byskov, A. G., and Grinsted, J. (1981). Feminizing effect of mesonephros on cultured differentiating mouse gonads and ducts. *Science* **212**, 817–818.

Byskov, A. G., and Saxén, L. (1976). Induction of meiosis in fetal mouse testis in vitro. *Dev. Biol.* **52**, 193–200.

Byskov, A. G., Höyer, P. E., Björkman, N., Mörk, A. B., Olsen, B., and Grinsted, J. (1986). Ultrastructure of germ cells and adjacent somatic cells correlated to initiation of meiosis in the fetal pig. *Anat. Embryol.* **175**, 57–67.

Cate, R. L., Mattaliano, R. J., Hession, C., Tizard, R., Farber, N. M., Cheung, A., Ninfa, E. G., Frey, A. Z., Gash, D. J., Chow, E. P., Fisher, R. A., Bertonis, J. M., Torres, G., Wallner, B. P., Ramachandran, K. L., Ragin, R. C., Manganaro, T. F., MacLaughlin, D. T., and Donahoe, P. K. (1986). Isolation of the bovine and human genes for Müllerian inhibiting substance and expression of the human gene in animal cells. *Cell* **45**, 685–698.

Cate, R. L., Donahoe, P. K., and MacLaughlin, D. T. (1990). Müllerian-inhibiting substance. *In* "Handbook of Experimental Pharmacology" (M. B. Sporn and A. B. Roberts, eds.), Vol. 95/II, pp. 179–210. Springer-Verlag, Berlin.

Chiquoine, A. D. (1954). The identification, origin, and migration of the primordial germ cells in the mouse embryo. *Anat. Rec.* **118**, 135–146.

Codesal, J., Regadera, J., Nistal, M., Regaderasejas, J., and Paniagua, R. (1990). Involution of human fetal Leydig cells—An immunohistochemical, ultrastructural and quantitative study. *J. Anat.* **172**, 103–114.

Cohen-Haguenauer, O., Picard, J. Y., Mattéi, M. G., Serero, S., Nguyen, V. C., de Tand, M. F., Guerrier, D., Hors-Cayla, M. C., Josso, N., and Frézal, J. (1987). Mapping of the gene for anti-Müllerian hormone to the short arm of human chromosome 19. *Cytogenet. Cell. Genet.* **44**, 2–6.

D'Agostino, A., Monaco, L., Stefanini, M., and Geremia, R. (1984). Study of the interaction between germ cells and Sertoli cells *in vitro*. *Exp. Cell Res.* **150**, 430–435.

Davis, R. L., Weintraub, H., and Lassar, A. B. (1987). Expression of a single transfected cDNA converts fibroblasts to myoblasts. *Cell* **51**, 987–1000.

Dechelotte, P., Chassagne, J., Labbe, A., Afane, M., Scheye, T., de Laguillaumie, B., and Boucher, D. (1989). Ultrastructural and immunohistochemical evidence of *in situ* differentiation of mononuclear phagocyte system cells in the interstitium of human fetal testis. *Early Hum. Dev.* **20**, 25–36.

de Felici, M., and Dolci, S. (1989). *In vitro* adhesion of mouse fetal germ cells to extracellular matrix components. *Cell Differ. Dev.* **26**, 87–96.

de Felici, M., and Siracusa, G. (1985). Adhesiveness of mouse primordial germ cells to follicular and Sertoli cell monolayers. *J. Embryol. Exp. Morphol.* **87**, 87–97.

de Felici, M., Boitani, C., and Cossu, G. (1985). Synthesis of glycoconjugates in mouse primordial germ cells. *Dev. Biol.* **109**, 375–380.

Dinges, H. P., Zatloukal, K., Schmid, C., Mair, S., and Wirnsberger, G. (1991). Coexpression of cytokeratin and vimentin in rete testis and epididymis—An immunohistochemical study. *Virchows Arch. (A)* **418**, 119–127.

Dolci, S., and de Felici, M. (1990). A study of meiosis in chimeric mouse fetal gonads. *Development* **109**, 37–40.

Dollé, P., and Duboule, D. (1989). Two gene members of the murine HOX-5 complex show regional and cell-type specific expression in developing limbs and gonads. *EMBO J.* **8**, 1507–1515.

Donahoe, P. K., Budzik, G. P., Trelstad, R., Mudgett-Hunter, M., Fuller, A., Jr., Hutson, J. M., Ikawa, H., Hayashi, A., and MacLaughlin, D. (1982). Müllerian-inhibiting substance: An update. *Rec. Prog. Horm. Res.* **38**, 279–330.

Donahoe, P. K., Hutson, J. M., Fallat, M. E., Kamagata, S., and Budzik, G. P. (1984). Mechanism of action of müllerian inhibiting substance. *Ann. Rev. Physiol.* **46**, 53–65.

Donovan, P. J., Stott, D., Cairns, L. A., Heasman, J., and Wylie, C. C. (1986). Migratory and postmigratory mouse primordial germ cells behave differently in culture. *Cell* **44**, 831–838.

Ekblom, M., Klein, G., Mugrauer, G., Fecker, L., Deutzmann, R., Timpl, R., and Ekblom, P. (1990). Transient and locally restricted expression of laminin A chain mRNA by developing epithelial cells during kidney organogenesis. *Cell* **60**, 337–346.

Engvall, E., Earwicker, D., Haaparanta, T., Ruoslahti, E., and Sanes, J. R. (1990). Distribution and isolation of four laminin variants: Tissue restricted distribution of heterotrimers assembled from five different subunits. *Cell Regul.* **1**, 731–740.

Escalante-Alcalde, D., and Merchant-Larios, H. (1992). Somatic and germ cell interactions during histogenetic aggregation of mouse fetal testes. *Exp. Cell Res.* **198**, 150–158.

Fauser, B. C. J. M., Galway, B., and Hsueh, A. J. W. (1988). Differentiation of ovarian and testicular cells: Intragonadal regulation by growth factors. *In* "Serono Symposia Publications: The Molecular and Cellular Endocrinology of the Testis" (B. A. Cooke and R. M. Sharpe, eds.), Vol. 50, pp. 281–296. Raven Press, New York.

Feldman, S. C., and Bloch, E. (1978). Developmental pattern of testosterone synthesis by fetal rat testes in response to luteinizing hormone. *Endocrinology* **102**, 999–1007.

Florini, J. R., Ewton, D. Z., and Magri, K. A. (1991). Hormones, growth factors, and myogenic differentiation. *Annu. Rev. Physiol.* **53**, 201–216.

Franchi, L. L., and Mandl, A. M. (1964). The ultrastructure of germ cells in foetal and neonatal male rats. *J. Embryol. Exp. Morphol.* **12**, 289–308.

Fridmacher, V., Locquet, O., and Magre, S. (1992). Differential expression of acidic cytokeratins 18 and 19 during sexual differentiation of the rat gonad. *Development* **115**, 503–517.

Fristrom, D. (1988). The cellular basis of epithelial morphogenesis. A review. *Tissue Cell* **20**, 645–690.

Fröjdman, K., and Pelliniemi, L. J. (1990). Morphogenesis and intermediate filaments in rat embryonic testis and mesonephros. *In* "Miniposters. 6th European Workshop on Molecular and Cellular Endocrinology of the Testis" (I. Huhtaniemi and M. Ritzén, eds.), p. C6. Mariehamns Tryckeri Ab, Mariehamn, Finland.

Fröjdman, K., Paranko, J., Kuopio, T., and Pelliniemi, L. J. (1989). Structural proteins in sexual differentiation of embryonic gonads. *Int. J. Dev. Biol.* **33**, 99–103.

Fröjdman, K., Paranko, J., Virtanen, I., and Pelliniemi, L. J. (1992a). Intermediate filaments and epithelial differentiation of male rat embryonic gonad. *Differentiation* **50**, 113–123.

Fröjdman, K., Malmi, R., and Pelliniemi, L. J. (1992b). Lectin-binding carbohydrates in sexual differentiation of rat male and female gonads. *Histochemistry* **97**, 469–477.

Gangnerau, M.-N., Funkenstein, B., and Picon, R. (1982). LH/hCG receptors and stimulation of testosterone biosynthesis in the rat testis: Changes during foetal development in vivo and in vitro. *Mol. Cell. Endocrinol.* **28**, 499–512.

Gelly, J. L., Richoux, J. P., Leheup, B. P., and Grignon, G. (1989). Immunolocalization of type IV collagen and laminin during rat gonadal morphogenesis and postnatal development of the testis and epididymis. *Histochemistry* **93**, 31–37.

Gilula, N. B., Fawcett, D. W., and Aoki, A. (1976). The Sertoli cell occluding junctions and gap junctions in mature and developing mammalian testis. *Dev. Biol.* **50**, 142–168.

Ginsburg, M., Snow, M. H. L., and McLaren, A. (1990). Primordial germ cells in the mouse embryo during gastrulation. *Development* **110**, 521–528.

Godin, I., and Wylie, C. C. (1991). TGFβ₁ inhibits proliferation and has a chemotropic effect on mouse primordial germ cells in culture. *Development* 113, 1451–1457.

Gondos, B. (1981). Cellular interrelationships in the human fetal ovary and testis. In "Eleventh International Congress of Anatomy: Advances in the Morphology of Cells and Tissues" (E. A. Vidrio and M. A. Galina, eds.), pp. 373–381. Liss, New York.

Gondos, B., and Byskov, A. G. (1981). Germ cell kinetics in the neonatal rabbit testis. *Cell Tissue Res.* 215, 143–151.

Gondos, B., and Conner, L. A. (1973). Ultrastructure of developing germ cells in the fetal rabbit testis. *Am. J. Anat.* 136, 23–42.

Gondos, B., and Hobel, C. J. (1971). Ultrastructure of germ cell development in the human fetal testis. *Z. Zellforsch.* 119, 1–20.

Gondos, B., Bhiraleus, P., and Hobel, C. J. (1971). Ultrastructural observations on germ cells in human fetal ovaries. *Am. J. Obstet. Gynecol.* 110, 644–652.

Gondos, B., Renston, R. H., and Conner, L. A. (1973). Ultrastructure of germ cells and Sertoli cells in the postnatal rabbit testis. *Am. J. Anat.* 136, 427–439.

Gould, R. P., and Haddad, F. (1978). Extratubular migration of gonocytes in the foetal rabbit testis. *Nature (London)* 273, 464–466.

Grinsted, J., Byskov, A. G., and Andreasen, M. P. (1979). Induction of meiosis in fetal mouse testis in vitro by rete testis tissue from pubertal mice and bulls. *J. Reprod. Fertil.* 56, 653–656.

Grove, B. D., and Vogl, A. W. (1989). Sertoli cell ectoplasmic specializations: A type of actin-associated adhesion junction? *J. Cell Sci.* 93, 309–323.

Grund, S. K., and Pelliniemi, L. J. (1987). Basal lamina as a criterion for ovarian cords and follicular differentiation. *Cell Diff.* 20, 83S.

Grund, S. K., Pelliniemi, L. J., Paranko, J., Müller, U., and Lakkala-Paranko, T. (1986). Reaggregates of cells from rat testis resemble developing gonads. *Differentiation* 32, 135–143.

Gubbay, J., Collignon, J., Koopman, P., Capel, B., Economou, A., Münsterberg, A., Vivian, N., Goodfellow, P., and Lovell-Badge, R. (1990). A gene mapping to the sex-determining region of the mouse Y chromosome is a member of a novel family of embryonically expressed genes. *Nature (London)* 346, 245–250.

Habert, R., Devif, I., Gangnerau, M. N., and Lecerf, L. (1991). Ontogenesis of the in vitro response of rat testis to gonadotropin-releasing hormone. *Mol. Cell. Endocrinol.* 82, 199–206.

Hall, J. L., and Wachtel, S. S. (1980). Primary sex determination: genetics and biochemistry. *Mol. Cell. Biochem.* 33, 49–66.

Hardy, M. P., Zirkin, B. R., and Ewing, L. L. (1989). Kinetic studies on the development of the adult population of Leydig cells in testes of the pubertal rat. *Endocrinology* 124, 762–770.

Hargitt, G. T. (1926). The formation of the sex glands and germ cells of mammals. II. The history of the male germ cells in the albino rat. *J. Morphol.* 42, 253–305.

Hayashi, M., Shima, H., Hayashi, K., Trelstad, R. L., and Donahoe, P. K. (1984). Immunocytochemical localization of Müllerian inhibiting substance in the rough endoplasmic reticulum and Golgi apparatus in Sertoli cells of the neonatal calf testis using a monoclonal antibody. *J. Histochem. Cytochem.* 32, 649–654.

Hilscher, B., Hilscher, W., Bülthoff-Ohnolz, B., Krämer, U., Birke, A., Pelzer, H., and Gauss, G. (1974). Kinetics of gametogenesis. I. Comparative histological and autoradiographic studies of oocytes and transitional prospermatogonia during oogenesis and prespermatogenesis. *Cell Tissue Res.* 154, 443–470.

Hölund, B., Clause, P. P., and Clemmensen, I. (1981). The influence of fixation and tissue preparation on the immunohistochemical demonstration of fibronectin in human tissue. *Histochemistry* 72, 291–299.

Holthöfer, H., Miettinen, A., Paasivuo, R., Lehto, V.-P., Linder, E., Alfthan, O., and Virtanen, I. (1983). Cellular origin and differentiation of renal carcinomas. A fluorescence microscopic study with kidney-specific antibodies, antiintermediate filament antibodies, and lectins. *Lab. Invest.* **49,** 317–326.

Huckins, C., and Clermont, Y. (1968). Evolution of gonocytes in the rat testis during late embryonic and early post-natal life. *Arch. Anat. Histol. Embryol.* **51,** 343–354.

Huhtaniemi, I., and Pelliniemi, L. J. (1992). Fetal Leydig cells: Cellular origin, morphology, life span, and special functional features. *Proc. Soc. Exp. Biol. Med.* **201,** 125–140.

Huhtaniemi, I. T., Korenbrot, C. C., and Jaffe, R. B. (1977). hCG binding and stimulation of testosterone biosynthesis in the human fetal testis. *J. Clin. Endocrinol. Metab.* **44,** 963–967.

Hutson, J. C. (1990). Changes in the concentration and size of testicular macrophages during development. *Biol. Reprod.* **43,** 885–890.

Josso, N. (1991). Anti-Müllerian hormone. *Baillieres Clin. Endocrinol. Metab.* **5,** 635–654.

Josso, N., Picard, J.-Y., and Tran, D. (1977). The antimüllerian hormone. *Rec. Prog. Horm. Res.* **33,** 117–167.

Josso, N., Picard, J.-Y., and Vigier, B. (1981). Purification de l'hormone antimüllérienne bovine à l'aide d'un anticorps monoclonal. *C. R. Acad. Sci. (III)* **293,** 447–450.

Jost, A. (1947). Recherches sur la différenciation sexuelle de l'embryon de lapin. III. Role des gonades foetales dans la différenciation sexuelle somatique. *Arch. Anat. Microsc. Morphol. Exp.* **36,** 271–315.

Jost, A. (1953). Problems of fetal endocrinology: The gonadal and hypophyseal hormones. *Rec. Prog. Horm. Res.* **8,** 379–418.

Jost, A. (1970). Hormonal factors in the sex differentiation of the mammalian foetus. *Phil. Trans. R. Soc. Lond. (Biol.)* **259,** 119–130.

Jost, A. (1988). Past, present and future of the research on gonadal differentiation and early function. *In* "Serono Symposia Review: Development and Function of the Reproductive Organs" (M. Parvinen, I. Huhtaniemi, and L. J. Pelliniemi, eds.), Vol. II, pp. 1–9, Ares-Serono Symposia, Rome.

Jost, A., Vigier, B., Prépin, J., and Perchellet, J.-P. (1973). Le développement de la gonade des freemartins. *Ann. Biol. Anim. Bioch. Biophys.* **13,** 103–114.

Jost, A., Valentino, O., Agelopoulou, R., and Magre, S. (1985). Animal morphogenesis — Action of an analogue of proline (L-azetidine-2-carboxylic acid) on the differentiation *in vitro* of the rat fetal testis. *C. R. Acad. Sci. (III)* **301,** 225–232.

Jost, A., Perlman, S., Valentino, O., Castanier, M., Scholler, R., and Magre, S. (1988). Experimental control of the differentiation of Leydig cells in the rat fetal testis. *Proc. Natl. Acad. Sci. U.S.A.* **85,** 8094–8097.

Kanai, Y., Kawakami, H., Kurohmaru, M., Hayashi, Y., Nishida, T., and Hirano, H. (1989a). Changes in lectin binding pattern of gonads of developing mice. *Histochemistry* **92,** 37–42.

Kanai, Y., Kurohmaru, M., Hayashi, Y., and Nishida, T. (1989b). Formation of male and female sex cords in gonadal development of C57BL/6 mouse. *Jpn. J. Vet. Sci.* **51,** 7–16.

Kanai, Y., Hayashi, Y., Kawakami, H., Takata, K., Kurohmaru, M., Hirano, H., and Nishida, T. (1991). Effect of tunicamycin, an inhibitor of protein glycosylation, on testicular cord organization in fetal mouse gonadal explants in vitro. *Anat. Rec.* **230,** 199–208.

Kanai, Y., Kawakami, H., Takata, K., Kurohmaru, M., Hirano, H., and Hayashi, Y. (1992). Involvement of actin filaments in mouse testicular cord organization *in vivo* and *in vitro*. *Biol. Reprod.* **46,** 233–245.

Kellokumpu-Lehtinen, P., and Pelliniemi, L. J. (1988). Hormonal regulation of differentiation of human fetal prostate and Leydig cells *in vitro*. *Folia Histochem. Cytobiol.* **26,** 113–118.

Kerr, J. B., and Donachie, K. (1986). Regeneration of Leydig cells in unilaterally cryptorchid rats: Evidence for stimulation by local testicular factors. *Cell Tissue Res.* **245,** 649–655.

Klein, G., Langegger, M., Timpl, R., and Ekblom, P. (1988). Role of laminin A chain in the development of epithelial cell polarity. *Cell* **55,** 331–341.

Klein, G., Ekblom, M., Fecker, L., Timpl, R., and Ekblom, P. (1990). Differential expression of laminin A and B chains during development of embryonic mouse organs. *Development* **110**, 823–837.

Kluin, P. M., Kramer, M. F., and de Rooij, D. G. (1984). Proliferation of spermatogonia and Sertoli cells in maturing mice. *Anat. Embryol.* **169**, 73–78.

Koji, T., Izumi, S., Tanno, M., Moriuchi, T., and Nakane, P. K. (1988). Localization *in situ* of c-*myc* mRNA and c-*myc* protein in adult mouse testis. *Histochem. J.* **20**, 551–557.

Koopman, P., Münsterberg, A., Capel, B., Vivian, N., and Lovell-Badge, R. (1990). Expression of a candidate sex-determining gene during mouse testis differentiation. *Nature (London)* **348**, 450–452.

Koopman, P., Gubbay, J., Vivian, N., Goodfellow, P., and Lovell-Badge, R. (1991). Male development of chromosomally female mice transgenic for Sry. *Nature (London)* **351**, 117–121.

Kormano, M. (1967). Dye permeability and alkaline phosphatase activity of testicular capillaries in the postnatal rat. *Histochemie* **9**, 327–338.

Kuopio, T., and Pelliniemi, L. J. (1989). Patchy basement membrane of rat Leydig cells shown by ultrastructural immunolabeling. *Cell Tissue Res.* **256**, 45–51.

Kuopio, T., Paranko, J., and Pelliniemi, L. J. (1989). Basement membrane and epithelial features of fetal-type Leydig cells in rat and human testis. *Differentiation* **40**, 198–206.

Kuroda, T., Lee, M. M., Haqq, C. M., Powell, D. M., Manganaro, T. F., and Donahoe, P. K. (1990). Müllerian inhibiting substance ontogeny and its modulation by follicle-stimulating hormone in the rat testes. *Endocrinology* **127**, 1825–1832.

LaQuaglia, M., Shima, H., Hudson, P., Takahashi, M., and Donahoe, P. K. (1986). Sertoli cell production of müllerian inhibiting substance in vitro. *J. Urol.* **136**, 219–224.

Leeson, T. S. (1975). Smooth muscle cells in the rat testicular capsule: A developmental study. *J. Morphol.* **147**, 171–186.

Leu, F. J., Engvall, E., and Damjanov, I. (1986). Heterogeneity of basement membranes of the human genitourinary tract revealed by sequential immunofluorescence staining with monoclonal antibodies to laminin. *J. Histochem. Cytochem.* **34**, 483–489.

Lording, D. W., and de Kretser, D. M. (1972). Comparative ultrastructural and histochemical studies of the interstitial cells of the rat testis during fetal and postnatal development. *J. Reprod. Fertil.* **29**, 261–269.

McGinley, D. M., and Posalaky, Z. (1986). Gap junctions between gonocytes and epithelial cells in the sex cords of the one day old rat. A freeze fracture study. *J. Submicrosc. Cytol. Pathol.* **18**, 75–84.

McKay, D. G., Hertig, A. T., Adams, E. C., and Danziger, S. (1953). Histochemical observations on the germ cells of human embryos. *Anat. Rec.* **117**, 201–219.

McLaren, A. (1990a). Of MIS and the mouse. *Nature (London)* **345**, 111.

McLaren, A. (1990b). What makes a man a man? *Nature (London)* **346**, 216–217.

McLaren, A. (1991a). Development of the mammalian gonad: The fate of the supporting cell lineage. *Bioessays* **13**, 151–156.

McLaren, A. (1991b). The making of male mice. *Nature (London)* **351**, 96.

Magre, S., and Jost, A. (1980). The initial phases of testicular organogenesis in the rat. An electron microscopy study. *Arch. Anat. Microsc.* **69**, 297–318.

Magre, S., and Jost, A. (1983). Early stages of the differentiation of the rat testis: relations between Sertoli and germ cells. *In* "Current Problems in Germ Cell Differentiation: Symposium of British Society for Developmental Biology" (A. McLaren and C. C. Wylie, eds.), pp. 201–214. Cambridge University Press, Cambridge.

Magre, S., and Jost, A. (1984). Dissociation between testicular organogenesis and endocrine cytodifferentiation of Sertoli cells. *Proc. Natl. Acad. Sci. U.S.A.* **81**, 7831–7834.

Magre, S., and Jost, A. (1991). Sertoli cells and testicular differentiation in the rat fetus. *Microsc. Res. Tech.* **19**, 172–188.

Malmi, R., Fröjdman, K., and Pelliniemi, L. J. (1990a). Lectin binding of glycoconjugates in differentiating rat testis. *In* "Miniposters. 6th European Workshop on Molecular and Cellular Endocrinology of the Testis" (I. Huhtaniemi and M. Ritzén, eds.), p. C5. Mariehamns Tryckeri Ab, Mariehamn, Finland.

Malmi, R., Fröjdman, K., and Söderström, K.-O. (1990b). Differentiation-related changes in the distribution of glycoconjugates in rat testis. *Histochemistry* **94**, 387–395.

Manova, K., and Bachvarova, R. F. (1991). Expression of c-*kit* encoded at the W locus of mice in developing embryonic germ cells and presumptive melanoblasts. *Dev. Biol.* **146**, 312–324.

Manova, K., Nocka, K., Besmer, P., and Bachvarova, R. F. (1990). Gonadal expression of c-*kit* encoded at the W locus of the mouse. *Development* **110**, 1057–1069.

Mason, A. J., Hayflick, J. S., Ling, N., Esch, F., Ueno, N., Ying, S.-Y., Guillemin, R., Niall, H., and Seeburg, P. H. (1985). Complementary DNA sequences of ovarian follicular fluid inhibin show precursor structure and homology with transforming growth factor beta. *Nature (London)* **318**, 659–663.

Massagué, J. (1990). The transforming growth factor-β family. *Ann. Rev. Cell Biol.* **6**, 597–641.

Matsui, Y., Zsebo, K. M., and Hogan, B. L. M. (1990). Embryonic expression of a haematopoietic growth factor encoded by the Sl locus and the ligand for c-*kit*. *Nature (London)* **347**, 667–669.

Mayerhofer, A., Seidl, K., Lahr, G., Bitter-Suermann, D., Christoph, A., Barthels, D., Wille, W., and Gratzl, M. (1992). Leydig cells express neural cell adhesion molecules *in vivo* and *in vitro*. *Biol. Reprod.* **47**, 656–664.

Meijer, D., Hermans, A., von Lindern, M., van Agthoven, T., de Klein, A., Mackenbach, P., Grootegoed, A., Talarico, D., Della Valle, G., and Grosveld, G. (1987). Molecular characterization of the testis specific c-abl mRNA in mouse. *EMBO J.* **6**, 4041–4048.

Merchant, H. (1975). Rat gonadal and ovarian organogenesis with and without germ cells. An ultrastructural study. *Dev. Biol.* **44**, 1–21.

Merchant-Larios, H. (1975). The onset of testicular differentiation in the rat: An ultrastructural study. *Am. J. Anat.* **145**, 319–330.

Merchant-Larios, H. (1979). Ultrastructural events in horse gonadal morphogenesis. *J. Reprod. Fertil.* **27**, (Suppl.), 479–485.

Meyers-Wallen, V. N., Manganaro, T. F., Kuroda, T., Concannon, P. W., MacLaughlin, D. T., and Donahoe, P. K. (1991). The critical period for Müllerian duct regression in the dog embryo. *Biol. Reprod.* **45**, 626–633.

Møller, C. J., Byskov, A. G., Roth, J., Celis, J. E., and Bock, E. (1991). NCAM in developing mouse gonads and ducts. *Anat. Embryol.* **184**, 541–548.

Moon, Y. S., Hardy, M. H., and Raeside, J. I. (1973). Biological evidence for androgen secretion by the early fetal pig testes in organ culture. *Biol. Reprod.* **9**, 330–337.

Münsterberg, A., and Lovell-Badge, R. (1991). Expression of the mouse anti-Müllerian hormone gene suggests a role in both male and female sexual differentiation. *Development* **113**, 613–624.

Nagano, T., and Suzuki, F. (1976). Freeze-fracture observations on the intercellular junctions of Sertoli cells and of Leydig cells in the human testis. *Cell Tissue Res.* **166**, 37–48.

Nagano, T., and Suzuki, F. (1978). Cell to cell relationships in the seminiferous epithelium in the mouse embryo. *Cell Tissue Res.* **189**, 389–401.

Narbaitz, R., and Adler, R. (1967). Submicroscopical aspects in the differentiation of rat fetal Leydig cells. *Acta Physiol. Lat. Am.* **17**, 286–291.

Necklaws, E. C., LaQuaglia, M. P., MacLaughlin, D., Hudson, P., Mudgett-Hunter, M., and Donahoe, P. K. (1986). Detection of Müllerian inhibiting substance in biological samples by a solid phase sandwich radioimmunoassay. *Endocrinology* **118**, 791–796.

Niemi, M., and Ikonen, M. (1961). Steroid-3β-ol-dehydrogenase activity in foetal Leydig's cells. *Nature (London)* **189**, 592–593.

Niemi, M., Ikonen, M., and Hervonen, A. (1967). Histochemistry and fine structure of the interstitial tissue in the human foetal testis. *In* "Ciba Foundation Colloquia on Endocrinology: Endocrinology of the Testis" (G. E. W. Wolstenholme and M. O'Connor, eds.), Vol. 16, pp. 31–55. Churchill, London.

Ohno, S. (1967). "Sex Chromosomes and Sex-Linked Genes," Monographs on endocrinology. Springer-Verlag, Berlin.

Ohno, S. (1980). The identification of testis-organizing H-Y antigen of man as hydrophobic polymers of M.W. 18,000 subunit. *Ann. Endocrinol.* **41**, 263–274.

Orth, J. M. (1982). Proliferation of Sertoli cells in fetal and postnatal rats: A quantitative autoradiographic study. *Anat. Rec.* **203**, 485–492.

Orth, J. M. (1984). The role of follicle-stimulating hormone in controlling Sertoli cell proliferation in testes of fetal rats. *Endocrinology* **115**, 1248–1255.

Orth, J. (1986). FSH-induced Sertoli cell proliferation in the developing rat is modified by β-endorphin produced in the testis. *Endocrinology* **119**, 1876–1878.

Orth, J. M., and Boehm, R. (1990). Endorphin suppresses FSH-stimulated proliferation of isolated neonatal Sertoli cells by a pertussis toxin-sensitive mechanism. *Anat. Rec.* **226**, 320–327.

Page, D. C., Mosher, R., Simpson, E. M., Fisher, E. M. C., Mardon, G., Pollack, J., McGillivray, B., de la Chapelle, A., and Brown, L. G. (1987). The sex-determining region of the human Y chromosome encodes a finger protein. *Cell* **51**, 1091–1104.

Palmer, M. S., Sinclair, A. H., Berta, P., Ellis, N. A., Goodfellow, P. N., Abbas, N. E., and Fellous, M. (1989). Genetic evidence that ZFY is not the testis-determining factor. *Nature (London)* **342**, 937–939.

Paranko, J. (1987). Expression of type I and III collagen during morphogenesis of fetal rat testis and ovary. *Anat. Rec.* **219**, 91–101.

Paranko, J., and Pelliniemi, L. J. (1992). Differentiation of smooth muscle cells in the fetal rat testis and ovary: Localization of alkaline phosphatase, smooth muscle myosin, F-actin, and desmin. *Cell Tissue Res.* **268**, 521–530.

Paranko, J., Pelliniemi, L. J., Vaheri, A., Foidart, J.-M., and Lakkala-Paranko, T. (1983). Morphogenesis and fibronectin in sexual differentiation of rat embryonic gonads. *Differentiation (Suppl.),* **23**, S72–S81.

Paranko, J., Kallajoki, M., Pelliniemi, L. J., Lehto, V.-P., and Virtanen, I. (1986). Transient coexpression of cytokeratin and vimentin in differentiating rat Sertoli cells. *Dev. Biol.* **117**, 35–44.

Paranko, J., Fröjdman, K., Grund, S. K., and Pelliniemi, L. J. (1988). Differentiation of epithelial cords in the fetal testis. *In* "Serono Symposia Review: Development and Function of the Reproductive Organs" (M. Parvinen, I. Huhtaniemi, and L. J. Pelliniemi, eds.), Vol. II, pp. 21–28. Ares-Serono Symposia, Rome.

Patsavoudi, E., Magre, S., Castanier, M., Scholler, R., and Jost, A. (1985). Dissociation between testicular morphogenesis and functional differentiation of Leydig cells. *J. Endocrinol.* **105**, 235–238.

Pelletier, J., Schalling, M., Buckler, A. J., Rogers, A., Haber, D. A., and Housman, D. (1991). Expression of the Wilms' tumor gene WT1 in the murine urogenital system. *Genes Dev.* **5**, 1345–1356.

Pelletier, R.-M., and Friend, D. S. (1983). The Sertoli cell junctional complex: Structure and permeability to filipin in the neonatal and adult guinea pig. *Am. J. Anat.* **168**, 213–228.

Pelliniemi, L. J. (1970). Fine structure of germ cords in human fetal testis. *In* "Advances in Andrology: Morphological Aspects of Andrology" (A. F. Holstein and E. Horstmann, eds.), Vol. 1, pp. 5–8. Grosse Verlag, Berlin.

Pelliniemi, L. J. (1975a). Ultrastructure of gonadal ridge in male and female pig embryos. *Anat. Embryol.* **147**, 19–34.

Pelliniemi, L. J. (1975b). Ultrastructure of the early ovary and testis in pig embryos. *Am. J. Anat.* **144**, 89–112.

Pelliniemi, L. J. (1976a). Human primordial germ cells during migration and entrance to the gonadal ridge. *J. Cell Biol.* **70**, 226a.

Pelliniemi, L. J. (1976b). Ultrastructure of the indifferent gonad in male and female pig embryos. *Tissue Cell* **8**, 163–174.

Pelliniemi, L. J. (1985). Sexual differentiation of the pig gonad. *Arch. Anat. Microsc. Morphol. Exp.* **74**, 76–80.

Pelliniemi, L. J., and Dym, M. (1993). The fetal gonad and sexual differentiation. *In* "Maternal–Fetal Endocrinology" (D. Tulchinsky and B. Little, eds.), 2d Ed. Saunders, Philadelphia.

Pelliniemi, L. J., and Lauteala, L. (1981). Development of sexual dimorphism in the embryonic gonad. *Hum. Genet.* **58**, 64–67.

Pelliniemi, L. J., and Niemi, M. (1969). Fine structure of the human foetal testis. I. The interstitial tissue. *Z. Zellforsch.* **99**, 507–522.

Pelliniemi, L. J., Paranko, J., Grund, S. K., Fröjdman, K., Foidart, J.-M., and Lakkala-Paranko, T. (1984a). Extracellular matrix in testicular differentiation. *Ann. N.Y. Acad. Sci.* **438**, 405–416.

Pelliniemi, L. J., Paranko, J., Grund, S. K., Fröjdman, K., Foidart, J.-M., and Lakkala-Paranko, T. (1984b). Morphological differentiation of Sertoli cells. *INSERM* **123**, 121–140.

Pelliniemi, L. J., Fröjdman, K., and Paranko, J. (1993). Embryological and prenatal development and function of Sertoli cells. *In* "The Sertoli Cell" (L. D. Russell and M. D. Griswold, eds.), pp. 87–113. Cache River Press, Clearwater, Florida.

Pepinsky, R. B., Siclair, L. K., Chow, E. P., Mattaliano, R. J., Manganaro, T. F., Donahoe, P. K., and Cate, R. L. (1988). Proteolytic processing of Müllerian inhibiting substance produces a transforming growth factor-beta-like fragment. *J. Biol. Chem.* **263**, 18961–18964.

Pfeiffer, D. C., and Vogl, A. W. (1991). Evidence that vinculin is co-distributed with actin bundles in ectoplasmic ("junctional") specializations of mammalian Sertoli cells. *Anat. Rec.* **231**, 89–100.

Picard, J.-Y., and Josso, N. (1976). Anti-Müllerian hormone: Estimation of molecular weight by gel filtration. *Biomedicine* **25**, 147–150.

Picard, J.-Y., and Josso, N. (1984). Purification of testicular anti-Müllerian hormone allowing direct visualization of the pure glycoprotein and determination of yield and purification factor. *Mol. Cell. Endocrinol.* **34**, 23–29.

Picard, J.-Y., Tran, D., and Josso, N. (1978). Biosynthesis of labelled anti-müllerian hormone by fetal testes: Evidence for the glycoprotein nature of the hormone and for its disulfide-bonded structure. *Mol. Cell. Endocrinol.* **12**, 17–30.

Picard, J.-Y., Benarous, R., Guerrier, D., Josso, N., and Kahn, A. (1986). Cloning and expression of cDNA for anti-Müllerian hormone. *Proc. Natl. Acad. Sci. U.S.A.* **83**, 5464–5468.

Picon, R. (1976a). Testicular inhibition of fetal müllerian ducts in vitro: Effect of dibutyryl cyclic AMP. *Mol. Cell. Endocrinol.* **4**, 35–42.

Picon, R. (1976b). Testosterone secretion by foetal rat testes *in vitro*. *J. Endocrinol.* **71**, 231–238.

Pointis, G., Latreille, M. T., and Cedard, L. (1984). Regulation of testosterone production in fetal testicular cells: effect of androgens. *Experientia* **40**, 756–757.

Pritchard-Jones, K., Fleming, S., Davidson, D., Bickmore, W., Porteous, D., Gosden, C., Bard, J., Buckler, A., Pelletier, J., Housman, D., van Heyningen, V., and Hastie, N. (1990). The candidate Wilms' tumour gene is involved in genitourinary development. *Nature (London)* **346**, 194–197.

Rabinovici, J., Goldsmith, P. C., Roberts, V. J., Vaughan, J., Vale, W., and Jaffe, R. B. (1991). Localization and secretion of inhibin/activin subunits in the human and subhuman primate fetal gonads. *J. Clin. Endocrinol. Metab.* **73**, 1141–1149.

Reyes, F. I., Winter, J. S. D., and Faiman, C. (1989). Endocrinology of the fetal testis. *In* "The Testis" (H. Burger and D. de Kretser, eds.), 2d Ed., pp. 119–142. Raven Press, New York.

Rivenson, A., Ohmori, T., Hamazaki, M., and Madden, R. (1981). Cell surface recognition: Spontaneous identification of mouse Leydig cells by lymphocytes, macrophages, and eosinophiles. *Cell. Mol. Biol.* **27**, 49–56.

Robertson, D. M., McLachlan, R. I., Burger, H. G., and de Kretser, D. M. (1989). Inhibin and inhibin-related proteins in the male. *In* "The Testis" (H. Burger and D. de Kretser, eds.), p. 231–254. Raven Press, New York.

Rodriguez-Martinez, H., Hurst, M., and Ekstedt, E. (1990). Localization of carbonic anhydrase activity in the developing boar testis. *Acta Anat.* **139,** 173–177.

Roosen-Runge, E. C. (1961). The rete testis in the albino rat: Its structure, development and morphological significance. *Acta Anat.* **45,** 1–30.

Roosen-Runge, E. C., and Anderson, D. (1959). The development of the interstitial cells in the testis of the albino rat. *Acta Anat.* **37,** 125–137.

Roosen-Runge, E. C., and Leik, J. (1968). Gonocyte degeneration in the postnatal male rat. *Am. J. Anat.* **122,** 275–300.

Sapsford, C. S. (1962). Changes in the cells of the sex cords and seminiferous tubules during the development of the testis of the rat and mouse. *Aust. J. Zool.* **10,** 178–192.

Schindler, A. M., Dayer, A., and Bischof, P. (1986). Immunohistochemical localization of pregnancy-associated plasma protein-A in the male genital tract. *Hum. Reprod.* **1,** 55–59.

Setchell, B. P., and Brooks, D. E. (1988). Anatomy, vasculature, innervation, and fluids of the male reproductive tract. *In* "The Physiology of Reproduction" (E. Knobil, J. Neill, L. L. Ewing, G. S. Greenwald, C. L. Markert, and D. W. Pfaff, eds.), pp. 753–836. Raven Press, New York.

Shaha, C., Morris, P. L., Chen, C.-L. C., Vale, W., and Bardin, C. W. (1989). Immunostainable inhibin subunits are in multiple types of testicular cells. *Endocrinology* **125,** 1941–1950.

Sinclair, A. H., Berta, P., Palmer, M. S., Hawkins, J. R., Griffiths, B. L., Smith, M. J., Foster, J. W., Frischauf, A.-M., Lovell-Badge, R., and Goodfellow, P. N. (1990). A gene from the human sex-determining region encodes a protein with homology to a conserved DNA-binding motif. *Nature (London)* **346,** 240–244.

Skinner, M. K. (1987). Cell–cell interactions in the testis. *Ann. N.Y. Acad. Sci.* **513,** 158–171.

Skinner, M. K., and Fritz, I. B. (1986). Identification of a non-mitogenic paracrine factor involved in mesenchymal–epithelial cell interactions between testicular peritubular cells and Sertoli cells. *Mol. Cell. Endocrinol.* **44,** 85–97.

Skinner, M. K., Fetterolf, P. M., and Anthony, C. T. (1988). Purification of a paracrine factor, P-mod-S, produced by testicular peritubular cells that modulates Sertoli cell function. *J. Biol. Chem.* **263,** 2884–2890.

Sokka, T., Hämäläinen, T., and Huhtaniemi, I. (1992). Functional LH receptor appears in the neonatal rat ovary after changes in the alternative splicing pattern of the LH receptor mRNA. *Endocrinology.* **130** 1738–1740.

Spiegelman, M., and Bennett, D. (1973). A light- and electron-microscopic study of primordial germ cells in the early mouse embryo. *J. Embryol. Exp. Morphol.* **30,** 97–118.

Steinberger, E., Steinberger, A., and Perloff, W. H. (1964). Studies on growth in organ culture of testicular tissue from rats of various ages. *Anat. Rec.* **148,** 581–590.

Stewart, D. W., and Raeside, J. I. (1976). Testosterone secretion by the early fetal pig testes in organ culture. *Biol. Reprod.* **15,** 25–28.

Stewart, T. A., Bellvé, A. R., and Leder, P. (1984). Transcription and promoter usage of the myc gene in normal somatic and spermatogenic cells. *Science* **226,** 707–710.

Suganuma, T., Muramatsu, H., Muramatsu, T., Ihida, K., Kawano, J., and Murata, F. (1991). Subcellular localization of N-acetylglucosaminide $\beta 1 \rightarrow 4$ galactosyltransferase revealed by immunoelectron microscopy. *J. Histochem. Cytochem.* **39,** 299–309.

Taketo, T., and Koide, S. S. (1981). *In vitro* development of testis and ovary from indifferent fetal mouse gonads. *Dev. Biol.* **84,** 61–66.

Taketo, T., Merchant-Larios, H., and Koide, S. S. (1984). Induction of testicular differentiation in the fetal mouse ovary by transplantation into adult male mice. *Proc. Soc. Exp. Biol. Med.* **176,** 148–153.

Tannahill, D., and Melton, D. A. (1989). Localized synthesis of the Vg1 protein during early *Xenopus* development. *Development* **106,** 775–785.

Tapanainen, J., Kuopio, T., Pelliniemi, L. J., and Huhtaniemi, I. (1984). Rat testicular endogenous steroids and number of Leydig cells between the fetal period and sexual maturity. *Biol. Reprod.* **31,** 1027–1035.

Torney, A. H., Robertson, D. M., Hodgson, Y. M., and de Kretser, D. M. (1990). *In vitro* bioactive and immunoactive inhibin concentrations in bovine fetal ovaries and testes throughout gestation. *Endocrinology* **127**, 2938–2946.

Torney, A. H., Robertson, D. M., and de Kretser, D. M. (1992). Characterization of inhibin and related proteins in bovine fetal testicular and ovarian extracts: Evidence for the presence of inhibin subunit products and FSH-suppressing protein. *J. Endocrinol.* **133**, 111–120.

Tran, D., and Josso, N. (1982). Localization of anti-Müllerian hormone in the rough endoplasmic reticulum of the developing bovine Sertoli cell using immunocytochemistry with a monoclonal antibody. *Endocrinology* **111**, 1562–1567.

Tran, D., Meusy-Dessolle, N., and Josso, N. (1977). Anti-Müllerian hormone is a functional marker of foetal Sertoli cells. *Nature (London)* **269**, 411–412.

Tung, P. S., and Fritz, I. B. (1980). Interactions of Sertoli cells with myoid cells *in vitro*. *Biol. Reprod.* **23**, 207–217.

Tung, P. S., and Fritz, I. B. (1986a). Cell–substratum and cell–cell interactions promote testicular peritubular myoid cell histotypic expression *in vitro*. *Dev. Biol.* **115**, 155–170.

Tung, P. S., and Fritz, I. B. (1986b). Extracellular matrix components and testicular peritubular cells influence the rate and pattern of Sertoli cell migration *in vitro*. *Dev. Biol.* **113**, 119–134.

Tung, P. S., and Fritz, I. B. (1987). Morphogenetic restructuring and formation of basement membranes by Sertoli cells and testis peritubular cells in coculture: Inhibition of the morphogenetic cascade by cyclic AMP derivatives and by blocking direct cell contact. *Dev. Biol.* **120**, 139–153.

Tung, P. S., Skinner, M. K., and Fritz, I. B. (1984). Cooperativity between Sertoli cells and peritubular myoid cells in the formation of the basal lamina in the seminiferous tubule. *Ann. N.Y. Acad. Sci.* **438**, 435–446.

Tung, P. S., Rosenior, J., and Fritz, I. B. (1987). Isolation and culture of ram rete testis epithelial cells: Structural and biochemical characteristics. *Biol. Reprod.* **36**, 1297–1312.

Ungar, F., Geiger, B., and Ben-Ze'ev, A. (1986). Cell contact- and shape-dependent regulation of vinculin synthesis in cultured fibroblasts. *Nature (London)* **319**, 787–791.

Upadhyay, S., and Zamboni, L. (1982). Ectopic germ cells: Natural model for the study of germ cell sexual differentiation. *Proc. Natl. Acad. Sci. U.S.A.* **79**, 6584–6588.

Vale, W., Rivier, C., Hsueh, A., Campen, C., Meunier, H., Bicsak, T., Vaughan, J., Corrigan, A., Bardin, W., Sawchenko, P., Petraglia, F., Yu, J., Plotsky, P., Spiess, J., and Rivier, J. (1988). Chemical and biological characterization of the inhibin family of protein hormones. *Rec. Prog. Horm. Res.* **44**, 1–34.

Vergouwen, R. P. F. A., Jacobs, S. G. P. M., Huiskamp, R., Davids, J. A. G., and de Rooij, D. G. (1991). Proliferative activity of gonocytes, Sertoli cells and interstitial cells during testicular development in mice. *J. Reprod. Fertil.* **93**, 233–243.

Viebahn, C., Lane, E. B., and Ramaekers, F. C. S. (1988). Keratin and vimentin expression in early organogenesis of the rabbit embryo. *Cell Tissue Res.* **253**, 553–562.

Vigier, B., Legeai, L., Picard, J.-Y., and Josso, N. (1982a). A sensitive radioimmunoassay for bovine anti-müllerian hormone, allowing its detection in male and freemartin fetal serum. *Endocrinology* **111**, 1409–1411.

Vigier, B., Picard, J.-Y., and Josso, N. (1982b). A monoclonal antibody against bovine anti-müllerian hormone. *Endocrinology* **110**, 131–137.

Vigier, B., Picard, J.-Y., Tran, D., Legeai, L., and Josso, N. (1984a). Production of anti-müllerian hormone: Another homology between Sertoli and granulosa cells. *Endocrinology* **114**, 1315–1320.

Vigier, B., Tran, D., Legeai, L., Bézard, J., and Josso, N. (1984b). Origin of anti-Müllerian hormone in bovine freemartin fetuses. *J. Reprod. Fertil.* **70**, 473–479.

Vigier, B., Picard, J.-Y., Campargue, J., Forest, M. G., Heyman, Y., and Josso, N. (1985). Secretion of anti-Müllerian hormone by immature bovine Sertoli cells in primary culture, studied by a competition-type radioimmunoassay: Lack of modulation by either FSH or testosterone. *Mol. Cell. Endocrinol.* **43**, 141–150.

Vigier, B., Watrin, F., Magre, S., Tran, D., and Josso, N. (1987). Purified bovine AMH induces a characteristic freemartin effect in fetal rat prospective ovaries exposed to it *in vitro.* *Development* **100**, 43–55.

Vihko, K., Penttilä, T.-L., Toppari, J., Parvinen, M., and Belin, D. (1987). Regulation and localization of plasminogen activators in the rat seminiferous epithelium. *Ann. N.Y. Acad. Sci.* **513**, 310–311.

Virtanen, I., Miettinen, M., Lehto, V.-P., Kariniemi, A.-I., and Paasivuo, R. (1985). Diagnostic application of monoclonal antibodies to intermediate filaments. *Ann. N.Y. Acad. Sci.* **455**, 635–648.

Virtanen, I., Kallajoki, M., Närvänen, O., Paranko, J., Thornell, L.-E., Miettinen, M., and Lehto, V.-P. (1986). Peritubular myoid cells of human and rat testis are smooth muscle cells that contain desmin-type intermediate filaments. *Anat. Rec.* **215**, 10–20.

Voutilainen, R., and Miller, W. L. (1987). Human Müllerian inhibitory factor messenger ribonucleic acid is hormonally regulated in the fetal testis and in adult granulosa cells. *Mol. Endocrinol.* **1**, 604–608.

Wachtel, S. S. (1983). "H–Y Antigen and the Biology of Sex Determination." Grune and Stratton, New York.

Warren, D. W., Haltmeyer, G. C., and Eik-Nes, K. B. (1972). Synthesis and metabolism of testosterone in the fetal rat testis. *Biol. Reprod.* **7**, 94–99.

Warren, D. W., Haltmeyer, G., and Eik-Nes, K. B. (1975). The effect of gonadotrophins on the fetal and neonatal rat testis. *Endocrinology* **96**, 1226–1229.

Warren, D. W., Huhtaniemi, I. T., Tapanainen, J., Dufau, M. L., and Catt, K. J. (1984). Ontogeny of gonadotropin receptors in the fetal and neonatal rat testis. *Endocrinology* **114**, 470–476.

Wartenberg, H. (1978). Human testicular development and the role of the mesonephros in the origin of a dual Sertoli cell system. *Andrologia* **10**, 1–21.

Wartenberg, H. (1980). Control of differentiation and proliferation of germ cells by somatic cells in the developing gonad. *Arch. Biol. Med. Exp. (Santiago)* **13**, 321–324.

Wartenberg, H., Rodemer-Lenz, E., and Viebahn, C. (1989). The dual Sertoli cell system and its role in testicular development and in early germ cell differentiation (prespermatogenesis). *In* "Reproductive Biology and Medicine" (A. F. Holstein, K. D. Voigt, and D. Grässlin, eds.), pp. 44–57. Diesbach Verlag, Berlin.

Witschi, E. (1948). Migration of the germ cells of human embryos from the yolk sac to the primitive gonadal folds. *Contrib. Embryol.* **32**, 67–82.

Yoshinaga, K., Nishikawa, S., Ogawa, M., Hayashi, S.-I., Kunisada, T., Fujimoto, T., and Nishikawa, S.-I. (1991). Role of c-*kit* in mouse spermatogenesis: Identification of spermatogonia as a specific site of c-*kit* expression and function. *Development* **113**, 689–699.

Zondek, L. H., and Zondek, T. (1976). Fetal hilar cells and Leydig cells in early pregnancy. *Biol. Neonate* **30**, 193–199.

3

Nuclear Morphogenesis during Spermiogenesis

MARVIN L. MEISTRICH

I. Introduction

The morphogenesis of the spermatid nucleus is important in the production of a spermatozoon that is transported to the ovum effectively and is capable of achieving successful fertilization. Defects in sperm nuclear morphogenesis, for example, abnormal shape or incomplete nuclear condensation, do result in lowered fertility.

In a spermatozoon, the nucleus constitutes most of the volume of the head. The acrosome, which covers the nucleus as a thin saccule in many species including mice, rats, and humans, is the other major component of the sperm head; its outline usually conforms to that of the nucleus.

The precision of sperm nuclear morphogenesis, particularly in rodents, is extremely high. In outbred and even some inbred strains of mice, over 95% of the sperm nuclei have the same shape. Indeed, the coefficients of variation of the nuclear length and width are only 4% (Wyrobek et al., 1976). The shape of the sperm nucleus is under genetic control; quantitative measurements of the dimensions can distinguish sperm from different inbred strains (Illison, 1969).

How the detailed shape of the sperm nucleus is produced is not known. The purpose of this chapter is to review succinctly the structures that could contribute to nuclear shaping and condensation (references will not be comprehensive but are chosen to provide an introduction to the more recent literature); to describe their temporal appearance, molecular composition, and biophysical properties; and to relate these characteristics to physical forces that are necessary to effect the stepwise changes in the shape of the nucleus to achieve its final configuration. The discussion emphasizes the mechanisms involved in morphogenesis of the mammalian sperm nucleus, particularly the falciform sperm of rodents. Selected information about other animal species will be related to or contrasted with that about mammals, but I do not suggest that the mechanisms proposed for mammalian nuclear shaping and condensation are applicable to nonmammalian species.

For the purposes of this discussion, nuclear shaping and chromatin condensation will be considered separate processes. In rats and mice, the steps of nuclear shaping are resolved fairly well from those in which nuclear condensation occurs. Although in other species [e.g., human, dog (Russell *et al.*, 1990)] condensation and shaping appear to occur concomitantly, I assume, based on observations in rodents, that they still are controlled separately.

II. Spermiogenesis

Spermiogenesis is the part of spermatogenesis that begins with the formation of the spermatid at the completion of the second meiotic division and ends with the release of the spermatozoon from the seminiferous epithelium. This phase is devoted to a remarkable series of morphogenic events that are important to producing a spermatozoon that is most suited to its task of fertilizing the ovum of that particular species.

Spermiogenesis can be divided into different steps, providing excellent landmarks to identify cells and arrange events sequentially. In the rat (Fig. 1), 19 steps require 21 days for completion; other species show similar events at analogous stages. The events of spermiogenesis have been best described in the rat and mouse, so those species are considered in this chapter. Although interspecies differences undoubtedly occur, some of the basic principles of morphogenesis should be applicable to all mammalian species.

Several morphogenic processes occur simultaneously during spermatid development, including the development of an acrosome, the development of a flagellum, nuclear shaping, nuclear condensation, mitochondrial reorganization, and elimination of the residual cytoplasm. A thorough description of these events is given by Russell *et al.* (1990), based on the earlier studies of Leblond and Clermont (1952) in the rat and Oakberg (1956) in the mouse. In this chapter, only the events of nuclear shaping and condensation are considered.

At the beginning of spermiogenesis the nucleus is spherical, contains decondensed chromatin typical of interphase cells, and is located roughly in the center of the cell (Fig. 1; Steps 1–7). During spermiogenesis, these three properties undergo dramatic changes. The nucleus moves to a position abutting the plasma membrane, which covers slightly less than half the nuclear

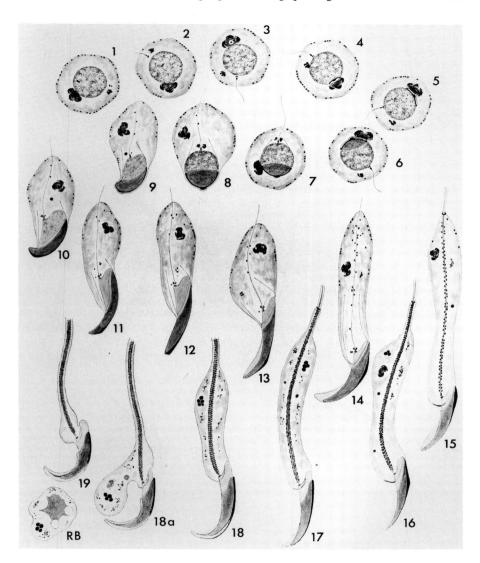

FIGURE 1   Cell structures for the 19 steps of rat spermiogenesis, showing the development of the nucleus and the relationship between the nuclear and cytoplasmic changes.   From Clermont and Rambourg (1978).   *Am. J. Anat.*   Copyright © 1978.   Reprinted by permission of Wiley-Liss, a Division of John Wiley and Sons, Inc.

surface (separated by only the acrosome); this event occurs at the beginning of Step 8 in the rat (Fig. 1) and mouse (Fig. 2). In rodents, the nuclear shape changes from a sphere to an asymmetric form that, when viewed in a longitudinal section, has a hook and specific curvatures and indentations but

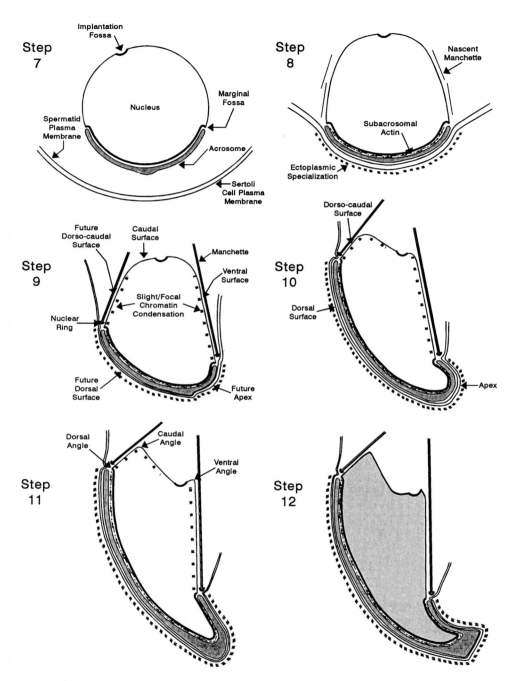

FIGURE 2 Nuclear shape, identifiable nuclear structures, and surrounding cytoskeletal and membranous structures for the steps of mouse spermiogenesis during which nuclear shaping occurs. Specific surfaces and angles are marked at stages at which they first become apparent.

is flattened in the perpendicular direction. This change in shape occurs primarily between Steps 9 and 13 in the rat (Fig. 1) and Steps 9 and 12 in the mouse (Fig. 2). Although some nuclear shape changes occur after Step 12 or 13, most of these are subtle and can be related to the chromatin condensation that occurs subsequently. In spatulate sperm of nonrodent species, the nuclear translocation, flattening, and condensation also occur; the major difference is that the hook does not form and the nucleus lacks dorso-ventral asymmetry.

Before the initiation of nuclear shaping at Step 8 in both the rat and the mouse, several morphological events occur that may be related to the determination of particular locations on the nuclear surface that later will have certain shapes. The position on the nucleus where the acrosomal granule makes contact in Step 3 will be the center of the anterior pole that becomes closely apposed to the plasma membrane at Step 8. At Step 7, the centrioles of the flagellum make contact with the nucleus at a site that is called the implantation fossa (Fig. 2). Initially, the site usually is displaced from the pole opposite from the acrosome, but by the beginning of Step 8 the contact site is at this pole. Note that, in the regions of the nucleus covered by the acrosome and at the implantation fossa, specific changes occur in the nuclear envelope, in conjunction with an increase in density of the immediately underlying nuclear material (Fig. 2; Step 7), which could represent specific chromatin structures. In the case of the acrosome, the altered membrane and the slight circular depression in the nuclear envelope at the periphery of the acrosomal sac are known as the modified nuclear envelope and the marginal fossa, respectively (Russell *et al.*, 1983).

III. Principles of Nuclear Shape Determination

Studies of somatic cells have shown that the structure responsible for the maintenance of the overall three-dimensional organization and shape of the nucleus is the nuclear matrix (Pienta *et al.*, 1989). This matrix is composed of a dynamic fibrous network of proteins running throughout the nucleus, nucleolar structures, RNA, polysaccharides, and a surrounding lamina that contains the nuclear pore complexes; the matrix is essentially devoid of histones and lipids and contains only 5% of the DNA in a protected environment. Although nuclei maintain their shape when isolated in suspension, their size may be altered. The nuclear matrix appears to show some degree of elasticity and varies in size to a limited degree *in vitro* in response to changes in salt (particularly divalent cation) concentrations (Anderson and Wilbur, 1952), presumably as a result of changes in the degree of chromatin condensation.

Although the basis for maintenance of nuclear shape has been well described, very little is known about the forces responsible for the determination of this shape, even in the case of somatic cells. One basic question is how much of the shape is contributed by internal forces rather than external forces.

Different cell types have distinctive shapes. About one-third of nuclei in different normal tissues are spherical or slightly ellipsoid (Rhodin, 1974). Other nuclei may be "irregular" or have a specific pattern to their shape. Some nonspherical nuclei, such as the flattened nuclei of endothelial cells and the triangular or polygonal outline of Sertoli cell nuclei when viewed in the cross section of a tubule (de Kretser and Kerr, 1988), appear to conform to the cell shape.

A spherical shape could result from sufficient surface tension at the nuclear periphery which, to reduce the energy state, would minimize the surface area for a given volume; however, no proof exists that this is the origin of spherical nuclei. Alternatively, spherical nuclei could be a result of symmetric shaping forces around the surface or in the interior of the nucleus. Internal forces involved in deviation from this spherical shape could be generated by local expansion or contraction of chromatin, the growth by polymerization of structural elements, or the activity of contractile elements. Although the nuclear matrix is not known to have contractile properties, accumulating evidence suggests the existence of intranuclear actin, microfilaments, and actin-binding proteins (Fukuda *et al.*, 1987).

External shaping forces could originate from or be transmitted by the actin microfilaments, intermediate filaments, and microtubules that compose the cytoskeleton. The nuclear lamina is linked to the cytoplasmic intermediate filaments that extend to the plasma membrane. Ultimately some of these filaments may be linked to the extracellular matrix or to adjacent cells, which may be the origin of forces on the nucleus (Pienta *et al.*, 1989).

IV. Elements Possibly Involved in Spermatid Nuclear Shaping

A. DNA and Chromosome Structure

Precedent exists in the structure of the chromosomes for DNA structure to contribute to morphogenesis. The length of each chromatid and the location of the centromere are determined largely by the length of the DNA and the position of the centromere-specific DNA sequences, respectively, and are modulated only slightly by the degree of condensation or packing. Fawcett *et al.* (1971) suggested that sperm "shape may be largely determined from within by a specific genetically controlled pattern of aggregation of DNA and protein." If DNA structure were important for sperm shape then, by analogy with chromosome morphology, specific sequences should be found in specific regions of the sperm nucleus. Also, sperm with abnormal DNA content or arrangement should have abnormal shapes.

The localization of specific DNA sequences within the sperm nucleus has been investigated using *in situ* hybridization. The Y chromosome revealed no obvious specificity with respect to position within the nucleus of human sperm (Wyrobek *et al.*, 1990). Centromeric DNA sequences hybridized to foci primarily in the equatorial region of bovine sperm (Powell *et al.*, 1990), but with no systematic distribution in rat elongated spermatids (Moens and Pearlman, 1989). Attempts to localize the centromeres also have been made by immunocytochemical labeling of kinetochore proteins with CREST antisera from patients with scleroderma, although the possibility that these proteins might become displaced from the centromere when histones are displaced in late spermatids must be considered (Courtens, 1982). Discrete foci of immunofluorescence were observed to be distributed randomly throughout the nucleus in both bovine and human sperm (Sumner, 1987; Palmer *et al.*, 1990). In mouse spermatids, the foci were clustered at Step 8 and later became dispersed and then disappeared (Brinkley *et al.*, 1986). Thus, the localization of the centromeres does not appear to be a general phenomenon.

The localization of DNA sequences also can be examined by their association with the nuclear annulus, a protein structure located at the implantation fossa of hamster sperm (Ward and Coffey, 1989). When the nuclear matrix is disrupted by decondensation of sperm *in vitro*, apparently all the DNA remains associated with the nuclear annulus, suggesting that it has at least one attachment site in each chromosome (Ward and Coffey, 1989). The DNA that remains associated with the nuclear annulus after shearing is a nonrandom representation of the total DNA (Ward and Cummings, 1991); however, no specific sequences associated with it have been identified to date. If specific sequences are found, they might have a role in the organization of the annulus and, hence, indirectly in sperm head morphogenesis.

The relationship between chromosomal abnormalities and sperm head shape has been tested using mice carrying Robertsonian, reciprocal, or other translocations. Morphologically normal sperm that contained an abnormal number of chromosome arms and abnormal amounts of DNA were observed (Stolla and Gropp, 1974). Further, no correlation was seen between the incidence of aneuploid spermatids and the incidence of abnormal sperm (Wyrobek *et al.*, 1975). Since sperm carrying translocated chromosomes or an aneuploid number of chromosomes can be morphologically normal, the sperm shape is not determined by the arrangement of DNA in chromosomes. Instead, other forces must be involved to cause the DNA and its associated protein to form the observed configurations.

Irrespective of whether DNA has an active role in nuclear shaping, it may act as a resistive force to shaping because of its viscous properties and its continuous linear structure, which resists twisting and the passage of strands through one another. The constraints of the linear structure can be relieved by topoisomerases. High levels of topoisomerase II are observed during nuclear shaping in chicken and rat spermatids (Roca and Mezquita, 1989; McPherson and Longo, 1992).

Analysis of the sperm nuclei in the rat and in several other mammalian species by freeze-fracture has revealed lamellar structures that run perpendicular to the narrowest dimension (Koehler *et al.*, 1983). These structures have been suggested to be a result of a sheet-like condensation of the nucleoprotein. Although a sheet-like arrangement of the DNA would favor a flattened structure for the sperm head, these lamellae have not been proved to represent the DNA and its associated protein, nor have they been documented to be formed during the time of nuclear shaping.

In summary, no convincing evidence exists that DNA and chromosome structure are principal determinants of the shape of the spermatozoon. However, DNA and chromatin structure may contribute to specific aspects of the shape.

B. Chromosomal Proteins

The basic chromosomal proteins constitute the bulk of the protein within the nucleus. A role for them in the shaping of the sperm head has been suggested (Fawcett *et al.*, 1971). However, because of the dramatic transitions that occur in these proteins during spermiogenesis, determining which ones are actually present during the time nuclear morphogenesis occurs is important.

In round spermatids of all species, the major basic chromosomal proteins are histones. Testis-specific histones have been identified in various mammals and, although their function is not known, they partially replace their somatic counterparts (Meistrich, 1989).

In contrast to the uniform composition of early spermatids, spermatozoa of different species contain very different basic proteins, including true protamines, keratinous protamines, somatic histones, sperm-specific histones, and intermediate sperm basic proteins (Kasinsky, 1989). Despite this heterogeneity, chromatin condensation is a consistent feature of spermatozoa from most organisms. Note, however, that sperm nuclear shapes are as diverse as their nuclear protein compositions. However, I do not believe that a relationship exists between the two.

The sperm of all eutherian mammals contain keratinous protamines encoded by two genes, *P1* and *P2* (Balhorn, 1989). Although most mammalian sperm have almost exclusively P1, the mouse, hamster, stallion, and human have a large proportion of P2 in their sperm.

Since in mammals the histones are not replaced directly by the protamines, an additional class of proteins called the transition proteins must be considered. These proteins are found only in elongated spermatids (and possibly elongating spermatids in some species) and are present after the loss of most of the histones but before the deposition of protamine. The major proteins of this class in mammals are TP1 and TP2 (Meistrich, 1989; Alfonso and Kistler, 1993). Other minor transition proteins appear to be present in these steps as well, for example, TP4 in the rat (Unni and Meistrich, 1992) and others in the ram (Dupressoir *et al.*, 1985) and boar (Akama *et al.*, 1990). Some nonmammalian species also have transition proteins, for example, S1 and S2 from dogfish (Chauviere *et al.*, 1989) and TH1 and TH2 from cricket (Kaye and McMaster-Kaye, 1982). Other proteins that are unique to late spermatids, for example, TP3 of the rat (E. Unni, Y. Zhang, M. L. Meistrich, and R. Balhorn, unpublished results) and T1 and T2 of cuttlefish (Wouters-Tyrou *et al.*, 1991), are actually precursors of protamines.

The question of a role for the transition proteins or protamines in rodent spermatid nuclear morphogenesis is resolved by a careful analysis of the timing of the deposition of these proteins, which is best studied in the rat (Fig. 3). TP1 and TP2 are first present in the nucleus at late Step 12 and are lost at Steps 15–16 (Heidaran *et al.*, 1988; Meistrich *et al.*, 1992; Alfonso and Kistler, 1993). The protamines, P1, and a precursor to P2, are first present at Step 16 (Grimes *et al.*, 1977; E. Unni, Y. Zhang, M. L. Meistrich, and R. Balhorn, unpublished results). In the mouse as well, the transition proteins and protamine do not appear until Step 12 or later (Mayer and Zirkin, 1979; Courtens and Loir, 1981; Balhorn *et al.*, 1984; Alfonso and Kistler, 1993). Since in both species the most significant parts of the shaping process occur between Steps 8 and 12 (Russell *et al.*, 1990), in rodents the transition proteins and protamines clearly cannot have a role in the major events of shaping the spermatid nucleus. Condensation does appear to be initiated by the deposition of the transition proteins at Step 12, and further compaction occurs later, at least in the mouse at Step 14 (Russell *et al.*, 1990), concurrent with the deposition of the protamines (Courtens and Loir, 1981). However, these events only produce an overall nuclear compression with subtle changes in shape.

Since the major chromosomal proteins during spermatid shaping are the histones, their possible role should be discussed also. Conceiving of a mechanism by which histones actively function in the shaping process is difficult. However, the histones must be in a state to allow the chromatin to be sufficiently plastic to permit shaping to occur. Further, these proteins could be involved in subsequent chromatin condensation, since histone modifications appear to play a role during chromatin condensation in metaphase as well as in spermiogenesis of species such as sea urchins with histone-containing sperm. However, this case is unlikely in rodent spermiogenesis since the major phase of chromatin condensation occurs as the histones are lost from the chromatin (Russell *et al.*, 1990; Meistrich *et al.*, 1992).

During the stages of spermatid nuclear morphogenesis, testis-specific histones, most of which were synthesized in spermatocytes with the excep-

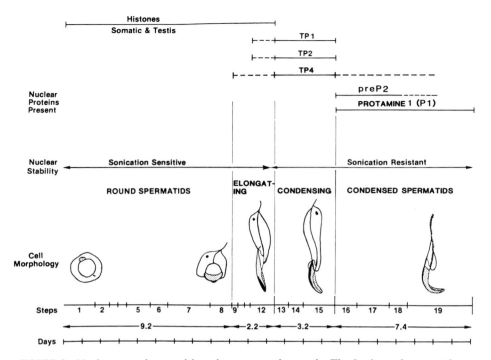

FIGURE 3 Nuclear protein transitions in rat spermiogenesis. The basic nuclear proteins present during various stages are correlated with cell morphology and kinetics. The definite presence of a nuclear protein at that stage is represented by a solid line. The dashed lines indicate accumulation of the protein (TP1, TP2), uncertainty as to the stage of appearance (TP4), or very low residual levels of the protein (TP4, preP2).

tion of one variant of H2B that is synthesized in spermatids (Moss *et al.,* 1989), constitute a significant percentage of the total histones. No evidence exists that they are localized to particular structures within the spermatid nucleus (Bhatnagar, 1985; Moss and Orth, 1992); some mammals exist in which testis-specific histone variants have not been identified (Meistrich, 1989). Thus, no role for the testis-specific histones in morphogenesis seems likely.

In addition, during spermiogenesis various posttranslational modifications of the histones (somatic and testis-specific) occur, the most well characterized of which is the acetylation of H4 in the rat (Meistrich *et al.,* 1992). H4 becomes highly acetylated during Steps 9–12. During Steps 9–11, this histone is present throughout the nucleus but in Step 12 it is associated primarily with the caudal part of the nucleus. The acetylation of H4 does cause decompaction of nucleosomes and chromatin and might facilitate chromatin rearrangement during nuclear shaping, but a more likely role would be facilitating histone displacement, allowing for replacement by the transition proteins which are most likely to cause the initial condensation of the chromatin.

C. Nuclear Matrix

The nuclear matrix must be considered, at least in the maintenance of the spermatid nuclear shape after the nucleus is molded into a given configuration by other forces. Assignment of an active role for the matrix in shaping is limited by the lack of known contractile or motor elements associated with it.

The most prominent and well-characterized proteins of somatic cell nuclear matrices are the lamins, which form the lamina that lines the inside surface of the inner nuclear membrane. Somatic cells contain lamins A, B, and C, which constitute a subfamily of intermediate filament proteins. Lamins A and C can form a fibrous meshwork; lamin B contains binding domains for membranes and for cytoplasmic intermediate filament proteins (Georgatos and Blobel, 1987).

In contrast to somatic cells, spermatid nuclei lack a morphologically distinct fibrous lamina structure (Fawcett, 1966; Ierardi *et al.*, 1983). Biochemical studies in rodents have shown that these cells also lack typical somatic cell lamins; for instance, no evidence exists for proteins related to lamins A and C (Moss *et al.*, 1987). However, these nuclei do contain a lamina composed of a protein closely related to lamin B (Sudhakar and Rao, 1990); this protein persists in late spermatids and epididymal spermatozoa (Sudhakar *et al.*, 1992).

Morphological studies of the nuclear matrix of mouse spermatozoa have revealed a matrix of interwoven fibrous material coursing through the interior of the nucleus (Bellve, 1982). However, instead of being surrounded by a distinct lamina, the fibers terminate at a perinuclear structure called the perinuclear theca.

Although an active role of the nuclear matrix in shaping the spermatid currently remains undefined, the properties of the matrix that would permit the shaping to occur and would maintain the shape will be considered. Swelling and condensation of nuclear matrices of spermatogenic cells (Ierardi *et al.*, 1983) and sperm (Longo *et al.*, 1987; Moss *et al.*, 1987) have been observed during preparation, indicating that the nuclear matrix is elastic to some extent. During spermiogenesis, the absence of a typical lamina containing a network of lamins A and C under the nuclear membrane may be necessary to allow the dramatic decrease in nuclear surface area that occurs during spermatid condensation. If the spermatid nuclear matrix is also slightly plastic, shaping forces can gradually and irreversibly alter the configuration of the nuclear matrix. In any case, the nuclear matrix has the function of organization of the DNA within the shape defined by these forces (Ward *et al.*, 1989).

D. Perinuclear Theca

Nuclei of spermatozoa are unusual because they lack a typical lamina but are surrounded by a dense cytoskeletal element referred to as the perinuclear theca (Lalli and Clermont, 1981). The perinuclear theca appears to be connected with the nuclear matrix. The maintenance of sperm nuclear size appears to be dependent on its integrity (Bellve, 1982; Longo *et al.*, 1987). The perinuclear theca displays two distinct regions, the subacrosomal portion and the postacrosomal dense lamina. In falciform spermatozoa of rodents, the subacrosomal portion is prominent and is called the perforatorium.

Whether the perinuclear theca plays any role in nuclear shaping depends on its time of appearance. In the rat, the postacrosomal dense lamina appears morphologically at Step 14 or 15 and the perforatorium appears during Step 19 as a result of condensation of material that has accumulated there during Steps 14–18 (Lalli and Clermont, 1981). In both cases, these structures appear after the major events of nuclear morphogenesis are complete. However, others indicate an accumulation of dense material within the subacrosomal space at earlier stages, particularly along the posterior margin of the acrosome and extending slightly into the postacrosomal region between the acrosome and the manchette (Czaker, 1985; Longo and Cook, 1991).

Further, the perinuclear thecal proteins, particularly those in the subacrosomal region, may appear at earlier stages; this question has been addressed using antibodies to specific proteins or groups of proteins derived from immunization with the perinuclear theca or nuclear matrix. One such group of proteins, an immunologically related heterogeneous group called the multiple band polypeptides, was identified throughout the perinuclear theca of spermatozoa (Longo *et al.*, 1987). The antibodies raised against these proteins reacted with material in the subacrosomal space of round, elongating, and early elongated spermatids of a variety of mammals (Longo and Cook, 1991). The multiple band polypeptides are not present in the postacrosomal region until later stages, during which the manchette migrates caudally away from the acrosome.

Another group of proteins, designated thecins, have been identified immunochemically by monoclonal antibodies and have been localized to the subacrosomal region of mouse testicular spermatids (Bellvé *et al.*, 1990). These proteins are present during the latter part of the round spermatid stage through the end of spermiogenesis and are localized to the area covered by the acrosome; they are absent from the postacrosomal segment.

An additional group of proteins recognized by antibodies raised against perforatorium preparations are localized specifically to the perforatorium of rat spermatozoa, but show a surprising distribution in spermiogenesis (Oko and Clermont, 1991). The perforatorium proteins first become prominent in round spermatids of the rat and are present in both the cytoplasm and the nucleus. Nuclear staining becomes more intense, reaches a peak at Step 9, and subsequently disappears during the condensation of the chromatin at Steps 12–13. Concomitant with the disappearance of nuclear staining is an increase in cytoplasmic labeling, particularly in the subacrosomal region. The intranuclear localization of these structural proteins during the steps of active nuclear morphogenesis raises the possibility that they could generate or transmit forces from within, as well as from the surroundings of the nucleus during nuclear shaping.

The presence of these groups of structural proteins, in addition to actin (discussed subsequently), in the subacrosomal space in close association with the nuclear membrane suggests that this region could have the machinery for force generation and nuclear shaping. Additional characterization of these proteins and their possible interactions with actin and motor proteins is needed to evaluate whether such a role is likely.

In contrast to the earlier appearance of structures and thecal proteins in the subacrosomal region, their appearance in the postacrosomal region does not occur until the manchette (see subsequent text) migrates caudally. Hence, these structures appear too late to have a role in the major events of nuclear morphogenesis. For example, the protein calicin has been identified and is localized in the postacrosomal region of the theca (Longo *et al.,* 1987) in spermatozoa of most species studied (Parnako *et al.,* 1988). However, during spermiogenesis calicin is found first in early elongating spermatids in a ring that appears to coincide with the caudal edge of the acrosome or the nuclear ring. In addition, other proteins (defined by their strong silver staining) that are components of the postacrosomal dense lamina (Czaker, 1985; Longo and Cook, 1991) actually are present earlier in the region at the caudal edge of the acrosome and between the acrosome and nuclear ring, and may link these structures during nuclear morphogenesis. The extension of these proteins to form the postacrosomal dense lamina as the nuclear ring–manchette system recedes is also consistent with a relationship between the acrosome and the nuclear ring.

E. Manchette and Nuclear Ring

The manchette is a microtubular structure that surrounds the caudal part of the nucleus of mammalian spermatids. The tubulins of the manchette microtubules are, at least in part, testis-specific and testis-enriched variants (Hecht *et al.,* 1988; Lewis and Cowan, 1988). Microtubule-associated proteins (MAP) identified in the manchette include MAP4 (Parysek *et al.,* 1984) and tau (Ashman *et al.,* 1992), which has been suggested to form connecting arms linking the microtubules. Further, in the rat, microtubule-dependent motor proteins have been identified on the manchette, using monoclonal antibodies (Hall *et al.*, 1992). Cytoplasmic dynein was observed on manchettes in Steps 15–17 (after the completion of nuclear shaping) whereas kinesin was present on manchettes in rat spermatids in Steps 10–18. However, the presence of other forms of these motors that are not recognized by these antibodies at earlier stages is not ruled out.

The nuclear ring is a structure to which the proximal tips of the manchette microtubules are attached (Rattner and Olson, 1973); this structure appears concurrent with manchette elaboration. The nuclear ring is a circular trough of dense material under a cup-shaped invagination in the plasma membrane just beyond the caudal edge of the acrosome, where the plasma and nuclear membranes come into close apposition. Since this structure maintains a constant relationship with the acrosomal complex during the steps of nuclear shaping, despite the dramatic changes that both undergo, the acrosome (or associated structures) is likely to be linked to the nuclear ring either directly or through the nucleus, the plasma membrane, or the ectoplasmic specialization of the Sertoli cell. The ring does appear to be linked with the plasma and nuclear membranes (Russell *et al.*, 1991).

The manchette is connected intimately to the nuclear ring (Cole *et al.*, 1988). The nuclear ring has been proposed to act as an organizing center for the manchette (Brinkley, 1985) and, hence, may be involved in determining the position of the manchette. The relationship between the nuclear ring and the manchette is demonstrated in spermatids from *azh* mutant mice, in which a nuclear ring-like structure extends along the plasma membrane, away from the nuclear membrane, and the manchette is correspondingly in an ectopic position (Meistrich *et al.*, 1990).

The association of the nuclear ring with the plasma membrane indicates that this structure also could interact with the membrane skeleton. Although no morphological evidence of a membrane skeleton has been reported in spermatids, fodrin—a major protein of the membrane skeleton in nonerythroid cells—is present along the plasma membrane of spermatids (De Cesaris *et al.*, 1989).

Currently no ultrastructural or immunochemical evidence exists for any cytoskeletal structures at the caudal end of the manchette. Thus, any external forces that deform the manchette must act on it at the anterior end through the nuclear ring.

A role for the manchette in nuclear shaping has long been debated (Fawcett *et al.*, 1971; Russell *et al.*, 1991). Several arguments have been presented to support a possible role for the manchette in nuclear shaping in mammals. The observation of links between the manchette and the nucleus that even extend through the nuclear membrane to the chromatin provides a mechanism to transmit forces (Courtens, 1982; Russell *et al.*, 1991). In a variety of genetic conditions and chemical treatments of rodents, manchette abnormalities are observed with nuclear abnormalities in elongating spermatids (Russell *et al.*, 1991).

Data from *Drosophila* provide strong genetic support that germ cell cytoplasmic microtubular structures, analogous to the manchette, are involved in nuclear elongation. In spermatids, a longitudinal bundle of microtubules forms along one side of the nucleus, which then elongates along the microtubules. Of the various mutations in the β2-tubulin gene, which shows germ cell-specific expression, only those that cause abnormalities in these microtubules result in failure of nuclear elongation (Fuller *et al.*, 1988). In contrast, observations on the leaf frog and scorpion have failed to show any microtubules associated with the nucleus at various steps during the shaping process (Asa and Phillips, 1988). Thus, microtubular structures may not be involved in shaping in all species.

A mechanism by which the manchette produces nuclear shaping is still needed. The most obvious way of exerting forces on the nucleus is through microtubule-based motors. Kinesin, which is first detected at Step 10 in rat spermiogenesis (Hall *et al.*, 1992), could be involved in the sliding of microtubules along each other to generate the asymmetry in this structure that occurs at that stage (see Fig. 2; Step 10). Other ways in which the manchette, which is rigid along the direction of the microtubules, could exert forces on the nucleus are by attaching the straight microtubules to the curved surface of the nucleus, transmitting forces originating in other regions of the cells through rigid microtubules with flexible linkers, and stimulating growth of microtubule bundles into the nucleus. In addition, one report has shown association of actin with the manchette in the opossum (Olson and Winfrey, 1991).

F. Subacrosomal Actin

Actin has been identified in the region between the nucleus and acrosome in round and elongating spermatids in all species studied (Russell *et al.*, 1986; Fouquet and Kann, 1992). Immunochemical evidence indicates that the actin in the subacrosomal space is a nonmuscle form of actin (Oko *et al.*, 1991).

During the development of the acrosome in round spermatids of the rat (Steps 4–7), actin filaments appear throughout the subacrosomal space. During nuclear shaping (Steps 9–13), they increase in number (Russell *et al.*, 1986). However, in later stages of spermiogenesis, the number of filaments and total amount of actin decrease. Similar observations have been reported in the mouse (Fouquet and Kann, 1992). A suggested role for actin in Steps 4–7 is the extension of the acrosomal cap over the nucleus; in Steps 8–13, its role could be the concordant shaping of the acrosome and the nucleus (Russell *et al.*, 1986). Although these authors considered the acrosome to follow the shape assumed by the nucleus, the reverse could be the case as well.

Since actin in association with other proteins can be involved in contractile elements, a possibility exists that the subacrosomal actin may be active in shaping the nucleus. In experimentally perturbed spermatogenesis, in which separations occur between the acrosome and nucleus, the actin remains closely apposed to the nuclear envelope, indicating that its primary interaction is with the nucleus (Fouquet *et al.*, 1989).

Actin filament-related contraction requires the presence of myosin, and only one report indicates the presence of myosin in the subacrosomal region of human sperm (Campanella *et al.*, 1979). However, the existence of other actin-based motor proteins cannot be excluded. For example, actin filaments can participate in motor functions solely with smaller proteins related to the head region of myosin (Titus *et al.*, 1989). Calmodulin is also present in the subacrosomal space, primarily during the stages of nuclear shaping in a variety of mammalian species (Kann *et al.*, 1991). Since calmodulin can regulate the polymerization state of actin through calcium levels and through the activity of myosin light-chain kinase, which initiates muscle contraction, its presence in the subacrosomal region is consistent with an active role for the actin present there.

Whether or not the subacrosomal actin has an active role in shaping, it may connect the acrosome tightly with the nucleus during nuclear morphogenesis. Spectrin-like proteins of the membrane skeleton of many cells bind actin and are important for anchoring actin to the membrane. Although spectrin-like proteins have been found primarily between the plasma membrane and the outer acrosomal membrane in elongating rabbit spermatids, they also are present in lower concentrations along the nuclear face of the inner acrosomal membrane (Camatini *et al.*, 1992).

G. Sertoli Cell Structures

The Sertoli cell contains numerous cytoskeletal elements that are in close contact with the region of its plasma membrane adjacent to the acrosome of each spermatid. In this region, the intercellular space between the Sertoli cell and the elongating spermatid is uniformly narrow (about 10 nm); connections between the plasma membrane of the Sertoli cell and that of the spermatid have been observed (Russell *et al.*, 1988). Further, whereas hypotonic treatment of tissue has caused separation of Sertoli cells from spermatids in other regions of contact, it has never increased the width of the intercellular space where the Sertoli cell faces the acrosomal system of spermatids (Russell, 1977).

Any forces generated (actively or passively) by Sertoli cell structures therefore can act directly on the acrosome, but only indirectly through the acrosome on the nucleus. Whether the acrosome is rigid enough to transmit forces from an external structure to the nucleus to contribute to its shaping is a topic of interest. The acrosome of elongating spermatids has a rigid matrix that can retain its structure after extraction with detergent and high salt

(Longo *et al.*, 1990), indicating that, during nuclear shaping, the acrosome may be capable of transmitting forces to the nucleus. However, additional studies are needed. In the meantime, possible forces generated by Sertoli cell elements should be considered.

Sertoli cells contain actin filaments, microtubules, and intermediate filaments in the region of the spermatid head. The actin is in the form of bundles of filaments, termed the ectoplasmic specialization (ES), that lie along the Sertoli cell plasma membrane adjacent to the acrosome of the elongating and elongated spermatid. Microtubules are aligned predominantly parallel to the Sertoli cell that surrounds elongating and elongated spermatids and are linked to the endoplasmic reticulum associated with the ES (Russell, 1977; Vogl, 1989). In addition, intermediate filaments of the vimentin type extend from the nucleus of the Sertoli cell to regions of the ES along the dorsal edge of each elongated spermatid in Steps 14–18 of the rat, where they are associated closely with the actin filaments (Amlani and Vogl, 1988).

Although short segments of ES occasionally face early spermatids, in rats and mice the ES first becomes prominent around Step 8 spermatids, coincidently with the translocation of the acrosome and spermatid nucleus to a position subjacent to the plasma membrane (Russell, 1977). Whether the ES directs the translocation of the spermatid nucleus to that position or whether it becomes enhanced in response to the presence of the spermatid nucleus at that region is not clear. Even when the acrosome is lacking, as is the case in drug-treated rats or mutant mice, nuclear translocation to the plasma membrane occurs over the region defined by the modified nuclear envelope, which is present even in the absence of the acrosome, but the development of the Sertoli cell ES in relation to the modified nuclear envelope appears variable (Russell *et al.*, 1983; Sotomayor and Handel, 1986).

After Step 8 in normal spermiogenesis, the ES intimately covers a large area of the plasma membrane overlying the acrosome of the spermatid. The actin filaments of the ES are associated with various actin-binding proteins including vinculin, a protein characteristic of actin-associated adhesion junctions, and appear to be linked to the Sertoli cell plasma membrane (Vogl, 1989). Separation of the spermatid and Sertoli cell membranes can be achieved only by disruption of the ES (Russell *et al.*, 1988). Further, when most of the membranes and cytoplasm from testis are solubilized by hypotonic detergent treatment, the ES (referred to as the mantle) still was observed to be attached (albeit loosely) to the elongating spermatid nucleus (Cole *et al.*, 1988). This evidence indicates that the ES is involved in adhesion between the Sertoli cell and the spermatid.

A role for the ES in generating forces for sperm nuclear morphogenesis remains speculative. Although in the ratfish myosin is present in the ES (Stanley and Lambert, 1985), in mammals myosin is absent so the ES cannot have the typical actomyosin sliding-filament contractile system (Vogl, 1989). However, the possibility of other force-generating or transmitting properties of actin in the ES cannot be ruled out. For example, the formation of actin fibers into bundles might increase their rigidity and linearity so they could exert a force tending to straighten the ES and the underlying acrosome.

In any case, the ES, acrosome, subacrosomal actin, and other perinuclear structures near the posterior margin of the acrosome (Czaker, 1985; Longo and Cook, 1991) are associated tightly and are referred to as the acrosomal complex. How forces are generated within this complex is not clear, but forces and shape changes may be transmitted through this complex.

V. Specific Forces Required in Spermatid Nuclear Shaping

From the intricate shape of the rodent spermatid nucleus, the several cytoskeletal elements surrounding the nucleus, and the limited area over which these elements extend, multiple elements that may act in concert appear to be responsible for the shaping. In the following discussion, I attempt to define specific nuclear shape changes in the steps of mouse spermiogenesis, to determine the location and directions of the forces necessary to produce these changes, and to speculate which of the known elements might be involved. This analysis was assisted by the use of a computer graphic representation (Harvard Graphics) of the structures, to evolve individual edges of the structures at each stage (Fig. 2) and to overlay the outlines of a given stage with the previous one (Fig. 4).

As discussed previously (Russell et al., 1991), nuclear shaping in rodents with asymmetric nuclei may be interpreted in two ways: (1) the structures surrounding the nucleus, the acrosome, the manchette, and the flagellum move over the nucleus during the most active steps of nuclear shaping or (2) these structures remain in fixed positions. For example, traditional histological descriptions indicate that the acrosome moves over the dorsal surface toward the caudal end at Step 10 of spermiogenesis (Oakberg, 1956). However, in the model used here, I assume that these structures remain in fixed positions from Step 8 onward and show how nuclear shapes may be generated. The existence of defined structures, such as the modified nuclear envelope and marginal fossa, argues against the sliding of the acrosome over the nuclear envelope. However, determining conclusively whether the fixed or mobile hypothesis is correct is not yet possible.

At Step 8 of mouse spermiogenesis, the nucleus first begins to deviate from its spherical shape (Fig. 2). A slight straightening of the acrosomal face and of the lateral surfaces occurs (Fig. 4). The straightening forces in the acrosomal region could be generated by any of the structures of the acrosomal complex, such as the ES or the subacrosomal actin. Although no marked changes occur in the subacrosomal actin filaments between Steps 7 and 8, the ES dramatically develops at this time.

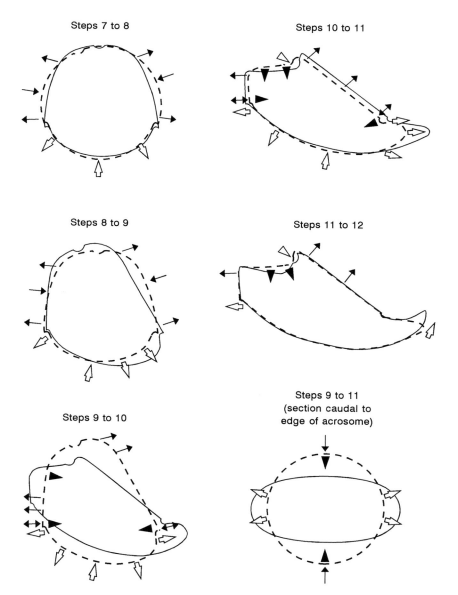

FIGURE 4 Changes in nuclear outline during spermiogenesis in the mouse from a given step (dashed line) to the next step (solid line) and indication of directions in which forces must act on the original nuclear outline to produce this change. *Arrows:* force vectors generated by or through the manchette; *double arrows:* inward force of the manchette against the acrosomal complex, but outward force pulling on the nucleus; *open arrows:* force vectors generated in or through the acrosomal region; *filled arrowheads:* force vectors transmitted because of strain on the nuclear contents or nuclear membrane, or generated by chromatin condensation; *open arrowheads:* force vectors acting through the flagellum. The lower right panel shows the nuclear outline in a transverse plane, perpendicular to the one shown in the other drawings.

Along the lateral surfaces, the manchette microtubules begin forming in Step 8, although at first the manchette does not appear to be very well organized. Manchette microtubules are quite rigid, relatively straight structures, and are linked to the nuclear envelope. We suggested that the initially spherical nuclear envelope could become linked progressively to the straight manchette microtubules, perhaps by a zipper mechanism that would be responsible for the forces that straighten these lateral faces at this and subsequent steps (Russell *et al.*, 1991).

By Step 9 of mouse spermiogenesis, the manchette is well formed and the nuclear ring, which appears to be an organizing center for the manchette, is prominent. As the manchette continues to become linked to the nuclear envelope, it straightens the outline of the lateral surfaces of the spermatid nucleus in longitudinal sections (Fig. 4). However, the nuclear surface is curved in a perpendicular plane, in which the microtubules are viewed in cross section (Fig. 4; lower right panel). Thus the linkers connecting the rigid straight tubules of the manchette appear to allow the tubules to form a curved cylindrical (or somewhat conical) surface.

The first indication of dorso-ventral asymmetry becomes apparent at Step 9 (Fig. 2). The surface over which the greater length of the manchette microtubules is linked to the nuclear envelope will be the future ventral surface; that over which the shortest length of microtubules is linked will be the dorso-caudal surface (Fig. 2). Note that spatulate sperm of nonrodents lack this dorso-ventral asymmetry; in these species, the manchette should be linked to the nucleus in a symmetrical manner.

As indicated earlier, some structure in the acrosomal complex appears to exert or transmit forces to straighten that surface. The straightening (increasing the radius of curvature) of the acrosome would act to increase its perimeter, which could be resisted by structures at the edge of the acrosome, within the nucleus, or within the nuclear ring–manchette system. One could envision that, although the straightening forces would increase the radius of curvature over most of the acrosomal complex, some weaker points in the structure would bend, forming surfaces with short radii of curvature (Fig. 4; Steps 9–10). The bends are located at the apex (Fig. 2) and the dorsal edge (not shown in plane of drawings). If the structures at these positions give way and allow bending, then additional straightening can occur on the short ventral portion, long dorsal portion, and lateral portions (not shown) of the acrosomal complex within the constraints of a fixed perimeter.

During Step 9, the outline of the mouse spermatid nucleus would be circular in a transverse cross section, but at Step 10 it becomes flattened into an ellipse or racetrack-shaped figure (Fig. 4; lower right panel). One mechanism by which this major change in nuclear configuration can be generated is that of straightening forces in the acrosomal region. If the forces are asymmetric and the displacements are constrained by a fixed perimeter of the caudal edge of the acrosomal complex, the complex will be straightened more in the plane shown in Fig. 2 than in the perpendicular plane, resulting in an elliptical outline. In addition, the length of the acrosomal complex increases in the plane shown with a concomitant shortening in the perpendicular direction.

The straightening forces from the acrosomal complex at the anterior end could be transmitted to the nuclear ring, via possible direct connections or indirectly via the nucleus, and then to the manchette. Although the manchette may be viewed as a scaffold of rigid microtubules, the linkers between the manchette microtubules must have some rotational freedom since the manchette cross section also becomes elliptical (Russell *et al.*, 1991). The manchette could transmit these flattening forces to the caudal part of the nucleus through the manchette–nuclear envelope links. A similar mechanism also could be responsible for flattening the nucleus of spatulate sperm; however, the straightening and extension of the acrosome are not as apparent as in rodent sperm.

At Step 10, the asymmetry of the caudal end of the mouse spermatid nucleus increases. The manchette may be envisioned to increase its linkage with the ventral surface of the nuclear envelope in a caudal direction while unzipping its links with the dorso-caudal surface, allowing the freed nuclear surface to become part of the caudal surface (Figs. 2, 4). The loss of caudal surface at the ventral portion, where it becomes linked to the manchette, and the gain at the dorsal portion may explain how the implantation fossa becomes situated close to the ventral surface although it did not begin in that position (Russell *et al.*, 1991).

The caudal surface of the nucleus also becomes straightened during Step 10. The cytoskeletal elements related to this surface are the flagellum and, around its perimeter, the manchette. If, as presumed earlier, the lateral, ventral, and dorso-caudal surfaces of the nucleus are being zippered close to the manchette, which is a sheath of fixed perimeter, and the area of the caudal surface does not change in this stage, the caudal surface would be pulled outward and the attempt to stretch it would cause a straightening of this surface. Also the flagellum, through its connections with the plasma membrane and membrane skeleton, may exert a force that indents the nucleus. At Step 11, these stretching forces also cause the edges between the caudal surface and those surfaces linked to the manchette to become sharper, as indicated at the ventral and caudal angles (Fig. 2).

Step 11 includes a continued straightening of the acrosomal complex, which results in a widening of the nucleus in the plane of the diagram and the further development of the apex (Fig. 4). Extension of the acrosomal complex in the plane of these diagrams, limited by a fixed perimeter at its caudal edge, could be the mechanism resulting in the growth of the apex.

Step 12 of mouse spermiogenesis is characterized by additional extension of the apex, increased concavity of the caudal surface, and the beginning of chromatin condensation (Mayer and Zirkin, 1979). The only known external structure that could exert the force needed to cause the concavity of the caudal surface is the flagellum. However, if chromatin condensation is beginning to decrease nuclear volume but the positions of the surfaces covered by the manchette and acrosome are fixed by these cytoskeletal structures, the only area that is free to undergo involution would be this caudal surface. Hence, indentation of the caudal surface could be a result of tensile forces produced by chromatin condensation.

The discussion presented here has indicated that nuclear morphogenesis in the mouse may be explained without invoking movement of the acrosome and manchette over the nucleus between Steps 8 and 12. After Step 12, the manchette undergoes a caudal migration over the nucleus and a dramatic remodeling (Cherry and Hsu, 1984; Cole *et al.*, 1988); however, little further nuclear morphogenesis occurs after Step 12 except that the nucleus becomes more condensed. With the manchette gone, the dorsal angle and dorso-caudal surface become more rounded as chromatin condensation proceeds. The caudal surface also becomes more concave, which could be attributed to chromatin condensation.

VI. Testing the Roles of Elements in Shaping

To test whether the elements listed in Section IV can exert the specific forces required for nuclear shaping as discussed in Section V, each element must be perturbed specifically and the effects observed. The effect of the perturbation on specific elements, as well as how it alters nuclear morphogenesis, should be characterized.

Having *in vitro* models to study sperm nuclear morphogenesis in mammals would be extremely useful since perturbations can be better controlled there than *in vivo*. Although some success has been had in achieving differentiation of mammalian germ cells in culture (Toppari and Parvinen, 1985), only fragmentary reports have been made to date of spermatids undergoing nuclear elongation in culture (Tres *et al.*, 1991). In contrast, newt spermatids undergo nuclear elongation in cell culture that is inhibited and reversed by antimicrotubule agents (Abe and Uno, 1984).

In the absence of a culture system, some information regarding the roles of elements in nuclear morphogenesis can be obtained by treatment of animals *in vivo* with agents that perturb spermatogenesis. Some agents may perturb certain cytoskeletal elements specifically, for example, cytochalasin D for microfilaments (Russell *et al.*, 1988) and vinblastine or taxol for microtubules (Russell *et al.*, 1981,1991); others, like procarbazine (Russell *et al.*, 1983), may act nonspecifically or indirectly. Although useful information has been gathered in those studies, for example, the role of Sertoli cell microtubules in maintaining cell integrity and, hence, the organization of germ cells and the ability of the manchette to distort the nucleus, this approach has

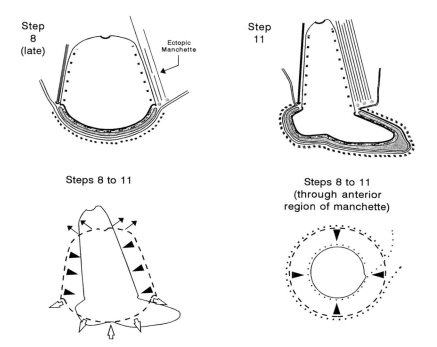

FIGURE 5 Abnormalities in nuclear morphogenesis in Steps 8 to 11 in spermatids from *azh* mutant mice, showing how forces generated by cytoskeletal elements can form the abnormal shapes observed. The lower right panel is a transverse cross section at the level of the anterior part of the manchette; the manchettes of the Step 8 and 11 spermatids are shown as dotted lines. (Arrows and arrowheads as in Fig. 4.)

several shortcomings. Even the specific agents given as examples perturb all the elements of the class. For instance, antimicrotubule agents will disrupt both the Sertoli cell microtubules and those of the manchette. Further, a variability of responses in different cells in the testis is seen often, particularly when an agent is injected intratesticularly.

The use of animals with single-gene mutations that affect spermatogenesis will produce a more homogeneous response, particularly if the mutation is in an inbred genetic background. In particular, mutations that affect only sperm nuclear morphogenesis and not spermatids at earlier stages or other structures within the spermatid will be most useful. The only mutation in mammals that fulfills these criteria is the *azh* mutation in the mouse, which produces normal numbers of sperm, all of which have abnormal shapes (Cole *et al.*, 1988).

The first defect observed in spermatids of *azh* mutants at Steps 8–9 is the abnormal position of the manchette at its formation (Fig. 5; upper left panel) (Meistrich *et al.*, 1990). At later stages, although the manchette largely encircles the nucleus, it appears to have gaps (Fig. 5; lower right panel). Other observations that should be explained by a consistent model are that the caudal part of the *azh* mutant sperm nucleus is never flat but always a cylinder or a tapered cylinder; that the manchette, although it is long, hardly extends beyond the caudal edge of the nucleus; and that the nucleus is longer in spermatids from *azh* mutants than in normal spermatids.

The model described in Section V for generation of forces during Steps 9–12 can be applied to the shaping of *azh* mutant spermatids. In normal spermatids, the zippering of the nucleus to the manchette is limited in the caudal direction by the tension exerted by the caudal surface of the nucleus, which is caused by the fixed perimeter of the intact manchette. However, the incomplete manchette of *azh* mutant spermatids imposes no such constraint and the zippering continues to the caudal end of the manchette (Fig. 5; upper right panel). The posterior portion of the nucleus must maintain its volume and, to compensate for the longitudinal extension, should decrease its cross-sectional area. The nucleus does so by becoming a longer, narrower, slightly conical cylinder (Fig. 5; lower left panel). As the underlying nucleus decreases its perimeter in cross section, more of the manchette may peel off and become ectopic (Fig. 5; lower right panel).

Meanwhile, some structure in the acrosomal complex should exert straightening forces on the anterior part of the nucleus and lengthening forces along its dorsal and ventral edges. As discussed in the next section, in the absence of a manchette these forces would result simply in a straightening of the anterior surface of the spermatid. However, in *azh* mutant spermatids, the manchette surrounds and is linked to most of the nucleus. Some structure, possibly the nuclear ring associated with the manchette or the nucleus itself (which is becoming a narrow cylinder in its posterior part and a smaller circle in a cross section just posterior to the acrosome), limits the perimeter of the caudal edge of the acrosomal complex. Hence the complex must bend at its weak points. Since the nuclear diameters at this level are smaller in *azh* mutant than in normal spermatids, the acrosome must form bends additional to those that normally occur at the apex and along the dorsal edge (Fig. 5; top right panel).

This discussion indicates that the mechanisms proposed for normal nuclear shaping in Section V can be applied to explain the abnormalities observed in the *azh* mutant. However, further observations and experiments are still required to prove the validity of this model.

Because of the limited numbers of naturally occurring mutations and the nonspecificity of chemical- or radiation-induced mutations, transgenic animals hold promise for specifically targeting mutations to spermatogenesis. Original approaches usually employed the addition of genes coupled with testis-specific promoters. More recently, the use of homologous recombination to knock out specific genes in embryonic stem cells offers a method for producing a defect in a specific gene that normally is expressed during spermiogenesis. Frequently, in the course of construction of transgenic animals for different purposes, recombination events may delete an unknown gene that is essential for normal spermiogenesis. One such line of transgenic mice is *sys*, which produces symplastic spermatids as the most apparent defect in the testis (MacGregor *et al.*, 1990).

Some of the symplastic spermatids in the *sys* mutant mice develop to Step 9 or 10; a few of these completely lack manchettes. These spermatids show a straightened nuclear surface where the acrosome is subjacent to the plasma membrane and the ES of the surrounding Sertoli cell (Fig. 6). The posterior portion of these nuclei is hemispherical and shows no lateral straightening or elongation.

As proposed in Section V, some structure in the acrosomal complex may produce straightening forces in this region (Fig. 6; lower panel). The opposing forces that would normally prevent this action are generated by the nuclear ring–manchette system, which constrains the posterior part of the nucleus. In the absence of a manchette, no such constraint exists and the posterior part of the nucleus should assume the most energetically favorable configuration. Since this area was initially part of a sphere, it becomes somewhat ovoid, a shape that may be enhanced if the acrosomal complex extends along one of its axes.

In summary, the results of studies of normal spermiogenesis and various perturbed systems are consistent with the following general concepts of

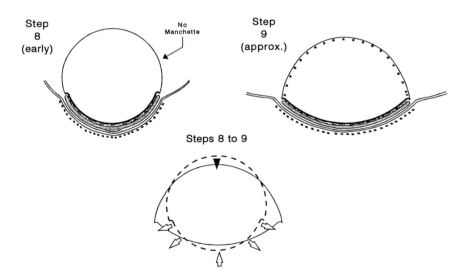

FIGURE 6 Abnormal nuclear morphogenesis in spermatids from *sys* mutant mice that lack manchettes, showing how forces generated by cytoskeletal elements in the acrosomal region can generate the shapes observed. (Arrows and arrowheads as in Fig. 4.)

nuclear morphogenesis in rodent spermatids. Forces are exerted in the part of the nucleus covered by the acrosome to straighten the surface but, because of constraints from other regions of the nucleus, specific sharp bends are produced also. The manchette, since it links with the nuclear envelope, causes the surfaces in the posterior part of the nucleus to lose their curvature along the direction of the microtubules. Although the nuclear ring–manchette system may not actively originate forces on its own, it acts as a scaffold— rigid in the longitudinal direction but flexible in the transverse plane—to transmit forces that it receives from its anterior end to the posterior part of the nucleus. The origin of the anterior forces may be peripheral to the spermatid nucleus or from within the nucleus. Forces from within the nucleus may not actually be generated from within but, because of its limited elasticity, the nucleus can transmit tensile forces exerted on one part of it to another region. In rodents, chromatin condensation occurs after the major part of nuclear shaping and can induce only subtle modifications to the shape already established by these other elements. Although generation of specific force vectors may be assigned to individual cytoskeletal or intranuclear elements, the formation of the shape of the sperm nucleus must involve interactions between the acrosomal complex, the nuclear ring—manchette system, the nucleus itself, and perhaps other as yet unidentified structures.

Acknowledgments

The research in the author's laboratory referred to in this chapter was supported by Grant HD-16843 from the National Institutes of Health. I thank Lonnie Russell for many interesting discussions of nuclear morphogenesis that contributed to ideas presented here and Emmanual Unni and Walter Pagel for editorial advice.

References

Abe, S.-I., and Uno, S. (1984). Nuclear elongation of dissociated newt spermatids *in vitro* and their shortening by antimicrotubule agents. *Exp. Cell Res.* **154**, 243–255.

Akama, K., Maruyama, R., Mochizuki, H., and Tobita, T. (1990). Boar transition protein 2 and 4 isolated from late spermatid nuclei by high-performance liquid chromatography. *Biochim. Biophys. Acta* **1041**, 264–268.

Alfonso, P. J., and Kistler, W. S. (1993). Immunohistochemical localization of spermatid nuclear transition protein 2 (TP2) in the testes of rats and mice. *Biol. Reprod.* **48**, 522–529.

Amlani, S., and Vogl, A. W. (1988). Changes in the distribution of microtubules and intermediate filaments in mammalian Sertoli cells during spermatogenesis. *Anat. Rec.* **220**, 143–160.

Anderson, N. G., and Wilbur, K. M. (1952). Studies on isolated cell components. IV. The effect of various solutions on the isolated rat liver nucleus. *J. Gen. Physiol.* **35**, 781–796.

Asa, C. S., and Phillips, D. M. (1988). Nuclear shaping in spermatids of the Thai leaf frog *Megophrys montana. Anat. Rec.* **220**, 287–290.

Ashman, J. B., Hall, E. A., Eveleth, J., and Boekelheide, K. (1992). Tau, the neuronal heat-stable microtubule-associated protein, is also present in the cross-linked microtubule network of the testicular spermatid manchette. *Biol. Reprod.* **46,** 120–129.

Balhorn, R. (1989). Mammalian protamines: Structure and molecular interactions. *In* "Molecular Biology of Chromosome Function" (K. W. Adolph, ed.), pp. 366–395. Springer-Verlag, New York.

Balhorn, R., Weston, S., Thomas, C., and Wyrobek, A. J. (1984). DNA packaging in mouse spermatids: Synthesis of protamine variants and four transition proteins. *Exp. Cell Res.* **150,** 298–308.

Bellve, A. R. (1982). Biogenesis of the mammalian spermatozoon. *In* "Prospects for Sexing Mammalian Sperm" (R. P. Amann and G. E. Seidel, Jr., eds.), pp. 69–102. Colorado Associated University, Boulder.

Bellve, A. R., Chandrika, R., and Barth, A. (1990). Temporal expression, polar distribution and transition of an epitope domain in the perinuclear theca during mouse spermatogenesis. *J. Cell Sci.* **96,** 745–756.

Bhatnagar, Y. M. (1985). Immunocytochemical localization of an H2A variant in the spermatogenic cells of the mouse. *Biol. Reprod.* **32,** 957–968.

Brinkley, B. R. (1985). Microtubule organizing centers. *Annu. Rev. Cell Biol.* **1,** 145–172.

Brinkley, B. R., Brenner, S. L., Hall, J. M., Tousson, A., Balczon, R. D., and Valdivia, M. M. (1986). Arrangements of kinetochores in mouse cells during meiosis and spermiogenesis. *Chromosoma* **94,** 309–317.

Camatini, M., Colombo, A., and Bonfanti, P. (1992). Cytoskeletal elements in mammalian spermiogenesis and spermatozoa. *Microsc. Res. Tech.* **20,** 232–250.

Campanella, C., Gabbiani, G., Baccetti, B., Burrini, A. G., and Pallini, V. (1979). Actin and myosin in the vertebrate acrosomal region. *J. Submicrosc. Cytol.* **11,** 53–71.

Chauviere, M., Martinage, A., Briand, G., Sautiere, P., and Chevaillier, P. (1989). Nuclear basic protein transition during sperm differentiation: Primary structure of the spermatid-specific protein S2 from the dog-fish *Scylliorhinus caniculus. Eur. J. Biochem.* **180,** 329–335.

Cherry, L. M., and Hsu, T. C. (1984). Antitubulin immunofluorescence studies of spermatogenesis in the mouse. *Chromosoma* **90,** 265–274.

Clermont, Y., and Rambourg, A. (1978). Evolution of the endoplasmic reticulum during rat spermiogenesis. *Am. J. Anat.* **151,** 191–212.

Cole, A., Meistrich, M. L., Cherry, L. M., and Trostle-Weige, P. K. (1988). Nuclear and manchette development in spermatids of normal and *azh/azh* mutant mice. *Biol. Reprod.* **38,** 385–401.

Courtens, J.-L. (1982). Roles indirects des microtubules dans la morphogenese nucleaire des spermatides. *Reprod. Nutr. Dev.* **22,** 825–840.

Courtens, J. L., and Loir, M. (1981). A cytochemical study of nuclear changes in boar, bull, goat, mouse, rat, and stallion spermatids. *J. Ultrastruc. Res.* **74,** 327–340.

Czaker, R. (1985). Morphogenesis and cytochemistry of the postacrosomal dense lamina during mouse spermiogenesis. *J. Ultrastr. Res.* **90,** 26–39.

De Cesaris, P., Filippini, A., Stefanini, M., and Ziparo, E. (1989). Spectrin, fodrin and protein 4.1-like proteins in differentiating rat germ cells. *Differentiation* **41,** 216–222.

de Kretser, D. M., and Kerr, J. B. (1988). The cytology of the testis. *In* "The Physiology of Reproduction" (E. Knobil and J. Neill, eds.), pp. 837–932. Raven Press, New York.

Dupressoir, T., Sautiere, P., Lanneau, M., and Loir, M. (1985). Isolation and characterization of the ram spermatidal nuclear proteins P1, 3 and T. *Exp. Cell Res.* **161,** 63–74.

Fawcett, D. W. (1966). On the occurrence of a fibrous lamina on the inner aspect of the nuclear envelope in certain cells of vertebrates. *Am. J. Anat.* **119,** 129–146.

Fawcett, D. W., Anderson, W. A., and Phillips, D. M. (1971). Morphogenetic factors influencing the shape of the sperm head. *Dev. Biol.* **26,** 220–251.

Fouquet, J.-P., and Kann, M.-L. (1992). Species-specific localization of actin in mammalian spermatozoa: Fact or artifact? *Microsc. Res. Tech.* **20,** 251–258.

Fouquet, J.-P., Kann, M.-L., and Dadoune, J.-P. (1989). Immunogold distribution of actin during spermiogenesis in the rat, hamster, monkey, and human. *Anat. Rec.* **223,** 35–42.

Fukuda, Y., Uchiyama, S., Masuda, Y., and Masugi, Y. (1987). Intranuclear rod-shaped actin filament bundles in poorly differentiated axillary adenosquamous cell carcinoma. *Cancer* **60,** 2979–2984.

Fuller, M. T., Caulton, J. H., Hutchens, J. A., Kaufman, T. C., and Raff, E. C. (1988). Mutations that encode partially functional β2 tubulin subunits have different effects on structurally different microtubule arrays. *J. Cell Biol.* **107,** 141–152.

Georgeatos, S. D., and Blobel, G. (1987). Lamin B constitutes an intermediate filament attachment site at the nuclear envelope. *J. Cell Biol.* **105,** 117–125.

Grimes, S. R., Meistrich, M. L., Platz, R. D., and Hnilica, L. S. (1977). Nuclear protein transitions in rat testis spermatids. *Exp. Cell Res.* **110,** 31–39.

Hall, E. S., Eveleth, J., Jiang, C., Redenbach, D. M., and Boekelheide, K. (1992). Distribution of the microtubule-dependent motors cytoplasmic dynein and kinesin in rat testis. *Biol. Reprod.* **46,** 817–828.

Hecht, N. B., Distel, R. J., Yelick, P. C., Tanhauser, S. M., Driscoll, C. E., Goldberg, E., and Tung, K. S. K. (1988). Localization of a highly divergent mammalian testicular α tubulin that is not detectable in brain. *Mol. Cell. Biol.* **8,** 996–1000.

Heidaran, M. A., Showman, R. M., and Kistler, W. S. (1988). A cytochemical study of the transcriptional and translational regulation of nuclear transition protein 1 (TP1), a major chromosomal protein of mammalian spermatids. *J. Cell Biol.* **106,** 1427–1433.

Ierardi, L. A., Moss, S. B., and Bellve, A. R. (1983). Synaptonemal complexes are integral components of the isolated mouse spermatocyte nuclear matrix. *J. Cell Biol.* **96,** 1717–1726.

Illison, L. (1969). Spermatozoal head shape in two inbred strains of mice and their F_1 and F_2 progenies. *Austr. J. Biol. Sci.* **22,** 947–963.

Kann, M.-L., Feinberg, J., Rainteau, D., Dadoune, J.-P., Weinman, S., and Fouquet, J.-P. (1991). Localization of calmodulin in perinuclear structures of spermatids and spermatozoa: A comparison of six mammalian species. *Anat. Rec.* **230,** 481–488.

Kasinsky, H. E. (1989). Specificity and distribution of sperm basic proteins. *In* "Histones and Other Basic Nuclear Proteins" (L. S. Hnilica, G. S. Stein, and J. L. Stein, eds.), pp. 73–163. CRC Press, Boca Raton, Florida.

Kaye, J. S., and McMaster-Kaye, R. (1982). Characterization of the unusual basic proteins of cricket spermatid nuclei on the basis of their molecular weights and amino acid compositions. *Biochim. Biophys. Acta* **696,** 44–51.

Koehler, J. K., Wurschmidt, U., and Larsen, M. P. (1983). Nuclear and chromatin structure in rat spermatozoa. *Gamete Res.* **8,** 357–370.

Lalli, M., and Clermont, Y. (1981). Structural changes of the head components of the rat spermatid during late spermiogenesis. *Am. J. Anat.* **160,** 419–434.

Leblond, C. P., and Clermont, Y. (1952). Spermiogenesis of rat, mouse, hamster and guinea pig as revealed by the "periodic acid-fuchsin sulfurous acid" technique. *Am. J. Anat.* **90,** 167–215.

Lewis, S. A., and Cowan, N. J. (1988). Complex regulation and functional versatility of mammalian α- and β-tubulin isotypes during the differentiation of testis and muscle cells. *J. Cell Biol.* **106,** 2023–2033.

Longo, F. J., and Cook, S. (1991). Formation of the perinuclear theca in spermatozoa of diverse mammalian species: Relationship of the manchette and multiple band polypeptides. *Mol. Reprod. Devel.* **28,** 380–393.

Longo, F. J., Krohne, G., and Franke, W. W. (1987). Basic proteins of the perinuclear theca of mammalian spermatozoa and spermatids: A novel class of cytoskeletal elements. *J. Cell Biol.* **105,** 1105–1120.

Longo, F. J., Cook, S., and Baillie, R. (1990). Characterization of an acrosomal matrix protein in hamster and bovine spermatids and spermatozoa. *Biol. Reprod.* **42**, 553–562.

MacGregor, G. R., Russell, L. D., van Beek, M. E. A. B., Hanten, G. R., Kovac, M. J., Kozak, C. A., Meistrich, M. L., and Overbeek, P. A. (1990). Symplastic spermatids (SYS), a recessive insertional mutation in mice causing a defect in spermatogenesis. *Proc. Natl. Acad. Sci. U.S.A.* **87**, 5016–5020.

McPherson, S. M., and Longo, F. J. (1992). Endogenous nicks in elongating spermatid DNA: Involvement of DNA topoisomerase II and protamine. *Mol. Biol. Cell* **3**, 102a.

Mayer, J. F., Jr., and Zirkin, B. R. (1979). Spermatogenesis in the mouse: I. Autoradiographic studies of nuclear incorporation of loss of ^{3}H-amino acids. *J. Cell Biol.* **81**, 403–410.

Meistrich, M. L. (1989). Histone and basic nuclear protein transitions in mammalian spermatogenesis. *In* "Histones and Other Basic Nuclear Proteins" (L. S. Hnilica, G. S. Stein, and J. L. Stein, eds.), pp. 165–182. CRC Press, Orlando, Florida.

Meistrich, M. L., Trostle-Weige, P. K., and Russell, L. D. (1990). Abnormal manchette development in spermatids of *azh/azh* mutant mice. *Am. J. Anat.* **188**, 74–86.

Meistrich, M. L., Trostle-Weige, P. K., Lin, R., Bhatnagar, Y. M., and Allis, C. D. (1992). Highly acetylated H4 is associated with histone displacement in rat spermatids. *Mol. Reprod. Devel.* **31**, 170–181.

Moens, P. B., and Pearlman, R. E. (1989). Satellite DNA I in chromatin loops of rat pachytene chromosomes and in spermatids. *Chromosoma* **98**, 287–294.

Moss, S. B., and Orth, J. M. (1993). Localization of a spermatid-specific histone 2B protein in mouse spermiogenic cells. *Biol. Reprod.* (In press).

Moss, S. B., Donovan, M. J., and Bellve, A. R. (1987). The occurrence and distribution of lamin proteins during mammalian spermatogenesis and early embryonic development. *Ann. N.Y. Acad. Sci.* **513**, 74–89.

Moss, S. B., Challoner, P. B., and Groudine, M. (1989). Expression of a novel histone 2B during mouse spermiogenesis. *Devel. Biol.* **133**, 83–92.

Oakberg, E. F. (1956). A description of spermiogenesis in the mouse and its use in analysis of the cycle of the seminiferous epithelium and germ cell renewal. *Am. J. Anat.* **99**, 391–413.

Oko, R., and Clermont, Y. (1991). Origin and distribution of perforatorial proteins during spermatogenesis of the rat: An immunocytochemical study. *Anat. Rec.* **230**, 489–501.

Oko, R., Hermo, L., and Hecht, N. B. (1991). Distribution of actin isoforms within cells of the seminiferous epithelium of the rat testis: Evidence for a muscle form of actin in spermatids. *Anat. Rec.* **231**, 63–81.

Olson, G. E., and Winfrey, V. P. (1991). Changes in actin distribution during sperm development in the opossum *Monodelphis domestica*. *Anat. Rec.* **230**, 209–217.

Palmer, D. K., O'Day, K., and Margolis, R. L. (1990). The centromere specific histone CENP-A is selectively retained in discrete foci in mammalian sperm nuclei. *Chromosoma* **100**, 32–36.

Paranko, J., Longo, F., Potts, J., Krohne, G., and Franke, W. W. (1988). Widespread occurrence of calicin, a basic cytoskeletal protein of sperm cells, in diverse mammalian species. *Differentiation* **38**, 21–27.

Parysek, L. M., Wolosewick, J. J., and Olmsted, J. B. (1984). MAP 4: A microtubule-associated protein specific for a subset of tissue microtubules. *J. Cell Biol.* **99**, 2287–2296.

Pienta, K. J., Partin, A. W., and Coffey, D. S. (1989). Cancer as a disease of DNA organization and dynamic cell structure. *Cancer Res.* **49**, 2525–2532.

Powell, D., Cran, D. G., Jennings, C., and Jones, R. (1990). Spatial organization of repetitive DNA sequences in the bovine sperm nucleus. *J. Cell Sci.* **97**, 185–191.

Rattner, J. B., and Olson, G. (1973). Observations on the fine structure of the nuclear ring of the mammalian spermatid. *J. Ultrastruc. Res.* **43**, 438–444.

Rhodin, J. A. G. (1974). "Histology: A Text and Atlas." Oxford, New York.

Roca, J., and Mezquita, C. (1989). DNA topoisomerase II activity in nonreplicating, transcriptionally inactive, chicken late spermatids. *EMBO J.* **8,** 1855–1860.

Russell, L. (1977). Observations on rat Sertoli ectoplasmic ('junctional') specializations in their association with germ cells of the rat testis. *Tissue Cell* **9,** 475–498.

Russell, L. D., Malone, J. P., and MacCurdy, D. S. (1981). Effect of the microtubule disrupting agents, colchicine and vinblastine, on seminiferous tubule structure in the rat. *Tissue Cell* **13,** 349–367.

Russell, L. D., Lee, I. P., Ettlin, R., and Peterson, R. N. (1983). Development of the acrosome and alignment, elongation and entrenchment of spermatids in procarbazine-treated rats. *Tissue Cell* **15,** 615–626.

Russell, L. D., Weber, J. E., and Vogl, A. W. (1986). Characterization of filaments within the subacrosomal space of rat spermatids during spermiogenesis. *Tissue Cell* **18,** 887–898.

Russell, L. D., Goh, J. C., Rashed, R. M. A., and Vogl, A. W. (1988). The consequences of actin disruption at Sertoli ectoplasmic specialization sites facing spermatids after *in vivo* exposure of rat testis to cytochalasin D. *Biol. Reprod.* **39,** 105–118.

Russell, L. D., Ettlin, R. A., Hikim, A. P. S., and Clegg, E. D. (1990). "Histological and Histopathological Evaluation of the Testis." Cache River Press, Clearwater, Florida.

Russell, L. D., Russell, J. A., MacGregor, G. R., and Meistrich, M. L. (1991). Linkage of manchette microtubules to the nuclear envelope and observations of the role of the manchette in nuclear shaping during spermiogenesis in rodents. *Am. J. Anat.* **192,** 97–120.

Sotomayor, R. E., and Handel, M. A. (1986). Failure of acrosome assembly in a male sterile mouse mutant. *Biol. Reprod.* **344,** 171–182.

Stanley, H. P., and Lambert, C. C. (1985). The role of a Sertoli cell actin-myosin system in sperm bundle formation in a ratfish, *Hydrolagus colliei* (Chrondrichthyes, Holocephali). *J. Morphol.* **186,** 223–236.

Stolla, R., and Gropp, A. (1974). Variation of the DNA content of morphologically normal and abnormal spermatozoa in mice susceptible to irregular meiotic segregation. *J. Reprod. Fertil.* **38,** 335–346.

Sudhakar, L., and Rao, M. R. S. (1990). Stage-dependent changes in localization of a germ cell-specific lamin during mammalian spermatogenesis. *J. Biol. Chem.* **265,** 22526–22532.

Sudhakar, L., Sivakumar, N., Behal, A., and Rao, M. R. S. (1992). Evolutionary conservation of a germ cell-specific lamin persisting through mammalian spermiogenesis. *Exp. Cell Res.* **198,** 78–84.

Sumner, A. T. (1987). Immunocytochemical demonstration of kinetochores in human sperm heads. *Exp. Cell Res.* **171,** 250–253.

Titus, M. A., Warrick, H. M., and Spudich, J. A. (1989). Multiple actin-based motor genes in *Dictyostelium. Cell Regul.* **1,** 55–63.

Toppari, J., and Parvinen, M. (1985). *In vitro* differentiation of rat seminiferous tubular segments from defined stages of the epithelial cycle: Morphologic and immunolocalization analysis. *J. Androl.* **6,** 334–343.

Tres, L. L., Smith, F. F., and Kierszenbaum, A. L. (1991). Spermatogenesis *in vitro*: Methodological advances and cellular functional parameters. *In* "Reproduction, Growth and Development" (A. Negro-Vilar and G. Perez-Palacios, eds.), Serono Symposia 71, pp. 115–125. Raven Press, New York.

Unni, E., and Meistrich, M. L. (1992). Purification and characterization of the rat spermatid basic nuclear protein TP4. *J. Biol. Chem.* **267,** 25359–25363.

Vogl, A. W. (1989). Distribution and function of organized concentrations of actin filaments in mammalian spermatogenic cells and Sertoli cells. *Int. Rev. Cytol.* **119,** 1–56.

Ward, W. S., and Coffey, D. S. (1989). Identification of a sperm nuclear annulus: A sperm DNA anchor. *Biol. Reprod.* **41,** 361–370.

Ward, W. S., and Coffey, D. S. (1990). Specific organization of genes in relation to the sperm nuclear matrix. *Biochem. Biophys. Res. Commun.* **173,** 20–25.

Ward, W. S., and Cummings, K. B. (1991). DNA associated with the sperm nuclear annulus is specific for that structure: Repeated 5S RNA gene as probe. *Surg. Forum* **42**, 677–679.

Ward, W. S., Partin, A. W., and Coffey, D. S. (1989). DNA loop domains in mammalian spermatozoa. *Chromosoma* **98**, 153–159.

Wouters-Tyrou, D., Chartier-Harlin, M.-C., Martin-Ponthieu, A., Boutillon, C., Van Dorsselaer, A., and Sautiere, P. (1991). Cuttlefish spermatid-specific protein T. *J. Biol. Chem.* **266**, 17388–17395.

Wyrobek, A. J., Heddle, J. A., and Bruce, W. R. (1975). Chromosomal abnormalities and the morphology of mouse sperm heads. *Can. J. Genet. Cytol.* **17**, 675–681.

Wyrobek, A. J., Meistrich, M. L., Furrer, R., and Bruce, W. R. (1976). The physical characteristics of mouse sperm nuclei. *Biophys. J.* **16**, 811–825.

Wyrobek, A. J., Alhborn, T., Balhorn, R., Stanker, L., and Pinkel, D. (1990). Fluorescence *in situ* hybridization to Y chromosomes in decondensed human sperm nuclei. *Mol. Reprod. Devel.* **27**, 200–208.

4

Hormonal Control of Spermatogenesis

GERHARD F. WEINBAUER & EBERHARD NIESCHLAG

I. Introduction

Mammalian spermatogenesis is regulated by pituitary hormones. A large body of literature provides compelling evidence that luteinizing hormone (LH) and follicle stimulating hormone (FSH) are the main regulators of the spermatogenic process in mammals. FSH acts directly on the seminiferous epithelium whereas LH exerts its effects on the spermatogenic process through testosterone produced by Leydig cells in response to LH (Fig. 1). The synthesis and release of the gonadotropic hormones LH and FSH are triggered by the hypothalamic peptide gonadotropin releasing hormone (GnRH) and are subject to feedback regulation mediated by testicular steroids (mainly testosterone) and peptides, inhibins, and presumably activins. Unlike the case for gonadotropic hormones, our knowledge about the involvement of pituitary factors other than gonadotropic hormones in mammalian spermatogenesis is much less comprehensive.

This chapter review considers several topics of hormonal control of male gametogenesis including the specific roles of LH, androgens, and FSH and their targets within the seminiferous epithelium; the importance of intratesticular androgen concentrations; the interactions between androgens and FSH; the relevance of pituitary hormones other than LH and FSH; and whether the duration of spermatogenesis is under hormonal control. The goal of this chapter is to provide a comparative assessment of the hormonal control of spermatogenesis in laboratory rodent models, in nonhuman primate models, and in men. Nonhuman primates are considered to represent an appropriate

ENDOCRINE BRAIN-TESTIS AXIS

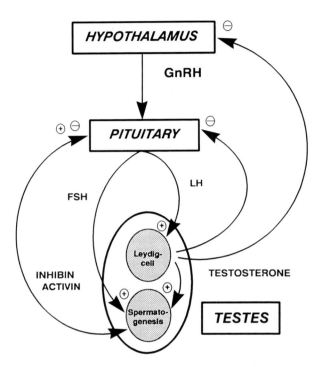

FIGURE 1 Representation of the hormonal interrelationships between the brain and the testis.

surrogate model for the study of reproductive processes in men (Nieschlag and Wickings, 1980). Finally, the implications of our advanced understanding of the regulation of spermatogenesis for the therapy of endocrine testicular disorders and for the development of an endocrine approach to male contraception are presented.

Since nonhuman primates are evolutionarily close to human primates and, in contrast to the rat and other rodent models, show many similarities to men in terms of regulation of testicular function, a distinction is made in some of the following sections of this chapter between primates and rodents.

II. Organization and Kinetics of Spermatogenesis

The process of spermatogenesis which takes place in the seminiferous tubules of the testis, encompasses a series of coordinated events that culminate in the formation and release of testicular spermatozoa. Spermatogenesis begins with the mitotic divisions of diploid A-type and B-type spermatogonia. The latter cells are named primary spermatocytes once they enter meiosis I and give rise to secondary spermatocytes. These cells contain a haploid chromosome number, but in duplicate sets. During meiosis II, haploid round spermatids evolve. The round spermatids do not divide but develop into elongated spermatids (testicular spermatozoa). This process, designated spermiogenesis, is followed by the release of the testicular spermatozoa (spermiation). Theoretically, one primary spermatocyte gives rise to four round spermatids. In the primate testis, two types of A spermatogonia have been identified, A-dark and A-pale spermatogonia. A-dark spermatogonia undergo divisions only after massive depletion of the A-pale spermatogonia (van Alphen et al., 1989) and are considered the testicular reserve stem cells. A-pale spermatogonia multiply regularly throughout the spermatogenic process and are considered the testicular renewing stem cells (Clermont, 1969; Clermont and Antar, 1973). In addition to the germ cells, somatic cells, that is, Leydig cells, peritubular cells, and Sertoli cells, play a key role in the spermatogenic process. For details of the cellular and molecular biology of these cells, the reader is referred to other chapters in this book.

When interpreting the effects of hormones on the spermatogenic process, a researcher must distinguish between three phases of germ cell development: initiation, maintenance, and reinitiation (Table 1). Initiation denotes the first completion of spermatogenesis, that is, the formation of testicular spermatozoa at puberty. Maintenance refers to the requirement of continuing spermatogenic events in the sexually mature organism. Reinitiation relates to the restart of gametogenesis once this process has been interrupted. To describe the latter phase, the term restoration is used also. Additionally, a clear distinction should be made between qualitatively normal spermatogenesis (presence of all germ cell types in reduced numbers) and quantitatively normal spermatogenesis (development of complete numbers of germ cells). Such distinction may be relevant to endocrine therapy of male infertility because, despite the presence of qualitatively normal spermatogenesis (production of subnormal numbers of spermatozoa), fertility might be impaired.

Comprehensive descriptions of the morphological and cytological characteristics of the Leydig cells, peritubular cells, Sertoli cells, and seminiferous

TABLE 1 Definitions of Certain Characteristics of the Spermatogenic Process

Initiation	First completion of spermatogenesis during puberty
Maintenance	Requirements during continued spermatogenesis
Reinitiation	Restart of the spermatogenic process
Qualitatively normal spermatogenesis	Presence of all germ cell types but at reduced numbers
Quantitatively normal spermatogenesis	Presence of complete numbers of all germ cell types

epithelium of rodents, nonhuman primates, and men are available (Fawcett, 1975; Dym and Cavicchia, 1978; Holstein, 1981, 1988; Nistal and Paniagua, 1984; Kerr, 1989; Russell *et al.*, 1990) and are discussed in other chapters of this book. The evolution and differentiation of germ cells—in the intact testis—progresses according to a remarkably stable and well-defined pattern. Early researchers recognized that the spermatogenic process could be categorized into stages, each of which contained a characteristic assembly of specific germ cells (cellular associations) (see Clermont,[1] 1972, for review). The classification of the spermatogenic stages is based on the morphology of the developing acrosome in spermatids (Leblond and Clermont, 1952a). Spermatogenesis has been divided into 14 stages in the rat (Leblond and Clermont, 1952b), into 8 stages in the cynomolgus monkey (Dang, 1971a), into 12 stages in the rhesus monkey and the stump-tailed macaque (Clermont and Antar, 1973; de Rooij *et al.*, 1986), and into 6 stages in man (Clermont, 1963).

With respect to the topographical arrangement of the spermatogenic stages along the length of the seminiferous tubule, important differences exist between rodents, nonhuman primates, and men. In the rat, a cross section of a seminiferous tubule yields a single stage; the stages follow a longitudinal arrangement in the seminiferous tubule. This spatial arrangement has been described as a spermatogenic wave (Perey *et al.*, 1961). Throughout the entire human testis (Heller and Clermont, 1964), in approximately one-third of the baboon testis (Chowdhury and Marshall, 1980), and in very small areas of the cynomolgus monkey testis (Dietrich *et al.*, 1986), several stages are present in cross sections of seminiferous tubules. This, at first glance irregular distribution of the spermatogenic stages, could be explained as a helical rather than longitudinal arrangement of the stages (Schulze, 1982; Schulze and Rehder, 1984). Thus, a spermatogenic wave also exists in men; the number of stages per tubular cross section reflects the extent of coiling of the spermatogenic helix. At present, whether this species-specific organization of spermatogenesis relates to differences in the hormonal requirements of the spermatogenic process remains unknown.

III. Initiation of Spermatogenesis by Testosterone

A. Rodents

The serum concentrations of FSH and testosterone attain peak levels 3–8 days postnatally during the prepubertal period of the rat (Lee *et al.*, 1975; Ketelslegers *et al.*, 1978). This activation of the hormonal axis is GnRH dependent and appears crucial to the initiation of spermatogenesis and the development of reproductive capacity, evidenced by immunization against

TABLE 2 Effects of LH/Androgens and FSH on the Qualitative and Quantitative Initiation, Maintenance, and Reinitiation of Spermatogenesis in Rats, Hamsters, Nonhuman Primates, and Men[a]

Phase	Rat	Nonhuman primate	Man	Hamster
Initiation				
Qualitatively normal	LH + FSH	LH or FSH	LH *FSH*	—[b]
Quantitatively normal	LH + FSH	*LH + FSH*	*LH + FSH*	—
Maintenance				
Qualitatively normal	LH or FSH	LH or FSH	LH *FSH*	—
Quantitatively normal	LH	*LH + FSH*	*LH + FSH*	—
Reinitiation				
Qualitatively normal	LH + *FSH*	LH or FSH	LH + FSH	FSH
Quantitatively normal	LH + FSH	*LH + FSH*	LH + FSH	LH + FSH

[a] *LH/FSH*: Requirement likely.
[b] Not studied.

GnRH (Bercu *et al.*, 1977) and administration of a GnRH antagonist causing permanent testicular dysfunction and impaired fertility (Huhtaniemi *et al.*, 1986; Kolho and Huhtaniemi, 1989).

Treatment of immature rats that had been hypophysectomized or exposed to estrogens with androgens, LH, human menopausal gonadotropin (hMG), or human chorionic gonadotropin (hCG) exerted a stimulatory effect on spermatogenic development (Lostroh *et al.*, 1963; Chowdhury and Steinberger, 1975; Chemes *et al.*, 1976, 1979; Kula, 1988). However, spermatogenesis failed to proceed beyond the meiotic stages or the appearance of early spermatids; the hypophysectomy-induced loss of germ cells was prevented only partially by LH (Russell *et al.*, 1987). The significance of the findings by Kula (1988) that estradiol stimulated the multiplication of the testicular reserve stem cells during the first spermatogenic cycle in the rat remains unclear at present. Among rodents, at least in the rat, testosterone alone does not appear to be capable of initiating qualitatively normal spermatogenesis (Table 2).

B. Nonhuman Primates and Men

Similar to the events in rats, a neonatal rise of gonadotropin and testosterone release also occurs in primates. Unlike in the rat, however, the neonatal phase precedes a period of quiescence in reproductive hormone secretion; germ cell proliferation is halted for several years before the juvenile phase and puberty are attained (Robinson and Bridson, 1978; Steiner and Bremner, 1981; Swerdloff and Heber, 1981; Plant, 1985). A neonatal peak of intratesticular testosterone level was noted in the human (Bidlingmeier *et al.*, 1983). Administration of ovine LH for induction of testicular development has been tried in several studies with varying success. Kar *et al.* (1966) reported mitotic figures in some germ cells within 5 days of LH treatment in 4- to 5-month-old rhesus monkeys. In 12- to 24-month-old animals of the same species, LH was ineffective but exerted a stimulatory effect in 30- to 42-month-old animals (Arslan *et al.*, 1981). The sensitivity of the nonhuman primate testis to LH has been suggested to depend on the age of the animal.

The results become more consistent and unequivocal when high doses of testosterone or long-acting testosterone esters are employed for longer time periods. In two of four 12-month-old cynomolgus monkeys, qualitatively normal spermatogenesis could be induced with testosterone alone; round spermatids were present in the remaining two animals (Marshall *et al.*, 1984). We were able to show that testosterone—given over a period of 12 weeks—stimulated the number of A-pale spermatogonia and Sertoli cells and induced the appearance of a few prophase I spermatocytes in 12- to 18-month-old rhesus monkeys (Fig. 2) (Arslan *et al.*, 1993). The numbers of A-dark spermatogonia did not change significantly under testosterone treatment when compared with vehicle-treated animals.

Although the observations were made under pathological conditions, evidence suggests that testosterone also is able to induce the spermatogenic process in boys (Steinberger *et al.*, 1973; Chemes *et al.*, 1982). Histological examination of testicular tissue containing a Leydig cell tumor and presumably high local concentrations of androgens revealed the presence of qualitatively normal spermatogenesis, but only in those seminiferous tubules located in the immediate vicinity of the tumor. Thus, testosterone can initiate qualitatively normal spermatogenesis in primates (Table 2).

IV. Maintenance and Reinitiation of Spermatogenesis by Testosterone

A. Rodents

Immunization against LH induces testicular involution in the rat (Dym and Raj, 1977), demonstrating the importance of androgens in maintaining spermatogenesis. Selective blockade of intratesticular testosterone action by a peripherally selective antiandrogen, Casodex, as an alternative to immunization against LH elicited only a small reduction of germ cell numbers (Chandolia *et al.*, 1991a). However, the antiandrogen may not compete effectively with the high concentrations of androgens present in the testis.

The importance of testosterone in the spermatogenic process also is supported by studies in hypophysectomized rats. Provision of exogenous testosterone, either on the day of surgery or shortly thereafter, prevented

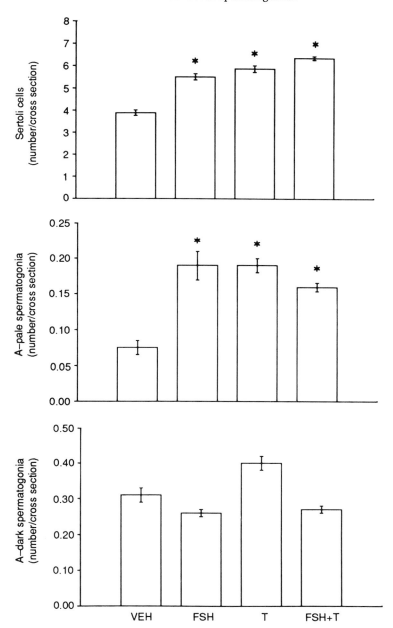

FIGURE 2 Effects of testosterone enanthate (T) (125 mg/week) and human FSH (2 × 3 IU/kg per day) alone or in combination on the numbers of Sertoli cells, A-pale spermatogonia, and A-dark spermatogonia in immature rhesus monkeys. Note the significant and comparable stimulation of the numbers of Sertoli cells and A-pale spermatogonia by testosterone and FSH. VEH, vehicle. Values are mean ± SEM of 4 animals/group. Modified from Arslan *et al.* (1993).

testicular involution and maintained spermatogenesis in a qualitative manner in several studies (Sun *et al.*, 1990; see Weinbauer and Nieschlag, 1990, for review). Similar findings were obtained in intact rats receiving high-dose testosterone treatment (Berndtson *et al.*, 1974; Huang and Nieschlag, 1986) and rats depleted of Leydig cells and supplemented with high doses of testosterone (Sharpe *et al.*, 1988a). Evidently, testosterone alone is capable of maintaining qualitatively normal spermatogenesis in the rat. Other androgens such as dihydrotestosterone (DHT) and 5-α-androstenediol, and even androgen precursors such as pregnenolone, progesterone, and 17-hydroxyprogesterone (probably acting through conversion to testosterone and DHT), were also effective in partially maintaining the spermatogenic process (Ahmad *et al.*, 1975; Chowdhury and Steinberger, 1975; Harris and Bartke, 1975).

Results are less uniform with respect to the ability of testosterone to maintain quantitatively normal spermatogenesis. In hypophysectomized rats, testosterone was unable to uphold completely the formation of testicular sperm (Sun *et al.*, 1989; Santulli *et al.*, 1990). The number of testicular spermatozoa was determined by enumeration of homogenization-resistant spermatids in these studies. The efficacy of testosterone in maintaining the spermatogenic process, however, was more pronounced in intact than in hypophysectomized rats, suggesting that the choice of the experimental model might be important. Quantitative maintenance of spermatogenesis by testosterone has been achieved in intact rats (Sharpe *et al.*, 1988b; Zirkin *et al.*, 1989), in GnRH-immunized rats (Awoniyi *et al.*, 1992a), and in estradiol-suppressed rats (Robaire *et al.*, 1979).

Findings comparable to those just described emerged with respect to the capacity of testosterone to reinitiate spermatogenesis in the rat model. In hypophysectomized animals, testosterone stimulated germ cell development only to a qualitative extent in the majority of investigations (see Weinbauer and Nieschlag, 1990, for review). Different results were obtained with other experimental approaches. In estradiol-suppressed or GnRH-immunized but otherwise intact animals, testosterone quantitatively restored spermatogenesis (Awoniyi *et al.*, 1989a,b). Serum concentrations of FSH were below the detection limit of the radioimmunoassay. This group of investigators also reported that, in the GnRH-immunized model, testosterone was even able to restore full fertility (Awoniyi *et al.*, 1992b). In contrast to the rat model, the Djungarian hamster *(Phodopus sungorus)*, a photosensitive species, responds to LH with only partially restored spermatogenesis although the intratesticular levels of androgens are stimulated markedly by exogenously administered LH (Niklowitz *et al.*, 1989).

Testosterone is able to maintain quantitatively normal spermatogenesis in the rat (Table 2), but only in the presence of an intact pituitary gland. A similar situation also prevails with respect to the reinitiation of spermatogenesis. Interestingly, LH is only marginally effective in restoring spermatogenesis in a seasonally breeding rodent, the Djungarian hamster.

B. Nonhuman Primates and Men

Immunization against testosterone was an obvious approach to study the selective role of testosterone in nonhuman primate spermatogenesis. This strategy, however, was not successful. Although testosterone could be neutralized outside the testis by active immunization against testosterone, testosterone production and intratesticular testosterone were not affected by the antibodies and actually were increased. Testicular size and sperm numbers remained unaffected (Wickings and Nieschlag, 1978). Current knowledge about the hormonal requirements of nonhuman primate spermatogenesis derives from studies in hypophysectomized or GnRH antagonist-treated animals. GnRH antagonists specifically suppress the synthesis and release of LH and FSH through competitive occupancy of the pituitary GnRH receptor (Marshall *et al.*, 1986a; Weinbauer *et al.*, 1992a) and induce testicular alterations comparable to those seen after hypophysectomy or transection of the pituitary stalk (see Weinbauer and Nieschlag, 1992, for review).

Administration of hMG or pregnant mare serum gonadotropin (PMSG) exerted only marginal effects on spermatogenesis in hypophysectomized monkeys, probably because of the formation of antibodies (Smith, 1942; Bennett *et al.*, 1973). In contrast, high-dose testosterone substitution initiated within 45 min of pituitary ablation had a positive effect on testicular size and maintained the spermatogenic process to a qualitatively normal extent in cynomolgus monkeys (Marshall *et al.*, 1986b). These findings were extended by the demonstration that testosterone, in a dose-dependent manner, could prevent the induction of azoospermia in GnRH antagonist-treated cynomolgus monkeys (Weinbauer *et al.*, 1988).

Most knowledge about testosterone and maintenance of spermatogenesis in humans is based on trials with androgens for male fertility regulation (Swerdloff *et al.*, 1979; Knuth and Nieschlag, 1989; Nieschlag *et al.*, 1992) and on studies using androgens and selective gonadotropin substitution for the study of the hormonal regulation of spermatogenesis (Matsumoto, 1989). In the latter type of study, young and healthy volunteers received high doses of testosterone or hCG for suppression of testicular function. In spite of a treatment duration of 9–16 months, azoospermia could not be achieved. From the results of these studies, testosterone appears to have the ability to maintain qualitative spermatogenesis in men in the absence of detectable serum concentrations of immunoactive FSH. This conclusion, however, might require revision, since testosterone or hCG did not abolish the bioactivity of FSH completely (Matsumoto and Bremner, 1990). On the other hand, clinical experience in hypogonadotropic hypogonadal patients clearly demonstrates that, once the formation of sperm has been induced, testosterone alone maintains qualitatively normal spermatogenesis and even fertility for some time (Johnsen, 1978).

Complete reinitiation of spermatogenesis, although not to a normal degree, by high doses of testosterone was achieved in hypophysectomized (Smith, 1944) and pituitary stalk-sectioned nonhuman primates (Marshall *et al.*, 1983). Administration of LH to hypophysectomized patients restimulated germ cell development up to the level of spermatocytes (Mancini, 1969). The induction of sperm formation in hypogonadotropic hypogonadal patients with testosterone or hCG has been achieved, but the numbers of sperm produced were rather low in the majority of patients (Johnson, 1978; D'Agata *et al.*, 1984; Finkel *et al.*, 1985; Burris *et al.*, 1988; Matsumoto, 1989; Vicari *et al.*, 1992). Moreover, the possibility must be considered that low amounts of FSH were present in these patients. Although the reinitiation of qualitatively normal spermatogenesis with testosterone alone is possible in primates (Table 2), testosterone appears to be considerably more potent in maintaining than in reinitiating the spermatogenic process.

The validity of this concept, reported earlier for the rat (Harris *et al.*, 1977), was substantiated in nonhuman primate studies. Amounts of testosterone that partially maintained germ cell formation in GnRH agonist- or GnRH antagonist-suppressed animals (Akhtar *et al.*, 1983a; Weinbauer *et al.*, 1988) did not reinitiate sperm production (Akhtar *et al.*, 1983b; Weinbauer *et al.*, 1989). These findings are highly relevant for the reinitiation of spermatogenesis in patients with secondary hypogonadism but also for design of contraceptive studies in men, and will be discussed further in Section VIII.

V. Intratesticular Testosterone and Spermatogenesis

In many species, the androgen concentrations in testicular tissue were noted to far exceed those present in circulation, leading to the concept that spermatogenesis requires high intratesticular androgen levels. However, this assumption had to be corrected when Cunningham and Huckins (1979) demonstrated in rats that the spermatogenic events took place in a qualitatively normal manner in the presence of testicular testosterone concentrations that were 5% or even less of normal. These findings were confirmed subsequently in many investigations, so obviously mammalian spermatogenesis can proceed with much lower amounts of androgen than produced by the Leydig cells and present intratesticularly (see Weinbauer and Nieschlag, 1990, for review). DHT is considered a major metabolite of testosterone within the testis (Payne *et al.*, 1973; Rivarola *et al.*, 1973). The affinity of DHT for the androgen receptor is about twofold greater than that of testosterone; testosterone dissociates about fivefold faster than DHT from the receptor (Grino *et al.*, 1990). Testosterone has been shown to overcome this effect by mass action. In the intact rat testis, the concentrations of testosterone exceed those of DHT approximately 20-fold (Arslan *et al.*, 1989).

Evident from the foregoing discussion on the importance of testosterone in different phases of spermatogenesis is the fact that testosterone is a key modulator of male gametogenesis. Considerable research efforts have been devoted to establish a quantitative relationship between intratesticular levels of testosterone and germ cell formation and to reveal how much testicular testosterone is needed for quantitatively normal spermatogenesis (see Sharpe, 1987; Rommerts, 1988; Weinbauer and Nieschlag, 1990, for details).

In brief, the precise relationships between testicular testosterone and quantitative aspects of germ cell development are still not fully understood. Many of the investigations have yielded puzzling findings. For example, spermatogenesis was disrupted in hypophysectomized rats in the presence of <5% of intratesticular androgen levels whereas the administration of exogenous testosterone, rendering testicular androgen levels unaltered compared with hypophysectomy alone, maintained spermatogenesis at approximately 80% of the control level (Bartlett *et al.*, 1989a). Neither qualitative nor quantitative spermatogenesis correlated with the concentrations of testosterone and DHT in the seminiferous tubules (Fig. 3). Similar to the case in rats, no direct relationship was found between testicular testosterone or DHT in the nonhuman primate model (Weinbauer *et al.*, 1988). In contrast, in the presence of FSH, the number of elongated spermatids correlated with the levels of testosterone in the interstitial fluid (Zirkin *et al.*, 1989). Possible explanations for the lack of close correlation between spermatogenesis and intratesticular androgen concentrations (in the absence of FSH) include pitfalls in the methodology for determination of testicular androgens (Sharpe *et al.*, 1988b) or the inability of current methods to distinguish between biologically relevant and nonrelevant fraction of testicular androgens.

How much testicular testosterone actually is needed to support quantitatively normal spermatogenesis? In rats with intact pituitaries (high-dose testosterone or estradiol treatment; immunization against LH or GnRH) but not in hypophysectomized animals, spermatogenesis proceeded in a quantitative manner in the presence of 20–40% of normal concentrations of testicular testosterone (Huang and Nieschlag, 1986; Sharpe *et al.*, 1988b; Awoniyi *et al.*, 1989b,1992a; Sun *et al.*, 1989; Santulli *et al.*, 1990; Roberts *et al.*, 1991). In the presence of circulating FSH, however, the androgen requirements for quantitatively normal spermatogenesis were much lower. The experimental paradigm of GnRH antagonist plus testosterone administration permitted this discovery, since testosterone or DHT but not estradiol (in the rat) selectively stimulates the pituitary synthesis and release of bioactive FSH (Rea *et al.*, 1986,1987; Arslan *et al.*, 1989; Sharma *et al.*, 1990). In this experimental situation, quantitatively normal spermatogenesis was achieved with only 10% of testicular testosterone compared with control animals. Unlike the case in rats, testosterone does not stimulate FSH secretion in the presence of a

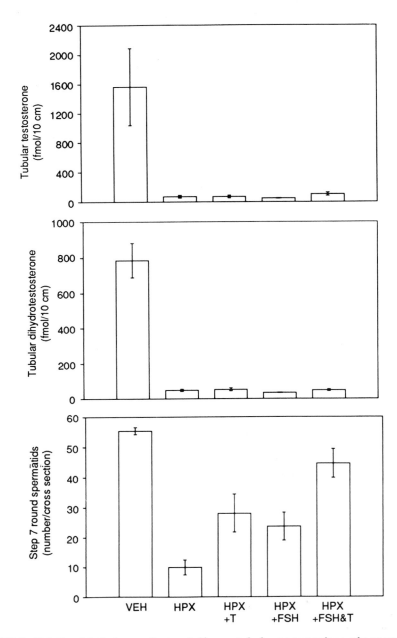

FIGURE 3 Relationship between the seminiferous tubule concentrations of testosterone and dihydrotestosterone and the number of round spermatids in hypophysectomized (HPX) and hypophysectomized and hormone-treated rats. T, 1.5-cm Silastic implant; human FSH, 2×15 IU per day for 14 days. Values are mean $\pm$ SEM of 3–6 animals/group. VEH, vehicle. The stimulatory effects of hormone administration on spermatid development were unrelated to the seminiferous androgen concentrations. Modified from Bartlett *et al.* (1989a).

GnRH antagonist in nonhuman primates and men (Bagatell *et al.*, 1989; Khurshid *et al.*, 1991). However, in the nonhuman primate spermatogenesis appears to be comparatively less dependent on androgens than in the rat, since azoospermia was achieved with a GnRH antagonist in the presence of 20–30% of baseline testicular androgen levels (Weinbauer *et al.*, 1988).

In conclusion, quantitatively normal spermatogenesis does not depend on the high intratesticular androgen levels present under normal conditions. The physiological importance of the high testicular androgen concentrations still remains unknown. However, the testis forms the reservoir from which the peripheral blood stream and target organs are provided with testosterone. Considering the relatively low blood flow through the testis, the reservoir must contain a massive testosterone concentration (maintained by continuous production) to insure adequate supply to the periphery. Thus, the intratesticular concentrations might only be high since testosterone has an endocrine function in addition to the local paracrine function. Rather than being perplexed by the question of how little testosterone may be minimally required for spermatogenesis (why should it be more than for other target organs?), one should wonder why the extremely high intratesticular testosterone concentrations are not harmful to spermatogenesis.

VI. Initiation of Spermatogenesis by FSH

A. Rodents

FSH is an important factor in the initiation of germ cell divisions in the immature rat testis. This role was proven clearly by the administration of FSH to hypophysectomized or estrogen-treated animals and by selective immunization against FSH (Lostroh *et al.*, 1963; Chemes *et al.*, 1976,1979; Madhwa Raj and Dym, 1976; Almiron *et al.*, 1984; Kerr and Sharpe, 1985). Studies using flow-cytometric quantification of testicular cells suggest that FSH is involved in meiotic divisions since the ratio of 4C cells (tetraploid cells, predominantly primary spermatocytes) to 1C cells (spermatids) was reduced to about 10% of control following immunization against FSH (Vaishnav and Moudgal, 1992). Notwithstanding the importance of FSH for the development of germ cells in the immature testis, the formation of testicular spermatozoa could not be achieved with FSH alone, thus questioning the ability of FSH alone to initiate qualitatively normal spermatogenesis (Table 2).

In addition to its action on germ cells, FSH apparently has a decisive role in the maturation of the testis (Heckert and Griswold, 1992) and in determining the size of the Sertoli cell population. In the fetal rat, FSH stimulates the division and numbers of Sertoli cells (Orth, 1984) and is involved in Sertoli cell proliferation after hemicastration (Orth *et al.*, 1984). The administration of an antimitotic drug to newborn rats decreases the number of Sertoli cells by 50%; during adulthood, the testicular weight of these animals was about 50% of that of age-matched controls (Orth *et al.*, 1988). Since Sertoli cells do not divide in the adult testis, the establishment of adequate numbers of Sertoli cells prior to completion of the first spermatogenic cycle is of pivotal importance to the achievement of normal testicular dimensions during adulthood.

B. Nonhuman Primates and Men

The ability of FSH to exert a stimulatory effect on the immature primate testis was first demonstrated in short-term studies (6 days) in the rhesus monkey (Arslan *et al.*, 1981). FSH increased the diameter of the seminiferous tubules and testicular weight. Following a 12-week period of FSH administration, the numbers of A spermatogonia and Sertoli cells increased significantly beyond that of an age-matched control group (Fig. 2) (Arslan *et al.*, 1993). However, meiotic divisions of germ cells could not be induced. Even pulsatile administration of GnRH for 12 weeks, initiating an adult-like pattern of gonadotropin and testosterone secretion, only induced the transition of a few spermatogonia into prophase I spermatocytes (Abeyawardenee *et al.*, 1989).

Whether FSH alone can induce complete spermatogenesis in the primate remains to be shown (Table 2). In view of the prolonged hiatus of gonadotropin secretion in primates, considerably longer periods of FSH administration than those tried to date are likely to be required to clarify this issue.

VII. Maintenance and Reinitiation of Spermatogenesis by FSH

A. Rodents

Contrasting results were obtained with respect to the involvement of FSH in the maintenance of adult spermatogenesis in the rat model. Immunization against FSH had no effect on spermatogenesis (Dym *et al.*, 1978). Administration of FSH to hypophysectomized rats failed to support the development (Woods and Simpson, 1961; von Berswordt-Wallrabe *et al.*, 1968a) or the survival of germ cells (Russell and Clermont, 1977; Toppari *et al.*, 1988).

On the other hand, clear-cut evidence for an involvement of FSH in the spermatogenic process of the adult rat was obtained in our laboratory. We used a highly purified urinary human FSH preparation, obviously devoid of intrinsic LH activity as demonstrated by the absence of stimulatory effects of this FSH preparation on peripheral and testicular testosterone concentrations. In hypophysectomized rats, germ cell development was supported qualitatively by FSH up to the level of step 7 (round spermatids; Bartlett *et al.*, 1989a) whereas, in GnRH antagonist-treated animals, the complete process of spermatogenesis (including step 19 spermatids) was preserved, albeit not to a quantitative extent (Chandolia *et al.*, 1991b). In hypophysectomized and Leydig cell-depleted rats, FSH failed to exert a stimulatory effect on spermatids (Kerr *et al.*, 1992). These findings suggest, analogous to the effects of testosterone on spermatogenesis (Awoniyi *et al.*, 1989a), that the choice of the experimental approach for induction of a hypogonadotropic state might influence considerably the hormonal responsiveness of the seminiferous epithelium. Testicular weights were reduced to 33 and 27% of control 2 weeks after hypophysectomy (Bartlett *et al.*, 1989a; Spiteri-Grech *et al.*, 1991) whereas, following the same period of exposure to a GnRH antagonist, the loss of testicular weight amounted only to 40–50% (Chandolia *et al.*, 1991b; Spiteri-Grech *et al.*, 1993).

That FSH plays a role in the spermatogenesis of the adult rat is actually not surprising, considering the fact that the seminiferous tubule is *the* target for FSH. We are not aware of studies that deal with the effects of highly purified FSH on the reinitiation of spermatogenesis in the rat model. The GnRH antagonist-induced suppression of spermatogenesis could be restored quantitatively by the administration of testosterone, which restimulated FSH secretion (Rea *et al.*, 1987). These observations provide indirect evidence for a role for FSH in the reinitiation of the spermatogenic process. Direct support in favor of a role for FSH during restoration of gametogenesis was obtained in another rodent model, the photosensitive Djungarian hamster *(Phodopus sungorus)*. In this species, shortening of the daylength leads to secondary hypogonadism and testicular involution. The histoarchitecture of fully regressed testes is similar to that seen in hypophysectomized hamsters (Niklowitz *et al.*, 1989). Interestingly, FSH in the absence of steroidogenically active Leydig cells reinitiates spermatogenesis to a qualitatively normal extent. We observed that the spermatozoa produced under these experimental conditions are released and reach the epididymis (Lerchl *et al.*, 1993). A small amount of exogenous testosterone was provided for induction of mating behavior. Mating studies suggested that the sperm produced actually might be capable of fertilizing and inducing normal pregnancies. A combination of hCG and PMSG quantitatively restored spermatogenesis in the seasonally breeding woodchuck *(Marmota monax)* (Sinha Hikim *et al.*, 1991).

In contrast to previous hypotheses, FSH apparently is involved in the maintenance of spermatogenesis in the adult rat (Table 2). Whether this involvement also occurs in reinitiation of the spermatogenic process is currently unknown. The Djungarian hamster is clearly different since testosterone is unable to restore spermatogenesis whereas FSH is highly effective in that respect. Whether this mode of hormonal regulation of spermatogenesis is specific to this hamster species or applies to photoperiodic species in general remains to be seen.

B. Nonhuman Primates and Men

Passive and active immunization against FSH of rhesus monkeys specifically neutralizes FSH but not LH; testosterone production overall remains unaltered (Wickings and Nieschlag, 1980; Wickings et al., 1980; Srinath et al., 1983). Therefore, the specific role of FSH in nonhuman primate spermatogenesis could be investigated in this experimental paradigm. Passive immunization against FSH provokes a 50% reduction of testicular size and a substantial decline of sperm production. Comparable effects on spermatogenesis are obtained following active immunization with FSH. Moreover, inactivation of circulating FSH by means of immunization even induces infertility in the bonnet monkey (Moudgal, 1981; Moudgal et al., 1992). Notably, spermatogenesis is severely compromised in these studies in spite of unaltered testosterone production. Subnormal serum levels of FSH also are able to maintain qualitatively normal sperm production in men (Matsumoto, 1989).

These findings provide compelling evidence for a crucial role for FSH in the maintenance of spermatogenesis in the primate. To clarify whether FSH is able to maintain testicular function in the presence of markedly lowered testosterone production, the GnRH antagonist-treated nonhuman primate model was used (Weinbauer et al., 1991). The effects of a highly purified human FSH preparation that did not stimulate serum testosterone levels were studied over a period of 8 weeks. FSH maintained testicular size and qualitatively normal spermatogenesis over the time interval studied. The numbers of A-pale spermatogonia, expressed per 10 Sertoli cells, was preserved quantitatively and the numbers of pachytene spermatocytes and round and elongated spermatids were maintained at ~50% of baseline (Fig. 4). Flow-cytometric analysis revealed that the proportion of the different types of testicular cells including the 1CC cells (elongated spermatids) were maintained by FSH, in contrast to treatment with GnRH antagonist alone (Fig. 5). This observation, in conjunction with the data obtained during germ cell enumeration, strongly suggests that FSH is able to maintain the spermatogenic process in this nonhuman primate model.

Reinitiation of the spermatogenic process by FSH also was investigated in the study just described over a period of 8 weeks (Weinbauer et al., 1991; Figs. 4, 5). Although FSH restimulated the appearance of small numbers of round spermatids, this hormone failed to restore the complete spermatogenic process. Again, the most pronounced stimulatory action of FSH occurred with respect to the numbers of A-pale spermatogonia. As for testosterone, these findings indicate that the ability of FSH to maintain spermatogenesis in

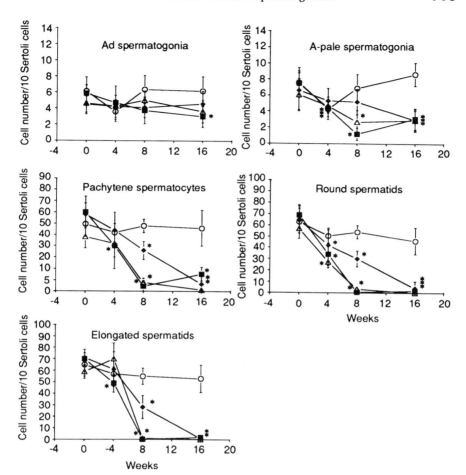

FIGURE 4   Effects of treatment with vehicle (open circles), GnRH antagonist for 16 weeks (open triangles), GnRH antagonist plus human FSH (2 × 15 IU per day) for the first 8 weeks (filled diamonds), or GnRH antagonist plus human FSH for the second 8 weeks (filled squares) on the numbers of germ cells in the cynomolgus monkey. Analysis was performed on 1-μm semithin sections of glutaraldehyde-fixed biopsy specimens. Values are mean ± SEM of 5 animals/group. The numbers of A-dark spermatogonia remained unaffected by any treatment whereas the A-pale spermatogonia were reduced under GnRH antagonist alone and stimulated by FSH. FSH partially maintained germ cell development but was much less effective in reinitiating the spermatogenic process. Modified from Weinbauer *et al.* (1991).

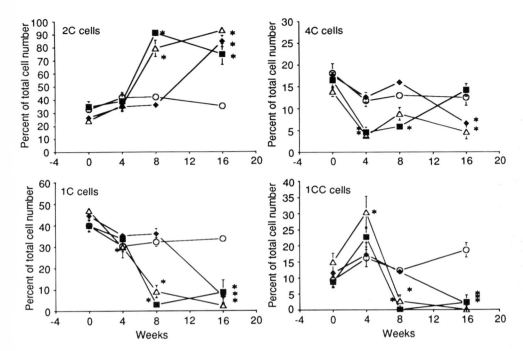

FIGURE 5 Effects of treatment with vehicle (open circles), GnRH antagonist for 16 weeks (open triangles), GnRH antagonist plus human FSH (2 × 15 IU per day) for the first 8 weeks (filled diamonds), or GnRH antagonist plus human FSH for the second 8 weeks (filled squares) on the numbers of 2C cells (diploid), 4C cells (tetraploid, mainly primary spermatocytes), 1C cells (haploid, round spermatids), and 1CC cells (elongated spermatids) in the cynomolgus monkey. Analysis was performed by flow cytometry of biopsies (2,118–14,732 cells/biopsy). Cell numbers are expressed as percentages of the total cell number/biopsy. The relative increase of 2C cells results from the proportional loss of 4C, 1C, and 1CC cells. FSH quantitatively maintained spermatogenesis but was considerably less active during reinitiation of spermatogenesis. Values are mean ± SEM of 5 animals/group. Modified from Weinbauer et al. (1991).

the primate is considerably more pronounced than its ability to reinitiate this process. Similar to the effects of testosterone, FSH increased the numbers of A-pale spermatogonia without significant effects on the numbers of A-dark spermatogonia (Weinbauer et al., 1991; Arslan et al., 1993), indicating that one of the important actions of FSH in the primate testis is the activation of the proliferation of A-pale spermatogonia, the renewing testicular stem cells. Interestingly, FSH also stimulated the spermatogenic process in intact non-human primates; the suggested target germ cell was the A-pale spermatogonia (van Alphen et al., 1988).

In conclusion, FSH plays an essential role in the maintenance and reinitiation of spermatogenesis in primates (Table 2). At present, FSH appears to maintain or reinitiate qualitatively normal spermatogenesis. However, only very few studies addressed this aspect of primate spermatogenesis, so future results may change the conclusions drawn here.

VIII. Hormonal Control of Male Fertility for Contraception

As already mentioned in Section IV,B, supranormal quantities of testosterone were administered to suppress spermatogenesis in an attempt to identify a hormonal method for male contraception. Advantage was taken of the negative feedback action of androgens on gonadotropin release. In Caucasian men, azoospermia was achieved in approximately two-thirds of the volunteers (see Nieschlag *et al.*, 1992, for review). Surprisingly, among Asian men, the rate of azoospermia approached 100% (World Health Organization, 1990; Pangkahila, 1991). Although several explanations for these observations are conceivable, speculating that the dependency of spermatogenesis on androgens (and FSH) or the kinetics of the spermatogenic process might vary among certain human races is tempting. That some rat strains differ in the duration of the spermatogenic process is well known (Clermont, 1972).

Another approach to endocrine male contraception is based on the suppressive effects of GnRH agonists and antagonists on gonadotropin secretion and testosterone production, and subsequently on the spermatogenic process. The clinical use of GnRH analogs for the purpose of male fertility regulation necessitates androgen substitution therapy to prevent the symptoms of androgen deficiency imposed by the GnRH analog-induced Leydig cell insufficiency. The agonistic analogs of GnRH, in conjunction with testosterone, fail to suppress sperm production consistently. This observation can be explained by the fact that GnRH agonists only incompletely inhibit FSH release (Huhtaniemi *et al.*, 1988; Pavlou *et al.*, 1988; Behre *et al.*, 1992). Thus, testosterone supplementation with incompletely suppressed FSH probably exerts a maintaining effect on germ cell development.

In the case of GnRH antagonists, the bioactivity of LH and FSH is suppressed (see Weinbauer and Nieschlag, 1992, for review). Concomitant administration of GnRH antagonist and testosterone prevents the induction of azoospermia (Weinbauer *et al.*, 1988), whereas a delay of testosterone substitution, at the same dose used during concomitant supplementation, by 6 weeks from the onset of GnRH antagonist treatment resulted in the development of azoospermia (Weinbauer *et al.*, 1989). In this study, the interval between the onset of GnRH antagonist treatment and the initiation of testosterone supplementation was 6 weeks. We found that a 2-week interval between GnRH antagonist administration and the testosterone substitution also induced azoospermia in the nonhuman primate model (Weinbauer *et al.*, 1992b). The reason for the reduced ability of delayed testosterone treatment to stimulate spermatogenesis, compared with concomitant testosterone administration, is not known. In terms of qualitative testicular histology and testicular cell numbers, no differences could be detected just prior to the onset of delayed (2 and 4 weeks) testosterone substitution when compared with administration of GnRH antagonist alone for the respective time periods. Testicular blood flow may be reduced in the absence of gonadotropic hormones and only a small portion of the substituted testosterone may arrive at the testis. In the absence of gonadotropic hormones, the testicular androgen receptor may undergo functional modifications that can be prevented by concomitant but not by delayed androgen supplementation.

The regimen of GnRH antagonist and a 2-week delay of testosterone substitution proved effective in clinical studies for male contraception (Pavlou et al., 1991; Tom et al., 1992). Of a total of 16 Caucasian men, 14 developed reversible azoospermia when subjected to GnRH antagonist and delayed (2 weeks) testosterone substitution. Currently, GnRH antagonists and delayed androgen supplementation and testosterone alone provide promising approaches to hormonal male contraception, particularly for men belonging to races that respond poorly to testosterone alone.

IX. Intratesticular Targets of Testosterone and FSH

Binding sites for testosterone have been found on Leydig cells (Nakhla et al., 1984; Namiki et al., 1991), peritubular cells (Nakhla et al., 1984), and Sertoli cells (Sanborn et al., 1975; Tindall et al., 1977; Nakhla et al., 1984). The androgen receptor has been localized immunohistochemically to these cells in both rats and men (Sar et al., 1990; Takeda et al., 1990; Ruizeveld de Winter et al., 1991). Germ cells apparently do not contain receptors for androgens (Grootegoed et al., 1977; Frankel et al., 1989). Receptors for FSH are present on Sertoli cells whereas spermatocytes and spermatids do not possess binding sites for FSH. Spermatogonia have been reported to contain FSH receptors on the basis of the ultrastructural distribution of labeled FSH (Orth and Christensen, 1978). On the other hand, using histochemical techniques at the light microscopic level, spermatogonia were found to be FSH negative in the rat and human testis (Wahlström et al., 1983). However, whether this approach permits a reliable assessment of the existence of bindings sites on spermatogonia remains questionable. Unequivocal proof for the existence of binding sites for FSH on spermatogonia is still lacking.

Consequently, the stimulatory effects of FSH are mediated through the Sertoli cells and eventually also through direct action on spermatogonia. The latter action, however, still awaits experimental proof. The effects of androgens are conveyed through the somatic cells of the seminiferous tubules (Table 3). Investigations in male mice chimeric for the normal and the androgen receptor-deficient genotype provided evidence that the response of

TABLE 3 Binding Sites for Androgens and FSH in Testicular Cells

Cell type	Androgens	FSH
Somatic cells		
Sertoli cell	Present	Present
Peritubular cell	Present	Absent
Leydig cell	Present	Absent
Germ cells		
Spermatogonia	Absent	Present[a]
Spermatocytes	Absent	Absent
Spermatids	Absent	Absent

[a] Presence likely but not unequivocally established.

the seminiferous epithelium to androgens indeed might be conveyed entirely by somatic cells (Lyon et al., 1975). Moreover, the peritubular cells have been shown to produce factors in response to testosterone that stimulate Sertoli cell function (Hutson et al., 1987; Skinner, 1991; Verhoeven et al., 1992).

Sertoli cells produce an impressive number of substances including binding and transport proteins, growth factors, proteases, and peptide and steroid hormones (see Jégou, 1992, for review). The synthesis and secretion of many of these factors are regulated by testosterone and FSH, whereas other Sertoli cell factors are believed to be influenced by the germ cells (Kangasniemi et al., 1990a; Sharpe et al., 1991; Jégou et al., 1992; Steinberger et al., 1992). To date, assigning a specific role to testosterone or FSH with respect to the entire spermatogenic process has been difficult. In many instances, both hormones were able to stimulate spermatogenesis to a similar extent (see Sections III–VII). Quantitative studies in the nonhuman primate model suggest that the stimulatory effects of both testosterone (Marshall et al., 1986b; Arslan et al., 1993) and FSH (van Alphen et al., 1988; Weinbauer et al., 1991) on germ cell development might be caused by a similar mechanism, namely an increased proliferation of the A-pale spermatogonia. Some investigators actually emphasized the lack of hormone specificity of the effects of testosterone and FSH on germ cells and suggested that both hormones exert a common effect on spermatogenesis, that is, increased germ cell viability and spermatogenic efficacy (Vernon et al., 1975; Russell et al., 1987).

On the other hand, data obtained in the rat model provide evidence that the various cellular associations of spermatogenesis differ in their sensitivity, requirements, or thresholds for androgens and FSH. Gonadotropin deprivation induces the loss of germ cells in the late stages of spermatogenesis rather than affecting all spermatogenic stages to a similar degree (Clermont and Morgenthaler, 1955; Russell and Clermont, 1977). Androgen receptor concentrations vary in a stage-specific manner throughout the spermatogenic cycle (Isomaa *et al.,* 1985); messenger RNA expression of the androgen receptor changes throughout the spermatogenic cycle, with highest expression in Stages XIII–I and lowest expression in Stages II–VI (Linder *et al.,* 1991). Stage-specific androgen-regulated proteins have been identified in the adult rat testis (Sharpe *et al.,* 1992).

The number of binding sites for FSH also varies in a stage-specific manner with the highest concentrations in Stages XIII–I and lowest concentrations in Stages VI–VIII in the testes of irradiated rats (Kangasniemi *et al.,* 1990b). Similar observations were reported in rats with vitamin A-induced synchronized spermatogenesis. Northern blot analysis revealed highest FSH receptor expression in Stages XIV–I and lowest expression in Stages VII–VIII (Heckert and Griswold, 1991). On the other hand, *in situ* hybridization analysis of FSH receptor mRNA expression in the testes of intact rats yielded opposing results (Kliesch *et al.,* 1992). The strongest signal was found in stages IX–X, decreased signal intensity was seen in Stages XI–XII, and no specific binding could be demonstrated in Stages XIII–VII (Fig. 6). The reasons for these contradictory findings, obtained in different animal models using different techniques, are currently unknown. Nonetheless, the expression of the FSH receptor obviously occurs in a stage-specific manner.

The possibility should be considered that modifications of the androgen and FSH receptor might permit modulations of hormone action within the gonads. Several LH/CG receptor transcripts resulting from alternative splicing have been described in the testis and ovary (Bernard *et al.,* 1990;

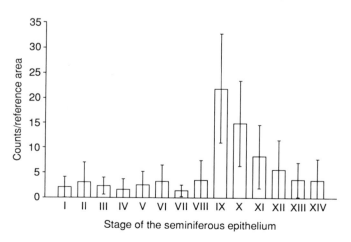

FIGURE 6 Localization of the testicular FSH receptor mRNA by *in situ* hybridization in the rat. A cRNA antisense probe of a nonhuman primate FSH receptor cDNA was used. One testis was analyzed. Values are mean ± SD derived from 10–15 measurements/spermatogenic stage using interactive image analysis. The FSH receptor is expressed in a stage-specific manner with the strongest signal in stages IX and X. Modified from Kliesch *et al.* (1992).

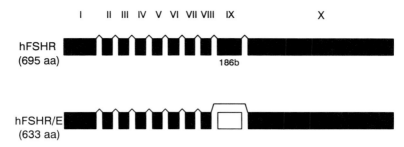

FIGURE 7 Alternative splicing of the human FSH receptor transcript through a cassette exon mode. The exons (I-IX) are represented by filled boxes; nucleotide sequences lost during the splicing process are indicated by open boxes. The lengths of the human FSH receptor (hFSHR) and of the isoform lacking exon IX (hFSHR/e) are indicated in amino acids (aa). Modified from Gromoll *et al.* (1992).

Minegish *et al.*, 1990; Segaloff *et al.*, 1990; Tsai-Morris *et al.*, 1990; Aatsiniki *et al.*, 1992). The smallest transcript presumably represents a truncated isoform of the LH/CG receptor that acts as a secreted hormone binding protein (see Hsueh and La Polt, 1992, for review). Similarly, multiple transcripts for the FSH receptor are expressed in the rat (Heckert and Griswold, 1991) and human testis (Gromoll *et al.*, 1992). In the human testis, a truncated isoform of the FSH receptor is detected, indicating that extensive splicing is not restricted to the testicular LH receptor. The truncation of the FSH receptor does not lead to a shift in the open reading frame but results in the deletion of 62 amino acids. This FSH receptor isoform is spliced through a cassette exon mode, thereby deleting exon IX (Fig. 7). Differential regulation of multiple receptor transcripts and secretion of truncated receptor isoforms with hormone binding activity might represent important mechanisms for regulating target cell responsiveness to gonadotropins. Whether the receptor isoforms indeed convey differential biological responses after ligand binding remains to be seen and may, in addition, play a role in certain forms of male infertility.

In conclusion, FSH acts on spermatogenesis via the Sertoli cells and eventually also by directly influencing spermatogonial divisions. Testosterone acts via the somatic cells of the seminiferous tubules, that is, the peritubular cells and the Sertoli cells. In addition, testosterone and FSH influence spermatogenesis in a stage-specific manner.

X. Collaborative Effects of Testosterone and FSH on Spermatogenesis

Testosterone and FSH generally are assumed to be required for quantitatively normal spermatogenesis. Experimental findings that either hormone alone under certain circumstances supports spermatogenesis in a quantitative manner (Sections III, IV, VI, and VII) are not contradictory to this assumption since, under normal conditions, both hormones are present. Consequently, the ability of a single hormone factor (testosterone or FSH) to stimulate quantitatively normal spermatogenesis by no means indicates that the other factor is irrelevant.

Many studies in rodents demonstrate that, in terms of stimulating testicular weight and germ cell numbers, the combination of testosterone and FSH is more effective than either hormone alone in immature rats (Russell et al., 1987) and in adult animals (Woods and Simpson, 1961; von Berswordt-Wallrabe and Neumann, 1968a,b; Sivelle et al., 1978; Bartlett et al., 1989a; Toppari et al., 1989; Spiteri-Grech et al., 1991; Kerr et al., 1992). This result is also seen in normal men (Matsumoto, 1989) and in patients with hypogonadotropic hypogonadism (Mancini, 1969; Johnsen, 1978; Finkel et al., 1985; Finkelstein et al., 1987).

Some work suggests that the stimulatory effects of FSH on spermatogenesis are influenced by the availability of testicular androgens. The ability of FSH to increase germ cell numbers in GnRH antagonist-suppressed rats is diminished significantly in the absence of Leydig cells (Spiteri-Grech et al., 1993) or when the action of intratesticular androgens has been blocked by the administration of an antiandrogen (Chandolia et al., 1991b). The lowering of the beneficial effects of FSH on germ cell development is particularly obvious with respect to the evolution of elongated spermatids (Fig. 8). These observations demonstrate that the beneficial action of FSH on spermatogenesis, at least in the rat model, is under the influence of testicular androgens. Based on studies in hypophysectomized and Leydig cell-depleted rats, Kerr et al. (1992) concluded that the role of FSH in spermatogenesis is to potentiate the effectiveness of androgens. Whatever the real situation is, androgens and FSH evidently interact with each other in their ability to stimulate the spermatogenic process.

How do androgens and FSH cooperate at the testicular level? One basis for the interactions of testosterone and FSH at the testicular level might be the regulation of the number and activity of the respective hormone receptors. Homologous regulation of the testicular receptors has been demonstrated for androgens (Guillou et al., 1987; Verhoeven and Cailleau, 1988; Blok et al., 1989,1992a) and FSH (O'Shaugnessy and Brown, 1978; Closset and Hennen, 1989; Themmen et al., 1991; Tsutsui, 1991; Blok et al., 1992b). However, FSH also can stimulate androgen receptor mRNA expression and the number of androgen receptors in Sertoli cells (Verhoeven and Cailleau, 1988; Blok et al., 1989, 1992b). Vice versa, a slight stimulatory effect of testosterone on the number of FSH receptors also has been reported (Tsutsui, 1991). In terms of the androgen receptor, the effects of FSH plus androgen were more pronounced than those of either hormone alone (Verhoeven and Cailleau, 1988; Blok et al., 1989; Sanborn et al., 1991). These findings raise

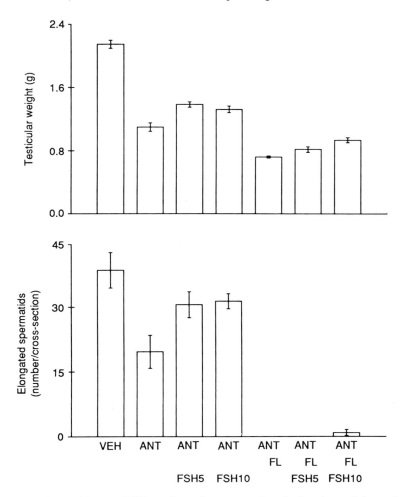

FIGURE 8 Effects of human FSH on the maintenance of testicular size and formation of elongated spermatids in GnRH antagonist (ANT)-suppressed rats in the absence and presence of the antiandrogen flutamide (FL). FSH5 and FSH10, 2×5 IU and 2×10 IU per day for 14 days. Values are mean $\pm$ SEM of 7 animals/group. The combination of ANT plus FL markedly enhanced spermatogenic involution. FSH was able to uphold the spermatogenic process qualitatively, but these effects were diminished markedly in the presence of an antiandrogen. VEH, vehicle. Modified from Chandolia *et al.* (1991b).

the possibility that testosterone and FSH interact at the receptor level by stimulating the receptor for the "other" hormone, thereby sensitizing the seminiferous epithelium to either hormone and resulting in a more pronounced biological effect compared with that of either hormone alone.

Some work indicates that the effects of testosterone and FSH on the respective receptor mRNA expression (Themmen *et al.*, 1991; Blok *et al.*, 1992b) and the effects of FSH on the androgen receptor mRNA expression (Blok *et al.*, 1992a) are related to the stability of the mRNA. This view is based on the observations that the transcription rate of mRNA is not influenced by the hormone, as demonstrated by translational run-on experiments (Blok *et al.*, 1992c). Note that androgens also might be involved in the manifestation of structural modifications of their testicular receptors (Blok *et al.*, 1992a). This assumption derives from the findings that androgen deprivation for 5 days fails to alter mRNA expression, whereas immunoprecipitation and Western blotting suggest the presence of a very low amount of total androgen protein. A possible explanation for this observation is a structural change of the androgen receptor that is induced during androgen deficiency.

In vitro studies suggest that testosterone and FSH might interact at the level of communication between spermatids and Sertoli cells. Sertoli cells possess a number of structural devices that anchor the spermatids and maintain metabolic exchange between Sertoli cells and maturing spermatids (see Jégou *et al.*, 1992, for review). The morphological basis for the interrelationships between developing spermatids and Sertoli cells has been studied in great detail (see Russell, 1991, for review), although the related physiological and biochemical events have not been revealed.

In coculture experiments using Sertoli cells from immature rat testis and round spermatids (pre-Step 9), neither testosterone nor FSH influences the spermatid density, that is, the number of spermatids per Sertoli cell (Cameron and Muffly, 1991). Similarly, pre-exposure to testosterone followed by FSH and vice versa has no effect. However, when testosterone and FSH are present concomitantly, spermatid density approximately doubles. FSH but not testosterone alters the distribution of Sertoli cell actin and vinculin—Sertoli cell–spermatid junction-related cytoskeletal proteins— from a stress fiber organization into a peripheral pattern within the cell. The authors suggest that FSH induces the spermatid-binding competency whereas testosterone enhances the binding of spermatids to the Sertoli cells. These results provide a first hint of how testosterone and FSH might cooperate in the promotion of spermatid development and provide clues to the relative role of either hormone in the spermatogenic process.

XI. Nongonadotropic Hormones and Spermatogenesis

That selective suppression of gonadotropin release by GnRH antagonists can be used to suppress spermatogenesis completely provides a strong indication of the crucial dependence of testicular functions on LH and FSH. For that reason, nongonadotropic hormones on their own are unlikely to be essential to the spermatogenic process in mammals. Nonetheless, the outcome of several investigations raises the possibility that nongonadotropic pituitary hormones might modulate the actions of gonadotropic hormones at the testicular level.

Rat Leydig cells possess receptors for prolactin (Aragona and Friesen, 1975), growth hormone (GH) (Odell and Swerdloff, 1976), and thyroid-stimulating hormone (TSH) (Amir *et al.*, 1978; Davies *et al.*, 1980). Adrenocorticotropic hormone (ACTH) has been found in Leydig cells (Tsong *et al.*, 1982). Both prolactin and GH are capable of increasing the number of LH binding sites on Leydig cells (Bex *et al.*, 1978; Morris and Saxena, 1980); these hormones—in combination—elicit a synergistic effect (Zipf *et al.*, 1978). In contrast, the possibility of regulation of Leydig cell LH receptors by TSH and ACTH has not been established clearly. In men suffering from disease as a result of elevated ACTH secretion, serum levels of testosterone are reduced without an accompanying diminution of LH secretion (Beitins *et al.*, 1973; Irvine *et al.*, 1974). However, whether ACTH and TSH can interfere directly with Leydig cell testosterone production still remains unknown. On the other hand, a lack of adrenal precursor steroids, for example, dehydroepiandrosterone, as in the case of patients with Addison's disease, has been suggested as a possible cause of diminished testosterone production capacity (Nieschlag and Kley, 1975).

With respect to testicular size and spermatogenesis, GH has been shown to potentiate the stimulatory effects of either testosterone or FSH in the rat model. In hypophysectomized animals, GH augments the response of the seminiferous epithelium to testosterone during reinitiation of spermatogenesis (Woods and Simpson, 1961; Boccabella, 1963). Similarly in hypophysectomized rats, GH synergizes with testosterone and FSH in the maintenance of testicular weight and spermatogenesis, whereas GH alone is ineffective (Woods and Simpson, 1961). Receptors for GH have been found on Sertoli cells by means of immunohistochemical techniques (Lobie *et al.*, 1990). Testicular concentrations of epidermal growth factor and insulin-like growth factor I are elevated when spermatogenesis has been disturbed severely by hypophysectomy plus ethane dimethane-sulfonate administration (Spiteri-Grech *et al.*, 1991). Testosterone plus FSH lowers the testicular levels of insulin-like growth factor I; the addition of GH slightly potentiates this effect. GH apparently is required for achievement of normal testicular size, since testicular weights are lower in a GH-deficient mutant rat strain, the Dwarf rat, compared with age-matched controls (Bartlett *et al.*, 1990a). Interestingly, however, the evolution of germ cells and testicular sperm is indistinguishable in Dwarf and normal animals. Since adult testicular size is dependent on the number of Sertoli cells (Section VI,A), GH might play a role in establishing quantitatively normal numbers of Sertoli cells in the developing testis.

For quite some time, clinical experience has shown that hypothyroidism retards pubertal development and causes macroorchidism (Laron *et al.*, 1970). Thyroid autoimmune disease also is associated with increased testis size (Hoffman *et al.*, 1991). For the study of the causes of increased testicular size in developmental hypothyroidism, a rat model has been established using the administration of methimazole or 6-propyl-2-thiouracil to pregnant rats or to neonatal animals. This treatment lowers the circulating concentrations of thyroxine and triiodothyronine (Francavilla *et al.*, 1991). During adulthood, these rats exhibit an average twofold elevation of testis weight (Cooke and Meisami, 1991) and sperm production (Cooke *et al.*, 1991).

Morphological and kinetic studies reveal that hypothyroidism leads to increased proliferation of Sertoli cells and that morphological maturation of the Sertoli cells is probably delayed (Francavilla *et al.*, 1991; van Haaster *et al.*, 1992). At the age of 36 days, the Sertoli cell numbers per testis were $11.4 \pm 0.9 \times 10^6$ compared with $6.2 \pm 0.6 \times 10^6$ in age-matched controls (van Haaster *et al.*, 1992). This consequence of subnormal thyroid function is believed to reflect a direct effect of thyroid hormone deficiency, since only the Sertoli cell has specific receptors for these hormones (Palmero *et al.*, 1989). The testicular triiodothyronine receptor concentrations are highest shortly after birth and decline at about the time that Sertoli cells normally cease proliferation (Jannini *et al.*, 1990). In the adult testis, these receptors are barely detectable. Therefore, thyroid hormones are quite likely to be important for pubertal testicular development and the attainment of normal testicular size in the adult stage.

In conclusion, GH and thyroid hormones seem to be required for the development of quantitatively normal numbers of Sertoli cells and spermatogenesis, at least in the rat. The action of these hormones, however, requires the presence of LH and androgens and/or FSH.

XII. Is the Duration of Spermatogenesis under Hormonal Control?

The time span required for a spermatogonium to develop into a testicular sperm has been studied by localizing the incorporation of [³H] thymidine in germ cells at various intervals after administration of the thymidine analog (see Clermont, 1972, for review). The duration of the spermatogenic process was observed to be species specific and remarkably constant within a species. The duration of the spermatogenic cycle, that is, one succession of all stages, was ~12.5 days in the rat (Clermont, 1972), 9.3–12.3 days in nonhuman primates (Dang, 1971b; Clermont and Antar, 1973; de Rooij *et al.*, 1986), and 16 days in men (Heller and Clermont, 1964). Since the evolution of testicular spermatozoa from spermatogonia requires 4.0–4.6 spermatogenic cycles, the total duration of spermatogenesis is estimated to be 51–53 days in the rat, 37–42 days in nonhuman primates, and 74 days in men.

Based on work by Clermont and colleagues, our current answer to the question posed in this section is that the timing of spermatogenesis is under genetic rather than under hormonal control. This view derives from observations that the duration of the spermatogenic cycle is similar in hypophysectomized rats and in hypophysectomized rats treated with testosterone or hMG (Clermont and Harvey, 1965). On the other hand, circumstantial evidence for an alteration of the duration of spermatogenesis derives from studies with drugs that disturb the spermatogenic process. Administration of 2,5-hexanedione (Chapin *et al.,* 1983), 1,2-dinitrobenzene (Hess *et al.,* 1988), or ethylene glycol monomethyl (Creasy *et al.,* 1985) induces significant changes in the frequency of certain spermatogenic stages. Treatment with procarbazine, a chemotherapeutic agent, induced asynchronous development of germ cells (Russell *et al.,* 1983). Since the frequency of a particular stage is believed to reflect its duration, the possibility of an alteration of spermatogenic timing should be considered.

Another piece of evidence, compatible with changes of spermatogenic timing, derives from studies using the vitamin A-deficient rat model. Vitamin A deficiency induced in the fetal or the early postnatal period arrests germ cell development, most likely at the level of the transition of A spermatogonia to B spermatogonia or spermatocytes (Huang *et al.,* 1990; van Pelt and de Rooij, 1990). One or two injections of high doses of vitamin A and resupplementation of the diet with vitamin A, however, initiates spermatogenesis to qualitatively normal levels (Bartlett *et al.,* 1989b). More importantly, under these conditions germ cell development becomes synchronized, resulting in testes with a predominance of certain stages, depending on the duration of vitamin A resubstitution (Morales and Griswold, 1987; Huang *et al.,* 1990). When vitamin A-synchronized animals were followed for a period of approximately 10 spermatogenic cycles, the surprising observation was made that the spermatogenic process desynchronized (Bartlett *et al.,* 1990a). The stage frequencies in these rats were similar to those of control animals (Fig. 9). Spermatogenesis, however, had not resumed in all seminiferous tubules and the serum concentrations of FSH were higher than in controls. Clearly, the possibility of an altered duration of spermatogenesis in situations of spermatogenic or endocrine disturbances should be considered. Recall, however, that this view is entirely speculative at present, since the actual duration of spermatogenesis has not been investigated in any of these studies.

That the duration of spermatogenesis is subject to alterations throughout life is suggested by observations that spermatogenesis proceeds faster in immature mice than in adult animals (Kluin *et al.,* 1982; Janca *et al.,* 1986). This view is supported by the time sequence of normal germ cell development in the immature rat. At day 1 of age, only gonocytes are present (Zhengwei *et al.,* 1990) whereas type A spermatogonia are recognized first on day 4 and testicular spermatozoa appear first at 45 days (Clermont and

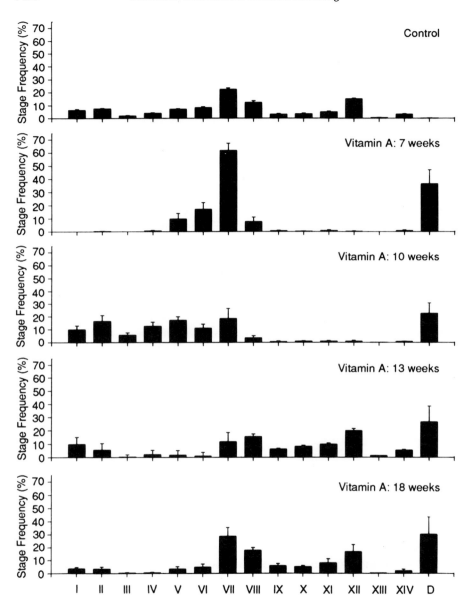

FIGURE 9 Frequency of the stages of the seminiferous epithelium in control rats and vitamin A-deficient rats after 7, 10, 13, and 18 weeks of resupplementation with vitamin A. Values are mean ± SEM of 5–7 animals/group. Per animal, 200 cross sections of seminiferous tubules were evaluated. Spermatogenesis is highly synchronized after 7 and 10 weeks of vitamin A replacement. By week 18, spermatogenesis has become markedly desynchronized. Modified from Bartlett *et al.* (1990b).

Perey, 1957). In contrast, the development of testicular spermatozoa requires more than 50 days in the adult rat.

XIII. Summary and Concluding Thoughts

LH, testosterone, and FSH are the key regulators of mammalian spermatogenesis and act synergistically to promote germ cell development. During the prepubertal period, GH and thyroid hormones are required *additionally* to establish quantitatively normal numbers of Sertoli cells and, subsequently, adult testicular size.

For quite some time, testosterone and FSH have been suggested to control specific events during spermatogenesis: FSH acts on spermatogoniogenesis whereas testosterone influences meiosis and, in particular, spermiogenesis. This view, however, is challenged by observations that testosterone and FSH alone are capable of stimulating the complete spermatogenic process in rats, nonhuman primates, and men. These findings indicate that testosterone and FSH exert a common beneficial effect on germ cell development. For example, both testosterone and FSH have been shown to stimulate the proliferation of the renewing testicular stem cells (A-pale spermatogonia) during initiation, maintenance, and reinitiation of spermatogenesis in nonhuman primates.

The failure to achieve complete spermatogenesis with testosterone and FSH simply might be related to the experimental regimen, that is, insufficient doses of hormones, short treatment length, and choice of the experimental model. This view is based on the following observations: the spermatogenic stages differ in sensitivity to the effects of testosterone and FSH; and the ability of FSH or testosterone to uphold or restore spermatogenesis is much more pronounced in GnRH antagonist-treated or GnRH-immunized rats (selective gonadotropin deprivation) than in hypophysectomized animals.

The precise mechanisms of action of testosterone and FSH on spermatogenesis have not been elucidated to date. Testosterone acts on the somatic cells of the seminiferous tubule (peritubular and Sertoli cells); FSH acts on Sertoli cells and possibly directly on spermatogonia. Obviously, testosterone and FSH interact at the receptor level through their capability to regulate not only the homologous receptor but also the receptor for the "other" hormone. Evidence has accumulated to suggest that isoforms of the testicular LH and FSH receptor exist that might be secreted and have hormone binding activity. This behavior would provide a means of regulating target cell sensitivity and might become relevant to our understanding of the yet unknown causes of male idiopathic infertility.

Interestingly, testosterone and FSH are far more effective in maintaining than in reinitiating the spermatogenic process. With respect to testosterone, a 2-week period of androgen deficiency already is sufficient to render testosterone ineffective. This reduced efficacy of testosterone remains unexplained but gives rise to a highly promising approach to male fertility regulation: a combined regimen of GnRH antagonist and delayed (2 weeks) testosterone administration. To date, reversible azoospermia has been achieved in 14 of 16 men using of this approach. Testosterone alone is also a potential candidate for male fertility regulation based on its negative feedback effect on gonadotropin secretion. The surprising observation was made that Asian men respond much better to testosterone in terms of inhibition of spermatogenesis than Caucasian men. Consequently, the possibility should be considered that certain human races might differ with respect to the hormonal control of spermatogenesis.

Finally, circumstantial evidence is available that suggests that the duration of the spermatogenic process, contrary to the present dogma, can be influenced by hormones and drugs. Synchronization of spermatogenesis in the vitamin A-deficient rat model is lost over a period of about 10 spermatogenic cycles, and certain drugs alter the frequency of the spermatogenic stages. Whether these changes in the relative duration of spermatogenic stages reflect an alteration of the duration of the entire spermatogenic process remains to be seen.

Acknowledgments

Our own work reported here was supported by grants from the Max Planck Society, the Deutsche Forschungsgemeinschaft, and the World Health Organization (Human Reproduction Programme).

References

Aatsinki, J. T., Pietilä, E. M., Lakkakorpi, J. T., and Rajaniemi, H. J. (1992). Expression of the LH/hCG receptor gene in rat ovarian tissue is regulated by an extensive alternative splicing of the primary transcript. *Mol. Cell. Endocrinol.* **84,** 127–135.

Abeyawardenee, S. A., Vale, W. W., Marshall, G. R., and Plant, T. M. (1989). Circulating inhibin concentrations in infant, prepubertal and adult male rhesus monkeys *(Macaca mulatta)* and in juvenile males during premature initiation of puberty with pulsatile gonadotropin-releasing hormone treatment. *Endocrinology* **125,** 250–256.

Ahmad, N., Haltmeyer, G. C., and Eik-Nes, K. (1975). Maintenance of spermatogenesis with testosterone or dihydrotestosterone in hypophysectomized rats. *J. Endocrinol.* **44,** 103–107.

Akhtar, F. B., Marshall, G. R., Wickings, E. J., and Nieschlag, E. (1983a). Reversible induction of azoospermia in rhesus monkeys by constant infusion of a GnRH agonist using osmotic minipumps. *J. Clin. Endocrinol. Metab.* **56,** 534–540.

Akhtar, F. B., Marshall, G. R., and Nieschlag, E. (1983b). Testosterone supplementation attenuates the antifertility effects of an LH-RH agonist in male monkeys. *Int. J. Androl.* **6,** 461–468.

Almiron, I., Domene, H., and Chemes, H. E. (1984). The hormonal regulation of premeiotic steps of spermatogenesis in the newborn rat. *J. Androl.* **5**, 235–242.

Amir, S. M., Sullivan, R. C., and Ingbar, S. H. (1978). Binding of bovine thyrotropin to receptors in rat testis and its interaction with gonadotropins. *Endocrinology* **103**, 101–111.

Aragona, L., and Friesen, H. G. (1975). Specific prolactin binding sites in the prostate and testis of rats. *Endocrinology* **97**, 677–684.

Arslan, M., Zaidi, P., Akhtar, F. B., Amin, S., Rana, T., and Qazi, M. H. (1981). Effects of gonadotrophin treatment *in vivo* on testicular function in immature rhesus monkeys (Macaca mulatta). *Int. J. Androl.* **4**, 462–474.

Arslan, M., Weinbauer, G. F., Khan, S. A., and Nieschlag, E. (1989). Testosterone and dihydrotestosterone, but not estradiol, selectively maintain pituitary and serum follicle-stimulating hormone in gonadotropin-releasing hormone antagonist treated male rats. *Neuroendocrinology* **49**, 395–401.

Arslan, M., Weinbauer, G. F., Schlatt, S., Shahab, M., and Nieschlag, E. (1993). Follicle-stimulating hormone and testosterone, alone or in combination, activate spermatogonial proliferation and stimulate testicular growth in the immature nonhuman primate (*Macaca mulatta*). *J. Endocrinol.* **136**, 235–243.

Awoniyi, C. A., Santulli, R., Sprando, R. L., Ewing, L. L., and Zirkin, B. B. (1989a). Restoration of advanced spermatogenic cells in the experimentally regressed rat testis: Quantitative relationship to testosterone concentration within the testis. *Endocrinology* **124**, 1217–1223.

Awoniyi, C. A., Santulli, R., Chandrashekar, V., Schanbacher, B. D., and Zirkin, B. R. (1989b). Quantitative restoration of advanced spermatogenic cells in adult male rats made azoospermic by active immunization against luteinizing hormone or gonadotropin-releasing hormone. *Endocrinology* **125**, 1303–1309.

Awoniyi, C. A., Zirkin, B. R., Chandrashekar, V., and Schlaff, W. D. (1992a). Exogenously administered testosterone maintains spermatogenesis quantitatively in adult rats actively immunized against gonadotropin-releasing hormone. *Endocrinology* **130**, 3283–3288.

Awoniyi, C. A., Kim, W. K., Hurst, B. S., and Schlaff, W. D. (1992b). Immunoneutralization of gonadotropin-releasing hormone and subsequent treatment with testosterone Silastic implants in rats: An approach toward developing a male contraceptive. *Fertil. Steril.* **58**, 403–408.

Bagatell, C. J., McLachlan, R. I., de Kretser, D. M., Burger, H. G., Vale, W. W., Rivier, J., and Bremner, W. J. (1989). A comparison of the suppressive effects of testosterone and a potent new gonadotropin-releasing hormone antagonist on gonadotropin and inhibin levels in normal men. *J. Clin. Endocrinol. Metab.* **69**, 43–48.

Bartlett, J. M. S., Weinbauer, G. F., and Nieschlag, E. (1989a). Differential effects of FSH and testosterone on the maintenance of spermatogenesis in the adult hypophysectomized rat. *J. Endocrinol.* **121**, 49–58.

Bartlett, J. M. S., Weinbauer, G. F., and Nieschlag, E. (1989b). Quantitative analysis of germ cell numbers and relation to intratesticular testosterone following vitamin A-induced synchronization of spermatogenesis in the rat. *J. Endocrinol.* **123**, 403–412.

Bartlett, J. M. S., Charlton, H. M., Robinson, I. C. A. F., and Nieschlag, E. (1990a). Pubertal development and testicular function in the male growth hormone-deficient rat. *J. Endocrinol.* **126**, 193–301.

Bartlett, J. M. S., Weinbauer, G. F., and Nieschlag, E. (1990b). Stability of spermatogenic synchronization achieved by depletion and restoration of vitamin A in rats. *Biol. Reprod.* **42**, 603–612.

Behre, H. M., Nashan, D., Hubert, W., and Nieschlag, E. (1992). Depot GnRH agonists blunts the androgen-induced suppression of spermatogenesis in a clinical trial for male contraception. *J. Clin. Endocrinol. Metab.* **74**, 84–90.

Beitins, I. Z., Baynard, F., Kowarski, A., and Migeon, C. J. (1973). The effect of ACTH administration on plasma testosterone and serum LH concentrations in normal men. *Steroids* **32**, 553–563.

Bennett, W. I., Dufau, M. L., Catt, K. J., and Tullner, W. W. (1973). Effect of human menopausal gonadotropin upon spermatogenesis and testosterone production in juvenile rhesus monkeys. *Endocrinology* **92**, 813–821.

Bercu, B. B., Jackson, I. M. D., Sawin, C. T., Safaii, H., and Reichlin, S. (1977). Permanent impairment of testicular development after transient immunological blockade of endogenous luteinizing hormone releasing hormone in the neonatal rat. *Endocrinology* **101**, 1871–1877.

Bernard, M. P., Myers, M. V., and Moyle, W. R. (1990). Cloning of rat lutropin (LH) receptor analogs lacking the soybean lectin domain. *Mol. Cell. Endocrinol.* **71**, R19–R23.

Berndtson, W. E., Desjardins, C., and Ewing, L. L. (1974). Inhibition and maintenance of spermatogenesis in rats implanted with polydimethylsiloxane capsules containing various androgens. *J. Endocrinol.* **62**, 125–135.

Bex, F. J., Bartke, A., Goldman, B. D., and Dalterio, S. (1978). Prolactin, growth hormone, and luteinizing hormone receptors and seasonal changes in testicular activity in the Golden hamster. *Endocrinology* **103**, 2069–2080.

Bidlingmaier, F., Dörr, H. G., Eisenmenger, W., Kuhnle, U., and Knorr, D. (1983). Testosterone and androstendione concentrations in human testis and epididymis during the first two years of life. *J. Clin. Endocrinol. Metab.* **57**, 311–315.

Blok, L. J., Mackenbach, P., Trapman, J., Themmen, A. P. N., Brinkmann, A. O., and Grootegoed, J. A. (1989). Follicle-stimulating hormone regulates androgen receptor mRNA in Sertoli cells. *Mol. Cell. Endocrinol.* **63**, 267–271.

Blok, L. J., Bartlett, J. M. S., Bolt-de-Vries, J., Themmen, A. P. N., Brinkmann, A. O., Weinbauer, G. F., Nieschlag, E., and Grootegoed, J. A. (1992a). Effect of testosterone deprivation on expression of the androgen receptor in rat prostate, epididymis and testis. *Int. J. Androl.* **15**, 182–198.

Blok, L. J., Hoogerbrugge, J. W., Themmen, A. P. N., Baarends, W. M., Post, M., and Grootegoed, A. J. (1992b). Transient down-regulation of androgen receptor messenger ribonucleic acid (mRNA) expression in Sertoli cells by follicle-stimulating hormone is followed by up-regulation of androgen receptor mRNA and protein. *Endocrinology* **131**, 1343–1349.

Blok, L. J., Themmen, A. P. N., Peters, A. H. F. M., Trapman, J., Baarends, W. M., Hoogerbrugge, J. W., and Grootegoed, J. A. (1992c). Transcriptional regulation of androgen receptor gene expression in Sertoli cells and other cell types. *Mol. Cell. Endocrinol.* **88**, 153–164.

Boccabella, A. V. (1963). Reinitiation and restoration of spermatogenesis with testosterone propionate and other hormones after long-term post-hypophysectomy regression period. *Endocrinology* **72**, 787–798.

Burris, A. S., Rodbard, H. W., Winters, S. J., and Sherins, R. J. (1988). Gonadotropin therapy in men with isolated hypogonadotropic hypogonadism: The response to human chorionic gonadotropin is predicted by initial testicular size. *J. Clin. Endocrinol. Metab.* **66**, 1144–1151.

Cameron, D. F., and Muffly, K. E. (1991). Hormonal regulation of spermatid binding. *J. Cell Sci.* **100**, 623–633.

Chandolia, R. K., Weinbauer, G. F., Behre, H. M., and Nieschlag, E. (1991a). Evaluation of a peripherally selective antiandrogen (Casodex) as a tool for studying the relationship between testosterone and spermatogenesis in the rat. *J. Steroid Biochem. Mol. Biol.* **38**, 367–375.

Chandolia, R. K., Weinbauer, G. F., Fingscheidt, U., Bartlett, J. M. S., and Nieschlag, E. (1991b). Effects of flutamide on testicular involution induced by an antagonist of gonadotrophin-releasing hormone and on stimulation of spermatogenesis by follicle-stimulating hormone in rats. *J. Reprod. Fertil.* **93**, 313–323.

Chapin, R. E., Morgan, K. T., and Bus, J. S. (1983). The morphogenesis of testicular degeneration induced in rats by orally administered 2,5-hexanedione. *Exp. Mol. Pathol.* **38**, 149–169.

Chemes, H. E., Podesta, E., and Rivarola, M. A. (1976). Action of testosterone, dihydrotestosterone, and 5-alpha androstane-3-alpha,17-beta-diol on spermatogenesis of immature rats. *Biol. Reprod.* **14**, 332–338.

Chemes, H. E., Dym, M., and Madhwa Raj, H. G. (1979). The role of gonadotropins and testosterone on initiation of spermatogenesis in the immature rat. *Biol. Reprod.* **21**, 241–249.

Chemes, H. E., Pasqualini, T., Rivarola, M. A., and Bergada, C. (1982). Is testosterone involved in the initiation of spermatogenesis in humans? A clinicopathological presentation and physiological considerations in four patients with Leydig cell tumors of the testis or secondary Leydig cell hyperplasia. *Int. J. Androl.* **5**, 299–245.

Chowdhury, A. K., and Marshall, G. (1980). Irregular patterns of spermatogenesis in the baboon *(Papio anubis)* and its possible mechanism *In* "Testicular Development, Structure and Function" (A. Steinberger and E. Steinberger, eds.), pp. 129–137. Raven Press, New York.

Chowdhury, A. K., and Steinberger, E. (1975). Effect of 5-alpha-reduced androgens on sex accessory organs, initiation and maintenance of spermatogenesis in the rat. *Biol. Reprod.* **12**, 609–617.

Clermont, Y. (1963). The cycle of the seminiferous epithelium in man. *Am. J. Anat.* **112**, 35–46.

Clermont, Y. (1969). Two classes of spermatogonial stem cells in the monkey *(Cercopithecus aethiops). Am. J. Anat.* **126**, 57–72.

Clermont, Y. (1972). Kinetics of spermatogenesis in mammals: Seminiferous epithelium cycle and spermatogonial renewal. *Physiol. Rev.* **52**, 198–236.

Clermont, Y., and Antar, M. (1973). Duration of the cycle of the seminiferous epithelium and the spermatogonial renewal in the monkey, *Macaca arctoides. Am. J. Anat.* **136**, 153–166.

Clermont, Y., and Harvey, S. C. (1965). Duration of the cycle of the seminiferous epithelium of normal, hypophysectomized and hypophysectomized-hormone treated albino rats. *Endocrinology* **76**, 80–89.

Clermont, Y., and Morgenthaler, H. (1955). Quantitative study of spermatogenesis in the hypophysectomized rat. *Am. J. Anat.* **57**, 369–382.

Clermont, Y., and Perey, B. (1957). Quantitative study of the cell population of the seminiferous tubules in immature rats. *Am. J. Anat.* **100**, 241–267.

Closset, J., and Hennen, G. (1989). Biopotency of highly purified procine FSH and human LH on gonadal function. *J. Endocrinol.* **120**, 89–96.

Cooke, P. S., and Meisami, E. (1991). Early hypothyroidism in rats causes increased adult testis and reproductive organ size but does not change testosterone levels. *Endocrinology* **129**, 237–243.

Cooke, P. S., Hess, R. A., Porcell, J., and Meisami, E. (1991). Increased sperm production in adult rats after transient neonatal hypothyroidism. *Endocrinology* **129**, 244–248.

Creasy, D. M., Flynn, J. C., Gray, T. J. B., and Butler, W. H. (1985). A quantitative study of stage-specific spermatocyte damage following administration of ethylene glycol monomethyl ether in the rat. *Exp. Mol. Pathol.* **43**, 321–336.

Cunningham, G. R., and Huckins, C. (1979). Persistence of complete spermatogenesis in the presence of low intratesticular concentrations of testosterone. *Endocrinology* **105**, 177–186.

D'Agata, R., Heindel, J. J., Vicari, E., Aliffi, A., Gulizia, S., and Polosa, P. (1984). hCG-induced maturation of the seminiferous epithelium in hypogonadotropic men. *Horm. Res.* **19**, 23–32.

Dang, D. C. (1971a). Stade dy cycle de l'épithélium séminifére du singe crabier, *Macaca fascicularis* (= Irus ou *Cynomolgus). Ann. Biol. Anim. Bioch. Biophys.* **11**, 363–371.

Dang, D. C. (1971b). Durée du cycle de l'épithélium séminifére du singe chrabier, Macaca fascicularis. *Ann. Biol. Anim. Bioch. Biophys.* **11**, 373–377.

Davies, T. F., Katikineni, M., Chan, V., Harwood, J. P., Dufau, M. L., and Catt, K. J. (1980). Lactogenic receptor regulation in hormone stimulated steroidogenic cells. *Nature (London)* **283**, 863–865.

de Rooij, D. G., van Alphen, M. M. A., and van de Kant, H. J. G. (1986). Duration of the cycle of the seminiferous epithelium and its stages in the rhesus monkey *(Macaca mulatta)*. *Biol. Reprod.* **35**, 587–591.

Dietrich, T., Schulze, W., and Riemer, M. (1986). Untersuchung zur Gliederung des Keimepithels beim Javaneraffen *(Macaca cynomolgus)* mittels digitaler Bildverarbeitung. *Urologe* **25**, 179–186.

Dym, M., and Cavicchia, J. C. (1978). Functional morphology of the testis. *Biol. Reprod.* **18**, 1–15.

Dym, M., and Raj, M. H. G. (1977). Response of adult rat Sertoli cells and Leydig cells to depletion of luteinizing hormone and testosterone. *Biol. Reprod.* **17**, 676–696.

Dym, M., Raj, H. G. M., Lin, Y. C., Chemes, H. E., Kotite, N. J., Nayfeh, S. N., and French, F. S. (1978). Is FSH required for maintenance of spermatogenesis in adult rats? *J. Reprod. Fertil. Suppl.* **26**, 175–181.

Fawcett, D. W. (1975). Ultrastructure and function of the Sertoli cell. *In* "Handbook of Physiology, Section 7: Endocrinology" (D. W. Hamilton and R. O. Greep, eds.), Vol. 5, pp. 21–55. American Physiology Society, Washington, D.C.

Finkel, D. M., Phillips, J. L., and Snyder, P. J. (1985). Stimulation of spermatogenesis by gonadotropins in men with hypogonadotropic hypogonadism. *N. Engl. J. Med.* **313**, 651–655.

Finkelstein, J. S., O'Dea, L. S. L., Spratt, D. I., and Crowley, W. F., Jr. (1987). Pulsatile GnRH therapy in men with idiopathic hypogonadotropic hypogonadism. *In* "Highlights on Endocrinology" (C. Christiansen and B. J. Riis, eds.). Proceedings of the First European Congress on Endocrinology, Copenhagen.

Francavilla, S., Cordeschi, G., Properzi, G., Di Cicco, L., Jannini, E. A., Palmero, S., Fugassa, E., Loras, B., and D'Armiento, M. (1991). Effect of thyroid hormones on the pre- and postnatal development of the rat testis. *J. Endocrinol.* **129**, 34–42.

Frankel, A. I., Chapman, J. C., and Wright, W. W. (1989). The equivocal presence of nuclear androgen binding proteins in mammalian spermatids and spermatozoa. *J. Steroid Biochem.* **33**, 71–79.

Grino, P. B., Griffin, J. E., and Wilson, J. D. (1990). Testosterone at high concentrations interacts with the human androgen receptor similarly to dihydrotestosterone. *Endocrinology* **126**, 1165–1172.

Gromoll, J., Gudermann, T., and Nieschlag, E. (1992). Molecular cloning of a truncated isoform of the human follicle-stimulating hormone receptor. *Biochem. Biophys. Res. Comm.* **188**, 1077–1083.

Grootegoed, J. A., Peters, M. J., Mulder, E., Rommerts, F. F. G., and van der Molen, H. J. (1977). Absence of nuclear androgen receptor in isolated germinal cells of rat testis. *Mol. Cell. Endocrinol.* **9**, 159–167.

Guillou, F., Marinat, N., and Combarnus, Y. (1987). Homologous desensitization of rat Sertoli cells by non-stimulating concentrations of follicle-stimulating hormone. *Biol. Cell* **59**, 227–232.

Harris, M. E., and Bartke, A. (1975). Maintenance of rete testis fluid testosterone and dihydrotestosterone levels by pregnenolone and other C_{21} steroids in hypophysectomized rats. *Endocrinology* **96**, 1396–1402.

Harris, M. E., Bartke, A., Weisz, J., and Watson, D. (1977). Effects of testosterone and dihydrotestosterone on spermatogenesis, rete testis fluid, and peripheral androgen levels in hypophysectomized rats. *Fertil. Steril.* **28**, 1113–1117.

Heckert, L. L., and Griswold, M. D. (1991). Expression of follicle-stimulating hormone receptor mRNA in rat testes and Sertoli cells. *Mol. Endocrinol.* **5**, 670–677.

Heckert, L., and Griswold, M. D. (1992). The changing functions of follicle-stimulating hormone in the testes of prenatal, newborn, immature, and adult rats. *In* "Follicle Stimulating Hormone — Regulation of Secretion and Molecular Mechanism of Action" (M. Hunzicker-Dunn and N. B. Schwartz, eds.), pp. 237–245. Springer Verlag, New York.

Heller, C. G., and Clermont, Y. (1964). Kinetics of the germinal epithelium in man. *Rec. Prog. Horm. Res.* **20**, 545–575.

Hess, R. A., Linder, R. E., Strader, L. F., and Perreault, S. D. (1988). Acute effects and long-term sequelae of 1,3-dinitrobenzene on male reproduction in the rat. I. Quantitative and qualitative histopathology of the testis. *J. Androl.* **9**, 327–342.

Hoffman, W. H., Kovacs, K. T., Gala, R. R., Keel, B. A., Jarrell, T. S., Ellegood, J. O., and Burek, C. L. (1991). Macroorchidism and testicular fibrosis associated with autoimmune thyroiditis. *J. Clin. Invest.* **14**, 609–616.

Holstein, A. F. (1981). "Atlas of Human Spermatogenesis." Grosse Verlag, Berlin.

Holstein, A. F. (1988). "Illustrated Pathology of Human Spermatogenesis." Grosse Verlag, Berlin.

Hsueh, A. J. W., and La Polt, P. S. (1992). Molecular basis of gonadotropin receptor regulation. *Trends Endocrinol. Metab.* **3**, 164–170.

Huang, H. F. S., and Nieschlag, E. (1986). Suppression of the intratesticular testosterone is associated with quantitative changes in spermatogonial populations in intact adult rats. *Endocrinology* **118**, 619–627.

Huang, H. F. S., Marshall, G. R., and Nieschlag, E. (1990). Enrichment of the stages of the seminiferous epithelium in vitamin A-replaced-vitamin A-deficient rats. *J. Reprod. Fertil.* **88**, 51–60.

Huhtaniemi, I. T., Nevo, N., Amsterdam, A., and Naor, Z. (1986). Effect of postnatal treatment with a gonadotropin-releasing hormone antagonist on sexual maturation of male rats. *Biol. Reprod.* **35**, 501–507.

Huhtaniemi, I. T., Dahl, K. D., Rannikko, S., and Hsueh, A. J. W. (1988). Serum bioactive and immunoreactive follicle-stimulating hormone in prostatic cancer patients during gonadotropin-releasing hormone agonist treatment and after orchidectomy. *J. Clin. Endocrinol. Metab.* **66**, 308–313.

Hutson, J. C., Yee, J. B., and Yee, J. A. (1987). Peritubular cells influence Sertoli cells at the level of translation. *Mol. Cell. Endocrinol.* **52**, 11–15.

Irvine, W. J., Toft, A. D., Nilson, K. S., Fraser, R., Wilson, A., Young, J., Hunter, W. M., Ismail, A. A. A., and Burger, P. E. (1974). The effect of synthetic corticotropin analogues onadrenocortical, anterior pituitary and testicular function. *J. Clin. Endocrinol. Metab.* **39**, 522–529.

Isomaa, V., Parvinen, M., Janne, O. A., and Bardin, C. W. (1985). Nuclear androgen receptors in different stages of the seminiferous epithelial cycle and the interstitial tissue of rat testis. *Endocrinology* **116**, 132–135.

Janca, F. C., Jost, L. K., and Evenson, D. P. (1986). Mouse testicular and sperm cell development characterized from birth to adulthood by dual parameter flow cytometry. *Biol. Reprod.* **34**, 613–623.

Jannini, E. A., Olivieri, M., Francavilla, S., Gulino, A., Ziparo, E., and D'Armiento, M. (1990). Ontogenesis of the nuclear 3,5,3'-triiodothyronine receptor in the rat testis. *Endocrinology* **126**, 2521–2526.

Jegou, B. (1992). The Sertoli cell. *Bailliere's Clin. Endocrin. Metab.* **6**, 273–311.

Jegou, B., Syed, V., Sourdaine, P., Byers, S., Gerard, N., Velez de la Calle, J., Pineau, C., Garnier, D. H., and Bauche F. (1992). The dialogue between late spermatids and Sertoli cells in vertebrates: A century of research. *In* "Spermatogenesis, Fertilization, Contraception" (E. Nieschlag and U. F. Habenicht, eds.), pp. 57–96. Springer Verlag, Berlin.

Johnsen, S. G. (1978). Maintenance of spermatogenesis induced by HMG treatment by means of continuous HCG treatment in hypogonadotrophic men. *Acta Endocrinol.* **89**, 763–769.

Kangasniemi, M., Kaipia, A., Toppari, J., Mali, P., Huhtaniemi, I., and Parvinen, M. (1990a).

Cellular regulation of basal and FSH-stimulated cyclic AMP production in irradiated testes. *Anat. Rec.* **227**, 32–36.

Kangasniemi, M., Kaipia, A., Toppari, J., Perheentupa, A., Huhtaniemi, I., and Parvinen, M. (1990b). Cellular regulation of follicle-stimulating hormone (FSH) binding in rat seminiferous tubules. *J. Androl.* **11**, 336–343.

Kar, A. B., Chandra, H., and Kamboj, V. P. (1966). Effect of non-primate gonadotrophins on the testis of prepuberal rhesus monkeys. *Acta Biol. Med. German.* **16**, 450–455.

Kerr, J. B. (1989). The cytology of the human testis. *In* "The Testis" (H. Burger and D. M. de Kretser, eds.), 2d Ed., pp. 197–229. Raven Press, New York.

Kerr, J. B., and Sharpe, R. M. (1985). FSH induction of Leydig cell maturation. *Endocrinology* **116**, 2592–2604.

Kerr, J. B., Maddocks, S., and Sharpe, R. M. (1992). Testosterone and FSH have independent, synergistic and stage-dependent effects upon spermatogenesis in the rat testis. *Cell Tissue Res.* **268**, 179–189.

Ketelslegers, J. M., Hetzel, W. D., Sherins, R. J., and Catt, K. J. (1978). Developmental changes in testicular gonadotropin receptors: Plasma gonadotropins and plasma testosterone in the rat. *Endocrinology* **103**, 212–222.

Khurshid, S., Weinbauer, G. F., and Nieschlag, E. (1991). Effects of testosterone and gonadotrophin-releasing hormone (GnRH) antagonist in basal and GnRH-stimulated gonadotrophin secretion in orchidectomized monkeys. *J. Endocrinol.* **129**, 363–370.

Kliesch, S., Penttilä, Gromoll, J., Saunders, P. T. K., Nieschlag, E., and Parvinen, M. (1992). FSH receptor mRNA is expressed stage-dependently during spermatogenesis. *Mol. Cell. Endocrinol.* **84**, R45–R49.

Kluin, P. H. M., Kramer, M. F., and de Rooij, D. G. (1982). Spermatogenesis in the immature mouse proceeds faster than in the adult. *Int. J. Androl.* **5**, 282–294.

Knuth, U. A., and Nieschlag, E. (1989). Male contraception based on androgen/gestagen combinations. *In* "The Molecular and Cellular Endocrinology of the Testis" (B. A. Cooke, and R. M. Sharpe, eds.), pp. 335–355. Raven Press, New York.

Kolho, K. L., and Huhtaniemi, I. (1989). Neonatal treatment of male rats with a gonadotropin-releasing hormone antagonist impairs ejaculation and fertility. *Physiol. Behav.* **46**, 373–377.

Kula, K. (1988). Induction of precocious maturation of spermatogenesis in infant rats by human menopausal gonaotropin and inhibition by simultaneous administration of gonadotropins and testosterone. *Endocrinology* **122**, 34–39.

Laron, Z., Karp, M., and Dolberg, L. (1970). Juvenile hypothyroidism with testicular enlargement. *Acta Pediat. Scand.* **59**, 317–322.

Leblond, C. P., and Clermont, Y. (1952a). Spermatogenesis of rat, mouse, hamster and guinea pig as revealed by the "periodic acid-fuchsin sulfurous acid" technique. *Am. J. Anat.* **90**, 167–215.

Leblond, C. P., and Clermont, Y. (1952b). Definition of the stages of the cycle of the seminiferous epithelium in the rat. *Ann. N.Y. Acad. Science.* **55**, 548–573.

Lee, V. W. K., de Kretser, D. M., Hudson, B., and Wang, C. (1975). Variations in serum FSH, LH and testosterone levels in male rats from birth to sexual maturity. *J. Reprod. Fertil.* **42**, 121–126.

Lerchl, A., Sotiriadou, S., Kliesch, S., Pierce, J., Behre, H. M., Weinbauer, G. F., Kliesch, S., and Nieschlag, E. (1993). FSH restores spermatogenesis and fertility in hypogonadotropic Djungarian hamsters. 37th Symposium of the German Endocrine Society, Berlin, March 3–6.

Linder, C. C., Heckert, L. L., Roberts, K. P., Kim, K. H., and Griswold, M. (1991). Expression of receptors during the cycle of the seminiferous epithelium. *Ann. N.Y. Acad. Sci.* **637**, 313–321.

Lobie, P. E., Breipohl, W., Aragon, J. G., and Waters, M. J. (1990). Cellular localization of the growth hormone receptor/binding protein in the male and female reproductive systems. *Endocrinology* **126**, 2214–2221.

Lostroh, A. L., Johnson, R., and Jordan, C. W., Jr. (1963). Effect of ovine gonadotrophins and antiserum to interstitial cell-stimulating hormone on the testis of the hypophysectomized rat. *Acta Endocrinol.* **44**, 536–544.

Lyon, M. F., Glenister, P. H., and Lamoreux, M. L. (1975). Normal spermatozoa from androgen-resistant germ cells of chimaeric mice and the role of androgen in spermatogenesis. *Nature (London)* **258**, 620–622.

Madhwa Raj, H. G., and Dym, M. (1976). The effects of selective withdrawal of FSH or LH on spermatogenesis in the immature rat. *Biol. Reprod.* **14**, 498–494.

Mancini, R. E. (1969). Effect of different types of gonadotrophins on the induction and restoration of spermatogenesis in the human testis. *Acta Europ. Fertil.* **1**, 401–429.

Marshall, G. R., Wickings, E. J., and Nieschlag, E. (1983). Stimulation of spermatogenesis in stalk-sectioned rhesus monkeys by testosterone alone. *J. Clin. Endocrinol. Metab.* **57**, 152–159.

Marshall, G. R., Wickings, E. J., and Nieschlag, E. (1984). Testosterone can initiate spermatogenesis in an immature nonhuman primate, *Macaca fascicularis. Endocrinology* **114**, 2228–2233.

Marshall, G. R., Akhtar, F. B., Weinbauer, G. F., and Nieschlag, E. (1986a). Gonadotrophin-releasing hormone (GnRH) overcomes GnRH antagonist-induced suppression of LH secretion in primates. *J. Endocrinol.* **110**, 145–150.

Marshall, G. R., Jockenhövel, F., Lüdecke, D., and Nieschlag, E. (1986b). Maintenance of complete but quantitatively reduced spermatogenesis in hypophysectomized monkeys by testosterone alone. *Acta Endocrinol.* **113**, 424–431.

Matsumoto, A. M. (1989). Hormonal control of spermatogenesis. *In* "The Testis" (H. Burger and D. M. de Kretser, eds.), 2d Ed., pp. 181–196. Raven Press, New York.

Matsumoto, A. M., and Bremner, W. J. (1991). Control of spermatogenesis in humans. *In* "Perspectives in Primate Reproductive Biology." (N. R. Moudgal, K. Yoshinaga, A. J. Rao, P. R. Adiga, eds.), pp. 173–180. New Delhi: Wiley Eastern Limited.

Minegish, T., Nakamura, K., Takakura, Y., Miyamoto, K., Hasegawa, Y., Ibuky, Y., and Igarashi, M. (1990). Cloning and sequencing of human LH/hCG receptor cDNA. *Biochem. Biophys. Res. Commun.* **172**, 1049–1054.

Morales, C., and Griswold, D. (1987). Retinol-induced stage synchronization in seminiferous tubules of the rat. *Endocrinology* **121**, 432–434.

Morris, P. L., and Saxena, B. B. (1980). Dose and age-dependent effects of prolactin (PRL) on LH and PRL-binding sites in rat Leydig cell homogenates. *Endocrinology* **107**, 1639–1645.

Moudgal, N. R. (1981). A need for FSH in maintaining fertility of adult subhuman primates. *Arch. Androl.* **7**, 117–125.

Moudgal, N. R., Ravindranath, N., Murthy, G. S., Dighe, R. R., Aravindan, G. R., and Martin, F. (1992). Long-term contraceptive efficacy of vaccine of ovine follicle-stimulating hormone in male bonnet monkeys *(Macaca radiata). J. Reprod. Fertil.* **96**, 91–102.

Nakhla, A. M., Mather, J. P., Jänne, O. A., and Bardin, C. W. (1984). Estrogen and androgen receptors in Sertoli, Leydig, myoid and epithelial cells: Effects of time in culture and cell density. *Endocrinology* **115**, 121–128.

Namiki, M., Yokokawa, K., Okuyama, A., Koh, E., Kioyhara, H., Nakao, M., Sakoda, S., Matsumoto, K., and Sonoda, T. (1991). Evidence for the presence of androgen receptors in human Leydig cells. *J. Steroid Biochem. Mol. Biol.* **38**, 79–82.

Nieschlag, E., and Kley, H. K. (1975). Possibility of adrenal-testicular interaction as indicated by plasma androgens in response to HCG in men with normal, suppressed and impaired adrenal function. *Horm. Metab. Res.* **7**, 326–330.

Nieschlag, E., and Wickings, E. J. (1980). Does the rhesus monkey provide a suitable model for human testicular function? *In* "Animal Models in Human Reproduction" (M. Serio and L. Martini, eds.), pp. 151–159. Raven Press, New York.

Nieschlag, E., Behre, H. M., and Weinbauer, G. F. (1992). Hormonal male contraception: A real chance? *In* "Spermatogenesis, Fertilization, Contraception" (E. Nieschlag and U. F. Habenicht, eds.), pp. 477–502. Springer Verlag, Berlin.

Niklowitz, P., Khan, S., Bergmann, M., Hoffman, K., and Nieschlag, E. (1989). Differential effects of follicle-stimulating hormone and luteinizing hormone on Leydig cell function and restoration of spermatogenesis in hypophysectomized and photoinhibited Djungarian hamsters (*Phodopus sungorus*). *Biol. Reprod.* **41**, 871–880.

Nistal, M., and Paniagua, R. (1984). "Testicular and Epididymal Pathology." Thieme-Stratton, New York.

Odell, W. D., and Swerdloff, R. S. (1976). Etiologies of sexual maturation: A model system based on the sexually maturing rat. *Rec. Prog. Horm. Res.* **32**, 245–263.

Orth, J. M. (1984). The role of follicle-stimulating hormone in controlling Sertoli cell proliferation in testes of fetal rats. *Endocrinology* **115**, 1248–1255.

Orth, J., and Christensen, A. K. (1978). Autoradiographic localization of specifically bound ^{125}I-labeled follicle-stimulating hormone on spermatogonia of the rat testis. *Endocrinology* **103**, 1944–1950.

Orth, J. M., Higginbotham, C. A., and Salisbury, R. L. (1984). Hemicastration causes and testosterone prevents enhanced uptake of [^{3}H]thymidine by Sertoli cells in testes of immature rats. *Biol. Reprod.* **30**, 263–270.

Orth, J. M., Gunsalus, G. L., and Lamperti, A. A. (1988). Evidence from the Sertoli cell-depleted rat indicates that spermatid number in adults depends on numbers of Sertoli cells produced during the perinatal development. *Endocrinology* **122**, 787–794.

O'Shaughnessy, P. J., and Brown, P. S. (1978). Reduction of FSH receptors in the rat testis by injection of homologous hormone. *Mol. Cell. Endocrinol.* **12**, 9–15.

Palmero, S., de Marchis, M., Gallo, G., and Fugassa, E. (1989). Thyroid hormone affects the development of Sertoli cell function in the rat. *J. Endocrinol.* **123**, 105–111.

Pangkahila, W. (1991). Reversible azoospermia induced by an androgen–progestin combination regimen in Indonesian men. *Int. J. Androl.* **44**, 248–256.

Pavlou, S. N., Dahl, K. D., Wakefield, G., Rivier, J., Vale, W., Hsueh, A. J. W., and Lindner, J. (1988). Maintenance of the ratio of bioactive to immunoreactive follicle-stimulating hormone in normal men during chronic luteinizing hormone-releasing hormone agonist administration. *J. Clin. Endocrinol. Metab.* **66**, 1005–1009.

Pavlou, S. N., Brewer, K., Farley, M. G., Lindner, J., Bastias, M. C., Rogers, B. J., Swift, L. L., Rivier, J. E., Vale, W., Conn, P. M., and Herbert, C. M. (1991). Combined administration of a gonadotropin-releasing hormone antagonist and testosterone in men induces reversible azoospermia without loss of libido. *J. Clin. Endocrinol. Metab.* **73**, 1360–1369.

Payne, A. H., Kawano, A., and Jaffe, R. B. (1973). Formation of dihydrotestosterone and other 5-alpha-reduced metabolites by isolated seminiferous tubules and suspension of interstitial cells in a human testis. *J. Clin. Endocrinol. Metab.* **37**, 448–453.

Perey, B., Clermont, Y., and Leblond, C. P. (1961). The wave of the seminiferous epithelium in the rat. *Am. J. Anat.* **108**, 47–77.

Plant, T. M. (1985). A study of the role of the postnatal testis in determining the ontogeny of gonadotropin secretion in the male rhesus monkey (*Macaca mulatta*). *Endocrinology* **116**, 1341–1350.

Rea, M. A., Marshall, G. R., Weinbauer, G. F., and Nieschlag, E. (1986). Testosterone maintains pituitary and serum FSH and spermatogenesis in gonadotrophin-releasing hormone antagonist-suppressed rats. *J. Endocrinol.* **108**, 101–107.

Rea, M. A., Weinbauer, G. F., Marshall, G. R., and Nieschlag, E. (1987). Testosterone stimulates pituitary and serum FSH in GnRH antagonist-suppressed rats. *Acta Endocrinol.* **113**, 487–492.

Rivarola, M. A., Podesta, E. J., Chemes, H. E., and Aguilar, D. (1973). *In vitro* metabolism of testosterone by whole human testis, isolated seminiferous tubules and interstitial tissue. *J. Clin. Endocrinol. Metab.* **37**, 454–463.

Robaire, B., Ewing, L. L., Irby, D. C., and Desjardins, C. (1979). Interactions of testosterone and estradiol-17-beta on the reproductive tract of the male rat. *Biol. Reprod.* **21**, 455–463.

Roberts, K. P., Awoniyi, C. A., Santulli, R., and Zirkin, B. R. (1991). Regulation of Sertoli cell transferrin and sulfated glycoprotein-2 messenger ribonucleic acid levels during the restoration of spermatogenesis in the adult hypophysectomized rat. *Endocrinology* **129**, 3417–3424.

Robinson, J. A., and Bridson, W. E. (1978). Neonatal hormone patterns in the macaque. I. Steroids. *Biol. Reprod.* **19**, 773–778.

Rommerts, F. F. G. (1988). How much androgen is required for maintenance of spermatogenesis? *J. Endocrinol.* **116**, 1–6.

Ruizeveld de Winter, J. A., Trapman, J., Vermey, M., Mulder, E., Zegers, N. D., and van der Kwast, T. H. (1991). Androgen receptor expression in human tissues: An immunohistochemical study. *J. Histochem. Cytochem.* **39**, 927–936.

Russell, L. D. (1991). The perils of sperm release—"Let my children go." *Int. J. Androl.* **14**, 307–311.

Russell, L. D., and Clermont, Y. (1977). Degeneration of germ cells in normal, hypophysectomized and hormone treated hypophysectomized rats. *Anat. Rec.* **187**, 347–366.

Russell, L. D., Lee, I. P., Ettlin, R., and Malone, J. P. (1983). Morphological pattern of response after administration of procarbazine: Alteration of specific cell associations during the cycle of the seminiferous epithelium. *Tiss. Cell* **15**, 391–404.

Russell, L. D., Alger, L. E., and Nequin, L. G. (1987). Hormonal control of pubertal spermatogenesis. *Endocrinology* **120**, 1615–1632.

Russell, L. D., Ettlin, R. A., Sinha Hikim, A. P., and Clegg, E. D. (1990). "Histological and Histopathological Evaluation of the Testis." Cache River Press, Clearwater, Florida.

Sanborn, B. M., Steinberger, A., Meistrich, M. L., and Steinberger, E. (1975). Androgen binding sites in testis cell fractions as measured by a nuclear exchange assay. *J. Steroid Biochem.* **6**, 1459–1465.

Sanborn, B. M., Caston, L. A., Chang, C., Kiao, S., Speller, R., Porter, L. D., and Ku, C. Y. (1991). Regulation of androgen receptor mRNA in rat Sertoli can peritubular cells. *Biol. Reprod.* **45**, 634–641.

Santulli, R., Sprando, R. L., Awoniyi, C. A., Ewing, L. L., and Zirkin, B. R. (1990). To what extent can spermatogenesis be maintained in hypophysectomized adult rat testis with exogenously administered testosterone? *Endocrinology* **126**, 95–102.

Sar, M., Lubahn, D. B., French, F. S., and Wilson, E. M. (1990). Immunohistochemical localization of the androgen receptor in rat and human tissues. *Endocrinology* **127**, 3180–3186.

Schulze, W. (1982). Evidence of a wave of spermatogenesis in human testis. *Andrologia* **14**, 200–207.

Schulze, W., and Rehder, U. (1984). Organization and morphogenesis of the human seminiferous epithelium. *Cell. Tissue Res.* **237**, 395–407.

Segaloff, D. L., Wang, H., and Richards, J. S. (1990). Hormonal regulation of luteinizing hormone/chorionic gonadotropin receptor mRNA in rat ovarian cells during follicular development and luteinization. *Mol. Endocrinol.* **4**, 1856–1865.

Sharma, O. P., Khan, S. A., Weinbauer, G. F., Arslan, M., and Nieschlag, E. (1990). Bioactivity and immunoreactivity of androgen stimulated pituitary FSH in GnRH antagonist-treated male rats. *Acta Endocrinol.* **122**, 168–174.

Sharpe, R. M. (1987). Testosterone and spermatogenesis. *J. Endocrinol.* **113**, 1–2.

Sharpe, R. M., Fraser, H. M., and Ratnasooriya, W. D. (1988a). Assessment of the role of Leydig cell products other than testosterone in spermatogenesis and fertility in adult rats. *Int. J. Androl.* **11**, 507–523.

Sharpe, R. M., Donachie, K., and Cooper, I. (1988b). Re-evaluation of the intratesticular level of testosterone required for quantitative maintenance of spermatogenesis in the rat. *J. Endocrinol.* **117**, 19–26.

Sharpe, R. M., Bartlett, J. M. S., and Allenby, G. (1991). Evidence for the control of testicular interstitial fluid volume in the rat by specific germ cell types. *J. Endocrinol.* **128**, 359–367.

Sharpe, R. M., Maddocks, S., Millar, M., Kerr, J. B., Saunders, P. T. K., and McKinnell, C. (1992).

Testosterone and spermatogenesis. Identification of stage-specific, androgen-regulated proteins secreted by adult rat seminiferous tubules. *J. Androl.* **13**, 172–184.

Sinha Hikim, A. P., Sinha Hikim, I., Amador, A. G., Bartke, A., Wolf, A., and Russell, L. D. (1991). Reinitiation of spermatogenesis by exogenous gonadotropins in a seasonal breeder, the woodchuck (*Marmota monax*), during gonadal inactivity. *Am. J. Anat.* **192**, 194–213.

Sivelle, P. C., McNeilly, A. S., and Collins, P. M. (1978). A comparison of the effectiveness of FSH, LH and prolactin in the reinitiation of testicular function of hypophysectomized and estrogen-treated rats. *Biol. Reprod.* **17**, 878–885.

Skinner, M. K. (1991). Cell–cell interactions in the testis. *Endocrinol. Rev.* **12**, 45–77.

Smith, P. E. (1942). Effect of equine gonadotropin on testes of hypophysectomized monkeys. *Endocrinology* **31**, 1–12.

Smith, P. E. (1944). Maintenance and restoration of spermatogenesis in hypophysectomized rhesus monkeys by androgen administration. *Yale J. Biol. Med.* **17**, 283–287.

Spiteri-Grech, J., Bartlett, J. M. S., and Nieschlag, E. (1991). Hormonal regulation of epidermal growth factor and insulin-like growth factor I in adult male hypophysectomized rats treated with ethance dimethane sulphonate. *J. Endocrinol.* **129**, 109–117.

Spiteri-Grech, J., Weinbauer, G. F., Bolze, P., Chandolia, R. K., Bartlett, J. M. S., and Nieschlag, E. (1993). Correlation between testicular insulin-like growth factor-I (IGF-I) content and specific germ cell populations in GnRH antagonist-treated rats. *J. Endocrinol.* **137**, 81–89.

Srinath, B. R., Wickings, E. J., Witting, C., and Nieschlag, E. (1983). Active immunization with follicle-stimulating hormone for fertility control: A 4.5-year study in male rhesus monkeys. *Fertil. Steril.* **40**, 110–117.

Steinberger, E., Root, A., Ficher, M., and Smith, K. D. (1973). The role of androgens in the initiation of spermatogenesis in man. *J. Clin. Endocrinol. Metab.* **37**, 746–751.

Steinberger, A., Janecki, A., and Jakubowiak, A. (1992). FSH actions on Sertoli cell secretions in stationary and superfused cultures. *In* "Follicle Stimulating Hormone—Regulation of Secretion and Molecular Mechanism of Action" (M. Hunzicker-Dunn and N. B. Schwartz, eds.), pp. 217–230. Springer Verlag, New York.

Steiner, R. A., and Bremner, W. J. (1981). Endocrine correlates of testicular development in the male monkey, *Macaca fascicularis. Endocrinology* **109**, 914–919.

Sun, Y. T., Irby, D. C., Robertson, D. M., and DeKretser, D. M. (1989). The effects of exogenously administered testosterone on spermatogenesis in intact and hypophysectomized rats. *Endocrinology* **125**, 1000–1010.

Sun, Y. T., Wreford, N. G., Robertson, D. M., and de Kretser, D. M. (1990). Quantitative cytological studies of spermatogenesis in intact and hypophysectomized rats: Identification of androgen-dependent stages. *Endocrinology* **127**, 1215–1223.

Swerdloff, R. S., and Heber, D. (1981). Endocrine control of testicular function from birth to puberty. *In* "The Testis" (H. Burger and D. M. de Kretser, eds.), pp. 107–126. Raven Press, New York.

Swerdloff, R. S., Campfield, L. A., Palacios, A., and McClure, R. D. (1979). Suppression of human spermatogenesis by depot androgen: Potential for male contraception. *J. Steroid Biochem.* **11**, 663–670.

Takeda, H., Chodak, G., Mutchnik, S., Nakamoto, T., and Chang, C. (1990). Immunohistochemical localization of androgen receptors with mono- and polyclonal antibodies to androgen receptor. *J. Endocrinol.* **126**, 17–25.

Themmen, A. P. N., Blok, L. J., Post, M., Baarends, W. M., Hoogerbrugge, J. W., Parmentier, M., Vassart, G., and Grootegoed, A. J. (1991). Follitropin receptor down-regulation involves a cAMP-dependent post-transcriptional decrease of receptor mRNA expression. *Mol. Cell. Endocrinol.* **78**, R7–R13.

Tindall, D. J., Miller, D. A., and Means, A. R. (1977). Characterization of androgen receptor in Sertoli cell-enriched testis. *Endocrinology* **101**, 13–23.

Tom, L., Bhasin, S., Salameh, W., Steiner, B., Peterson, M., Sokol, R. Z., Rivier, J., Vale, W., and

Swerdloff, R. S. (1992). Induction of azoospermia in normal men with combined Nal-Glu gonadotropin-releasing hormone antagonist and testosterone enanthate. *J. Clin. Endocrinol. Metab.* **75**, 476–483.

Toppari, J., Tsutsumi, I., Campeau, J. D., Ahmad, N., and diZerega, G. S. (1988). Plasminogen activator secretion in the rat seminiferous epithelium after hypophysectomy and gonadotropin treatment. *Arch. Androl.* **20**, 219–227.

Toppari, J., Tsutsumi, I., Bishop, P. C., Parker, J. W., Ahmad, N., Tsang, C., Campeau, J. D., and diZerega, G. S. (1989). Flow cytometry quantification of rat spermatogenic cells after hypophysectomy and gonadotropin treatment. *Biol. Reprod.* **40**, 623–634.

Tsai-Morris, C. H., Buczko, E., Wang, W., and Dufau, M. L. (1990). Intronic nature of the rat luteinizing hormone receptor gene defines a soluble receptor subspecies with hormone binding activity. *J. Biol. Chem.* **265**, 385–388.

Tsong, S. D., Phillips, D., Halmi, N., Liotta, A. S., Margioris, A., Bardin, C. W., and Krieger, D. T. (1982). ACTH and β-endorphin-like peptides are present in multiple sites in the reproductive tract of the male rat. *Endocrinology* **110**, 2204–2206.

Tsutsui, K. (1991). Pituitary and gonadal hormone-dependent and -independent induction of follicle-stimulating hormone receptors in the developing testis. *Endocrinology* **128**, 477–487.

Vaishnav, M. Y., and Moudgal, R. N. (1992). Effects of specific FSH deprivation on testicular germ cell transformations and on LDH-X hyaluronidase activity of immature and adult rats. In "Follicle Stimulating Hormone—Regulation of Secretion and Molecular Mechanism of Action" (M. Hunzicker-Dunn and N. B. Schwartz, eds.), pp. 364–368. Springer Verlag, New York.

van Alphen, M. M. A., van de Kant, H. J. G., and de Rooij, D. G. (1988). Follicle-stimulating hormone stimulates spermatogenesis in the adult monkey. *Endocrinology* **123**, 1449–1455.

van Alphen, M. M. A., van de Kant, H. J. G., and de Rooij, D. G. (1989). Protection from radiation-induced damage of spermatogenesis in the rhesus monkey *(Macaca mulatta)* by follicle-stimulating hormone. *Cancer Res.* **49**, 533–536.

van Haaster, L. H., de Jong, F. H., Docter, R., and de Rooij, D. G. (1992). The effect of hypothyroidism on Sertoli cell proliferation and hormone levels during testicular development in the rat. *Endocrinology* **131**, 1574–1576.

van Pelt, A. M. M., and de Rooij, D. G. (1990). The origin of the synchronization of the seminiferous epithelium in vitamin-deficient rats after vitamin A replacement. *Biol. Reprod.* **42**, 677–682.

Verhoeven, G., and Cailleau, J. (1988). Follicle-stimulating hormone and androgens increase the concentration of the androgen receptor in Sertoli cells. *Endocrinology* **122**, 1541–1550.

Verhoeven, G., Swinnen, K., Cailleau, J., Deboel, L., Rombauts, L., and Heyns, W. (1992). The role of cell–cell interactions in androgen action. *J. Steroid Biochem. Mol. Biol.* **41**, 487–494.

Vernon, R. G., Go, V. L. W., and Fritz, I. B. (1975). Hormonal requirements of different cycles of the seminiferous epithelium during reinitiation of spermatogenesis in long-term hypophysectomized rats. *J. Reprod. Fertil.* **42**, 77–94.

Vicari, E., Mongioi, A., Calogero, A. E., Moncada, M. L., Sidoti, G., Polosa, P., and D'Agata, R. (1992). Therapy with human chorionic gonadotrophin alone induces spermatogenesis in men with isolated hypogonadotrophic hypogonadism—long-term follow-up study. *Int. J. Androl.* **15**, 320–329.

von Berswordt-Wallrabe, R., Steinbeck, H., and Neumann, F. (1968a). Effect of FSH on the testicular structure of rats. *Endocrinologia* **53**, 35–42.

von Berswordt-Wallrabe, R., and Neumann, F. (1968b). Successful reinitiation and restoration of spermatogenesis in hypophysectomized rats with pregnant mare's serum after a long-term regression period. *Experientia* **24**, 499–501.

Wahlström, T., Huhtaniemi, I., Hovatta, O., and Seppälä, M. (1983). Localization of luteinizing hormone, follicle-stimulating hormone, prolactin, and their receptors in human and rat testis using immunohistochemistry and radioreceptor assay. *J. Clin. Endocrinol. Metab.* **57**, 825–830.

Weinbauer, G. F., and Nieschlag, E. (1990). The role of testosterone in spermatogenesis. *In* "Testosterone—Action, Deficiency, Substitution" (E. Nieschlag and H. M. Behre, eds.), pp. 23–50. Springer Verlag, Berlin.

Weinbauer, G. F., and Nieschlag, E. (1992). LH-RH antagonists: State of the art and future perspectives. *Rec. Results Cancer Res.* **124**, 113–136.

Weinbauer, G. F., Göckeler, E., and Nieschlag, E. (1988). Testosterone prevents complete suppression of spermatogenesis in the gonadotropin-releasing hormone (GnRH) antagonist-treated non-human primate *(Macaca fascicularis). J. Clin. Endocrinol. Metab.* **67**, 284–290.

Weinbauer, G. F., Khurshid, S., Fingscheidt, U., and Nieschlag, E. (1989). Sustained inhibition of sperm production and inhibin secretion induced by a gonadotrophin-releasing hormone antagonist and delayed testosterone substitution in non-human primates *(Macaca fascicularis). J. Endocrinol.* **123**, 303–310.

Weinbauer, G. F., Behre, H. M., Fingscheidt, U., and Nieschlag, E. (1991). Human follicle-stimulating hormone exerts a stimulatory effect on spermatogenesis, testicular size, and serum inhibin levels in the gonadotropin-releasing hormone antagonist-treated nonhuman primate *(Macaca fascicularis). Endocrinology* **129**, 1831–1839.

Weinbauer, G. F., Hankel, P., and Nieschlag, E. (1992a). Exogenous gonadotrophin-releasing hormone (GnRH) stimulates LH secretion in male monkeys *(Macaca fascicularis)* treated chronically with high doses of a GnRH antagonist. *J. Endocrinol.* **133**, 439–445.

Weinbauer, G. F., Behre, H. M., and Nieschlag, E. (1992b). Concomitant but not delayed androgen supplementation prevents complete testicular involution in GnRH antagonist-treated nonhuman primates. 74th Annual Meeting of the Endocrinology Society, San Antonio, Abstr. 1075.

Wickings, E. J., and Nieschlag, E. (1978). The effect of active immunization with testosterone on pituitary-gonadal feedback in the male rhesus monkey *(Macaca mulatta). Biol. Reprod.* **18**, 602–607.

Wickings, E. J., Usadel, K. H., Dathe, G., and Nieschlag, E. (1980). The role of follicle stimulating hormone in testicular function of the mature rhesus monkey. *Acta Endocrinol.* **95**, 117–128.

Woods, M. C., and Simpson, M. E. (1961). Pituitary control of the testis of the hypophysectomized rat. *Endocrinology* **69**, 91–125.

World Health Organization (1990). WHO Task Force on Methods for the Regulation of Male Fertility. Contraceptive efficacy of testosterone-induced azoospermia in normal men. *Lancet* **336**, 955–959.

Zhengwei, Y., Wreford, N., and de Kretser, D. M. (1990). A quantitative study of spermatogenesis in the developing rat testis. *Biol. Reprod.* **43**, 629–635.

Zipf, W. B., Payne, A. H., and Kelch, R. P. (1978). Prolactin, growth hormone and luteinizing hormone in the maintenance of testicular LH receptors. *Endocrinology* **103**, 595–560.

Zirkin, B. R., Santulli, R., Awoniyi, C. A., and Ewing, L. L. (1989). Maintenance of advanced spermatogenic cells in the adult rat testis: Quantitative relationship to testosterone concentrations within the testis. *Endocrinology* **124**, 3043–3049.

5

Patterns of Expression and Potential Functions of Proto-oncogenes during Mammalian Spermatogenesis

MARTIN A. WINER & DEBRA J. WOLGEMUTH

I. Introduction

Major developmental processes are involved in spermatogenesis, including mitotic and meiotic division, genetic recombination, and morphological differentiation. Studies of the molecular genetic basis of differentiation in this defined cell lineage have revealed several genes, the expression patterns of which suggest a role in this process (reviewed by Willison and Ashworth, 1987; Erickson, 1990; Hecht, 1990; Wolgemuth and Watrin, 1991). Some ubiquitously expressed genes exhibit particularly abundant expression in specific subsets of germ cells, often of uniquely sized transcripts. The stage-specific expression of different sized transcripts suggests that different isoforms of the same gene may be involved in discrete stages of germ cell development. Similarly, testis-specific genes can be expressed in spermatogenic cells in a stage-specific manner.

The genomes of several RNA tumor viruses contain genes, the so-called oncogenes, that cause the host cell to undergo transformation (reviewed by Bishop, 1983; Herrlich and Ponta, 1989). These viral oncogenes frequently have homologs in cellular DNA that are known as proto-oncogenes or cellular oncogenes and are thought to play critical roles in normal growth and development. A number of the genes that exhibit unique patterns of expression during spermatogenesis are proto-oncogenes.

Proto-oncogenes can be divided into eight general groups that participate in a wide range of cellular functions, including development and differentiation. These groups include (1) growth factors; (2) growth factor receptor tyrosine protein kinases, which become activated on binding their respective ligands; (3) nonreceptor tyrosine-specific protein kinases; (4) cytoplasmic serine- and threonine-specific protein kinases; (5) plasma membrane-associated GTP-binding and -hydrolyzing proteins; (6) nuclear proto-oncogenes that affect subsequent development by interacting with transcription regulatory elements; (7) tumor suppressor genes; and (8) proto-oncogenes that produce proteins with functions that remain to be elucidated.

II. Proto-oncogenes Known to Be Expressed in the Mouse Testis

With the exception of the cellular oncogenes that encode growth factors, members of all the classes of proto-oncogenes just noted are known to be expressed in testis. Some are expressed at levels higher than those in somatic tissues, some are expressed at different levels at specific stages of development, and some yield novel-sized transcripts in reproductive tissues. A reasonably comprehensive but by no means exhaustive compilation of these genes is presented in Table 1. In addition to genes expressed in a developmentally regulated pattern, several proto-oncogenes are expressed at all stages of germ cell development, as well as in testicular somatic cells. Such genes represent constitutively expressed "housekeeping" genes that are responsible for basic cellular functions not directly related to the specialized processes involved in spermatogenesis. In the sections that follow, selected examples of proto-oncogenes are discussed with respect to their possible function at different stages of spermatogenesis.

A. Proto-oncogenes that Might Play a Role in Fetal Development of the Testis

1. Growth factor receptor tyrosine protein kinases

The cellular homolog of the acute transforming feline retroviral oncogene v-*kit* (Besmer *et al.*, 1986), called c-*kit*, is a transmembrane receptor tyrosine kinase that is related to the platelet-derived growth factor (PDGF) receptor (Besmer *et al.*, 1986; Yarden *et al.*, 1987; Qui *et al.*, 1988) produced by the dominant *white spotted (W)* locus in the mouse (Geissler *et al.*, 1988; Nocka *et al.*, 1989). The pleiotrophic phenotypes of *W* mutants include white coat color, sterility in both sexes, and anemia, and have been attributed to the failure of stem cells to migrate or proliferate during development (Mintz and Russell, 1957). Analysis of the expression of c-*kit* during embryonic and adult development has shown that the gene is expressed in a wide variety of tissues, including sites in which no obvious phenotype exists in animals bearing the mutant alleles (Manova *et al.*, 1990; Orr-Urtreger *et al.*, 1990;

TABLE 1 **Proto-oncogenes and Related Genes Expressed in Mouse Testis**

Gene	Transcript size (kb)	Reference
Growth Factor receptor tyrosine protein kinases		
Flt3[a]	3.3[b,c]	Rosnet et al. (1991)
	2.6[b,c]	Rosnet et al. (1991)
c-kit[a]	5.5	Sorrentino et al. (1991)
	3.5[c,d]	Sorrentino et al. (1991)
	2.3[c,d]	Sorrentino et al. (1991)
Plasma membrane associated protein kinases		
c-abl[e]	8.0	Ponzetto and Wolgemuth (1985); Propst et al. (1987)
	6.2	Ponzetto and Wolgemuth (1985); Propst et al. (1988)
	4.7[c,d]	Ponzetto and Wolgemuth (1985); Goldman et al. (1987); Propst et al. (1988)
ferT[f]	2.4[b,c]	Fischman et al. (1990)
Cytoplasmic serine/threonine protein kinases		
c-mos	1.7[c,d]	Goldman et al. (1987); Mutter and Wolgemuth (1987); Propst et al. (1988)
c-raf-1	3.1[c]	Sorrentino et al. (1988); Wolfes et al. (1989); Storm et al. (1990); Wadewitz et al. (1993)
A-raf	4.3	Wadewitz et al. (1993)
	2.6	Storm et al. (1990); Wadewitz et al. (1993)
B-raf	4.0[c,d]	Storm et al. (1990); Wadewitz et al. (1993)
	2.6[c,d]	Storm et al. (1990); Wadewitz et al. (1993)
pim-1	2.8	Sorrentino et al. (1988)
	2.4[c,d]	Sorrentino et al. (1988)
Plasma membrane associated GTP binding proteins		
H-ras	1.3[c]	Leon et al. (1987); Sorrentino et al. (1988); Wolfes et al. (1989)
	1.1[c]	Leon et al. (1987); Sorrentino et al. (1988); Wolfes et al. (1989)
K-ras	2.2[c]	Leon et al. (1987); Sorrentino et al. (1988); Wolfes et al. (1989)
N-ras	5.0[c]	Leon et al. (1987); Sorrentino et al. (1988); Wolfes et al. (1989)
	3.5[c,d]	Leon et al. (1987); Wolfes et al. (1989)
	1.9[c,d]	Leon et al. (1987); Wolfes et al. (1989)
	1.4[c]	Leon et al. (1987); Wolfes et al. (1989)
rab1[g]	3.3	Olofsson et al. (1988)
	1.9	Olofsson et al. (1988)
rab2[g]	2.5	Olofsson et al. (1988)
	1.45	Olofsson et al. (1988)
rab4[g]	1.8	Olofsson et al. (1988)
RalA[g]	2.8	Olofsson et al. (1988)
	1.2	Olofsson et al. (1988)
Rho12[g]	2.3	Olofsson et al. (1988)
	1.75	Olofsson et al. (1988)

(continues)

TABLE 1   *(continued)*

Gene	Transcript size (kb)	Reference
Nuclear proto-oncogenes		
c-*cbl*	11	Langdon *et al.* (1989)
	5.6	Langdon *et al.* (1989)
	3.5	Langdon *et al.* (1989)
ets-1	5.3	Bhat *et al.* (1987)
ets-2	3.5	Bhat *et al.* (1987)
elk-1[h]	4.5[b]	Rao *et al.* (1989)
	2.8[b]	Rao *et al.* (1989)
	1.7	Rao *et al.* (1989)
c-*fos*	2.2[c]	Wolfes *et al.* (1989)
c-*myc*	2.4[c]	Taylor *et al.* (1986); Wolfes *et al.* (1989)
N-*myc*	3.2	Jakobovits *et al.* (1985)
c-*jun*	3.2[c]	Alcivar *et al.* (1990)
	2.7[c]	Hirai *et al.* (1989); Wolfes *et al.* (1989); Alcivar *et al.* (1990)
*jun*B	2.1[c]	Hirai *et al.* (1989); Alcivar *et al.* (1990)
*jun*D	1.8[c]	Hirai *et al.* (1989); Alcivar *et al.* (1991)
	1.6[c,d]	Alcivar *et al.* (1991)
Tumor suppressor genes		
RB	4.7	Bernards *et al.* (1989)
	2.8[b,c]	Bernards *et al.* (1989)
p53	2.0	Rogel *et al.* (1985)
WT1	3.1[c]	Pelletier *et al.* (1991)
	2.5[b,c]	Pelletier *et al.* (1991)
Other proto-oncogenes		
wnt-1	2.6	Schackleford and Varmus (1987)
W*nt*5b	3.2	Gavin *et al.* (1990)
W*nt*6	2.2[b]	Gavin *et al.* (1990)
	1.8[b]	Gavin *et al.* (1990)

[a] Member of PDGF receptor family.
[b] Testis-specific transcript.
[c] Developmentally regulated transcript.
[d] Germ cell-specific transcript.
[e] c-*abl* products also may be present as nuclear or cytoplasmic tyrosine kinases.
[f] *fps*/*fes* proto-oncogene homolog.
[g] *ras* proto-oncogene homolog.
[h] *ets* proto-oncogene homolog.

Keshet *et al.*, 1991). *In situ* analysis reveals the expression of c-*kit* in mouse primordial germ cells during days 9–10 of fetal development as these cells migrate from the yolk sac to the gonadal ridge (Orr-Urtreger *et al.*, 1990; Keshet *et al.*, 1991). In c-*kit* mutations such as W^v, primordial germ cells migrate to the seminiferous tubules but fail to proliferate or differentiate beyond the spermatogonial stage (Coloumbre and Russell, 1954), suggesting that the c-*kit* gene product also may play a role in the subsequent stages of male germ cell development.

The ligand for the c-*kit* receptor has been identified as the product of the *Steel (Sl)* gene (Copeland *et al.*, 1990; Flanagan and Leder, 1990; Huang *et al.*, 1990; D. E. Williams *et al.*, 1990; Zsebo *et al.*, 1990a,b). *Sl* is expressed abundantly in the stromal cells of the mesoderm along the migratory path of the primordial germ cells during the period of fetal development in which the germ cells express c-*kit*, and is down-regulated after the cells have migrated (Keshet *et al.*, 1991). *Sl* also is expressed in the genital ridge (Matsui *et al.*, 1990b; Keshet *et al.*, 1991). However, the *Sl* product does not appear to act as a chemotropic factor to guide the migration of the primordial germ cells, nor does it affect their proliferation. Instead, this protein seems to be required to maintain the viability of the cells (Dolci *et al.*, 1991; Godin *et al.*, 1991). Only the transmembrane form of the *Sl* gene product is effective at maintaining primordial germ cells (Dolci *et al.*, 1991).

The intracellular effects induced by *Sl* binding to the c-*kit* receptor have not been examined extensively, and have not been examined at all with respect to germ cell development. However, in human fibroblast and myeloid cell lines, *Sl* binding stimulates the tyrosine autophosphorylation of the c-*kit* receptor, as well as tyrosine phosphorylation of GTPase activating protein (GAP; Herbst *et al.*, 1991; Miyazawa *et al.*, 1991), the 42-kDa mitogen activated protein (MAP) kinase, GAP-associated proteins of 62 and 190 kDa (Miyazawa *et al.*, 1991), and phospholipase C (PLCγ; Herbst *et al.*, 1991). Intriguingly, the product of another proto-oncogene, the cytoplasmic serine/threonine kinase c-*raf*-1, co-immunoprecipitates with the c-*kit* gene product (Herbst *et al.*, 1991). Phosphorylation and subsequent activation of c-*raf*-1 kinase activity also is induced by *Sl*, albeit on serine residues, suggesting the presence of at least one intermediate step in signal transduction (Herbst *et al.*, 1991; Miyazawa *et al.*, 1991). Thus, the ligand-activated c-*kit* tyrosine kinase may exert its effects via a number of different signaling pathways.

2. Tumor suppressor genes

In addition to the proto-oncogenes involved in the regulated activation of cell proliferation, another class of genes, the tumor suppressor genes, normally functions to inhibit proliferation (Weinberg, 1991). Loss-of-function mutations of these genes result in deregulation of the cell cycle or a block to the cell cycle exit (Cooper and Whyte, 1989). One such putative tumor suppressor gene is *WT1*, the Wilms' Tumor gene, a nuclear proto-oncogene encoding a transcription regulatory factor (Madden *et al.*, 1991). The *WT1* product binds to DNA at the *egr*-binding motif and represses transcription. *WT1* is expressed in human fetal testis as an abundant 3.2-kb somatic transcript and a less abundant testis-specific 2.7-kb form. *In situ* hybridization analysis of embryonic gonads localized *WT1* to the sex cords of the gonadal ridge, but not to gonocytes, suggesting expression primarily in the somatic tissues of the developing human testis (Pritchard-Jones *et al.*, 1990).

B. Proto-oncogenes that Might Be Important in Stem Cell Stages

During spermatogenesis, a subset of the diploid primitive type A spermatogonia commit to the process of terminal differentiation and progress through intermediate and type B spermatogonia before entering meiosis. The remainder of the primitive type A spermatogonia remain as a pool of mitotically proliferating stem cells to insure the constant production of spermatozoa. A complex network of protein phosphorylation and dephosphorylation regulates entry of cells into mitosis and meiosis, affecting such events as nuclear membrane breakdown, chromosome condensation, spindle formation, and activation of microtubule-organizing centers (discussed by Llamazares *et al.*, 1991). Some of these phosphorylation events have been postulated to be mediated directly by the 34-kDa serine/threonine protein kinase product of the *cdc2* gene (p34^{cdc2}; Moreno and Nurse, 1990); several proto-oncogene products have consensus phosphorylation sequences for p34^{cdc2} (Shalloway and Shenoy, 1991). Numerous potential substrates are likely to reflect a diversity of phosphorylating enzymes, some of which may be the products of various proto-oncogenes. Some proto-oncogenes are expressed only in stem cells (or at unusually high levels in such cells, relative to later stages of germ cell development), suggesting that they may play crucial roles in the commitment to terminal differentiation. Several articles have discussed the potential roles of proto-oncogenes in regulatory cascades and signal transduction, particularly as they relate to the regulation of mitosis and meiosis (Storm and Bose, 1989; Pelech *et al.*, 1990; Shalloway and Shenoy, 1991).

1. Growth factor receptor tyrosine protein kinases

c-kit is expressed in adult as well as fetal testes, occurring as an abundant 5.5-kb transcript (Sorrentino *et al.*, 1991) encoding a 145-kDa tyrosine kinase (p145^{c-kit}; Majumder *et al.*, 1988). *In situ* and Northern hybridization analyses indicated that highest levels of the 5.5-kb *c-kit* transcript are expressed in type A spermatogonia (Manova *et al.*, 1990). Germ cell expression of *c-kit* decreases in intermediate and type B spermatogonia but still exhibits diffuse labeling in preleptotene spermatocytes. All subsequent stages of meiotic and postmeiotic germ cell development are unlabeled.

The pattern of p145^{c-kit} expression parallels that of the *c-kit* mRNA. In neonatal testes, immunoreactive *c-kit* protein is expressed very weakly in gonocytes and disappears as these begin their first wave of proliferation and differentiation into primitive type A spermatogonia. Undifferentiated (stem cell) type A spermatogonia do not express immunoreactive p145$^{c\text{-}kit}$. p145$^{c\text{-}kit}$ is detected at the cell surface of differentiating type A, intermediate, and type B spermatogonia and in the earliest preleptotene spermatocytes, but not at any subsequent stage of germ cell development, suggesting that c-*kit* is down-regulated on entry into meiosis. Intraperitoneal administration of monoclonal antibodies against p145$^{c\text{-}kit}$, which block binding of the *Sl* ligand, have no effect on the mitotic proliferation of undifferentiated type A spermatogonial stem cells or on the progression from intermediate to type B spermatogonia or type B to preleptotene spermatocytes (Yoshinaga *et al.*, 1991). However, antibody treatment does block the mitotic proliferation and differentiation of type A spermatogonia as they enter the spermatogenic pathway.

The role of c-*kit* in stem cell proliferation and differentiation in the testis has been examined further by surgically induced cryptorchidism. This process depletes the testes of all stages of germ cells except type A spermatogonia; in normal mice, reversal of the cryptorchid condition results in the resumption of spermatogenesis. However, mice heterozygous for the *W* mutation (*W*/+) fail to undergo spermatogenesis after the cryptorchid condition is reversed, although the proliferation of primitive type A spermatogonia is unimpaired (Koshimizu *et al.*, 1991). Similar studies with mice heterozygous for the *Sl* mutation also exhibit type A spermatogonia that proliferate mitotically at a rate similar to wild-type, but still have fewer type A cells, suggesting an increased rate of cell death. Spermatogenesis after reversal of cryptorchidism does occur in these *Sl* heterozygotes, but the transition from type A to intermediate/type B spermatogonia is impaired (Nishimune *et al.*, 1980; Tajima *et al.*, 1991b). Thus, although c-*kit* is not essential for the mitotic proliferation of the undifferentiated primitive type A spermatogonia, a functional *Sl*–c-*kit* interaction is required for maintenance of spermatogonial viability and commitment to the spermatogenic pathway. Further, two normal alleles of these otherwise apparently recessive mutations appear to be required for normal spermatogenesis.

2. Nonreceptor tyrosine protein kinases

The cellular homolog, c-*abl*, of the transforming region of the Abelson murine leukemia virus encodes a cytoplasmic-class tyrosine kinase and is present as a single copy gene in the mouse genome, spanning over 100 kb (Wang *et al.*, 1984) and producing a 150-kDa protein (p150$^{c\text{-}abl}$; Witte *et al.*, 1979). This gene is expressed in all murine tissues examined, including neonatal and adult testes, yielding transcripts 5.3 and 6.5 kb in size (Wang and Baltimore, 1983; Ponzetto and Wolgemuth, 1985; Ben-Neriah *et al.*, 1986). The c-*abl* protein encoded by the larger somatic transcript (type IV; Ben-Neriah *et al.*, 1986) initiates with the sequence Met–Gly–Gln, the myristylation site in the N terminus of both the *gag* v-*abl* and the related v-*src* and c-*src* proteins, suggesting that the protein may be anchored in the plasma membrane. However, a large fraction of type IV c-*abl* protein overexpressed in nontransformed 3T3 fibroblasts is localized in the nucleus (Van Etten *et al.*, 1989). Amino acid sequence analysis indicates that the type IV c-*abl* protein has a nuclear localization signal similar to that of SV40 large T antigen. Mammalian p150^{c-abl} contains a DNA binding domain in its C-terminal region that is capable of binding EP (Dikstein *et al.*, 1992; Kipreos and Wang, 1992), an element in the enhancer region of a number of genes that acts to integrate enhancer binding factors into functional units (Dikstein *et al.*, 1990). p150^{c-abl} binding to DNA stimulates its autophosphorylating activity. Such phosphorylation subsequently may recruit other proteins (p150^{c-abl} substrates) to the nucleus and phosphorylate them, inducing gene transcription (Dikstein *et al.*, 1992).

In vitro assays have demonstrated that p150^{c-abl} is phosphorylated on three sites during interphase and on seven additional sites during mitosis. p34^{cdc2} is capable of phosphorylating two of the interphase sites and all the mitotic sites; p34^{cdc2} isolated from mitotic cells is more efficient than p34^{cdc2} from interphase cells (Kipreos and Wang, 1990). Phosphorylation by p34^{cdc2} has no apparent effect on p150^{c-abl} tyrosine kinase activity, but does regulate the subcellular distribution of p150^{c-abl} between the cytoplasmic and nuclear compartments of the cell during mitosis (Shalloway and Shenoy, 1991). Phosphorylation by p34^{cdc2} during mitosis abolishes DNA binding; p150^{c-abl} apparently is dephosphorylated during interphase (Kipreos and Wang, 1990). However, only partial dephosphorylation is required for DNA binding, suggesting that the other phosphorylation sites regulate other functions of the gene product (Kipreos and Wang, 1992). c-*abl* therefore may play a crucial role during the mitotic proliferation of spermatogonia.

3. Cytoplasmic serine/threonine protein kinases

a. c-*raf*-1

c-*raf*-1, the cellular homolog of the murine sarcoma virus gene v-*raf* (Rapp *et al.*, 1983; Bonner *et al.*, 1985), is a member of the cytoplasmic serine/threonine protein kinase family, producing a 3.1-kb transcript encoding a 74-kDa protein (p74$^{c-raf-1}$). c-*raf*-1 mRNAs have been detected in germ cells from type A and B spermatogonia through the round spermatid stage, but not in residual bodies; the highest levels have been observed in pachytene spermatocytes (Wolfes *et al.*, 1989). *In situ* hybridization analysis of mouse testes reveals that c-*raf*-1 is expressed at low levels in spermatogonia (Wadewitz *et al.*, 1993).

c-*raf*-1 belongs to a family of related genes (c-*raf*-1, A-*raf*, and B-*raf*), all of which are known to be expressed in the testis (Storm *et al.*, 1990). All three *raf* family genes encode protein kinases that share a high degree of sequence homology at both the nucleotide and the amino acid level (Sithanandam *et al.*, 1990). Three regions in particular are highly conserved: a cysteine-rich putative zinc-finger region, a serine/threonine rich domain that may serve as a site of autophosphorylation or activation, and the kinase domain itself.

Elucidation of the role of *raf* family genes in testicular function is dependent on identification of the factors that influence *raf* production and activity, as well as *raf* protein substrates (reviewed by Rapp *et al.*, 1988; Rapp, 1991; Heidecker *et al.*, 1992). Mouse NIH 3T3 cells transformed with viral oncogenes or stimulated with hormones or mitogenic growth factors exhibit increased phosphorylation of p74$^{c\text{-}raf\text{-}1}$, primarily on serine and threonine residues (Morrison *et al.*, 1988; Baccarini *et al.*, 1990; Blackshear *et al.*, 1990; Kovacina *et al.*, 1990; Carroll *et al.*, 1991). Subsequent studies indicate that p74$^{c\text{-}raf\text{-}1}$ associates with membrane-bound receptor tyrosine kinases *in vitro* (Morrison *et al.*, 1989) and *in vivo* (App *et al.*, 1991).

Despite evidence for direct interactions with membrane-bound receptor tyrosine kinases, extracellular signals are likely to activate *raf* proteins *in vivo* by stimulating intervening cytoplasmic factors such as *raf* kinases (Lee *et al.*, 1991), protein kinase C (PKC; Siegel *et al.*, 1990), or the products of the *ras* family genes (Troppmair *et al.*, 1992; Wood *et al.*, 1992), although full activation by *ras* gene products requires coexpression of an as yet unidentified tyrosine kinase (N. G. Williams *et al.*, 1992).

The *in vivo* functions of *raf* proteins during spermatogenesis have not been elucidated. However, evidence in other systems suggests that *raf* proteins regulate both the expression (Jamal and Ziff, 1990; Qureshi *et al.*, 1991) and the activity (Wasylyk *et al.*, 1989; Bruder *et al.*, 1992) of transcription regulatory factors in the nucleus. Alternatively, both H1 histone (App *et al.*, 1991) and MAP kinase kinase (Dent *et al.*, 1992; Howe *et al.*, 1992; Kyriakis *et al.*, 1992) serve as substrates for p74$^{c\text{-}raf\text{-}1}$ *in vitro* and *in vivo*, respectively. p74$^{c\text{-}raf\text{-}1}$ may represent the initial step in a regulatory cascade wherein activated MAP kinase kinase phosphorylates MAP kinases, which in turn activate c-*jun* and ribosomal S6 kinase (RSK), leading to alterations in gene expression (Kyriakis *et al.*, 1992). Intriguingly, both MAP-1 and MAP-2 kinases phosphorylate p74$^{c\text{-}raf\text{-}1}$, but apparently do not stimulate its kinase activity (Lee *et al.*, 1992). Phosphorylation by such downstream target elements may represent part of a negative regulatory mechanism.

b. c-*mos*

As the cellular homolog of transforming sequences in the Moloney murine sarcoma virus, c-*mos* encodes a cytoplasmic serine/threonine protein kinase (Moloney, 1966). Relatively little is known about the function of c-*mos* during spermatogenesis. However, evidence from functional and immunoadsorption studies in vertebrate oocytes suggests that the 39-kDa c-*mos* product (p39$^{c\text{-}mos}$) is the cytostatic factor (CSF) responsible for arresting cells at the metaphase II stage of meiosis (Sagata *et al.*, 1989b; O'Keefe *et al.*, 1991; X. Zhao *et al.*, 1991). CSF has been proposed to block the proteolytic degradation of cyclin, thereby stabilizing the p34^{cdc2}–cyclin complex. A role for functional c-*mos* protein in mitosis seems implicated by the fact that 3T3 cells that overexpress a mutant c-*mos* protein with a deletion in the ATP binding domain are incapable of completing mitosis (Freeman *et al.*, 1989). c-*mos* protein may have a similar function during the mitotic proliferation of spermatogonia.

4. GTP binding proteins

The mammalian *ras* family has three members, designated H-*ras*, K-*ras*, and N-*ras*, that were isolated originally from rat Harvey and Kirsten sarcomas and from a human neuroblastoma cell line, respectively (Barbacid, 1987). All three genes encode very similar 21-kDa proteins (p21^{ras}) that can bind and hydrolyze GTP. Transcripts identical in size to the somatic transcripts of H-*ras* (1.3 and 1.1 kb), N-*ras* (5.0 kb), and low levels of K-*ras* (2.2 kb) are present in type A and B spermatogonia (Leon *et al.*, 1987; Wolfes *et al.*, 1989). p21^{ras} proteins are detected at both stages, being more abundant in type B spermatogonia and decreasing at later stages of development (Wolfes *et al.*, 1989).

Some evidence indicates a potential role for *ras* or *ras*-related proteins in mitosis. Injection of antibodies against p21^{ras} into axolotl embryos blocks cell division (Baltus *et al.*, 1988); expression of p21^{ras} is required for entry into the S phase of the cell cycle in serum-stimulated NIH 3T3 cells (Mulcahy *et al.*, 1985). H-*ras* initiates a cascade that results in the phosphorylation and initiation of the *trans*-activating function of the nuclear proto-oncogene c-*jun* (Binétruy *et al.*, 1991). p21^{ras} injected into 3T3 cells also induces rapid transient expression of c-*fos* protein (Stacey *et al.*, 1987).

p21^{ras} functions in a number of key regulatory pathways (reviewed by Lowy *et al.*, 1991). Normal *ras* expression is required for binding of a 120-kDa GAP (p120 GAP) to the PDGF receptor, where it subsequently is phosphorylated (Kaplan *et al.*, 1990). Phosphorylation of p120 GAP in turn stimulates the GTPase activity of p21^{ras} (Molloy *et al.*, 1989). Oncogenically active p21^{ras} can phosphorylate and activate MAP-2 kinase in 3T3 cells in a PKC-dependent and -independent manner (Leevers and Marshall, 1992). Mouse NIH 3T3 cells transformed with v-H-*ras* exhibit increased phosphorylation of p74$^{c\text{-}raf\text{-}1}$ on serine and threonine residues (Morrison *et al.*, 1988) and subsequent activation of its kinase activity (Troppmair *et al.*, 1992; N. G. Williams *et al.*, 1992; Wood *et al.*, 1992).

5. Nuclear proto-oncogenes

A number of nuclear proto-oncogenes, the so-called "early response genes," have been shown to be expressed in the testis. In response to mitogenic signals, these genes bind to specific DNA domains and *trans*-activate transcription of other genes (Faisst and Meyer, 1992). The expression of these nuclear proto-oncogenes in type B spermatogonia in particular suggests that their products may mediate changes in gene expression that are related to the regulation of mitotic cell division.

a. c-*myc*

c-*myc*, the cellular homolog of the transforming region of the avian myelocytomatosis virus, encodes a 58-kDa protein that has been implicated in both proliferation and differentiation of cells (Cole, 1986). The highest levels of c-*myc* expression in the testis occur in type B spermatogonia as the 2.4-kb (somatic-type) transcript (Wolfes *et al.*, 1989). c-*myc* protein has been detected in the nuclei of spermatogonia (Koji *et al.*, 1988). Immunolocalization of c-*myc* protein in human hematopoietic tumor cell lines (Bading *et al.*, 1989) or in monkey fibroblasts that overexpress c-*myc* protein (Henriksson *et al.*, 1988) indicates an exclusively nuclear localization of c-*myc* protein in cells in interphase. During mitosis, c-*myc* protein distributes evenly throughout the cytoplasm but is never associated with condensed chromosomes (Bading *et al.*, 1989). Instead, it colocalizes with small nuclear ribonucleoprotein particles, suggesting that c-*myc* protein may be involved in RNA processing (Spector *et al.*, 1987). In telophase, c-*myc* protein relocalizes to the nucleus (Bading *et al.*, 1989).

The precise role of c-*myc* during mitosis remains to be elucidated. However, c-*myc* is required for entry into S phase but not for the $G_0 - G_1$ transition in mitogen-stimulated lymphocytes (Heikkila *et al.*, 1987). c-*myc* may exert its effects on the cell cycle either directly or indirectly via the $p34^{cdc2}$ kinase, since prior induction of c-*myc* is required for expression of $p34^{cdc2}$ at the $G_1 - S$ transition in human T lymphocytes (Furukawa *et al.*, 1990). Blocking $p34^{cdc2}$ with antisense oligonucleotides of *cdc2* has no effect on c-*myc* expression.

b. c-*fos*

c-*fos* mRNA is expressed at highest levels during spermatogenesis in type B spermatogonia as a 2.2-kb transcript (Hall *et al.*, 1988; Wolfes *et al.*, 1989). Immunocytochemical labeling with specific antibodies has demonstrated a uniform distribution of the 62-kDa c-*fos* protein ($p62^{c-fos}$) in the nuclei of spermatogonia of mice (Wolfes *et al.*, 1989) and rats (Pelto-Huikko *et al.*, 1991). In a number of cultured cell lines, specific $p62^{c-fos}$ immunoreactivity is distributed uniformly in the nuclei during G_0, G_1, and S phase; no cytoplasmic staining is observed. During mitosis, $p62^{c-fos}$ immunostaining is reciprocal to DNA staining (Rahm *et al.*, 1990). In prophase, no $p62^{c-fos}$ is present on condensed chromatin; cells exhibit diffuse cytoplasmic stain during metaphase and anaphase. During telophase, the cytoplasmic $p62^{c-fos}$ begins to disappear whereas nuclear stain increases as $p62^{c-fos}$ reassociates with the chromatin. Thus, $p62^{c-fos}$ appears to be associated with loosely packed euchromatin.

Because of its inability to form homodimers (Halazonetis *et al.*, 1988), $p62^{c-fos}$ does not bind to any specific DNA sequences. However, the protein interacts via its leucine zipper with members of the *jun* family of genes to form the AP-1 transcription activator, which stimulates genes containing the tetradecanoyl phorbol acetate (TPA) responsive element in their promoter region (Curran and Franza, 1988; Ransone and Verma, 1990). Intriguingly, $p62^{c-fos}$ is capable of *trans*-repressing expression of c-*fos*, even in the absence of *jun* gene products (Lucibello *et al.*, 1989).

c. c-*jun* and *jun*-related genes

Encoding a component of the AP-1 transcription activator (Rauscher *et al.*, 1988), c-*jun* is expressed at highest levels during spermatogenesis in type B spermatogonia as a 2.7-kb transcript (Hirai *et al.*, 1989; Wolfes *et al.*, 1989). Transcription of c-*jun* mRNA occurs in serum-stimulated mouse NIH 3T3 fibroblasts during the $G_0 - G_1$ transition (Ryseck *et al.*, 1988). As for the other nuclear proto-oncogenes, stimulation of the *trans*-activating effects of c-*jun* protein by mitogens is likely to be mediated by regulatory cascades involving $p21^{ras}$ (Binétruy *et al.*, 1991) or $p74^{c-raf-1}$ (Kyriakis *et al.*, 1992) stimulation of MAP kinases, which in turn phosphorylate and activate c-*jun* protein (Pulverer *et al.*, 1991). Intriguingly, c-*fos* – c-*jun* protein complexes can *trans*-activate the ovalbumin gene by interacting with its estrogen-responsive element, even in the absence of functional estrogen receptor (Gaub *et al.*, 1990). Whether c-*fos* – c-*jun* protein complexes are capable of playing a similar role in the activation of other steroid-responsive genes in the testis is unknown.

Two other members of the *jun* family, *jun*B (2.1 kb) and *jun*D (1.8 kb), also are expressed in mouse testis (Hirai *et al.*, 1989). *jun* gene family members are capable of forming homo- and heterodimers with each other and with the c-*fos* gene product (Nakabeppu *et al.*, 1988). These dimers have different affinities for DNA and subsequent variations in their abilities to initiate gene transcription (Nakabeppu *et al.*, 1988; Schütte *et al.*, 1989). *jun*B, for example, can act as a negative regulator of the *trans*-activating activity of c-*jun* (Chiu *et al.*, 1989; Schütte *et al.*, 1989), possibly by competitive inhibition of c-*jun* binding to AP-1 recognition sites. *jun*D is expressed more abundantly than c-*jun* or *jun*B in adult mouse testis. However, the pattern of gene expression appears to be regulated developmentally, since all three genes are expressed abundantly in type B spermatogonia of 8-day-old prepubertal mice (Alcivar *et al.*, 1990,1991). Coexpression of c-*jun*-related genes in spermatogenic cells may provide specific regulatory mechanisms that affect the stage specificity of the expression of their target genes throughout spermatogenesis.

6. Tumor suppressor genes

a. *RB*

The best characterized member of the tumor suppressor class of proto-oncogenes is the retinoblastoma susceptibility gene (*RB*; Weinberg, 1991). In the mouse, *RB* is expressed as a 4.7-kb transcript in all adult tissues examined (Bernards *et al.*, 1989). The protein is synthesized throughout the cell cycle, but is present in an unphosphorylated form in G_0 and G_1 and is highly phosphorylated on serine and threonine residues during S, G_2, and M phases (Buchkovich *et al.*, 1989). The unphosphorylated form is the active growth suppressor form and induces quiescence in G_0 or blocks exit from G_1(DeCaprio *et al.*, 1989; Mihara *et al.*, 1989). Mouse RB is hyperphosphorylated at the G_1–S transition (Chen *et al.*, 1989). Dephosphorylation of RB precedes differentiation or entry into the G_0 or G_1 nonproliferative phase.

Immunocytochemical localization of RB protein indicates that the distribution of RB is constant during G_1, S, and G_2 phases. At interphase, RB is localized to the nuclei in compartments with low DNA density. During mitosis, RB protein apparently is released into the cytoplasm as the nuclear membranes break down. Cells in mitotic prophase exhibit RB protein in a perinuclear location, but the protein is not present in the newly condensed chromosomes. Cytoplasmic staining remains apparent during metaphase and anaphase. RB becomes reassociated with the recondensing daughter nuclei (Szekely *et al.*, 1991).

RB is a substrate for p34^{cdc2} or a p34^{cdc2}-like kinase (B. T.-Y. Lin *et al.*, 1991) and, thus, may be regulated by any of the previously discussed regulatory cascades that affect p34^{cdc2} function. In serum-induced or normal cycling NIH 3T3 cells, RB down-regulates expression of c-*fos* and subsequent AP-1 activity (Robbins *et al.*, 1990). The biological effects of RB in the testis may also be mediated in part by its interaction with tissue-specific a′ and c forms of the differentiation-regulated transcription factor 1 (Partridge and La Thangue, 1991), which bind RB and are expressed abundantly in mouse testis.

b. *p53*

The product of *p53*, originally discovered to complex with the SV40 large T antigen (Lane and Crawford, 1979), has been detected in numerous actively proliferating nontransformed cells but is undetectable or present at low levels in resting cells. Adult testis expresses relatively high levels of *p53* compared with most somatic tissues (Rogel *et al.*, 1985). p53 is present in normal mouse testes as the product of both the regular and an alternatively spliced form of mRNA. The alternatively spliced transcript produces a protein that is 9 amino acids shorter and differs over 25 amino acids at the C terminus (Han and Kulesz-Martin, 1992). Levels of p53 vary with the cell cycle; the protein is absent in cells that have completed division and accumulates in an unphosphorylated form during the G_1 phase (Bischoff *et al.*, 1990).

C. Proto-oncogenes that Could Be Implicated in Meiosis

As discussed previously, several proto-oncogenes from different classes are expressed at low levels in meiotic cells (e.g., c-*abl*, Ponzetto and Wolgemuth, 1985; Iwaoki *et al.*, 1993; c-*fos*, Sorrentino *et al.*, 1988; Pelto-Huikko *et al.*, 1991) as well as in spermatogonia, and may maintain general cell functions throughout development. However, genes such as c-*raf*-1 (Sorrentino *et al.*, 1988; Wolfes *et al.*, 1989; Wadewitz *et al.*, 1993), c-H-*ras* (Wolfes *et al.*, 1989), and K-*ras* (Sorrentino *et al.*, 1988; Wolfes *et al.*, 1989) appear to be induced specifically during the early meiotic stages and are expressed at highest levels in early pachytene spermatocytes. Although some of these genes may play an active role in meiosis itself, others may be translated later, playing specific roles relevant to haploid development of the germ cells.

Few studies address the function of proto-oncogenes during meiosis in male germ cells, partly because spermatocytes proceed rapidly through the diplotene and secondary spermatocyte stages of meiosis. However, the roles of some of these genes in oocyte meiosis have been studied extensively. If similar patterns of proto-oncogene expression are observed in pachytene spermatocytes and oocytes, such genes may play comparable roles in both systems.

1. Growth factor receptor tyrosine protein kinases

a. c-*kit*

In addition to the 5.5-kb c-*kit* transcript observed in spermatogonia through the late pachytene stage, a novel testis-specific c-*kit* transcript of 3.5 kb appears at low levels in late pachytene spermatocytes (Sorrentino *et al.*, 1991). Whether the product of the 3.5-kb transcript actually plays an integral role in the final stages of meiosis, or whether translational control mechanisms limit its expression to postmeiotic stages, is unknown. However, in *Sl*/+ heterozygous mice, meiotic division is impaired during recovery after reversal of surgically induced cryptorchidism (Nishimune *et al.*, 1980; Tajima *et al.*, 1991b), indicating a necessity of a functional *Sl* – c-*kit* axis at this stage of development.

b. c-*trk*

c-*trk* encodes a 140-kDa protein (p140$^{c\text{-}trk}$) that is a component of the high affinity receptor complex for nerve growth factor (NGF). Although germ cells do not express c-*trk*, the gene is expressed in Sertoli cells where it exhibits coordinated expression with the low affinity NGF receptor at the onset of meiosis (Parvinen *et al.*, 1992). The proposed role of c-*trk* in meiosis is discussed in greater detail in Section II,E,1.

2. Cytoplasmic serine/threonine protein kinases

In mouse and rat testis, c-*mos* is expressed most abundantly as a testis-specific 1.7-kb transcript in early spermatids (Goldman *et al.*, 1987; Mutter and Wolgemuth, 1987; Propst *et al.*, 1987; Sorrentino *et al.*, 1988; van der Hoorn *et al.*, 1991), although lower levels also are detected in pachytene spermatocytes (Mutter and Wolgemuth, 1987). Surprisingly, antibodies specific for the 43-kDa testis-specific c-*mos* gene product (p43$^{c\text{-}mos}$) demonstrate highest levels of the protein in purified populations of pachytene spermatocytes (Herzog *et al.*, 1989; van der Hoorn *et al.*, 1991).

The promoter of the 1.7-kb rat testicular c-*mos* transcript spans 250 bp, approximately 0.56 kb upstream of the coding region. Within this region are potential binding sites for AP-1 and helix–loop–helix proteins, as well as a testis-specific protein-binding site, D_1, that is also observed in the promoter regions of other testis-specific genes (van der Hoorn, 1992). Thus, regulation of c-*mos* expression during spermatogenesis may be under the control of factors unique to the testis. c-*mos* transcripts are associated with both polysomes and monosomes, suggesting that the rate at which the message is translated may be particularly slow, accounting for the elevated levels of transcript seen in late spermatids (Propst *et al.*, 1987). Therefore, p43^{c-mos} expression seems likely to be regulated in a stage-specific fashion at the level of translation as well as transcription.

Treatment with progesterone induces production of p39^{c-mos} in *Xenopus* oocytes, subsequently stimulating activation of the p34^{cdc2}–cyclin complex and germinal vesicle breakdown, and triggers the transition from G_2 to M phase (Sagata *et al.*, 1989a). A functional ATP binding domain is required for c-*mos* protein-mediated effects on meiosis in *Xenopus* oocytes (Freeman *et al.*, 1989).

3. GTP binding proteins

In addition to the 5.0-kb transcripts of N-*ras* expressed in spermatogonia, a second N-*ras* transcript of 1.4 kb is also present in spermatocytes (Leon *et al.*, 1987; Wolfes *et al.*, 1989). p21^{ras} is present in abundance through the leptotene and zygotene stages, and decreases somewhat in pachytene spermatocytes (Wolfes *et al.*, 1989). In vertebrate oocytes, injection of p21^{ras} induces meiotic maturation (Birchmeier *et al.*, 1985; Allende *et al.*, 1988) whereas injection of monoclonal antibodies against p21^{ras} inhibits insulin-induced maturation (Desphande and Kung, 1987; Korn *et al.*, 1987). p21^{ras} stimulates both RSK and p34^{cdc2} kinase. p21^{ras} is likely to function upstream of c-*mos*, since injecting antisense oligonucleotides of c-*mos* blocks p21^{ras} induction of oocyte maturation (Barrett *et al.*, 1990). However, injection of c-*mos* antisense oligonucleotides does not block all the p21^{ras}-induced stimulation of RSK, suggesting that p21^{ras} also may act via a c-*mos* protein-independent pathway. Whether p21^{ras} plays a similar role in the meiotic cycle of spermatocytes is unknown.

4. Nuclear proto-oncogenes

a. c-*jun* and *jun*-related genes

As discussed previously, the levels of all three *jun* family genes are regulated developmentally. In testes from 17-day-old mice, mRNA for all three members of the *jun* family are highest in preleptotene spermatocytes, decrease in leptotene/zygotene, and virtually undetectable in pachytene and later stages (Alcivar *et al.*, 1990,1991). *jun*D in particular appears to be loaded more efficiently on polysomes during the early phases (leptotene/zygotene) of meiosis than during the pachytene phase (Alcivar *et al.*, 1991).

b. *ets*

ets-2 is a nuclear proto-oncogene expressed as an abundant 3.5-kb (somatic-type) transcript in prepubertal but not adult mouse testes, encoding a 56-kDa protein also present in young but not adult mice (Bhat *et al.*, 1987). *ets*-2 expression appears to be involved in meiotic maturation in *Xenopus* oocytes (Chen *et al.*, 1990); whether a similar requirement for *ets* expression exists in the progression of meiosis in spermatocytes is unknown.

The related gene *ets*-1 also is expressed more abundantly in immature testes than in adult as a 5.3-kb (somatic-type) transcript (Bhat *et al.*, 1987). *elk*-1, an *ets*-related gene, also is expressed abundantly in mouse testis as a 1.7-kb (somatic) transcript, in conjunction with lower levels of the 4.5- and 2.8-kb forms (Rao *et al.*, 1989). Intriguingly, testes contain TIN-1, a 43- to 45-kDa testis-specific factor that inhibits transcription of phosphoglycerate kinase-1 (the somatic PGK) mRNA in a dose-dependent fashion by binding to an *ets*-binding motif located between 400 and 200 bp upstream of *PGK*-1 (Goto *et al.*, 1991). Thus, expression of an *ets* family gene may be required for maintenance of cellular function during the premeiotic and meiotic stages of germ cell development.

c. c-*erb*A

c-*erb*A is a member of a class of genes encoding transcription regulatory receptors for thyroid hormones (the c-*erb*A gene product) and steroids (Weinberger *et al.*, 1986). c-*erb*A-α-2 is present in rat testis as a 2.6-kb transcript encoding a 55-kDa protein (Lazar *et al.*, 1988). Rat *erb*A-α-2 protein binds to DNA at a putative triiodothyronine response element, but not to the hormone itself. The *erb*A-α-2 product may act as a competitive antagonist to thyroid hormone stimulation. Alternatively, as yet unidentified ligand of the steroid/thyroid hormone family may stimulate *erb*A-α-2 protein at the thyroid hormone-responsive DNA binding site (Lazar *et al.*, 1988).

In addition to the somatic isoform of c-*erb*A, a testis-specific gene has been isolated from a human testis cDNA library (Benbrook and Pfahl, 1987). c-*erb*A-T produces a testis-specific 55-kDa protein with high homology to chicken and human c-*erb*A and v-*erb*A, especially in the DNA and steroid/ thyroid hormone binding domains. c-*erb*A-T is related to but distinct from human c-*erb*A-β, having both different binding affinities for triiodothyronine and different DNA binding domains, suggesting different target genes (Benbrook and Pfahl, 1987).

The role of c-*erb*A in meiosis, if any, has not yet been elucidated. However, thyroid hormone receptors are present in rat Sertoli cells (Jannini *et al.*, 1990) and treatment of Sertoli cells from immature rats *in vitro* with thyroid hormone stimulates production of insulin-like growth factor I (IGF-I;

Palmero *et al.*, 1990). Since late pachytene spermatocytes from immature rats in culture bind IGF-I (Tres *et al.*, 1986), c-*erb*A or *erb*-related genes may play a significant role in meiosis as part of a hormonal regulatory cascade.

D. Proto-oncogenes that Could Regulate Spermiogenesis

After meiotic reduction, the germ cells undergo a series of dramatic morphological changes. A marked increase is seen in the synthesis of new proteins required for flagellar and acrosomal formation, inactivation and condensation of nuclear material, and reorganization of organelles within the cytoplasm. Immunocytochemical studies have indicated the presence of c-*fos* (Pelto-Huikko *et al.*, 1991) and c-*erb*A (Tagami *et al.*, 1990) proteins in the nuclei of elongating spermatids. Whether these proteins are the products of postmeiotic transcription and translation or left over from the meiotic stages is not apparent. However, haploid-specific expression of proto-oncogenes such as the 2.8-kb testis-specific form of *RB* (Bernards *et al.*, 1989) and the 1.6-kb *jun*D transcript (Alcivar *et al.*, 1991) does occur exclusively in the postmeiotic stages, strongly suggesting that such genes are involved in the terminal differentiation of the germ cells.

1. Growth factor receptor tyrosine protein kinases

The 3.5-kb c-*kit* transcript that first appears in late pachytene spermatocytes also is expressed abundantly in postmeiotic round spermatids. In addition, a haploid-specific 2.3-kb transcript is expressed at this stage (Sorrentino *et al.*, 1991). Since the open reading frame for c-*kit* is 2.9 kb, this latter transcript is likely to encode a truncated form of the c-*kit* product. Studies with mice chimeric for the *Sl* mutation have testes in which the seminiferous tubules have sperm that are differentiated, undifferentiated, or intermediate between the two conditions. Intermediate tubules contain both normally developing germ cells and others that appear to have undergone degeneration at varying stages of development, suggesting that a functional *Sl*–c-*kit* interaction is essential for maintenance of spermiogenesis (Nakayama *et al.*, 1988).

2. Nonreceptor tyrosine protein kinases

In addition to the larger somatic transcripts of c-*abl*, adult mouse testes also contain high levels of a unique 4.7-kb transcript (Müller *et al.*, 1982). This transcript is present only in haploid germ cells and is regulated developmentally (Ponzetto and Wolgemuth, 1985). This testis-specific transcript apparently has a 5′ sequence identical to that of the larger somatic transcript (Meijer *et al.*, 1987; Oppi *et al.*, 1987), although the 3′ end lacks ~1 kb of 3′ untranslated sequences (Duggal *et al.*, 1987; Meijer *et al.*, 1987) and uses an alternative polyadenylation site that lacks an AU-rich destabilization sequence. This absence may account for the increased stability of the message, which is observed at highest levels in elongating spermatids (Meijer *et al.*, 1987; Iwaoki *et al.*, 1993). Only a portion of the testis-specific 4.7-kb c-*abl* transcript is associated with polysomes; the remainder is associated with ribonucleoprotein particles and monosomes, suggesting that the remainder is stored for translation at a later stage of development (Zakeri *et al.*, 1988). The predicted amino acid sequence from cDNA cloning (Meijer *et al.*, 1987; Oppi *et al.*, 1987), as well as studies on c-*abl* proteins (Meijer *et al.*, 1987; Ponzetto *et al.*, 1989), indicates that the 4.7-kb transcript produces a protein that is similar in size ($\sim$ 150 kDa) to that resulting from the type IV somatic transcript. The testis-specific transcript appears to be similar to the type IV c-*abl* transcript.

Intriguingly, immunoprecipitation studies indicate that a 74-kDa phosphoprotein (p74) coprecipitates with p150^{c-abl} in extracts from postmeiotic spermatids and residual bodies, but not from earlier stages of development. Similarly, p150^{c-abl} is present in testicular cell lysates from both immature and mature testes, whereas p74 is present at detectable levels only in extracts of mature testes (Ponzetto *et al.*, 1989). Whether the p150^{c-abl} present in the immunoprecipitates is the product of the somatic form or of the 4.7-kb haploid-specific form of the c-*abl* gene is unclear. Phosphoamino acid analysis indicates that both p150^{c-abl} and p74 contain predominantly phosphoserine (Wadewitz and Wolgemuth, 1991). Since the phosphorylating activity of p150^{c-abl} is tyrosine specific, an additional serine/threonine protein kinase must be present in the immunoprecipitates, possibly the coprecipitating p74. Interestingly, the 74-kDa protein PKC phosphorylates p150^{c-abl} proteins on serine *in vitro*; treatment of NIH 3T3 cells with activators of PKC stimulates p150^{c-abl} kinase activity (Pendergast *et al.*, 1987). This result suggests the intriguing possibility that the coprecipitating p74 may be PKC. Alternatively, p74 may be the product of the proto-oncogene serine/threonine protein kinase c-*raf*-1.

3. Cytoplasmic serine/threonine protein kinases

B-*raf* first was identified in a human testis cDNA library (Ikawa *et al.*, 1988). This gene appears in the mouse testis as abundant testis-specific transcripts of 2.6 and 4.0 kb, along with lower levels of the 10- and 13-kb transcripts expressed in somatic tissues (Storm *et al.*, 1990). B-*raf* encodes a major 75-kDa protein as well as a less abundant 77-kDa protein that may be the result of alternative splicing or a posttranslational modification of the 75-kDa species (Sithanandam *et al.*, 1990). B-*raf* is expressed only in germ cells, and is particularly abundant in early spermatids. Further, the two B-*raf* transcripts are expressed in a stage-specific manner. Low levels of the 4.0-kb transcript are observed first in meiotic cells, whereas the 2.6-kb form is not expressed until the postmeiotic stages (Wadewitz *et al.*, 1993). Intriguingly, unlike A-*raf* or c-*raf*-1, the B-*raf* gene product contains a p34^{cdc2} kinase consensus phosphorylation sequence that precedes its putative zinc-finger region (Sithanandam *et al.*, 1990). Since the mouse cyclin B1 also is expressed at high levels in round spermatids (Chapman and Wolgemuth, 1992), the cyclin–p34^{cdc2} complex could regulate B-*raf* function during spermiogenesis.

4. GTP binding proteins

H-*ras* transcripts identical in size to the somatic transcripts (1.3 and 1.1 kb) are present throughout spermatid development and even have been detected in the residual bodies of late stage spermatids (Wolfes *et al.*, 1989). In addition to the ubiquitously expressed H-*ras* and the 5.0-kb form of N-*ras*, three additional N-*ras* transcripts of 1.4, 1.9, and 3.5 kb are present in postmeiotic germ cells. The 1.9- and 3.5-kb transcripts appear to be male germ cell-specific (Leon *et al.*, 1987; Wolfes *et al.*, 1989). In contrast, the levels of K-*ras* expression decrease dramatically in round spermatids relative to the high levels seen in pachytene spermatocytes, and are undetectable in residual bodies (Wolfes *et al.*, 1989). p21^{ras} is lower but still present in the early postmeiotic stages, but is virtually undetectable in residual bodies (Wolfes *et al.*, 1989).

5. Other proto-oncogenes

wnt-1 is a member of the *wnt* family of genes (reviewed by Nusse and Varmus, 1992). *wnt* genes are unusual among cellular oncogenes because they produce secretory glycoproteins thought to be involved in paracrine and autocrine regulatory mechanisms. The *wnt-1* gene product is approximately 36–44 kDa (Papkoff *et al.*, 1987). Unlike most proto-oncogenes, which often are expressed in several cell types and tissues, the expression of *wnt-1* is restricted to the neural tube of midgestational embryos and postmeiotic male germ cells (Shackleford and Varmus, 1987). Murine *wnt-1* transcripts of 2.6 kb are expressed abundantly in round spermatids and at sharply lower levels in elongating spermatids, but not at earlier stages of development. Further, *wnt-1* expression apparently occurs during late spermatid differentiation, since the transcripts are not detected until day 26 of postnatal development, well after the earliest appearance of round spermatids (Shackleford and Varmus, 1987).

E. Expression of Cellular Oncogenes in Nongerminal Compartments of the Testis

Proto-oncogene expression in the nongerminal cells of the testis has been documented much less extensively than has expression in germ cells, and the observations are based primarily on *in vitro* systems.

1. Growth factor receptor tyrosine protein kinases

a. c-*kit*

The 5.5-kb somatic c-*kit* transcript is expressed in Leydig cells (Manova *et al.*, 1990; Sorrentino *et al.*, 1991). Immunocytochemical labeling with antibodies against the c-*kit* product indicates that Leydig cells express abundant amounts of c-*kit* from day 3 of postnatal development onward. Intriguingly, treatment of mice with antibodies that block *Sl* protein binding to c-*kit* protein results in an increase in serum testosterone, suggesting that the c-*kit*–*Sl* protein complex may act as a negative regulator of Leydig cell function (Yoshinaga *et al.*, 1991).

Although Sertoli cells do not express c-*kit*, they do produce the *Sl* ligand as a membrane associated protein (Motro *et al.*, 1991; Rossi *et al.*, 1991; Tajima *et al.*, 1991a). *Sl* expression is higher in the Sertoli cells of testes from germ cell-deficient mice, suggesting that the presence of germ cells in the seminiferous epithelium interacts with the Sertoli cells to down-regulate *Sl* expression (Motro *et al.*, 1991). Expression of *Sl* mRNA is stimulated by cAMP, an effector that acts downstream of FSH in Sertoli cells (Rossi *et al.*, 1991). The presence of c-*kit* genes encoding the receptor for the *Sl* product throughout the germ cell lineage suggests that c-*kit* and *Sl* may be a crucial mechanism for coordinating the interaction between Sertoli and germ cells.

b. c-*trk*

In NGF-responsive cells, high affinity NGF binding and subsequent transmembrane signaling requires coexpression of p140^{c-trk} and the low affinity NGF receptor (p75^{LNGFR}; Hempstead *et al.*, 1991). NGF binding to the high affinity complex stimulates tyrosine phosphorylation of p140^{c-trk} and activates its tyrosine kinase activity (Kaplan *et al.*, 1991; Klein *et al.*, 1991).

Although c-*trk* is not expressed in germ cells, the protein nonetheless is likely to play an integral role in spermatogenesis, since NGF mRNA and protein are present in germ cells from mid-pachytene spermatocytes through the late spermatid stage in both mouse (Ayer-LeLievre *et al.*, 1988) and rat (Parvinen *et al.*, 1992). Rat c-*trk* mRNA is expressed in Sertoli cells throughout the spermatogenic cycle, but is expressed at slightly higher levels during Stages VII–VIII, corresponding to the onset of both meiosis and spermiation. These stages are the only ones that express the low affinity NGF receptor mRNA (Persson *et al.*, 1990; Parvinen *et al.*, 1992). Immunolocalization with monoclonal antibodies against the p75^{LNGFR} shows labeling in Sertoli cells in Stages VII–XI. Expression of LNGFR message at later stages apparently is down-regulated by a peak of inhibitory testosterone in the seminiferous tubules during Stage VIII (Persson *et al.*, 1990; Parvinen *et al.*, 1992). Treatment with NGF stimulates DNA synthesis in preleptotene spermatocytes at the onset of meiosis in a dose-dependent manner at Stages VIII–IX of the spermatogenic cycle (Parvinen *et al.*, 1992). Thus, c-*trk* production by Sertoli cells may be required for NGF produced by germ cells at the onset of meiosis to interact with a testosterone-regulated stage-specific Sertoli cell $p75^{LNGFR}$. Subsequent production of unspecified Sertoli cell factors in response to NGF stimulation may be required for completion of meiosis.

c. c-erbB and bek/flg

c-erbB and the *bek/flg* genes are proto-oncogenes encoding the receptors for epidermal growth factor (EGF) (Downward *et al.*, 1984) and basic fibroblast growth factor (bFGF) (Basilico and Moscatelli, 1992), respectively. Although their patterns of expression in testicular tissues have not been examined extensively, the ability of these growth factors to affect Sertoli (Jaillard *et al.*, 1987) and Leydig (Verhoeven and Cailleau, 1986; Sordoillet *et al.*, 1988) cell function *in vitro*, as well as to stimulate the expression of early response nuclear proto-oncogenes in these cells (Smith *et al.*, 1989a,b; Hall *et al.*, 1991), provides compelling evidence for expression of the receptor products of these proto-oncogenes.

2. Cytoplasmic serine/threonine protein kinases

A-*raf* is expressed at low levels in the mouse testis as a 2.6-kb transcript encoding a 68-kDa protein (Beck *et al.*, 1987). Low levels of an additional 4.3-kb transcript also have been observed in mouse testis (Wadewitz *et al.*, 1993). Among the somatic cells, the androgen-producing Leydig cells express A-*raf* abundantly. The 4.3-kb transcript is likely to be specific for somatic cells, since it is virtually undetectable in wild-type mice or heterozygous littermates of germ cell-deficient mutants in which the bulk of the mRNA is derived from germ cells. Conversely, the 2.6-kb transcript is expressed more abundantly in germ cell-containing testes. Further, *in situ* hybridization studies suggest that expression of A-*raf* may be enhanced in Leydig cells from testes containing germ cells (Wadewitz *et al.*, 1993), suggesting a paracrine regulatory mechanism on the part of the cells in the seminiferous tubules that affect Leydig cell expression of A-*raf*. Beck *et al.* (1990) have reported a potential hormone response element in the human A-*raf* promoter region. Tissue survey studies have demonstrated that highest levels of A-*raf* expression occur in the epididymis, seminal vesicle, and prostate (Storm *et al.*, 1990). These observations suggest that A-*raf* may be involved in regulating the hormonal responsiveness of these androgen-dependent tissues. Native A-*raf* activates the AP-1 transcription regulatory factor (Wasylyk *et al.*, 1989).

3. Nuclear proto-oncogenes

a. c-myc

c-*myc* expression in mouse Leydig cells is regulated developmentally, being highest during prepubertal stages when these cells are actively proliferating. Likewise, c-*myc* expression is highest in seminiferous tubules during the period of Sertoli cell proliferation and decreases at approximately day 12 of postnatal development when proliferation stops (Stewart *et al.*, 1984). By adulthood, no c-*myc* mRNA is detectable in mouse Leydig or Sertoli cells by *in situ* hybridization (Koji *et al.*, 1988). Despite these observations, c-*myc* is likely to play a significant role in Leydig cell physiology, even in the adult. hCG administered to rats *in vivo* causes a transient expression of c-*myc* mRNA (T. Lin *et al.*, 1988). Stimulation of a murine Leydig cell tumor line with hCG causes a transient increase in both c-*myc* mRNAs at an ED_{50} comparable to that required for maximal hCG binding and steroid production by Leydig cells. Dibutyryl cyclic AMP mimics these effects, suggesting that hCG stimulates expression of c-*myc* via a cAMP-dependent pathway (Czerwiec *et al.*, 1989). In cultured porcine Leydig cells, expression of c-*myc* is stimulated by hCG, IGF-I, bFGF, and EGF; EGF is the most effective inducer (Hall *et al.*, 1991).

b. c-*fos*

Immunocytochemical labeling with specific antibodies has demonstrated a uniform distribution of c-*fos* protein present in the nuclei of rat Sertoli cells (Pelto-Huikko *et al.*, 1991). Treatment of primary Sertoli cell cultures with FSH induces expression of c-*fos* (Hall *et al.*, 1988). Both bFGF and a Sertoli cell bFGF-like factor stimulate a transient increase in the levels of c-*fos* expression in cultured rat Sertoli cells, whereas PDGF, IGF-I, and EGF have minimal effects. FGF does not increase cAMP levels, calcium, or inositol phosphate, indicating that it acts via a cAMP- and PKC-independent pathway (Smith *et al.*, 1989a,b).

Although immunocytochemistry does not detect c-*fos* protein in the nuclei of Leydig cells (Pelto-Huikko *et al.*, 1991), stimulation of a murine Leydig cell tumor line with hCG causes a transient increase in c-*fos* mRNA at an ED_{50} comparable to that required for maximal hCG binding and steroid production by Leydig cells. Dibutyryl cyclic AMP mimics these effects, suggesting that hCG acts on c-*fos* via a cAMP-dependent pathway (Czerwiec *et al.*, 1989). In contrast, hCG administered to rats *in vivo* has no effect on Leydig cell expression of c-*fos* (Lin *et al.*, 1988). In cultured porcine Leydig cells, expression of c-*fos* is stimulated by hCG, IGF-I, bFGF, and EGF; hCG is the most effective inducer (Hall *et al.*, 1991).

c. c-*jun* and *jun*-related genes

In cultured porcine Leydig cells, expression of c-*jun* is stimulated by bFGF and EGF but not by hCG or IGF-I. In contrast, increased expression of *jun*B is stimulated by all four effectors; hCG is the most stimulatory (Hall *et al.*, 1991). *jun*D also is expressed abundantly in Leydig cells from prepubertal and adult mouse testes (Alcivar *et al.*, 1991), but the hormonal and cellular mechanisms regulating its expression have not yet been elucidated.

4. Tumor suppressor genes

WT1 is expressed exclusively in the nuclei of mouse Sertoli cells (Pelletier *et al.*, 1991). In addition to the somatic 3.1-kb transcript of *WT1*, a testis-specific 2.5-kb species is also present. The 3.1-kb form peaks in its expression at day 8 of postnatal development, then decreases. The 2.5-kb transcript increases to its maximum at 15 days of postnatal development, then remains constant in the adult testis. The *WT1* product may act to keep Sertoli cells in a nonproliferative state after puberty.

F. Is There Any Evidence that Proto-oncogenes Might Function in the Mature Sperm, Particularly during Fertilization?

Activation of proteins by phosphorylation may play an important role during fertilization, suggesting a potential function for proto-oncogene kinases. Mature caudal epididymal mouse sperm contain a 95-kDa protein phosphorylated on tyrosine in the acrosomal region of the sperm head (Leyton and Saling, 1989). This protein may be involved in sperm binding to the zona pellucida and subsequent induction of the binding-dependent acrosome reaction (Bleil and Wasserman, 1983). Phosphorylation of this protein is enhanced both by capacitation and by the binding of specific zona pellucida glycoproteins (Leyton and Saling, 1989), suggesting that the zona pellucida receptor kinase is capable of autophosphorylation on binding its ligand, although it also may serve as a substrate for other tyrosine kinases within the cell. Antibodies against specific tyrosine kinases have demonstrated the expression of the 62-kDa product of c-*yes*, a nonreceptor proto-oncogene tyrosine kinase of the *src* family, in the acrosomal region of mature rat spermatozoa (Y.-H. Zhao *et al.*, 1990). p62^{c-yes} can be activated by ligand-stimulated PDGF receptor (Kypta *et al.*, 1990). Proto-oncogene products therefore may be crucial for spermatozoal functions during fertilization.

Recent immunological studies have identified proto-oncogene products in mature spermatozoa from a number of mammalian species. EGF receptors (the product of the c-*erb*B gene) have been detected in human, mouse, rabbit, and rat spermatozoa (Naz and Ahmad, 1992). Such receptors bind ligand and exhibit tyrosine kinase activity, but binding/activation had no apparent effect on the fertilizing potential of the spermatozoa. In contrast, monoclonal antibodies against v-H-*ras*, which detected a 21 kDa *ras*-like protein in the acrosomal region of human spermatozoa, significantly reduced hyperactivated motility, as well as the ability of spermatozoa to undergo acrosome reactions or penetrate zonae (Naz *et al.*, 1992). This suggests that *ras*-like proteins may function in signaling cascades involved in capacitation. Naz *et al.* (1991) have also detected an immunoreactive doublet of a 62–64 kDa c-*myc* or *myc*-like protein occurring over the acrosomes of sperm from mice, human, and rabbits. Antibodies against c-*myc* protein block zona binding/

penetration and some aspects of hyperactivated motility. It is unclear what, if any, role the product of a nuclear proto-oncogene may play during fertilization.

III. Expression of Proto-oncogene-Related Genes in the Testis

In addition to the proto-oncogenes previously discussed, several related genes also are known to be expressed in the testis, although the cellular and developmental patterns of expression have not yet been elucidated. These genes include known nuclear proto-oncogenes (L-*myc*, Saksela *et al.*, 1989; Robertson *et al.*, 1991; N-*myc*, Jakobovits *et al.*, 1985), putative *trans*-activating genes (c-*cbl*, Langdon *et al.*, 1989), and proto-oncogenes of unknown function such as *Wnt5b* and *Wnt6* (Gavin *et al.*, 1990) and c-*dbl* (Ron *et al.*, 1988; Galland *et al.*, 1991).

Testes express a number of genes with homologies to known proto-oncogenes. For example, *Flt3*, an *fms*-like tyrosine kinase with homology to members of the *kit*/PDGF receptor family (Rosnet *et al.*, 1991), and *fer*, which shares homology to the *fps*/*fes* cytoplasmic tyrosine kinase proto-oncogenes (Fischman *et al.*, 1990; Keshet *et al.*, 1990), produce testis-specific transcripts that are expressed in a developmentally regulated stage-specific fashion in mouse testis.

Other *ras*-related genes also are expressed at high levels in testis, including *RalA*, *Rho12*, and the *rab* family of genes (Olofsson *et al.*, 1988). Similarly, the *ras*-related *smg*p21A (*rap*1A/K-*rev*-1; Matsui *et al.*, 1990a) and *smg*p21B (*rap*1B) have been detected by immunocytochemistry in the cytoplasm of all spermatogenic cell types in the rat, including spermatogonia (Kim *et al.*, 1990). In contrast, no p21[ras] was detected in any rat germ cells, a result that differs from the results of Wolfes *et al.*, (1989), who detected *ras* protein products in mouse spermatogonia. This discrepancy may reflect species differences or the relative sensitivities of the methods employed in these studies. *smg*p21A binds to GAP in a GTP-dependent fashion but is not activated, and competitively inhibits GAP binding to p21[ras] (French *et al.*, 1990), thereby suppressing the transformation phenotypes of v-Ki-*ras*-transformed NIH 3T3 cells (Kitayama *et al.*, 1989). Whether the products of these *ras*-related genes compete with *ras* gene products or act via independent mechanisms is unknown. The latter case is suggested by the observation that *ral*A interacts with a GAP present in mouse testis cytosol that is distinct from the *ras*-encoded p120 GAP (Emkey *et al.*, 1991).

Although not a proto-oncogene in the classical sense of the term, the transforming potential of *pim*-1, another gene encoding a serine/threonine protein kinase, can be activated by inserting the Moloney murine leukemia virus in the *pim*-1 locus (Selten *et al.*, 1985). *pim*-1 is expressed predominantly in somatic tissues as a 2.8-kb transcript and in mouse testis as an additional abundant transcript of 2.4 kb. This testis-specific transcript is expressed only in postmeiotic spermatids (Sorrentino *et al.*, 1988). As with c-*abl* (Duggal *et al.*, 1987; Meijer *et al.*, 1987), the testis-specific *pim*-1 transcript apparently uses an alternate polyadenylation signal, which results in the loss of an AU-rich destabilization sequence, relative to the longer transcript (Wingett *et al.*, 1992). This apparently results in a more stable transcript being produced in the postmeiotic germ cells, which may be available for translation at more advanced stages of development. The function of *pim*-1 during male germ cell differentiation remains to be elucidated. However, its presence in hematopoietic organs as well suggests that it may be crucial for the terminal differentiation of stem cells.

The Sertoli cell product inhibin belongs to a family of heterodimeric proteins with homology to the TGF-β family of growth factors and inhibits production of FSH by the pituitary. It may also serve paracrine/autocrine regulatory functions within the testes. A recent study by Matzuk *et al.* (1992) suggested that inhibin may have proto-oncogenic potential: null mutations which are deficient for the α-subunit of the inhibin heterodimer develop gonad-specific tumors of stromal cell origin. If the inhibin gene is indeed a tumor suppressor proto-oncogene, it would also represent the first example of a growth factor-type proto-oncogene present in the testis.

IV. Summary and Future Directions

Coordinated expression of gene products at specific stages of development is crucial for normal germ cell development in the testis. Certain general patterns of proto-oncogene expression have been observed during spermatogenesis. c-*kit*, for example, appears to be required throughout the spermatogenic process to maintain cell viability. In contrast, the "early response" nuclear proto-oncogenes are expressed in greatest abundance in type B spermatogonia, suggesting an important role in the initiation of meiosis. Genes encoding components of the signal transduction pathway are expressed at higher levels in late pachytene spermatocytes and early postmeiotic round spermatids. These gene products are likely to be involved in transducing external hormonal signals from the somatic cells of the testis to the germ cells to regulate terminal differentiation. Perhaps the most intriguing findings relate to the expression of haploid-specific transcripts of ubiquitously expressed genes (e.g., c-*kit*, c-*abl*, N-*ras*, *jun*D). The role of such genes in the final stages of germ cell development remain to be elucidated, as do the specific regulatory mechanisms involved in initiating their stage-specific expression. Finally, the role of proto-oncogenes in the coordinated interaction of the germ cells with the Sertoli and Leydig cells remains largely unexplored.

Additional descriptive studies are needed to establish the stage and developmental specificity of expression of a number of proto-oncogenes in the testis, both by *in situ* hybridization and by analysis of gene expression in germ cell-deficient mutants compared with their fertile littermates, as well as by analysis of age-dependent changes in mice at different stages of postnatal development and studies on enriched populations of purified germ cells. Functional studies are also essential to elucidate the role of such genes in spermatogenesis. Exogenous proto-oncogenes can be introduced into the germ line of transgenic animals and their expression can be regulated precisely under the direction of heterologous, spermatogenic, stage-specific promoters. Finally, the use of targeted homologous recombination in embryonic stem cells followed by generation of chimeras will permit disruption of endogenous genes and analysis of subsequent effects on the function of proto-oncogenes *in vivo*. As new proto-oncogenes are discovered and RNA from reproductive tissues is examined, other germ cell-specific transcripts or patterns of proto-oncogene expression are likely to be identified.

Acknowledgments

The authors would like to thank J. Zengotita, and T. Mathew, for their assistance and D. Weymann-Session for helpful comments during the preparation of this manuscript. This work was supported by NIH Grant P50 HD05077.

References

Alcivar, A. A., Hake, L. E., Hardy, M. P., and Hecht, N. B. (1990). Increased levels of *junB* and c-*jun* mRNAs in male germ cells following testicular cell dissociation. *J. Biol. Chem.* **266,** 20160–20165.

Alcivar, A. A., Hake, L. E., Kwon, Y. K., and Hecht, N. B. (1991). *jun*D mRNA expression differs from c-*jun* and *jun*B mRNA expression during male germinal cell differentiation. *Mol. Reprod. Devel.* **30,** 187–193.

Allende, C., Hinrichs, M. V., Santos, E., and Allende, J. E. (1988). Oncogenic Ras protein induces meiotic maturation of amphibian oocytes in the presence of protein synthesis inhibitors. *FEBS Lett.* **234,** 426–430.

App, H., Hazan, R., Zilberstein, A., Ullrich, A., Sclessinger, J., and Rapp, U. (1991). Epidermal growth factor (EGF) stimulates association and kinase activity of Raf-1 with the EGF receptor. *Mol. Cell. Biol.* **11,** 913–919.

Ayer-LeLievre, C., Olson, L., Ebendal, T., Hallbook, F., and Persson, H. (1988). Nerve growth factor mRNA and protein in the testis and epididymis of mouse and rat. *Proc. Natl. Acad. Sci. U.S.A.* **85,** 2628–2632.

Baccarini, M., Sabatini, D. M., App, H., Rapp, U. R., and Stanley, E. R. (1990). Colony stimulating factor-1 (CSF-1) stimulates temperature dependent phosphorylation and activation of the RAF-1 proto-oncogene product. *EMBO J.* **9,** 3649–3657.

Bading, H., Rauterberg, E. W., and Moelling, K. (1989). Distribution of c-*myc*, c-*myb*, and Ki-67 antigens in interphase and mitotic human cells evidenced by immunofluorescence staining technique. *Exp. Cell Res.* **185,** 50–59.

Baltus, E., Hanocq-Quertier, J., Hanocq, F., and Brachet, J. (1988). Injection of an antibody against a p21 c-Ha-*ras* protein inhibits cleavage in axolotl eggs. *Proc. Natl. Acad. Sci. U.S.A.* **85**, 502–506.

Barbacid, M. (1987). Ras proteins. *Ann. Rev. Biochem.* **56**, 779–827.

Barrett, C. B., Schroetke, R. M., Van der Hoorn, F. A., Nordeen, S., and Maller, J. L. (1990). Ha *ras*val-12,thr-59 activates S6 kinase and p34cdc2 kinase in *Xenopus* oocytes: Evidence for c-*mos*xe-dependent and -independent pathways. *Mol. Cell. Biol.* **10**, 310–315.

Basilico, C., and Moscatelli, D. (1992). The FGF family of growth factors and oncogenes. *Adv. Cancer Res.* **59**, 115–165.

Beck, T. W., Huleihel, M., Gunnell, M., Bonner, T. I., and Rapp, U. R. (1987). The complete coding sequence of the human A-*raf*-1 oncogene and transforming activity of a human A-*raf* carrying retrovirus. *Nucl. Acids Res.* **15**, 595–609.

Beck, T. W., Brennscheidt, U., Sithanandam G., Cleveland, J., and Rapp, U. R. (1990). Molecular organization of the human *raf*-1 promoter region. *Mol. Cell. Biol.* **10**, 3325–3333.

Benbrook, D., and Pfahl, M. (1987). A novel thyroid hormone receptor encoded by a cDNA clone from a human testis library. *Science* **238**, 788–791.

Ben-Neriah, Y. A., Bernards, A., Paskind, M., Daley, G. Q., and Baltimore, D. (1986). Alternative 5' exons in c-*abl* mRNA. *Cell* **44**, 577–586.

Bernards, R., Schackleford, G. M., Gerber, M. R., Horowitz, J. M., Friend, S. H., Schartl, M., Bogenmann, E., Rapaport, J. M., McGee, T., Dryja, T. P., and Weinberg, R. A. (1989). Structure and expression of the murine retinoblastoma gene and characterization of its encoded protein. *Proc. Natl. Acad. Sci. U.S.A.* **86**, 6474–6478.

Besmer, P., Murphy, P. C., George, P. C., Qui, F., Bergold, P. J., Lederman, L., Snyder, H. W., Brodeur, D., Zuckerman, E. E., and Hardy, W. D. (1986). A new acute transforming feline retrovirus and relationship of its oncogene v-*kit* with the protein kinase family. *Nature (London)* **320**, 415–421.

Bhat, N. K., Fisher, R. J., Fujiwara, S., Ascione, R., and Papas, T. S. (1987). Temporal and tissue-specific expression of mouse *ets* genes. *Proc. Natl. Acad. Sci. U.S.A.* **84**, 3161–3165.

Binétruy, B., Smeal, T., and Karin, M. (1991). Ha-Ras augments c-Jun activity and stimulates phosphorylation of its activation domain. *Nature (London)* **351**, 122–127.

Birchmeier, C., Broek, D., and Wigler, M. (1985). Ras proteins can induce meiosis in *Xenopus* oocytes. *Cell* **43**, 615–621.

Bischoff, J. R., Friedman, P. N., Marshak, D. R., Prives, C., and Beach, D. (1990). Human p53 is phosphorylated by p60-cdc2 and cyclin B-cdc2. *Proc. Natl. Acad. Sci. U.S.A.* **87**, 4766–4770.

Bishop, J. M. (1983). Cellular oncogenes and retroviruses. *Ann. Rev. Biochem.* **52**, 301–354.

Blackshear, P. J., Haupt, D. M., App, H., and Rapp, U. R. (1990). Insulin activates the Raf-1 protein kinase. *J. Biol. Chem.* **265**, 12131–12134.

Bleil, J. D., and Wasserman, P. M. (1983). Sperm-egg interaction in the mouse: Sequence of events and induction of the acrosome reaction by a zona binding glycoprotein. *Dev. Biol.* **95**, 317–324.

Bonner, T. I., Kerby, S. B., Sutrave, P., Gunnell, M. A., Mark, G., and Rapp, U. R. (1985). Structure and biological activity of human homologs of the *raf*/*mil* oncogene. *Mol. Cell. Biol.* **5**, 1400–1407.

Bruder, J. T., Heidecker, G., and Rapp, U. R. (1992). Serum-, TPA-, and Ras-induced expression from Ap-1/Ets-driven promoters requires Raf-1 kinase. *Genes Dev.* **6**, 545–556.

Buchkovich, K., Duffy, L. A., and Harlow, E. (1989). The retinoblastoma protein is phosphorylated during specific phases of the cell cycle. *Cell* **58**, 1097–1105.

Carroll, M. P., Spivak, J. L., McMahon, M., Weich, N., Rapp, U. R., and May, W. S. (1991). Erythropoietin induces Raf-1 activation and Raf-1 is required for erythropoietin-mediated proliferation. *J. Biol. Chem.* **266**, 14964–14969.

Chapman, D. L., and Wolgemuth, D. J. (1992). Identification of a mouse B-type cyclin which exhibits developmentally regulated expression in the germ line. *Mol. Reprod. Devel.* **33**, 259–269.

Chen, P.-L., Scully, P., Shew, J.-Y., Wang, J. Y. J., and Lee, W.-H. (1989). Phosphorylation of the retinoblastoma gene product is modulated during the cell cycle and cellular differentiation. *Cell* **58**, 1193–1198.

Chen, Z. Q., Burdett, L. A., Seth, A. K., Lautenberger, J. A., and Papas, T. S. (1990). Requirement of *ets*-2 expression for *Xenopus* oocyte maturation. *Science* **250**, 1416–1418.

Chiu, R., Angel, P., and Karin, M. (1989). Jun-B differs in its biological properties from, and is a negative regulator of, c-Jun. *Cell* **59**, 979–986.

Cole, M. D. (1986). The *myc* oncogene: Its role in transformation and differentiation. *Ann. Rev. Genet.* **13**, 361–384.

Cooper, J. A., and Whyte, P. (1989). RB and the cell cycle: Entrance or exit. *Cell* **58**, 1009–1011.

Copeland, N. G., Gilbert, D. J., Cho, B. C., Donovan, P. J., Jenkins, N. A., Cosman, D., Anderson, D., Lynman, S. D., and Williams, D. E. (1990). Mast cell growth factor maps near the *steel* locus on mouse chromosome 10 and is deleted in a number of *steel* alleles. *Cell* **63**, 175–183.

Coulombre, J. L., and Russell, E. S. (1954). Analysis of the pleiotropism at the W-locus in the mouse. The effects of W and W^v substitution upon postnatal development of germ cells. *J. Exp. Zool.* **126**, 277–295.

Curran, T., and Franza, B. R. (1988). Fos and jun: The AP-1 connection. *Cell* **55**, 395–397.

Czerwiec, F. S., Melner, M., and Puett, D. (1989). Transiently elevated levels of c-*fos* and c-*myc* oncogene messenger ribonucleic acids in cultured murine Leydig tumor cells after addition of human chorionic gonadotropin. *Mol. Endocrinol.* **3**, 105–109.

DeCaprio, J. A., Ludlow, J. W., Lynch, D., Furukawa, Y., Griffin, J., Piwnica-Worms, H., Huang, C.-M., and Livingston, D. M. (1989). The product of the retinoblastoma susceptibility gene has properties of a cell cycle regulatory element. *Cell* **58**, 1085–1095.

Dent, P., Haser, W., Haystead, T. A. J., Vincent, L. A., Roberts, T. M., and Sturgill, T. W. (1992). Activation of mitogen-activated protein kinase kinase by v-Raf in NIH 3T3 cells and *in vitro*. *Science* **257**, 1404–1407.

Desphande, A. T., and Kung, H.-F. (1987). Insulin induction of *Xenopus* oocyte maturation is inhibited by monoclonal antibody against p21 proteins. *Mol. Cell. Biol.* **7**, 1285–1288.

Dikstein, R., Faktor, O., Ben-Levy, R., and Shaul, Y. (1990). Functional organization of the hepatitis B enhancer. *Mol. Cell. Biol.* **10**, 3683–3689.

Dikstein, R., Heffetz, D., Ben-Neriah, Y., and Shaul, Y. (1992). c-Abl has a sequence-specific enhancer binding activity. *Cell* **69**, 751–757.

Dolci, S., Williams, D. E., Ernst, M. K., Resnick, J. L., Brannan, C. I., Lock, L. F., Lyman, S. D., Boswell, H. S., and Donovan, P. J. (1991). Requirement for mast cell growth factor for primordial germ cell survival in culture. *Nature (London)* **352**, 809–811.

Downward, J., Yarden, Y., Mayes, E., Screace, G., Totty, N., Stockwell, P., Ullrich, A., Schlessinger, J., and Waterfield, M. D. (1984). Close similarity of epidermal growth factor receptor and v-*erb*-B oncogene protein sequences. *Nature (London)* **307**, 521–527.

Duggal, R. N., Zakeri, Z. F., Ponzetto, C., and Wolgemuth, D. J. (1987). Differential expression of the c-*abl* proto-oncogene and the homeobox-containing gene *Hox* 1.4 during mouse spermatogenesis. *In* "Cell Biology of the Testis and Epididymis" (M. C. Orgebin-Crist, ed.), pp. 112–127. New York Academy of Sciences, New York.

Emkey, R., Freedman, S., and Feig, L. A. (1991). Characterization of a GTPase-activating protein for the Ras-related Ral protein. *J. Biol. Chem.* **266**, 9703–9706.

Erickson, R. P. (1990). Post-meiotic gene expression. *Trends Genet.* **6**, 264–268.

Faisst, S., and Meyer, S. (1992). Compilation of vertebrate encoded transcription factors. *Nucl. Acids Res.* **20**, 3–26.

Fischman, K., Edman, J. C., Shackleford, G. M., Turner, J. A., Rutter, W. J., and Nir, U. (1990). A murine *fer* testis-specific transcript *(ferT)* encodes a truncated Fer protein. *Mol. Cell. Biol.* **10**, 146–153.

Flanagan, J. G., and Leder, P. (1990). The Kit ligand: A cell surface molecule altered in *steel* mutant fibroblasts. *Cell* **63**, 185–194.

Freeman, R. S., Pickham, K. M., Kanki, J. P., Lee, B. A., Pena, S. V., and Donoghue, D. J. (1989).

Xenopus homolog of the *mos* protooncogene transforms mammalian fibroblasts and induces maturation of *Xenopus* oocytes. *Proc. Natl. Acad. Sci. U.S.A.* **86,** 5805–5809.

Frech, M., John, J., Pizon, V., Chardin, P., Tavitian, A., Clark, R., McCormick, F., and Wittinghofer, A. (1990). Inhibition of GTPase activating protein stimulation of Ras-p21 GTPase by the K-*rev*-1 gene product. *Science* **249,** 169–171.

Furukawa, Y., Piwnica-Worms, H., Ernst, T. J., Kanakura, Y., and Griffin, J. D. (1990). *cdc* gene expression at the G₁ to S transition in human T lymphocytes. *Science* **250,** 805–808.

Galland, F., Pirisi, V., deLapeyriere, O., and Birnbaum, D. (1991). Restriction and complexity of *Mcf2* proto-oncogene expression. *Oncogene* **6,** 833–839.

Gaub, M.-P., Bellard, M., Scheuer, I., Chambon, P., and Sassone-Corsi, P. (1990). Activation of the ovalbumin gene by the estrogen receptor involves the Fos-Jun complex. *Cell* **63,** 1267–1276.

Gavin, B. J., McMahon, J. A., and McMahon, A. P. (1990). Expression of multiple novel Wnt-1/int-1-related genes during fetal and adult mouse development. *Genes Dev.* **4,** 2319–2332.

Geissler, E. N., Ryan, M. A., and Housman, D. E. (1988). The dominant-white spotting (W) locus of the mouse encodes the c-*kit* proto-oncogene. *Cell* **55,** 185–192.

Godin, I., Deed, R., Cooke, J., Zsebo, K., Dexter, M., and Wylie, C. C. (1991). Effects of the *steel* gene product on mouse primordial germ cells in culture. *Nature (London)* **352,** 807–809.

Goldman, D. S., Kiessling, A. A., Millette, C. F., and Cooper, G. M. (1987). Expression of c-*mos* RNA in germ cells of male and female mice. *Proc. Natl. Acad. Sci. U.S.A.* **84,** 4509–4513.

Goto, M., Tamura, T.-A., Mikoshiba, K., Masamune, Y., and Nakanishi, Y. (1991). Transcription inhibition of the somatic-type phosphoglycerate kinase 1 gene *in vitro* by a testis-specific factor that recognizes a sequence similar to the binding site for Ets oncoproteins. *Nucl. Acids Res.* **19,** 3959–3963.

Halazonetis, T. D., Georgopoulos, K., Greenberg, M., and Leder, P. (1988). c-Jun dimerizes with itself and with c-Fos, forming complexes of different DNA binding affinities. *Cell* **55,** 917–924.

Hall, S. H., Joseph, D. R., French, F. S., and Conti, M. (1988). Follicle-stimulating hormone induces transient expression of the proto-oncogene c-*fos* in primary Sertoli cell cultures. *Mol. Endocrinol.* **2,** 55–61.

Hall, S. M., Berthelon, M.-C., Avallet, O., and Saez, J. M. (1991). Regulation of c-*fos*, c-*jun*, *jun*-B, and c-*myc* messenger ribonucleic acids by gonadotropin and growth factors in cultured pig Leydig cell. *Endocrinol.* **129,** 1243–1249.

Han, K. A., and Kulesz-Martin, M. F. (1992). Alternatively spliced *p53* RNA in transformed and normal cells of different tissue types. *Nucl. Acids Res.* **20,** 1979–1981.

Hecht, N. B. (1990). Regulation of "haploid expressed genes" in male germ cells. *J. Reprod. Fertil.* **88,** 679–693.

Heidecker, G., Kölch, W., Morrison, D. K., and Rapp, U. R. (1992). The role of Raf-1 phosphorylation in signal transduction. *Adv. Cancer Res.* **58,** 53–73.

Heikkila, R., Schwab, G., Wickstrom, E., Loke, S. L., Pluznik, D. H., Watt, R., and Neckers, L. M. (1987). A c-*myc* antisense oligodeoxynucleotide inhibits entry into S phase but not progress from G₀ to G₁. *Nature (London)* **328,** 445–449.

Hempstead, B. L., Martin-Zanca, D., Kaplan, D. R., Parada, L. F., and Chao, M. V. (1991). High-affinity NGF binding requires coexpression of the *trk* proto-oncogene and the low-affinity NGF receptor. *Nature (London)* **350,** 678–683.

Henriksson, M., Classon, M., Ingvarsson, S., Koskinen, P., Sümegi, J., Klein, G., and Thyberg, J. (1988). Elevated expression of c-*myc* and N-*myc* produces distinct changes in nuclear fine structure and chromatin organization. *Oncogene* **3,** 587–593.

Herbst, R., Lammers, R., Schlessinger, J., and Ullrich, A. (1991). Substrate phosphorylation specificity of the human c-*kit* receptor tyrosine kinase. *J. Biol. Chem.* **266,** 19908–19916.

Herrlich, P., and Ponta, H. (1989). "Nuclear" oncogenes convert extracellular stimuli into changes in the genetic program. *Trends Genet.* **5,** 112–116.

Herzog, N. K., Ramagli, L. S., and Arlinghaus, R. B. (1989). Somatic cell expression of the c-*mos* protein. *Oncogene* **4**, 1307–1315.

Hirai, S.-I., Ryseck, R.-P., Mechta, F., Bravo, R., and Yaniv, M. (1989). Characterization of *jun*D: A new member of the *jun* proto-oncogene family. *EMBO J.* **8**, 1433–1439.

Howe, L. R., Leevers, S. J., Gomez, N., Nakielny, S., Cohen, P., and Marshall, C. J. (1992). Activation of the MAP kinase pathway by the protein kinase Raf. *Cell* **71**, 335–342.

Huang, E., Nocka, K., Beier, D. R., Chu, T.-Y., Buck, J., Lahm, H.-W., Wellner, D., Leder, P., and Besmer, P. (1990). The hematopoietic growth factor KL is encoded by the *Sl* locus and is the ligand of the c-*kit* receptor, the gene product of the *W* locus. *Cell* **63**, 225–233.

Ikawa, S., Fukui, M., Ueyama, Y., Tamaoki, N., Yamamoto, T., and Toyoshima, K. (1988). B-*raf*, a new member of the *raf* family, is activated by DNA rearrangement. *Mol. Cell. Biol.* **8**, 2651–2654.

Iwaoki, Y., Matsuda, H., Mutter, G. L., Watrin, F., and Wolgemuth, D. J. (1993). Differential expression of the proto-oncogenes c-*abl* and c-*mos* in developing germ cells. *Exp. Cell Res.* **209**, 212–219.

Jaillard, C., Chatelain, P. G., and Saez, J. M. (1987). In vitro regulation of pig Sertoli cell growth and function: Effects of fibroblast growth factor and somatomedin-C. *Biol. Reprod.* **37**, 665–674.

Jakobovits, A., Schwab, M., Bishop, J. M., and Martin, G. R. (1985). Expression of N-*myc* in teratocarcinoma stem cells and mouse embryos. *Nature (London)* **318**, 188–191.

Jamal, S., and Ziff, E. (1990). Transactivation of c-fos and β-actin genes by *raf* as a step in early response to transmembrane signals. *Nature (London)* **344**, 463–466.

Jannini, E. A., Olivieri, M., Francavilla, S., Gulino, A., Ziparo, E., and D'Armiento, M. (1990). Ontogenesis of the nuclear 3,5,3' triiodothyronine receptor in the rat testis. *Endocrinol.* **126**, 2521–2526.

Kaplan, D. R., Morrison, D. K., Wong, G., McCormick, F., and Williams, L. T. (1990). PDGF β-receptor stimulates tyrosine phosphorylation of GAP and association of GAP with a signaling complex. *Cell* **61**, 125–133.

Kaplan, D. R., Martin-Zanca, D., and Parada, L. F. (1991). Tyrosine phosphorylation and tyrosine kinase activity of the *trk* proto-oncogene product induced by NGF. *Nature (London)* **350**, 158–160.

Keshet, E., Itin, A., Fischman, K., and Nir, U. (1990). The testis-specific transcript (*ferT*) of the tyrosine kinase FER is expressed during spermatogenesis in a stage-specific manner. *Mol. Cell. Biol.* **10**, 5021–5025.

Keshet, E., Lyman, S. D., Williams, D. E., Anderson, D. M., Jenkins, N. A., Copeland, N. G., and Parada, L. F. (1991). Embryonic expression patterns of the c-*kit* receptor and its cognate ligand suggest multiple functional roles in mouse development. *EMBO J.* **10**, 2425–2435.

Kim, S., Mizoguchi, A., Kikuchi, A., and Takai, Y. (1990). Tissue and subcellular distributions of the *smg*-21/*rap*1/*Krev*-1 proteins which are partly distinct from those of c-*ras* p21s. *Mol. Cell. Biol.* **10**, 2645–2652.

Kipreos, E. T., and Wang, J. Y. J. (1990). Differential phosphorylation of c-*abl* in cell cycle determined by *cdc*2 kinase and phosphatase activity. *Science* **248**, 217–220.

Kipreos, E. T., and Wang, J. Y. J. (1992). Cell cycle-regulated binding of c-Abl tyrosine kinase to DNA. *Science* **256**, 382–385.

Kitayama, H., Sugimoto, Y., Matsuzaki, T., Ikawa, Y., and Noda, M. (1989). A *ras*-related gene with transformation suppressor activity. *Cell* **56**, 77–84.

Klein, R., Jing, S., Nanduri, V., O'Rourke, E., and Barbacid, M. (1991). The *trk* proto-oncogene encodes a receptor for nerve growth factor. *Cell* **65**, 189–197.

Koji, T., Izumi, S., Tanno, M., Moriuchi, T., and Nakane, P. K. (1988). Localization *in situ* of c-*myc* mRNA and c-*myc* protein in adult mouse testis. *Histochem. J.* **20**, 551–557.

Korn, L. J., Siebel, C. W., McCormick, F., and Roth, R. A. (1987). *Ras* p21 as a potential mediator of insulin action in *Xenopus* oocytes. *Science* **236**, 840–842.

Koshimizu, U., Sawada, K., Tajima, Y., Watanabe, D., and Nishimune, Y. (1991). White-spotting mutations affect the regenerative differentiation of testicular germ cells: Demonstration by experimental cryptorchidism and its surgical reversal. *Biol. Reprod.* **45,** 642–648.

Kovacina, K. S., Yonezawa, K., Brautigan, D. L., Tonks, N. K., Rapp, U. R., and Roth, R. A. (1990). Insulin activates the kinase activity of the Raf-1 proto-oncogene by increasing its serine phosphorylation. *J. Biol. Chem.* **265,** 12115–12118.

Kypta, R. M., Goldberg, Y., Ulug, E. T., and Courtneidge, S. A. (1990). Association between the PDGF receptor and members of the *src* family of tyrosine kinases. *Cell* **62,** 481–492.

Kyriakis, J. M., App, H., Zhang, X. F., Banerjee, P., Brautigan, D. L., Rapp, U. R., and Avruch, J. (1992). Raf-1 activates MAP kinase-kinase. *Nature (London)* **358,** 417–421.

Lane, D. P., and Crawford, L. V. (1979). T antigen is bound to a host protein in SV40-transformed cells. *Nature (London)* **278,** 261–263.

Langdon, W. Y., Hyland, C. D., Grumont, R. J., and Morse, H. C., III. (1989). The c-*cbl* proto-oncogene is preferentially expressed in thymus and testis tissue and encodes a nuclear protein. *J. Virol.* **63,** 5420–5424.

Lazar, M. A., Hodin, R. A., Darling, D. S., and Chin, W. W. (1988). Identification of rat c-erbAα-related protein which binds deoxyribonucleic acid but does not bind thyroid hormone. *Mol. Endocrinol.* **2,** 893–901.

Lee, R. M., Rapp, U. R., and Blackshear, P. J. (1991). Evidence for one or more Raf-1 kinase kinase(s) activated by insulin and polypeptide growth factors. *J. Biol. Chem.* **266,** 10351–10357.

Lee, R. M., Cobb, M. H., and Blackshear, P. J. (1992). Evidence that extracellular signal-regulated kinases are the insulin-activated Raf-1 kinase kinases. *J. Biol. Chem.* **267,** 1088–1092.

Leevers, S. J., and Marshall, C. J. (1992). Activation of extracellular signal-related kinase, ERK2 by p21*ras* oncoprotein. *EMBO J.* **11,** 569–574.

Leon, J., Guerrero, I., and Pellicer, A. (1987). Differential expression of the *ras* gene family in mice. *Mol. Cell. Biol.* **7,** 1535–1540.

Leyton, L., and Saling, P. (1989). 95 kd sperm proteins bind ZP3 and serve as tyrosine kinase substrates in response to zona binding. *Cell* **57,** 1123–1130.

Lin, B. T.-Y., Gruenwald, S., Morla, A. O., Lee, W. H., and Wang, J. Y. (1991). Retinoblastoma cancer suppressor gene product is a substrate of the cell cycle regulator cdc2 kinase. *EMBO J.* **10,** 857–864.

Lin, T., Blaisdell, J., Barbour, K. W., and Thompson, E. A. (1988). Transient activation of c-*myc* proto-oncogene expression in Leydig cells by human chorionic gonadotropin. *Biochem. Biophys. Res. Commun.* **157,** 121–126.

Llamazares, S., Moreira, A., Tavares, A., Girdham, C., Spruce, B. A., Gonzalez, C., Karess, R. E., Glover, D. M., and Sunkel, C. E. (1991). *polo* encodes a protein kinase homolog required for mitosis in *Drosophila. Genes Dev.* **5,** 2153–2165.

Lowy, D. R., Zhang, K., Declue, J. E., and Willumsen, B. M. (1991). Regulation of p21^ras^ activity. *Trends Genet.* **7,** 346–351.

Lucibello, F. C., Lowag, C., Neuberg, M., and Müller, R. (1989). *Trans*-repression of the mouse c-*fos* promoter: A novel mechanism of fos-mediated *trans*-regulation. *Cell* **59,** 999–1007.

Madden, S. L., Cook, D. M., Morris, J. F., Gashler, A., Sukhatme, V. P., and Rauscher, F. J., III. (1991). Transcriptional repression mediated by the WT1 Wilms tumor gene product. *Science* **253,** 1550–1553.

Majumder, S., Brown, K., Qiu, F. H., and Besmer, P. (1988). c-Kit protein, a transmembrane kinase: Identification in tissues and characterization. *Mol. Cell. Biol.* **8,** 4896–4903.

Manova, K., Nocka, K., Besmer, P., and Bachvarova, R. F. (1990). Gonadal expression of c-*kit* encoded at the W locus of the mouse. *Development* **110,** 1057–1069.

Matsui, Y., Kikuchi, A., Kawata, M., Kondo, J., Teranishi, Y., and Takai, Y. (1990a). Molecular cloning of *smg* p21B and identification of *smg* p21 purified from bovine brain and human platelets as *smg* p21B. *Biochem. Biophys. Res. Commun.* **166,** 1010–1016.

Matsui, Y., Zsebo, K. M., and Hogan, B. L. M. (1990b). Embryonic expression of a haematopoie-tic growth factor by the *Sl* locus and the ligand for c-*kit*. *Nature (London)* **347**, 667–669.

Matzuk M. M., Finegold, M. J., Su, J.-G. J., Hsueh, A. J. W., and Bradley, A. (1992). α-Inhibin is a tumour-suppressor gene with gonadal specificity in mice. *Nature (London)* **360**, 313–319.

Meijer, D., Hermans, A., Von Lindern, M., Van Agthoven, T., De Klein, A., Mackenbach, P., Grootegoed, A., Talarico, D., Della Valle, G., and Grosveld, G. (1987). Molecular charac-terization of the testis specific c-*abl* mRNA in mouse. *EMBO J.* **6**, 4041–4048.

Mihara, K., Cao, X. R., Yen, A., Chandler, S., Driscoll, B., Murphree, A. L., T'Ang, A., and Fung, Y. K. T. (1989). Cell cycle-dependent regulation of phosphorylation of the human retino-blastoma gene product. *Science* **246**, 1300–1303.

Mintz, B., and Russell, E. S. (1957). Gene-induced embryological modifications of primordial germ cells in the mouse. *J. Exp. Zool.* **134**, 207–230.

Miyazawa, K., Hendrie, P. C., Mantel, C., Wood, K., Ashman, L. K., and Broxmeyer, H. E. (1991). Comparative analysis of signaling pathways between mast cell growth factor (c-*kit* ligand) and granulocyte-macrophage colony-stimulating factor in a human factor-dependent myeloid cell line involves phosphorylation of Raf-1, GTPase-activating pro-tein and mitogen-activated protein kinase. *Exp. Hematol.* **19**, 1110–1123.

Moloney, J. B. (1966). A virus-induced rhabdomyosarcoma of mice. *Natl. Cancer Inst. Monogr.* **22**, 139–142.

Molloy, C. J., Bottaro, D. P., Fleming, T. P., Marshall, M. S., Gibbs, J. B., and Aaronson, S. A. (1989). PDGF induction of tyrosine phosphorylation of GTPase activating protein. *Nature (London)* **342**, 711–714.

Moreno, S., and Nurse, P. (1990). Substrates for p34^{cdc2}: In vivo veritas? *Cell* **61**, 549–551.

Morrison, D. K., Kaplan, D. R., Rapp, U., and Roberts, T. M. (1988). Signal transduction from membrane to cytoplasm: Growth factors and membrane-bound oncogene products in-crease Raf-1 phosphorylation and associated protein kinase activity. *Proc. Natl. Acad. Sci. U.S.A.* **85**, 8855–8859.

Morrison, D. K., Kaplan, D. R., Escobedo, J. A., Rapp, U. R., Roberts, T. M., and Williams, L. T. (1989). Direct activation of the serine/threonine kinase activity of Raf-1 through tyrosine phosphorylation by the PDGF β-receptor. *Cell* **58**, 649–657.

Motro, B., Van Der Kooy, D., Rossant, J., Reith, A., and Bernstein, A. (1991). Contiguous patterns of c-*kit* and *Steel* expression: Analysis of mutations at the W and *Sl* Loci. *Development* **113**, 1207–1221.

Mulcahy, L. S., Smith, M. R., and Stacey, D. W. (1985). Requirement for *ras* proto-oncogene function during serum-stimulated growth of NIH 3T3 cells. *Nature (London)* **313**, 241–243.

Müller, R., Slamon, D. J., Tremblay, J. M., Cline, M. J., and Verma, I. M. (1982). Differential expression of cellular oncogenes during pre- and postnatal development of the mouse. *Nature (London)* **299**, 640–644.

Mutter, G. L., and Wolgemuth, D. J. (1987). Distinct developmental patterns of c-*mos* proto-on-cogene expression in female and male mouse germ cells. *Proc. Natl. Acad. Sci. U.S.A.* **84**, 5301–5305.

Nakabeppu, Y., Ryder, K., and Nathans, D. (1988). DNA binding activities of three murine Jun proteins: Stimulation by Fos. *Cell* **55**, 907–915.

Nakayama, H., Kuroda, H., Onoue, H., Fujita, J., Nishimune, Y., Matsumoto, K., Nagano, T., Suzuki, F., and Kitamura, Y. (1988). Studies of *Sl/Sl*^{d} ↔ +/+ mouse aggregation chi-maeras. II. Effect of the *Steel* locus on spermatogenesis. *Development* **102**, 117–126.

Naz, R. K., and Ahmad, K. (1992). Presence of expression products of c-*erb*B-1 and c-*erb*B-2/HER2 genes on mammalian sperm cell, and effects of their regulation on fertilization. *J. Reprod. Immunol.* **21**, 223–239.

Naz, R. K., Ahmad, K., and Kumar, G. (1991). Presence and role of c-*myc* proto-oncogene product in mammalian sperm cell function. *Biol. Reprod.* **44,** 842–850.

Naz, R. K., Ahmad, K., and Kaplan, P. (1992). Expression and function of *ras* proto-oncogene proteins in human sperm cells. *J. Cell Sci.* **102,** 487–494.

Nishimune Y., Haneji, T., and Kitamura Y. (1980). The effects of *steel* mutation on testicular germ cell differentiation. *J. Cell Phys.* **105,** 137–141.

Nocka, K., Majumder, S., Chabot, B., Ray, P., Cervone, M., Bernstein, A., and Besmer, P. (1989). Expression of c-*kit* gene products in known cellular targets of W mutations in normal and W mutant mice—Evidence for an impaired c-*kit* kinase in mutant mice. *Genes Dev.* **3,** 816–826.

Nusse, R., and Varmus, H. E. (1992). Wnt genes. *Cell* **69,** 1073–1087.

O'Keefe, S. J., Kiessling, A. A., and Cooper, G. M. (1991). The c-*mos* gene product is required for cyclin B accumulation during meiosis of mouse eggs. *Proc. Natl. Acad. Sci. U.S.A.* **88,** 7869–7872.

Olofsson, B., Chardin, P., Touchot, N., Zahraoul, A., and Tavitian, A. (1988). Expression of the *ras*-related *ral*A, *rho*12 and *rab* genes in adult mouse tissues. *Oncogene* **3,** 231–234.

Oppi, C. S., Shore, K., and Reddy, P. (1987). Nucleotide sequence of testis-derived c-*abl* cDNAs: Implications for testis-specific transcription and *abl* oncogene activation. *Proc. Natl. Acad. Sci. U.S.A.* **84,** 8200–8204.

Orr-Urtreger, A., Aviv, A., Zimmer, Y., Givol, D., Yarden, Y., and Lonai, P. (1990). Developmental expression of c-*kit*, a proto-oncogene encoded by the W locus. *Development* **109,** 911–923.

Palmero, S., Prati, M., Barreca, A., Minuto, F., Giordano, G., and Fugassa, E. (1990). Thyroid hormone stimulates the production of insulin-like growth factor I (IGF-I) by immature rat Sertoli cells. *Mol. Cell. Endocrinol.* **68,** 61–65.

Papkoff, J., Brown, A. M. C., and Varmus, H. E. (1987). The *int*-1 proto-oncogene products are glycoproteins that appear to enter the secretory pathway. *Mol. Cell. Biol.* **7,** 3978–3984.

Partridge, J. F., and La Thangue, N. B. (1991). A developmentally regulated and tissue dependent transcription factor complexes with the retinoblastoma gene product. *EMBO J.* **10,** 3819–3827.

Parvinen, M., Pelto-Huikko, M., Söder, O., Schultz, R., Kaipia, A., Mali, P., Toppari, J., Hakovirta, H., Lönnerberg, P., Ritzén, E. M., Ebendal, T., Olson, L., Hökfelt, T., and Persson, H. (1992). Expression of β-nerve growth factor and its receptor in rat seminiferous epithelium: Specific function at the onset of meiosis. *J. Cell Biol.* **117,** 629–641.

Pelech, S. L., Sanghera, J. S., and Daya-Makin, M. (1990). Protein kinase cascades in meiotic and mitotic cell cycle control. *Biochem. Cell Biol.* **68,** 1297–1330.

Pelletier, J., Schalling, M., Buckler, A. J., Roger, A., Haber, D. A., and Housman, D. (1991). Expression of the Wilms' tumor gene WT1 in the murine urogenital system. *Genes Dev.* **5,** 1345–1346.

Pelto-Huikko, M., Schultz, R., Koistinaho, J., and Hokfelt, T. (1991). Immunocytochemical demonstration of c-Fos protein in Sertoli cells and germ cells in rat testis. *Acta Physiol. Scand.* **141,** 283–284.

Pendergast, A. M., Traugh, J. A., and Witte, O. N. (1987). Normal cellular and transformation-associated *abl* proteins share common sites for protein kinase C phosphorylation. *Mol. Cell. Biol.* **7,** 4280–4289.

Persson, H., Lievre, C. A. L., Söder, O., Villar, M. J., Metsis, M., Olson, L., Ritzen, M., and Hökfelt, T. (1990). Expression of β-nerve growth factor receptor mRNA in Sertoli cells downregulated by testosterone. *Science* **247,** 704–707.

Ponzetto, C., and Wolgemuth, D. J. (1985). Haploid expression of a unique c-*abl* transcript in the mouse male germ line. *Mol. Cell. Biol.* **5,** 1791–1794.

Ponzetto, C., Wadewitz, A. G., Pendergast, A. M., Witte, O. N., and Wolgemuth, D. J. (1989). P150^{c-abl} is detected in mouse male germ cells by an *in vitro* kinase assay and is associated with stage-specific phosphoproteins in haploid cells. *Oncogene* **4**, 685–690.

Pritchard-Jones, K., Fleming, S., Davidson, D., Bickmore, W., Porteous, D., Gosden, C., Bard, J., Buckler, A., Pelletier, J., Housman, D., van Heyningen, V., and Hastie, N. (1990). The candidate Wilms' tumour gene is involved in genitourinary development. *Nature (London)* **346**, 194–197.

Propst, F., Rosenberg, M. P., Iyer, A., Kaul, K., and Vande Woude, G. (1987). c-*mos* proto-oncogene RNA transcripts in murine tissues: Structural features, developmental regulation, and localization in specific cell types. *Mol. Cell. Biol.* **7**, 1629–1637.

Propst, F., Rosenberg, M. P., Oskarsson, M. K., Russell, L. B., Nguyen-Huu, M. C., Nadeau, J., Jenkins, N. A., Copeland, N. G., and Vande Woude, G. F. (1988). Genetic analysis and developmental regulation of testis-specific RNA expression of *mos, abl,* actin and *Hox*-1.4. *Oncogene* **2**, 227–233.

Pulverer, B. J., Kyriakis, J. M., Avruch, J., Nikolakaki, E., and Woodgett, J. R. (1991). Phosphorylation of c-*jun* mediated by MAP kinases. *Nature (London)* **353**, 670–674.

Qui, F., Ray, P., Brown, K., Barker, P. E., Jhanwar, S., Ruddle, F. H., and Besmer, P. (1988). Primary structure of c-*kit*: Relationship with the CSF-1/PDGF receptor kinase family—Oncogenic activation of v-*kit* involves deletion of extracellular domain and C terminus. *EMBO J.* **7**, 1003–1011.

Qureshi, S. A., Rim, M., Bruder, J., Kölch, W., Rapp, U., Sukhatme, V. P., and Foster, D. A. (1991). An inhibitory mutant of c-Raf-1 blocks v-Src-induced activation of the Egr-1 promoter. *J. Biol. Chem.* **266**, 20594–20597.

Rahm, M., Hultgårdh-Nilsson, A., Jiang, W.-Q., Sejersen, T., and Ringertz, N. R. (1990). Intracellular distribution of the c-Fos antigen during the cell cycle. *J. Cell Physiol.* **143**, 475–482.

Ransone, L. J., and Verma, I. M. (1990). Nuclear proto-oncogenes *fos* and *jun. Ann. Rev. Cell Biol.* **6**, 539–557.

Rao, V. N., Huebner, K., Isobe, M., Ar-Rushdi, A., Croce, C. M., Shyam, E., and Reddy, P. (1989). *elk,* tissue-specific *ets*-related genes on chromosomes X and 14 near translocation breakpoints. *Science* **244**, 66–70.

Rapp, U. R. (1991). Role of *Raf*-1 serine/threonine protein kinase in growth factor signal transduction. *Oncogene* **6**, 495–500.

Rapp, U. R., Goldsborough, M. D., Mark, G. E., Bonner, T. I., Groffen, J., Reynolds, F. H., and Stephenson, J. R. (1983). Structure and biological activity of v-*raf,* a unique oncogene transduced by a retrovirus. *Proc. Natl. Acad. Sci. U.S.A.* **80**, 4218–4222.

Rapp, U. R., Heidecker, G., Huleihel, M., Cleveland, J. L., Choi, W. C., Pawson, T., Ihle, J. N., and Anderson, W. B. (1988). *raf* family serine/threonine protein kinases in mitogen signal transduction. *Cold Spring Harbor Symp. Quant. Biol.* **LIII**, 173–184.

Rauscher, F. J., Cohen, D. R., Curran, T., Bos, T. J., Vogt, P. K., Bohmann, D., Tjian, R., and Franza, B. R. (1988). Fos-associated protein p39 is the product of the *jun* proto-oncogene. *Science* **240**, 1010–1016.

Robbins, P. D., Horowitz, J. M., and Mulligan, R. C. (1990). Negative regulation of human c-*fos* expression by the retinoblastoma gene product. *Nature (London)* **346**, 668–671.

Robertson, N. G., Pomponio, R. J., Mutter, G. L., and Morton, C. C. (1991). Testis-specific expression of the human MYCL2 gene. *Nucl. Acids Res.* **19**, 3129–3137.

Rogel, A., Popliker, M., Webb, C. G., and Oren, M. (1985). p53 cellular tumor antigen: Analysis of mRNA levels in normal adult tissues, embryos, and tumors. *Mol. Cell. Biol.* **5**, 2851–2855.

Ron, D., Tronick, S. R., Aaronson, S. A., and Eva, A. (1988). Molecular cloning and characterization of the human *dbl* proto-oncogene: Evidence that its overexpression is sufficient to transform NIH/3T3 cells. *EMBO J.* **8**, 2465–2473.

Rosnet, O., Marchetto, S., deLapeyriere, O., and Birnbaum, D. (1991). Murine *Flt3,* a gene encoding a novel tyrosine kinase receptor of the PDGFR/CSF1R family. *Oncogene* **6**, 1641–1650.

Rossi, P., Albanesi, C., Grimaldi, P., and Geremia, R. (1991). Expression of the mRNA for the ligand of C-kit in mouse Sertoli cells. *Biochem. Biophys. Res. Commun.* **176**, 910–914.

Ryseck, R. P., Hirai, S. I., Yaniv, M., and Bravo, R. (1988). Transcriptional activation of c-*jun* during the G_0/G_1 transition in mouse fibroblasts. *Nature (London)* **334**, 535–537.

Sagata, N., Daar, I., Oskarsson, M., Showalter, S. D., and Vande Woude, G. F. (1989a). The product of the *mos* proto-oncogene as a candidate "initiator" for oocyte maturation. *Science* **245**, 643–645.

Sagata, N., Watanabe, N., Vande Woude, G. F., and Ikawa, Y. (1989b). The c-*mos* proto-oncogene product is a cytostatic factor responsible for meiotic arrest in vertebrate eggs. *Nature (London)* **342**, 512–518.

Saksela, K., Mäkelä, T. P., and Alitalo, K. (1989). Oncogene expression in small-cell lung cancer cell lines and a testicular germ-cell tumor: Activation of the N-*myc* gene and decreased *RB* mRNA. *Int. J. Cancer* **44**, 182–185.

Schütte, J., Viallet, J., Nau, M., Segal, S., Fedorico, J., and Minna, J. (1989). *jun*-B inhibits and c-*fos* stimulates the transforming and trans-activating activities of c-*jun*. *Cell* **59**, 987–997.

Selten, G., Cuypers, H. T., and Berns, A. (1985). Proviral activation of the putative oncogene *Pim*-1 in MuLV induced T-cell lymphomas. *EMBO J.* **4**, 1793–1798.

Shackleford, G. M., and Varmus, H. E. (1987). Expression of the proto-oncogene *int*-1 is restricted to postmeiotic male germ cells and the neural tube of mid-gestational embryos. *Cell* **50**, 89–95.

Shalloway, D., and Shenoy, S. (1991). Oncoprotein kinases in mitosis. *Adv. Cancer Res.* **57**, 185–225.

Siegel, J. N., Klausner, R. D., Rapp, U. R., and Samelson, L. E. (1990). T cell antigen receptor engagement stimulates c-*raf* phosphorylation and induces c-*raf*-associated kinase activity via a protein kinase C-dependent pathway. *J. Biol. Chem.* **265**, 18472–18480.

Sithanandam, G., Kölch, W., Duh, F. M., and Rapp, U. R. (1990). Complete coding sequence of a human B-raf cDNA and detection of B-raf protein kinase with isozyme specific antibodies. *Oncogene* **5**, 1775–1780.

Smith, E. P., Hall, S. H., Monaco, L., French, F. S., Wilson, E. M., and Conti, M. (1989a). A rat Sertoli cell factor similar to basic fibroblast growth factor increases c-*fos* messenger ribonucleic acid in cultured Sertoli cells. *Mol. Endocrinol.* **3**, 954–961.

Smith, E. P., Hall, S. H., Monaco, L., French, F. S., Wilson, E. M., and Conti, M. (1989b). Regulation of c-*fos* messenger ribonucleic acid by fibroblast growth factor in cultured Sertoli cells. *Ann. N.Y. Acad. Sci.* **564**, 132–139.

Sordoillet, C., Chauvin, M. A., Revol, A., Morera, A. M., and Benahmed, M. (1988). Fibroblast growth factor is a regulator of testosterone secretion in cultured immature Leydig cells. *Mol. Cell. Endocrinol.* **58**, 283–286.

Sorrentino, V., McKinney, M. D., Giorgi, M., Geremia, R., and Fleissner, E. (1988). Expression of cellular proto-oncogenes in the mouse male germ line: A distinctive 2.4-kilobase *pim*-1 transcript is expressed in haploid post-meiotic cells. *Proc. Natl. Acad. Sci. U.S.A.* **85**, 2191–2195.

Sorrentino, V., Giorgi, M., Geremia, R., Besmer, P., and Pellegrino, R. (1991). Expression of the c-*kit* proto-oncogene in the murine male germ cells. *Oncogene* **6**, 149–151.

Spector, D. L., Watt, R. A., and Sullivan, N. F. (1987). The v- and c-*myc* oncogene proteins colocalize *in situ* with small nuclear ribonucleoprotein particles. *Oncogene* **1**, 5–12.

Stacey, D. W., Watson, T., Kung, H.-F., and Curran, T. (1987). Microinjection of transforming *ras* protein induces c-*fos* expression. *Mol. Cell. Biol.* **7**, 523–527.

Stewart, T. A., Bellvé, A. R., and Leder, P. (1984). Transcription and promoter usage of *myc* gene in normal somatic and spermatogenic cells. *Science* **226**, 707–710.

Storm, S. M., Cleveland, J. L., and Rapp, U. R. (1990). Expression of *raf* family proto-oncogenes in normal mouse tissues. *Oncogene* **5**, 345–351.

Storms, R. W., and Bose, H. R., Jr. (1989). Oncogenes, protooncogenes, and signal transduction: Toward a unified theory? *Adv. Virus Res.* **37**, 1–34.

Szekely, L., Uzvolgyi, E., Jiang, W.-Q., Durko, M., Wiman, K. G., Klein, G., and Sumegi, J. (1991). Subcellular localization of the retinoblastoma protein. *Cell Growth Diff.* **2**, 287–295.

Tagami, T., Nakamura, H., Sasaki, S., Mori, T., Yoshioka, H., Yoshida, H., and Imura, H. (1990). Immunohistochemical localization of nuclear 3,5,3'-triiodothyronine receptor proteins in rat tissues studied with antiserum against c-ERB A/T_3 receptor. *Endocrinol.* **127**, 1727–1734.

Tajima, Y., Onoue, H., Kitamura, Y., and Nishimune, Y. (1991a). Biologically active kit ligand growth factor is produced by mouse Sertoli cells and is defective in Sl^d mutant mice. *Development* **113**, 1031–1035.

Tajima, Y., Sakamaki, K., Watanabe, D., Koshimizu, U., Matsuzawa, T., and Nishimune, Y. (1991b). Steel-dickie (Sl^d) mutation affects both maintenance and differentiation of testicular germ cells in mice. *J. Reprod. Fertil.* **91**, 441–449.

Tres, L. L., Smith, E. P., Van Wyk, J. J., and Kierszenbaum, A. L. (1986). Immunoreactive sites and accumulation of somatomedin-C in rat Sertoli-spermatogenic cell co-culture. *Exp. Cell Res.* **162**, 33–50.

Troppmair, J., Bruder, J. T., App, H., Cai, H., Liptak, L., Szeberényi, J., Cooper, G. M., and Rapp, U. R. (1992). Ras controls coupling of growth factor receptors and protein kinase C in the membrane to Raf-1 and B-Raf protein serine kinases in the cytosol. *Oncogene* **7**, 1867–1874.

van der Hoorn, F. A., Spiegel, J. E., Maylié-Pfenninger, M.-F., and Nordeen, S. K. (1991). A 43 kD c-*mos* protein is only expressed before meiosis during rat spermatogenesis. *Oncogene* **6**, 929–932.

van der Hoorn, F. A. (1992). Identification of the testis c-*mos* promoter: Specific activity in seminiferous tubule-derived extract and binding of a testis-specific nuclear factor. *Oncogene* **7**, 1093–1099.

Van Etten, R. A., Jackson, P., and Baltimore, D. (1989). The mouse type IV c-*abl* gene product is a nuclear protein, and activation of transforming ability is associated with cytoplasmic localization. *Cell* **58**, 669–678.

Verhoeven, G., and Cailleau, J. (1986). Stimulatory effects of epidermal growth factor on steroidogenesis in Leydig cells. *Mol. Cell. Endocrinol.* **47**, 99–106.

Wadewitz, A. G., and Wolgemuth, D. J. (1991). Characteristics of the expression of the c-*abl* and c-*mos* proto-oncogenes in mouse germ cells. *In* "Growth Factors in Fertility Regulation" (F. Haseltine and J. Findlay, eds), pp. 35–49. Cambridge University Press, Cambridge.

Wadewitz, A. G., Winer, M. A., and Wolgemuth, D. J. (1993). Developmental and cell-lineage specificity of raf family gene expression in mouse testis. *Oncogene* **8**, 1055–1062.

Wang, J., and Baltimore, D. (1983). Cellular RNA homologous to the Abelson murine leukemia virus transforming gene: Expression and relationship to the viral sequence. *Mol. Cell. Biol.* **3**, 773–779.

Wang, J. Y. J., Ledley, F., Goff, S., Lee, R., Groner, Y., and Baltimore, D. (1984). The mouse c-*abl* locus: Molecular cloning and characterization. *Cell* **36**, 349–356.

Wasylyk, C., Wasylyk, B., Heidecker, G., Huleihel, M., and Rapp, U. R. (1989). Expression of raf oncogenes by the PEA1 transcription factor motif. *Mol. Cell. Biol.* **9**, 2247–2250.

Weinberg, R. A. (1991). Tumor suppressor genes. *Science* **254**, 1138–1146.

Weinberger, C., Thompson, C. C., Ong, E. S., Lebo, R., Gruol, D. J., and Evans, R. M. (1986). The c-*erb*-A gene encodes a thyroid hormone receptor. *Nature (London)* **324**, 641–646.

Williams, D. E., Eisenman, J., Baird, A., Rauch, C., Van Ness, K., March, C. J., Park, L. S., Martin, U., Mochizuki, D. Y., Boswell, H. S., Burgess, G. S., Cosman, D., and Lyman, S. D. (1990). Identification of a ligand for the c-*kit* proto-oncogene. *Cell* **63**, 167–174.

Williams, N. G., Roberts, T. M., and Li, P. (1992). Both $p21^{ras}$ and $pp60^{v-src}$ are required, but neither alone is sufficient, to activate the Raf-1 kinase. *Proc. Natl. Acad. Sci. U.S.A.* **89**, 2922–2926.

Willison, K., and Ashworth, A. (1987). Mammalian spermatogenic gene expression. *Trends Genet.* **3,** 351–355.

Wingett, D., Reeves, R., and Magnuson, N. S. (1992). Characterization of the testes-specific *pim*-1 transcript in rat. *Nucl. Acids Res.* **20,** 3183–3189.

Witte, O. N., Rosenberg, N. E., and Baltimore, D. (1979). A normal cell protein cross reactive to the major Abelson murine leukemia virus gene product. *Nature (London)* **281,** 396–398.

Wolfes, H., Kogawa, K., Millette, C. F., and Cooper, G. M. (1989). Specific expression of nuclear proto-oncogenes before entry into meiotic prophase of spermatogenesis. *Science* **245,** 740–743.

Wolgemuth, D. J., and Watrin, F. (1991). List of cloned mouse genes with unique expression patterns during spermatogenesis. *Mammalian Genome* **1,** 283–288.

Wood, K. W., Sarnecki, C., Roberts, T. M., and Blenis, J. (1992). *ras* mediates nerve growth factor receptor modulation of three signal-transducing protein kinases: MAP kinase, Raf-1, and RSK. *Cell* **68,** 1041–1050.

Yarden, Y., Kuang, W. J., Yang-Feng, T., Coussens, L., Munemitsu, S., Dull, T. J., Chen, E., Schlessinger, J., Francke, U., and Ullrich, A. (1987). Human proto-oncogene c-*kit*: A new cell surface receptor tyrosine kinase for an unidentified ligand. *EMBO J.* **6,** 3341–3351.

Yoshinaga, K., Nishikawa, S., Ogawa, M., Hayashi, S. I., Kunisada, T., Fujimoto, T., and Nishikawa, S. I. (1991). Role of c-*kit* in mouse spermatogenesis: Identification of spermatogonia as a specific site of c-*kit* expression and function. *Development* **113,** 689–699.

Zakeri, Z. F., Ponzetto, C., and Wolgemuth, D. J. (1988). Translational regulation of the novel haploid-specific transcripts for the c-*abl* proto-oncogene and a member of the 70 kDa heat-shock protein gene family in the male germ line. *Devel. Biol.* **125,** 417–422.

Zhao, X., Singh, B., and Batten, B. E. (1991). The role of c-*mos* proto-oncoprotein in mammalian meiotic maturation. *Oncogene* **6,** 43–49.

Zhao, Y.-H., Krueger, J. G., and Sudol, M. (1990). Expression of cellular-*yes* protein in mammalian tissues. *Oncogene* **5,** 1629–1635.

Zsebo, K. M., Williams, D. A., Geissler, E. N., Broudy, V. C., Martin, F. H., Atkins, H. L., Hsu, R.-Y., Birkett, N. C., Okino, K. H., Murdock, D. C., Jacobsen, F. W., Langley, K. E., Smith, K. A., Takeishi, T., Cattanach, B. M., Galli, S. J., and Suggs, S. V. (1990a). Stem cell factor is encoded at the *Sl* locus of the mouse and is the ligand for the c-*kit* tyrosine kinase receptor. *Cell* **63,** 213–224.

Zsebo, K. M., Wypych, J., McNiece, I. K., Lu, H. S., Smith, K. A., Karkare, S. B., Sachdev, R. K., Yuschenkoff, V. N., Birkett, N. C., Williams, L. R., Satyagal, V. N., Tung, W., Bosseiman, R. A., Mendiaz, E. A., and Langley, K. E. (1990b). Identification, purification, and biological characterization of hematopoietic stem cell factor from buffalo rat liver-conditioned medium. *Cell* **63,** 195–201.

6

Gene Expression during Spermatogenesis

E. M. EDDY, JEFFREY E. WELCH,
& DEBORAH A. O'BRIEN

I. Introduction

To understand gene expression during spermatogenesis fully, researchers must know which proteins are responsible for the unique structural and functional characteristics of spermatogenic cells and spermatozoa, must determine which genes encode these proteins and when they are transcribed, and must identify the intrinsic and extrinsic mechanisms that regulate the expression of these genes. The process of spermatogenesis is a contradiction because it produces a cell that is developmentally totipotent but morphologically and functionally highly differentiated. Although exciting progress has been made in learning about the regulation of gene expression during spermatogenesis, significant gaps remain in the knowledge required to understand the overall process.

Spermatogenesis is a developmental program that occurs in mitotic, meiotic, and postmeiotic phases. In the mitotic phase, spermatogonia proliferate to expand the quantity of germ cells; in the second phase, spermatocytes accomplish chromosomal synapsis and genetic recombination before undergoing two reductional divisions; and in the third phase, the haploid spermatids are remodeled dramatically to fashion male gametes.

This chapter focuses on recent findings on intrinsic mechanisms that appear to be involved in regulating spermatogenic cell gene expression by considering three aspects of gene expression during spermatogenesis. The first is the regulation of gene expression, with a focus on promoter and enhancer sequence elements 5' to the protein coding region of spermatogenic cell-specific genes, the transcription factors that bind to these sequences, and the role of the 5' and 3' untranslated regions of the transcript in regulating message stability and translation. The second considers studies with transgenic mice. These animals have been used to identify sequences in the flanking regions that regulate cell- and stage-specific expression of spermatogenic cell genes and to produce new models of male infertility caused by transgene expression or insertional mutation. The third aspect concerns the processes of genomic imprinting and gene methylation that occur during spermatogenesis and how they may be involved in gene regulation. We also review the chromosomal location of spermatogenic cell-specific genes to determine their distribution in the mouse genome. These areas of research on spermatogenesis are still emerging, so the full significance of many of the observations reported in this chapter remains to be determined.

Other intriguing aspects of gene expression in spermatogenic cells have been reviewed and will not be discussed in detail in this chapter. For example, an impressive and continuously growing list of genes is expressed uniquely in spermatogenic cells. Many of these genes have been found to be transcribed from the haploid genome. Other genes on this list are cognates of genes expressed in somatic cells whose protein products are quite similar, but whose gene regulatory sequences are quite different. In addition, alternative transcription start or splice sites and polyadenylation signals frequently are used to produce unique transcripts in spermatogenic cells. Finally, the expression of many genes is regulated developmentally in spermatogenic cells. For additional information on these topics, see the reviews by Willison and Ashworth (1987), Erickson (1990), Hecht (1990,1992), Wolgemuth and Watrin (1991), and Eddy *et al.* (1991).

II. Regulation of Gene Expression in Spermatogenic Cells

A. Transcription Factors

Synthesis of messenger RNA in eukaryotic cells requires the interaction of RNA polymerase II with additional nuclear proteins to initiate synthesis at a specific position in a gene. These nuclear proteins or transcription factors, in addition to RNA polymerase II, are necessary for basal levels of transcription and activation of tissue-specific gene expression to occur in a proper temporal context (Mermelstein *et al.*, 1989; Sawadogo and Sentenac, 1990; Wingender, 1990). The DNA-binding domains of transcription factors often have an organized structure that is predictable from the amino acid sequence (see reviews by Johnson and McKnight, 1989; Mitchell and Tjian, 1989; Wingender, 1990). Common motifs include zinc finger, homeobox, helix–

loop–helix, leucine zipper, and high mobility group (HMG) box sequences, all of which can be found in transcription factors present in spermatogenic cells (Table 1).

1. Zinc fingers

Zinc finger structures first were identified in the transcription factor TFIIIA, a protein that binds the *Xenopus* 5S ribosomal RNA promoter (Miller *et al.*, 1985). To form the finger, a pair of cysteine residues and a pair of histidine residues interact with a zinc ion. As a result, the amino acids located between the cysteine and histidine pairs are pulled into a loop or "finger." The transcription factor SP1 is a member of the zinc finger protein (ZFP) family. SP1 binds to the GC box, a GC-rich promoter element (Kadonaga *et al.*, 1987), and is believed to play an integral role in the transcription of genes lacking a TATA motif (Dynan, 1986; Pugh and Tjian, 1990). Limiting amounts of SP1 have been proposed as a regulatory mechanism in the expression of GC box-containing genes. However, SP1 concentrations do not appear to be limiting in spermatogenic cells and high levels of SP1 protein are detected in spermatids (Saffer *et al.*, 1991).

The human ZFY gene encodes a ZFP and originally was discovered while searching the Y chromosome for the testis determining factor (Page *et al.*, 1987; Mardon and Page, 1989; Koopman *et al.*, 1991). Although the single human ZFY gene and two mouse (*Zfy-1, Zfy-2*) genes have been demonstrated not to have a primary role in testis formation (Koopman *et al.*, 1989; Palmer *et al.*, 1989; Gubbay *et al.*, 1990), the protein structure is indicative of a transcription factor (Mardon *et al.*, 1990). ZFP genes related to ZFY have been detected on the X chromosome of the human *(ZFX)* (Schneider-Gädicke *et al.*, 1989) and the mouse *(Zfx)* (Mardon *et al.*, 1990)) and on an autosome of the mouse *(Zfa)* (Ashworth *et al.*, 1990). The autosomal *Zfa* gene appears to represent a functional retroposon derived from *Zfx*, a gene subject to X chromosome inactivation.

Expression of the human ZFY and ZFX genes and the mouse *Zfx* gene has been detected in a wide variety of fetal and adult tissues (Koopman *et al.*, 1991). In contrast, adult expression of the mouse *Zfy-1, Zfy-2,* and *Zfa* genes appears to be restricted to male germ cells (Nagamine *et al.*, 1990; Koopman *et al.*, 1991). First detected in primordial germ cells, *Zfy-1* mRNA is present throughout male germ cell development and appears to increase in amount with the onset of meiosis. Although some *Zfy-2* expression may occur in primordial germ cells, *Zfy-2* transcripts only become abundant in the testis between 7 and 14 days of age, again coincident with the onset of meiosis.

Zfy genes do not encode identical proteins. Whereas the ZFY-1 protein contains an acidic domain and 13 zinc fingers (Mardon and Page, 1989), the ZFY-2 protein contains an amino acid substitution disrupting the third zinc finger and the ZFA protein is missing a portion of the acidic domain (Ash-

TABLE 1 Transcription Factors in Germ Cells

Transcription factor	Reference	Transcription factor	Reference
Primordial germ cells		**Postmeiotic cells**	
Zinc finger		Zinc finger	
ZFY-1	Koopman *et al.* (1989,1991)	ZFY-1	Nagamine *et al.* (1990)
		ZFY-2	Nagamine *et al.* (1990)
Homeobox		ZFP-29	Denny and Ashworth (1991)
OCT-1	Schöler *et al.* (1989)		
OCT-4A	Rosner *et al.* (1990); Schöler *et al.* (1990); Schöler (1991)	TSGA	Höög *et al.* (1991)
		SP1	Saffer *et al.* (1991)
		RARα	Eskid *et al.* (1991)
Premeiotic cells		Homeobox	
Zinc finger		HOX 1.4	Rubin *et al.* (1986); Wolgemuth *et al.* (1987)
ZFY-1	Koopman *et al.* (1991)		
Leucine zipper			
jun B	Alcivar *et al.* (1990)	Leucine zipper	
c-*jun*	Wolfes *et al.* (1989); Alcivar *et al.* (1990)	*jun* D	Alcivar *et al.* (1991a)
		CREB	Waeber *et al.* (1991)
jun D	Alcivar *et al.* (1991a)	CREMτ (agonist form)	Foulkes *et al.* (1992)
c-*fos*	Wolfes *et al.* (1989)		
CREM (antagonist form)	Foulkes *et al.* (1992)	**Spermatogenic cells, unknown**	
		Zinc finger	
Helix–loop–helix		ZFA	Ashworth *et al.* (1990); Koopman *et al.* (1991)
c-*myc*	Wolfes *et al.* (1989)		
Meiotic cells		Homeobox	
Zinc finger		MiniOCT 2	Schöler *et al.* (1991)
ZFY-1	Nagamine *et al.* (1990)	HMG box	
ZFY-2	Nagamine *et al.* (1990); Koopman *et al.* (1991)	SRY	Koopman *et al.* (1990)
		Adult testis, unknown cell type	
ZFP-29	Denny and Ashworth (1991)	Zinc finger	
		MKR3	Chowdhury *et al.* (1988)
ZFP-35	Cunliffe *et al.* (1990a)		
REX-1	Rogers *et al.* (1991)	MKR4	Chowdhury *et al.* (1988)
TSGA	Höög *et al.* (1991)		
RARα	Kim and Griswold (1990)	MKR5	Chowdhury *et al.* (1988)
		MOK2	Ernoult-Lange *et al.* (1990)
Homeobox			
HOX 1.4	Rubin *et al.* (1986); Wolgemuth *et al.* (1987)	KROX-8	Chavrier *et al.* (1988a)
		KROX-20	Chavrier *et al.* (1988b)
		ZFP	Nelki *et al.* (1990)
Leucine zipper		c-*erb*A	Benbrook and Pfahl (1987)
jun D	Alcivar *et al.* (1991a)		
CREB	Ruppert *et al.* (1992); Waeber *et al.* (1991)	Homeobox	
		HOX 1.1	Coleberg-Poley *et al.* (1985)
CREMτ (agonist form)	Foulkes *et al.* (1992)	HOX 1.3	Odenwald *et al.* (1987)

(continues)

TABLE 1 *(continued)*

Transcription factor	Reference	Transcription factor	Reference
HOX 2.3	Meijlink *et al.* (1987)	OCT-8	Schöler *et al.* (1989)
HOX 2.6	Graham *et al.* (1988)	OCT-9	Schöler *et al.* (1989)
HOX 5.1	Featherstone *et al.* (1988)	OCT-11	Goldsborough *et al.* (1990)
HOX 6.1	Sharpe *et al.* (1988)	BRN-3	Goldsborough *et al.* (1990)
OCT-1	Schöler *et al.* (1989)		
OCT-2A	Hatzopoulos *et al.* (1990)	Helix–loop–helix MYCL2	Robertson *et al.* (1991)
OCT-2B	Hatzopoulos *et al.* (1990)	Other *ets*	Rao *et al.* (1989)
OCT-4A	Rosner *et al.* (1990)	Retinoblastoma protein	Bernards *et al.* (1989)
OCT-6	He *et al.* (1989); Suzuki *et al.* (1990)		

worth *et al.*, 1990; Koopman *et al.*, 1991). These mutations presumably alter DNA binding and gene activation functions. The mouse *Srx'* mutant strain has a deletion on the mouse Y chromosome and is missing portions of the *Zfy-1* and *Zfy-2* genes (Koopman *et al.*, 1991). Spermatogonial stem cells are present in *Srx'* mice, but proliferation of spermatogonia beyond early stages is blocked. The mutation is suggested to cause the deletion of a putative *Spy* "spermatogenesis gene" (Levy and Burgoyne, 1986; Sutcliffe and Burgoyne, 1989). The relationship between *Zfy* genes, *Spy*, and this mutation is not known, but these studies suggest that ZFPs have a role during early spermatogenesis.

Genes for numerous other mouse ZFPs have been cloned by hybridization with probes derived from either the *Drosophila Krüppel* gene or a *Zfp* consensus DNA sequence (Chavrier *et al.*, 1988a,b; Cunliffe *et al.*, 1990a; Ernoult-Lange *et al.*, 1990; Denny and Ashworth, 1991). Several of these *Zfp* genes are expressed in spermatogenic cells. Present in 14.5-day mouse embryos, *Zfp-29* mRNA is found only in the testis of adult mice (Denny and Ashworth, 1991). *Zfp-29* transcripts are found in both spermatocytes and spermatids; maximal expression occurs in round spermatids. Another *Zfp* gene, *Zfp-35*, originally was detected by screening a urogenital ridge cDNA library derived from 11.5-day fetal mice, but also is expressed highly in pachytene spermatocytes. Little *Zfp-35* mRNA is found in spermatids, suggesting that *Zfp-35* expression in the adult mouse is restricted to meiotic germ cells (Cunliffe *et al.*, 1990a,b). The *Rex-1* gene encodes a retinoic acid-regulated ZFP and is expressed in spermatocytes, F9 cells, and embryonic tissues (Rogers *et al.*, 1991). Other *Zfp* genes are expressed in the testis and include *mKr3, mKr4, mKr5* (Chowdhury *et al.*, 1988), *MOK2* (Ernoult-Lange *et al.*,

1990), *Krox-8* (Chavrier *et al.*, 1988a), *Krox-20* (Chavrier *et al.*, 1988b), and *TSGA* (Höög *et al.*, 1991). Additional *Zfp* mRNAs (Nelki *et al.*, 1990) have been detected in the testis, but cell-specific expression has not been determined.

Transcription factors that use the zinc finger motif for DNA-binding include members of the hormone receptor superfamily. This superfamily includes receptors for steroid hormones, T_3 thyroid hormone, vitamin D_3, and retinoic acid (for review, see Carson-Jurica *et al.*, 1990). Although expressed in reproductive tissues, most of these receptors have not been found in spermatogenic cells. The androgen receptor (Anthony *et al.*, 1989; Sar *et al.*, 1990) and the estrogen receptor (West and Brenner, 1990) are undetectable in spermatogenic cells. Glucocorticoid receptor mRNA is present in rat round spermatids and the protein has been detected in rat epididymal sperm (Kaufmann *et al.*, 1992). However, the functional glucocorticoid receptor concentration in seminiferous tubules decreases with the appearance and differentiation of germ cells (Levy *et al.*, 1989). No high-affinity binding of glucocorticoids was detected in epididymal sperm (Kaufman *et al.*, 1992).

Expression of the c-*erb*A gene, which encodes a high-affinity receptor for T_3 thyroid hormone (Sap *et al.*, 1986; Weinberger *et al.*, 1986), has been detected in the human testis (Benbrook and Pfahl, 1987). However, receptor assays on adult rat testis have detected little or no T_3 binding (Oppenheimer *et al.*, 1974; Jannini *et al.*, 1990) and the c-*erb*A protein from testis (c-*erb*A-α-2) fails to bind T_3 (Lazar *et al.*, 1988,1989; Schueler *et al.*, 1990). Although it does not function as a receptor, c-*erb*A-α-2 still can bind to the T_3 response element DNA sequence and may serve as a constitutive repressor of T_3 responsive genes (Lazar *et al.*, 1989; Koenig *et al.*, 1989). Which cell types in the testis contain T_3 hormone receptors or c-*erb*A has not been reported.

The retinoic acid receptors (RAR) are additional members of the hormone receptor superfamily (Petkovich *et al.*, 1987). The mRNAs for *Rara* and *Rarb* are found in Sertoli cells, whereas only *Rara* mRNA has been detected in spermatogenic cells (Kim and Griswold, 1990; Eskild *et al.*, 1991). In rats and mice, retinol deprivation has a dramatic effect on the testis and blocks the proliferation of type A spermatogonia as well as entry into meiotic prophase (Griswold *et al.*, 1989; Ismail *et al.*, 1990; van Pelt and de Rooij, 1990a,b,1991). This block can be released by the administration of retinol (Siiteri *et al.*, 1992) or retinoic acid (van Pelt and de Rooij, 1991). Retinol has been used to induce synchronized spermatogenesis (Morales and Griswold, 1987). Whether retinol, retinoic acid, or their metabolites release this block is unclear (Griswold *et al.*, 1989), but retinoid treatment does cause a rapid induction of *Rara* mRNA (Kim and Griswold, 1990; van Pelt *et al.*, 1992).

An increase in *Rara* mRNA levels also is found in synchronized rat testes during stages VII–IX of spermatogenesis, suggesting that the activation of RAR-responsive genes is a prerequisite for both proliferation of type A spermatogonia and entry into meiosis (Kim and Griswold, 1990). Autoinduction of *Rarb* expression has been reported to occur in other cell types through an upstream retinoic acid response element (de Thé *et al.*, 1990; Sucov *et al.*, 1990). The induction of *Rara* in spermatogenic cells could occur by a similar mechanism, whereby autoregulation of *Rara* levels precedes retinoid-induced gene expression and entry into mitosis or meiosis (Kim and Griswold, 1990).

2. Homeoboxes

The homeobox domain first was recognized as a highly conserved region present in *Drosophila* genes that regulate axial development, but now has been identified in organisms ranging from nematodes to humans (Levine and Hoey, 1988). Although transcripts for *Hox 1.1*, *Hox 1.2* (Coleberg-Poley *et al.*, 1985), *Hox 1.3* (Odenwald *et al.*, 1987), *Hox 2.3* (Meijlink *et al.*, 1987), *Hox 2.6* (Graham *et al.*, 1988), *Hox 5.1* (Featherstone *et al.*, 1988), and *Hox 6.1* (Sharpe *et al.*, 1988) are present in the testis, only the *Hox 1.4* gene is known to be expressed in spermatogenic cells. Formerly designated *HBT-1* (Wolgemuth *et al.*, 1986), *Hox-1-3* (Duboule *et al.*, 1986), and *MH-3* (Rubin *et al.*, 1986), *Hox 1.4* transcripts are detected first in fetal mice during development of the central nervous system and in mesodermal structures. In adult mice, *Hox 1.4* expression appears to be restricted to the testis, where transcripts are found in spermatocytes, round spermatids, and elongating spermatids (Rubin *et al.*, 1986; Wolgemuth *et al.*, 1987). The *Hox 1.4* mRNA transcribed in spermatogenic cells is approximately 200 bases shorter than the transcript produced in fetal tissues (1.45 kb). The spermatogenic cell mRNA still contains the 180-base region encoding the homeobox (Galliot *et al.*, 1989), but preliminary reports suggest that the transcripts in fetal tissues and germ cells may encode different N-terminal sequences (Wolgemuth *et al.*, 1991).

The *Hox 1.4* gene is a member of the HOX 1 cluster located on mouse chromosome 6 (Bucan *et al.*, 1986). The DNA region between *Hox 1.3* and *Hox 1.4* has been shown to direct transcription in the testis and gut of transgenic mice (Wolgemuth *et al.*, 1989). Two putative transcription factors present in the testis may be involved in regulating *Hox 1.4* expression. First, the zinc finger protein KROX-20 recognizes two *cis*-elements in the 5′ promoter region of the *Hox 1.4* gene (Chavrier *et al.*, 1990). Although *Krox-20* and *Hox 1.4* expression do not overlap during fetal development, *Krox-20* is expressed in the testis (Chavrier *et al.*, 1988b) and expression of this ZFP may represent part of a gene activation cascade. Second, transfection studies of the *Hox 1.4* promoter region have shown that a 2.75-kb fragment containing 2.0-kb of sequence 5′ to the *Hox 1.4* gene can confer retinoic acid responsiveness on a β-globin reporter gene. The expression of *Rara* transcripts in germ cells (Kim and Griswold, 1990; Eskild *et al.*, 1991) and the responsiveness of *Hox 1.4* expression to retinoic acid suggests the possible involvement of retinoic acid receptors in *Hox 1.4* gene expression during spermatogenesis.

A second group of closely related homeobox-containing genes encodes the POU domain family of transcription factors (Herr *et al.*, 1988; Sturm *et al.*, 1988). Included in the POU domain family are the octamer-binding (OCT) proteins. The OCT proteins recognize an 8-base *cis* element (ATGCAAAT; the octamer motif) and are involved in both ubiquitous and tissue-specific expression (Schöler, 1991). Evidence suggests that at least eight members of the *Oct* gene family are expressed in the mouse testis. *Oct-1*, a ubiquitously expressed gene, has been found in testis and spermatozoa (Schöler *et al.*, 1989). Expression of the *Oct-2* gene in testis produces transcripts encoding OCT-2A, OCT-2B (Hatzopoulos *et al.*, 1990), and miniOCT-2 (Stoykova *et al.*, 1992) proteins. Present in spermatogenic cells, miniOCT-2 is a truncated version of OCT-2, which may function to supress gene activation by other OCT transcription factors (Stoykova *et al.*, 1992). OCT-2 protein also has been detected in sperm (Schöler *et al.*, 1989). Like *Hox 1.4*, some *Oct* genes may be expressed in both embryonic and adult tissues. The *Oct-4* gene, located on mouse chromosome 17, is expressed in primordial germ cells and in oocytes of the adult female (Rosner *et al.*, 1990; Schöler *et al.*, 1990). Evidence also exists for the expression of *Oct-4* mRNA in the testis (Rosner *et al.*, 1990). The *Oct-6* gene (referred to as *tst-1* or *SCIP* in the rat) is expressed in primordial germ cells and adult testis (He *et al.*, 1989; Suzuki *et al.*, 1990). The discovery of two additional octamer-binding proteins in sperm—OCT-8 and OCT-9 (Schöler *et al.*, 1989)—and the cloning of cDNAs for *Oct-11* and *Brn-3* from the testis (Goldsborough *et al.*, 1990) indicate that at least four additional *Oct* genes are expressed in the testis.

3. Leucine zipper

The leucine zipper motif was identified originally using protein sequence comparisons of the oncogenes *fos*, *jun*, and *myc* (see Chapter 5) and the yeast regulatory protein GCN4 (Landschulz *et al.*, 1988). Consisting of leucine residues arranged in a coiled-coil style heptad repeat within an α-helical domain, the leucine zipper permits the protein–protein associations required to form the functional DNA binding domain.

The AP-1 transcription factor functions as a heterodimer formed by products of *jun* and *fos* genes (Curran and Franza, 1988). Several transcription factors using the leucine zipper domain are expressed in the testis. The mRNAs for c-*jun*, *jun*B, and *jun*D are present in mouse spermatogenic cells (Wolfes *et al.*, 1989; Alcivar *et al.*, 1990,1991a). Although initially high levels of c-*jun* and *jun*B mRNA appeared to be present constitutively in isolated spermatogenic cells (Wolfes *et al.*, 1989), elevated expression of these genes was found to be induced by the cell isolation procedure. Only low amounts of c-*jun* and *jun*B mRNA were found in the intact testis (Alcivar *et al.*, 1990). However, *jun*D is expressed in all spermatogenic cells and is not affected by cell dissociation. *Jun*D mRNA was most prevalent in meiotic and postmeiotic cells, whereas uninduced *jun*B and c-*jun* transcripts were at their highest levels in type B spermatogonia (Alcivar *et al.*, 1991a). Just as the levels of *jun* mRNAs differ in spermatogenic cells, so does their translational activity (see Section III,C).

Coordinate expression of c-*fos* and *jun* is required for the formation of an active AP-1 transcription factor complex. Expression of c-*fos* was reported earlier to occur primarily in the premeiotic phase of spermatogenesis (Wolfes *et al.*, 1989). However, more recent studies have demonstrated that c-*fos* expression, like c-*jun* and *jun*B, also can be induced in meiotic and postmeiotic germ cells by cell isolation procedures (Tsuruta and O'Brien, 1992). These studies show that c-*fos* mRNA levels return to baseline after isolated cells are cultured for 5 hr, and can be restimulated by treatment with fetal bovine serum or phorbol esters. Although the association of c-*fos* mRNA with polysomes has not been investigated in spermatogenic cells, the expression of c-*fos* correlates well with the appearance of *jun* transcripts during spermatogenesis.

Other leucine zipper proteins are involved in modulating the cyclic AMP response. During this response, increases in the level of cAMP activate the catalytic subunit of cAMP-dependent protein kinase A. This kinase then phosphorylates transcription factors capable of binding to the cAMP responsive element (CRE) and activating transcription. Through this phosphorylation event, the cAMP responsive element binding (CREB) protein becomes a transcriptional activator (Gonzalez and Montminy, 1989; Gonzalez *et al.*, 1989). Although at least four isoforms of CREB are expressed in spermatogenic cells (Ruppert *et al.*, 1992), the ability of CREB to modulate the cAMP response in these cells has not been shown. In spermatogenic cells, alternative splicing events during mRNA processing produce two CREB isoforms that lack the leucine zipper domain (Ruppert *et al.*, 1992) and the nuclear translocation signal (Waeber *et al.*, 1991). CREB isoforms missing these essential domains are unlikely to function as transcriptional activators. Dimerization of functional and nonfunctional CREB isoforms also may modulate CREB protein action by sequestration of functional transcription factors into inactive complexes.

A closely related transcription factor, the cyclic AMP responsive element modulator (CREM) (Foulkes *et al.*, 1992), originally was described as an antagonist to the cAMP transcriptional response. CREM is expressed in premeiotic germ cells in the antagonist form but, once the cells have entered the pachytene spermatocyte stage, an agonist isoform of CREM (CREMτ) is produced exclusively (Foulkes *et al.*, 1992). CREMτ is likely to function as a transcriptional activator since it has leucine zipper and nuclear localization signal domains, as well as two glutamine-rich transcriptional activation domains acquired by alternative splicing.

The appearance of CREMτ in pachytene spermatocytes coincides with the loss of CREB, suggesting that the cAMP transcriptional response in now modulated by CREMτ. The presence of male germ cell-specific proteins and/or mRNAs of adenylate cyclase (Braun, 1974; Adamo *et al.*, 1980; Gordeladze and Hansson, 1981a,b), cAMP phosphodiesterase (Welch *et al.*, 1992), and the catalytic and regulatory subunits of protein kinase A (Oyen *et al.*, 1990; Lönnerberg *et al.*, 1992) is consistent with the presence of a unique cAMP transcriptional response pathway capable of activating spermatogenic cell-specific genes.

4. Basic helix – loop – helix

The helix – loop – helix domain first was identified by sequence homology between the κ-E2 immunoglobulin enhancer, the *Drosophila daughterless* gene, *MyoD*, and *myc* (Murre *et al.*, 1989). One member of the helix – loop – helix family that is found in spermatogenic cells is c-*myc* (Stewart *et al.*, 1984; Koji *et al.*, 1988; Wolfes *et al.*, 1989). c-*myc* contains both a helix – loop – helix domain and the leucine zipper motif discussed earlier. c-*myc* expression is restricted to spermatogonia, in which the mRNA and the protein have been detected (Koji *et al.*, 1988; Wolfes *et al.*, 1989). Transcripts for another member of the *myc* family, *MYCL2*, also have been detected in adult human testis, but the cell type containing them is unknown (Robertson *et al.*, 1991).

5. High mobility group box

The HMG proteins are a family of nonhistone chromosomal proteins found in all eukaryotes (reviewed by Einck and Bustin, 1985). HMG proteins are associated with transcriptionally active genes and probably serve to hold DNA in a conformation suitable for transcription factor binding and formation of the transcription complex. Although some HMG proteins bind to AT-rich sequences (Wright and Dixon, 1988) and may be involved in facilitating transcription by RNA polymerase II (Singh and Dixon, 1990), the HMG proteins themselves do not appear to be sequence-specific transcription factors (Landsman and Bustin, 1991).

Several transcription factors have been discovered that have a DNA binding motif that is also present in the HMG proteins. Designated the HMG box, this sequence is present in the eukaryotic upstream binding factor (UBF) (Jantzen *et al.*, 1990), in mitochondrial transcription factor 1 (Parisi and Clayton, 1991), and in the insulin response element binding protein (Nasrin *et al.*, 1991). Of particular interest here is the *Sry* gene product, which contains an HMG box capable of sequence-specific DNA binding (Sinclair *et al.*, 1990; Harley *et al.*, 1992). The *Sry* gene is believed to encode the testis determining factor (TDF) that is active during embryonic development. In the adult, *Sry* expression has been detected only in spermatogenic cells (Koopman *et al.*, 1990). Genes activated by Sry have not been identified, although *cis* elements capable of binding the Sry HMG box have been identified upstream of both the mouse *Sry* and the human *SRY* genes (Harley *et al.*, 1992), suggesting an autoregulatory function for these genes.

6. Other factors

In addition to the major transcription factors, other less well-character-ized transcription factors are present in the testis. The CCAAT box-binding proteins represent a large family of proteins binding to the CCAAT *cis* element motif (Santoro *et al.*, 1988). A testis-specific CCAAT-binding protein capable of interacting with the sperm histone 2B promoter has been found in sea urchin testis (Barberis *et al.*, 1987). CCAAT boxes are present in histone genes expressed specifically in rat spermatogenic cells (Kim *et al.*, 1987; Huh *et al.*, 1991; Grimes *et al.*, 1992b). Other potential transcription factors present in testis or spermatogenic cells include members of the *ets* proto-on-cogene family (Rao *et al.*, 1989; Goto *et al.*, 1991) and c-*cbl* (Langdon *et al.*, 1989). Expression of the retinoblastoma gene in testis (Bernards *et al.*, 1989) also may regulate transcription since the retinoblastoma protein has been shown to bind the transcription factor E2F and prevent activation of target genes involved in cell proliferation (Hiebert *et al.*, 1992).

B. Spermatogenic Cell-Specific Promoter Regions

In addition to the transcription factors first identified in somatic cells and now known to be expressed in spermatogenic cells (Table 2), several studies suggest that unique transcription factors and *cis*-elements are involved in regulating spermatogenic cell-specific gene expression. These studies used DNA footprinting, gel retardation, and *in vitro* transcription assays as well as transgenic mice to dissect the promoter regions of genes and to identify *cis* elements and associated transcription factors that do not correspond to any described previously in somatic cells. The elements include both positive regulatory factors present only in the testis and negative factors thought to be involved in suppressing inappropriate transcription in both somatic and spermatogenic cells.

Comparison of the promoter regions from the mouse protamine 1 *(Prm-1)* and protamine 2 *(Prm-2)* genes expressed during the postmeiotic phase iden-tified several regions of homologous sequence (elements D, E, and B) that are potentially involved in their regulation (Johnson *et al.*, 1988). Progressive deletion of distal 5′ sequence elements of *Prm-2* including the D element (Bunick *et al.*, 1990) increased the level of transcription in heterologous nuclear extracts (i.e., HeLa cells and mouse brain). This result suggested that the upstream region contained negative regulatory elements for *Prm-2* ex-pression in somatic cells. The deletions did not affect *in vitro* expression of the *Prm-2* gene in testis nuclear extracts until removal of the E and B

TABLE 2 Spermatogenic Cell-Specific Gene *cis*-Elements

Gene	Source	Domain	Regu-lation	Sequence	Reference
Prm-1	Mouse	Tet-1	+	TGACTTCATAA	Tamura *et al.* (1992)
Prm-2	Mouse	B	+	TGGGCAGG	Bunick *et al.* (1990)
Prm-2	Mouse	E	+	CCTCACAGA	Bunick *et al.* (1990)
Prm-2	Mouse	D	−	TGTCACAGTCAAGAAGTGA	Bunick *et al.* (1990)
PGK-2	Human	E1/E2	+	CCTAGCAACACAGTTGAC	Gebara and McCarrey (1992)
PGK-2	Human	E4	−	CAGATCGAGATTGACAG	Gebara and McCarrey (1992)
RT7	Rat	T	+	GGTCACAGAACACAAGAA	Van Der Hoorn and Tarnasky (1992)
RT7	Rat	D1	+	GGCCTTAGGG	Van Der Hoorn and Tarnasky (1992)
c-mos	Rat	D1	+	GGACTTAGGA	Van Der Hoorn (1992)
TH2B	Rat	E′	−	CTCTCTTCTTGGCCTTCAAG	Lim and Chae (1992)
H1t	Rat	ND[a]	ND	GAGGCGCCTAGGGATGCA	Grimes *et al.* (1992a,b)

[a] ND, Not determined.

sequence elements, located within the first 150 bp upstream of the transcription initiation site (Bunick *et al.*, 1990).

DNase protection assays were used to demonstrate binding of a potential transcription factor, Tet-1, to the proximal promoter region of *Prm-1*, approximately 20 bp downstream of the E element (Tamura *et al.*, 1992). Although Tet-1 binding to the promoter region of rat and mouse *Prm-1* genes was demonstrated, this factor only weakly activated transcription in testis nuclear extracts. Tet-1 therefore may enhance transcription of *Prm-1* once the gene is activated by other factors.

Like the protamines, the rat *RT7* gene is expressed only during the haploid phase of spermatogenesis (van der Hoorn *et al.*, 1990). Analysis of the *RT7* promoter region demonstrated specific DNA–protein interactions adjacent to activating transcription factor (ATF)/CREB transcription factor binding sites (T region) and near a sequence element similar to the mouse *Prm-1* and *Prm-2* D element (van der Hoorn and Tarnasky, 1992). The T region of the *RT7* promoter has been shown to bind a major 39-kDa protein present in seminiferous tubule (ST) nuclear extracts. However, *in vitro* transcription analysis suggests that the T region serves as a nonspecific positive regulator of *RT7* promoter activity present in both somatic and ST nuclear extracts. Deletion of the D element region severely reduced *RT7* promoter activity in ST extracts, but not in liver extracts, a result that is in apparent contrast to that for the *Prm-2* promoter, for which the D element may serve as a negative regulator of expression in somatic cells (Bunick *et al.*, 1990). ST extracts protected a specific *RT7* site (D_1) immediately upstream of the D element, whereas liver and total testis extracts protected an area immediately downstream (D_2). An ST extract-specific factor, TTF-D, has been found to bind the D_1 upstream element, possibly serving as a seminiferous tubule-specific enhancer of transcription. Similarly, binding of a second factor to the downstream element may block transcription in somatic cells (van der Hoorn and Tarnasky, 1992). The failure to detect binding to the upstream element in total testis extracts may be the result of binding of testicular somatic factors to adjacent sequence elements, resulting in the displacement of the TTF-D factor (van der Hoorn and Tarnasky, 1992).

The spermatogenic cell-specific promoter controlling expression of the c-*mos* gene has been shown to contain potential binding sites for AP-1, CCAAT box-binding transcription factors, and helix–loop–helix transcription factors (van der Hoorn, 1992). This promoter also contains a sequence 80% homologous with the D_1 promoter element of *RT7*. Binding of a testis-specific factor to the c-*mos* D_1 element can be competed for by the D_1 element of *RT7*, suggesting that both genes can interact with the testis-specific DNA binding protein TTF-D (van der Hoorn, 1992).

The use of transgenic mice (see Section III,A) has allowed elements specifying spermatogenic cell-specific expression of the human phosphoglycerate kinase-2 gene *(PGK-2)* to be mapped to a 327-bp enhancer region located upstream of the 180-bp core promoter region (Robinson *et al.*, 1989). Although the TATA-less core promoter contains CCAAT-binding transcription factor (CTF) and SP1 binding sites and permits general expression of *PGK-2*, the enhancer region is required for spermatogenic cell-specific expression. The E1 and E2 sequences, located within a 17-bp segment of the enhancer region, bind proteins present only in the adult testis and may serve as positive enhancer elements (Gebara and McCarrey, 1992). Sequence element E4 in the enhancer region binds a protein present only in somatic cells and may be a silencer of *PGK-2* expression in somatic cells.

The comparison of spermatogenic cell-specific histone genes with their somatic counterparts has shown many promoter elements to be conserved between the two groups. Unlike somatic cells, spermatogenic cells express these cognate genes in a replication-independent manner, predominantly during meiotic prophase (Meistrich *et al.*, 1985). Nevertheless, the rat germ cell-specific *H1t* gene is expressed in pachytene spermatocytes and round spermatids and shows conservation of SP1, H1/CCAAT, and TATA elements when compared with somatic *H1* gene promoter sequences (Grimes *et al.*, 1992a,b). Transfection of *H1t* promoter constructs into somatic cells even results in gene replication-dependent transcription of the reporter gene (Kremer and Kistler, 1992). A testis-specific *cis* element has been identified in *H1t* between the SP1 and CCAAT elements (Grimes *et al.*, 1992a). This

18-bp sequence is recognized by a binding factor specific to primary spermatocytes and early spermatids (Grimes *et al.*, 1992b). *H1t* mRNA is detectable by Northern blotting only in primary spermatocytes, but preliminary evidence obtained from run-off transcription experiments suggests that *H1t* transcription occurs in early spermatids. However, instability of *H1t* mRNA in haploid germ cells may prevent accumulation to levels detectable by Northern analysis (Grimes *et al.*, 1992b). Closely linked to the *H1t* gene, the histone *H4t* gene is expressed in pachytene spermatocytes and in the liver (Wolfe and Grimes, 1991). Although germ cell-specific protein–DNA interactions have been demonstrated in the *H4t* promoter region, the *cis* elements involved in germ cell expression have not been identified.

Analysis of the germ cell-specific *TH2A* (Huh *et al.*, 1991) and *TH2B* (Kim *et al.*, 1987) histone genes has shown that a high percentage of homology also exists between the promoter regions of these genes and their somatic counterparts. CCAAT and OCT transcription factor binding sites are present in promoter sequences of both somatic and testis-specific genes. In addition, the *TH2A* and *TH2B* genes share a functional S phase-specific regulatory element (Huh *et al.*, 1991). Since the expression of these histones occurs primarily in spermatocytes independent of DNA replication, the stringent conservation of testis and somatic promoter elements is puzzling. The *TH2B* gene in somatic cells is suggested to be held in a transcriptionally inactive state by DNA methylation (Choi and Chae, 1991) (see Section IV,D). Germ cells maintain the *TH2B* gene in a hypomethylated state throughout spermatogenesis, which might allow the gene to be transcribed, in contrast to the expression of *TH2B*, which occurs only in spermatocytes. Lim and Chae (1992) demonstrated the presence of a repressor protein that binds to the *TH2B* promoter and prevents transcription. This repressor binds to a *cis* element (designated E′ to distinguish it from the protamine E element) located between the TATA box and the CAP site where it presumably blocks initiation of transcription. The absence of *TH2B* expression in premeiotic germ cells may be the result of the action of this repressor during early spermatogenesis (Lim and Chae, 1992).

A complex system of gene regulation involving numerous transcription factors and *cis* elements is responsible for controlling gene expression in spermatogenic cells. Transcription factors representing most known groups have been shown to be expressed in the testis. Positive and negative factors specific to spermatogenic cells also have been shown to be involved in regulating transcription during spermatogenesis, possibly by competing for binding sites. Negative regulation of ubiquitous transcription factors apparently can involve binding to *cis* elements by transcription factor isoforms that are incapable of gene activation (e.g., c-*erb*A, miniOCT-2). Although many transcription factors have been identified in the testis, most have not been localized specifically to spermatogenic cells. For the majority, the mRNA but not the protein has been shown to be present. Functional studies must examine those transcription factors that are predicted to have unusual activity (i.e., c-*erb*A, miniOCT-2, CREB, CREM).

C. Translational Regulation

The expression of many eukaryotic genes is regulated at the level of translation. Sequences in both the 5' and 3' untranslated regions (UTRs) of mRNAs have been shown to modulate translational efficiency; both positive and negative control mechanisms have been defined. In addition, processes that alter the stability of a specific transcript clearly can affect how much of the protein product is synthesized.

The rate-limiting initiation step of translation involves formation of a specific complex between multiple proteins (eIF–4F) and the 5' cap structure of mRNA. Therefore, factors that affect the structure or accessibility of the cap influence translation. Phosphorylation of specific eIF–4F constituents in response to stimuli such as growth factors have been shown to activate protein synthesis, whereas phosphorylation of other initiation and elongation factors inhibits translation (reviewed by Hershey, 1991; Thach, 1992). In addition, interactions between proteins and specific 5' UTR sequences can block translation. Ferritin synthesis, for example, is regulated by binding of a repressor protein to the 5' iron-responsive element, which must be close to the cap to inhibit translation (Goossen *et al.*, 1990).

Sequences in the 3' UTR also have been implicated in the regulation of protein synthesis. The translational repression of several mRNAs is mediated by 3' UTR sequences, although mechanisms for how 3' repressors modulate initiation events at the 5' end of the transcript have not been defined (Thach, 1992). The poly(A) tail complexed with poly(A)-binding protein (PABP) increases mRNA stability *in vitro* (Bernstein *et al.*, 1989; reviewed by Bernstein and Ross, 1989). In addition, the PABP–poly(A) tail complex can increase translational efficiency (Gallie, 1991). Genetic studies in yeast suggest that PABP may be required for translational initiation (Sachs and Davis, 1989; reviewed by Jackson and Standart, 1990).

Several studies have shown that cytoplasmic elongation of the poly(A) tail is necessary and sufficient to stimulate the translation of specific transcripts during meiotic maturation in both *Xenopus* and mouse oocytes (Vassalli *et al.*, 1989; reviewed by Richter, 1991). AU-rich sequences in these transcripts have been identified as cytoplasmic polyadenylation elements (CPEs). These same sequences appear to be responsible for the reversible deadenylation and translational repression of these transcripts during earlier stages of oocyte growth (Huarte *et al.*, 1992). In addition, a shorter AU-rich sequence (AUUUA) has been identified as a destabilizer motif that is present in the 3' UTR of several lymphokine and proto-oncogene transcripts and may regulate the selective degradation of these transiently expressed mRNAs (Shaw and Kamen, 1986).

During spermatogenesis, the need for translational regulatory mechanisms is evident during the postmeiotic phase in elongating spermatids, in which protein synthesis continues for several days following nuclear condensation and the cessation of transcription. Translational regulation also occurs during the meiotic phase and may be particularly important in the switch from somatic to germ cell-specific isoforms of multiple proteins. In several cases, alterations in transcript size, which are common during spermatogenesis, have been correlated with changes in translation.

1. Genes transcribed during the meiotic phase

Cytochrome c_T is a mitochondrial electron transport protein of spermatogenic cells. This protein is a cognate of cytochrome c_S in somatic cells, but is encoded by a unique gene that is expressed during the meiotic and postmeiotic phases of spermatogenesis (Goldberg *et al.*, 1977; Virbasius and Scarpulla, 1988). Heterogenous cytochrome c_T transcripts with poly(A) tails 50–200 nucleotides in length were detected in spermatogenic cells isolated from prepubertal and adult mice (Hake *et al.*, 1990). To determine which transcripts are associated with polysomes and apparently are translated, RNAs were isolated from cell fractions separated by density gradient centrifugation. Northern analysis of RNA from these gradients indicated that the longer transcripts in pachytene spermatocytes (0.7–0.9 kb) are associated with polysomes whereas the shorter transcripts (0.6–0.75 kb) are present predominantly in nonpolysomal fractions. Transcripts in round spermatids are less heterogeneous and shorter (0.7 kb) and are present primarily in nonpolysomal fractions. The cytochrome c_T mRNA contains two sequences near the poly(A) addition signal that have homology to the AU-rich CPE identified in several oocyte transcripts (Hake *et al.*, 1990; Richter, 1991). As occurs during oocyte maturation, cytoplasmic elongation of the poly(A) tail could activate translation of cytochrome c_T in pachytene spermatocytes.

An increase in poly(A) tail length also is correlated with translation of lactate dehydrogenase (LDH) C, another isozyme found only in spermatogenic cells. Two mouse *Ldh-c* transcripts, a 1.2-kb mRNA predominant in pachytene spermatocytes and a 1.3-kb transcript abundant in round spermatids, differ only in the lengths of their poly(A) tracts (Fujimoto *et al.*, 1988). Polysome gradients prepared from isolated spermatogenic cells suggested that the longer 1.3-kb transcripts present in both pachytene spermatocytes and round spermatids are associated preferentially with polysomes (Alcivar *et al.*, 1991b). Like cytochrome c_T, the *Ldh-c* mRNA contains an AU-rich sequence near its poly(A) addition signal that may serve as a CPE. Additional studies are needed to determine whether these longer transcripts result from new transcription or from cytoplasmic polyadenylation of existing mRNA and whether these variations in poly(A) tail length regulate translation.

Sequences 5′ to the coding region also may regulate polyadenylation and translational recruitment. During spermatogenesis in *Drosophila melanogas-*

ter, transcription ceases before the meiotic divisions. *Mst87F* and related germ cell-specific genes in *Drosophila* contain a conserved sequence element in the 5' UTR that is required for delayed translation (Schafer *et al.*, 1990). Further, this sequence element mediates elongation of the poly(A) tail that is coincident with translation of the stored mRNAs exclusively during the postmeiotic phase. To date, similar 5' sequence elements have not been identified in genes expressed during mammalian spermatogenesis.

The translational activation of germ cell-specific transcripts is not always correlated with longer poly(A) tails. PGK-2, a glycolytic isozyme restricted to spermatogenic cells, is transcribed first in pachytene spermatocytes but is not associated with polysomes during the meiotic phase of spermatogenesis in the mouse (Gold *et al.*, 1983). After meiosis, *Pgk-2* mRNA levels increase and translation is initiated without a change in transcript size (Erickson *et al.*, 1980; Gold *et al.*, 1983).

Similarly, no change in transcript length accompanies the translation of *jun* mRNAs. *Jun*D mRNA is found associated mostly with polysomes in meiotic cells, whereas *jun*B mRNA is present in both polysomal and nonpolysomal fractions (Alcivar *et al.*, 1991a). Although found at low levels in spermatogenic cells, most c-*jun* mRNA is present on polysomes (Alcivar *et al.*, 1991a). These findings suggest that *jun* proteins are synthesized in spermatogenic cells and that efficient translation of c-*jun* compensates for the low amount of mRNA.

Several transcripts present in the meiotic and postmeiotic phases of spermatogenesis are not translated efficiently. For example, transcripts for the more widely expressed cytochrome c_S and LDH A persist after their corresponding germ cell-specific cognates (cytochrome c_T and LDH C) begin to be synthesized, and are still present in early spermatids. Multiple cytochrome c_S transcripts have been identified in spermatogenic cells, including a longer 1.7-kb mRNA that appears in round spermatids (Hake *et al.*, 1990). Polysome gradients indicated that the cytochrome c_S transcripts in adult pachytene spermatocytes and round spermatids are present primarily in nonpolysomal fractions. Similarly, longer LDH A transcripts (1.8–2 kb) accumulate during the later stages of spermatogenesis but are present mostly in nonpolysomal fractions (Thomas *et al.*, 1990; Alcivar *et al.*, 1991b). Mechanisms for the translational repression of these longer transcripts have not been identified. The larger size of the 1.7-kb cytochrome c_S mRNA is solely the result of polyadenylation, suggesting that additional sequences in this transcript may inhibit translation.

Longer germ cell transcripts that are not translated efficiently have been identified also for proenkephalin (Penk) and ornithine decarboxylase (Odc) in rat and mouse testes. Although a 1.7-kb *Pek* transcript is abundant in pachytene spermatocytes and round spermatids, low levels of enkephalin-related peptides have been detected in these cells (Kilpatrick *et al.*, 1985,1987; Garrett *et al.*, 1989). Compared with transcripts in somatic cells, this spermatogenic cell-specific mRNA has a longer 5′ UTR with multiple short open reading frames, which are likely to inhibit translation from the downstream *Penk* initiation codon (Garrett *et al.*, 1989; Kilpatrick *et al.*, 1990). Similarly, *Odc* transcripts (2.2 and 2.7 kb) accumulate during meiotic and postmeiotic stages, although testicular polysome gradients suggested that less than 10% of the transcripts are associated with polysomes (Alcivar *et al.*, 1989). Structural features responsible for the longer transcripts in spermatogenic cells or the apparent translational repression in these cells have not been identified. However, other studies indicate that 5′ UTR secondary structure may be responsible for translational inhibition of *Odc* in somatic cells (Manzella *et al.*, 1991; Thach, 1992).

2. Genes transcribed only during the postmeiotic phase

Transcripts for several germ cell-specific proteins are synthesized in round spermatids during the early postmeiotic phase of spermatogenesis. Subsequently, RNA synthesis is terminated as the haploid nucleus condenses. Later in the postmeiotic phase, protein synthesis continues. Elongating spermatids undergo marked structural changes as they assume the polarized architecture of the mature gamete. Proteins synthesized during these terminal stages must be derived from transcripts that are stable for at least several days (reviewed by Hecht, 1990,1992; Eddy *et al.*, 1991).

Spermatids isolated from adult mice have been used to monitor the translational activation of several haploid-expressed genes, including transition protein 1 *(Tp-1)*, transition protein 2 *(Tp-2)*, protamine 1 *(Prm-1)*, protamine 2 *(Prm-2)*, and the seleno-protein *(Msp)* of the mitochondrial capsule (Kleene *et al.*, 1984,1990; Kleene, 1989; Yelick *et al.*, 1989). The transition proteins and protamines are structural proteins of the nucleus that sequentially replace histones during nuclear condensation and inactivation of the genome. The seleno-protein is a structural component of the rigid mitochondrial capsule that is characteristic of mammalian sperm. Although the mRNAs are present in round spermatids, Northern analysis of polysome gradients has shown that they are not associated with polysomes in these cells. The mRNAs become translationally active in the elongating spermatid stages, in which 30–40% of each mRNA is associated with polysomes.

Unlike the increase in poly(A) tail length associated with cytochrome c_T and LDH C synthesis during meiosis, translation of *Tp-1, Tp-2, Prm-1, Prm-2,* and *Msp* mRNA during the postmeiotic phase is correlated with a marked reduction in the size of the poly(A) tract from an initial length of 120–180 nucleotides to ~30 nucleotides (Kleene, 1989). Whether this poly(A) shortening plays a regulatory role in initiating protein synthesis or is a consequence of translation that ultimately leads to mRNA degradation has not yet been determined. The correlation between poly(A) shortening and translation may be characteristic of transcripts from haploid genes expressed during the terminal stages of differentiation or it may be a modification common to transcripts for structural proteins of the developing gamete.

Another shortened transcript that is produced specifically in spermatids is an ~ 4-kb mRNA for the c-*abl* proto-oncogene (Ponzetto and Wolgemuth, 1985). In this case, the shorter transcript results from the use of an unusual upstream polyadenylation signal (Meijer *et al.*, 1987; Oppi *et al.*, 1987). The truncated c-*abl* transcript persists in elongating spermatids and appears to be more stable than its counterparts in somatic cells (Meijer *et al.*, 1987). Some of the transcripts are associated with polysomes (Zakeri *et al.*, 1988) and may encode a 150-kDa phosphoprotein that can be immunoprecipitated from spermatid lysates (Ponzetto *et al.*, 1989). Meijer and associates (1987) suggested that elimination of the downstream portion of the 3' UTR present in somatic c-*abl* transcripts may increase the stability of the shorter spermatid transcript. This downstream region contains AU-rich segments that are similar, but not identical, to the destabilizer motif identified by Shaw and Kamen (1986).

Although mechanisms for translational activation during the late postmeiotic phase have not been defined completely, studies using transgenic mice have shown that 3' UTR sequences near the polyadenylation signal are responsible for the delayed translation of mouse *Prm-1* (Braun *et al.*, 1989a). Sequence comparisons have shown that this region is highly conserved in *Prm-1* and *Prm-2* among several mammalian species (Johnson *et al.*, 1988; Hecht, 1990).

A more detailed analysis of the 3' UTRs of mammalian protamines and transition proteins has identified three conserved sequences denoted Y, H, and Z (Kwon and Hecht, 1991a). The Y and H *cis* elements of *Prm-2* mRNA are linked closely and are within 60 nucleotides upstream of the polyadenylation signal. Gel retardation assays have demonstrated that these elements form specific complexes with cytoplasmic proteins isolated from the testis or from round spermatids, but not with proteins from elongating spermatids (Kwon and Hecht, 1991a,b). An 18-kDa protein that binds specifically to the Y element has been identified in these extracts using ultraviolet (uv) crosslinking and Northwestern blotting techniques. This binding is inhibited when the 18-kDa protein is treated first with phosphatase (Kwon and Hecht, 1991b). Further, cytoplasmic extracts containing the 18-kDa protein inhibited *in vitro* translation of human growth hormone only when its 3' UTR was replaced with the 3' UTR from *Prm-2* (Kwon and Hecht, 1991b). These studies have provided the first detailed model for translational regulation during spermatogenesis, in which a specific repressor binding to the Y element inhibits translation during the early postmeiotic phase. In elongating spermatids, binding of the repressor apparently is abolished, perhaps as a result of dephosphorylation, allowing translation to begin.

Two additional RNA binding proteins similar to such proteins in *Xenopus* oocytes have been identified in mouse spermatogenic cells, in which they appear to bind to RNA in a sequence-independent fashion (Kwon and Hecht, 1991b; Kwon *et al.*, 1992). These proteins are present at highest levels in round spermatids and in nonpolysomal fractions of testis extracts, and may play a role in the temporary storage of mRNAs.

Clearly the regulation of gene expression during spermatogenesis involves translational as well as transcriptional control mechanisms that include the translational activation of germ cell-specific genes in a defined temporal sequence and the repression of somatic cognate genes. Further, mechanisms for the stabilization and storage of mRNAs are essential during germ cell differentiation. Additional studies are needed to identify common translational regulatory mechanisms and to define their molecular constituents.

III. Transgenic Mice

The regulation of gene expression has been more difficult to study in spermatogenic cells than in some other cell types because of the lack of permanent germ cell lines for *in vitro* transfection assays. However, this drawback has been overcome partially by using cell-free systems (Section II,B). Another experimental approach that has proven to be effective for studying spermatogenic cell gene regulation is the use of transgenic mice. Promoter–reporter transgene constructs have been used to test flanking regions of spermatogenic cell-specific genes for their roles in determining tissue-specific and developmentally restricted gene expression (Section III,A). Additional studies with transgenic mice have produced unexpected new genetic models of male infertility. Some of these models have occurred because expression of the reporter gene disrupts spermatogenic cell development (Section III,C). Other transgenes cause male infertility by producing an insertional mutation in an endogenous gene that is required for spermatogenesis (Section III,D). The studies described in this section demonstrate that transgenic mice have provided new insights into understanding regulation of gene expression in spermatogenic cells and suggest that these animals are likely to continue to be used for this purpose in the future.

A. Spermatogenic Cell-Specific Genes

Transgenic mice have been used to identify both 5′ and 3′ flanking regions that regulate tissue- and stage-specific gene expression during spermatogenesis. Transgene constructs used in these studies generally have flanking sequences from genes expressed exclusively in male germ cells fused with reporter genes, for example, those encoding bacterial chloramphenicol acetyltransferase (CAT) or β-galactosidase (β-gal), both of which are not nor-

TABLE 3 5' Regulatory Regions Directing Germ Cell- and Stage-Specific Expression of Transgenes

Promoter	5' Region (bp)	From ATG	Reference
Meiotic phase			
PGK-2, human	327	−515 to −188	Robinson *et al.* (1989)
Acr, rat	2255	−2300 to −45	Nayernia *et al.* (1992)
Postmeiotic phase			
Prm-1, mouse	425	−557 to −132	Peschon *et al.* (1989)
Prm-2, mouse	920	−962 to −42	Stewart *et al.* (1988)
Ace$_t$, mouse	702	−698 to +4	Langford *et al.* (1991)

mally expressed in the testis. Specific 5' flanking regions required for spermatogenic cell-specific transcription have been identified for several genes by analyzing the synthesis and distribution of the heterologous mRNAs or proteins in these transgenic animals (Table 3).

1. Gene expression initiated during the meiotic phase

The spermatogenic cell PGK-2 isozyme is expressed during both the meiotic and the postmeiotic phase (Gold *et al.*, 1983; McCarrey and Thomas, 1987). Fusion constructs containing successive deletions in the 5' region of the human *PGK-2* gene have been used to produce several lines of transgenic mice (Robinson *et al.*, 1989). These studies have identified a 327-bp region required for tissue-specific and developmentally regulated transcription. Transgenes are expressed appropriately when either a 1.4- or a 0.5-kb sequence 5' to the ATG initiation codon is fused to a reporter gene containing CAT or luciferase coding sequences and simian virus 40 (SV40) poly(A) addition signals. In contrast, a transgene containing 192 bp of 5' sequence is not expressed in any tissues in three transgenic lines, although this 5' sequence contains minimal promoter elements (CAAT and GC box sequences). Subsequent studies have focused on the 327-bp tissue-specific enhancer region, using *in vitro* methods to examine DNA–protein interactions associated with spermatogenic cell expression of *Pgk-2* (Gebara and McCarrey, 1992; see Section II,B).

Although the human *PGK-2* transgenes and the endogenous mouse *Pgk-2* gene are transcribed coordinately, the transgenes do not exhibit the delayed translation that occurs for mouse *Pgk-2*. Therefore, the *PGK-2* coding region or 3' UTR sequences not present in the fusion constructs may contain additional regulatory sequences that are required for proper translational control (Robinson *et al.*, 1989).

Proacrosin, an acrosomal serine protease, is another spermatogenic cell-specific gene that is expressed first during meiosis (Kashiwabara *et al.*, 1990;

Anakwe *et al.*, 1991; Escalier *et al.*, 1991). A rat proacrosin-CAT/SV40 fusion gene, containing 2.3 kb of 5' proacrosin sequence from 45 to 2300 bp upstream of the translation initiation codon, is transcribed appropriately in pachytene spermatocytes and early spermatids and is not expressed in somatic tissues (Nayernia *et al.*, 1992). Additional studies with transgenes containing further deletions in this 5' region should identify the germ cell-specific regulatory elements more precisely. Although proacrosin has been identified in pachytene spermatocytes and spermatids of mice and of other species by immunological methods (Anakwe *et al.*, 1991; Escalier *et al.*, 1991), CAT activity was detected only in spermatids of transgenic mice. Further, the fusion protein is not confined to the acrosome, perhaps because the transgene lacks the proacrosin signal sequence (Nayernia *et al.*, 1992).

2. Gene expression initiated during the postmeiotic phase

Genes that are expressed first during the haploid stages of spermatogenesis also have 5' regulatory sequences that direct appropriate spermatogenic cell-specific and developmentally regulated transcription. Transgenic mice have been used to identify these 5' regulatory regions for three mouse genes encoding Prm-1, Prm-2, and a unique isozyme for angiotensin-converting enzyme (ACE) (Table 3).

Both the somatic ACE and the spermatogenic cell-specific isozyme (ACE_t) are encoded by a single gene, although transcription of the smaller spermatogenic cell isozyme begins approximately 7 kb downstream from the somatic *ACE* promoter, within a region that is treated as an intron by somatic tissues (Howard *et al.*, 1990). Three lines of transgenic mice were produced using a genomic fragment containing 698 bp of 5' flanking sequence immediately upstream from the transcription start site for ACE_t linked to a β-gal reporter gene. This transgene is expressed only in the testis, with high levels of β-gal activity restricted to elongating spermatids and spermatozoa. Although the 698-bp 5' region exhibits few structural features typically associated with a RNA polymerase II promoter, this intragenic region appears to contain a unique promoter for ACE_t that is recognized during spermatogenesis (Langford *et al.*, 1991).

A 920-bp segment of the *Prm-2* gene containing 5' flanking sequences and 61 bp of the 5' untranslated region has been used to prepare fusion genes expressed in transgenic mice. This segment directs testis-specific transcription when linked to exons 2 and 3 of the mouse c-*myc* gene (Stewart *et al.*, 1988). However, anomalous transcription occurs in somatic tissues when this *Prm-2* 5' sequence is fused with an SV40 T-antigen reporter (Stewart *et al.*, 1988; see Section III,B). Transcription of the transgene containing c-*myc* or SV40 sequences is restricted in the testis to round spermatids, in which endogenous protamines are transcribed. Although the T-antigen was not detected in the *Prm-2*/SV40 transgenic mice (Stewart *et al.*, 1988), the 0.9-kb *Prm-2* 5' segment linked to boar proacrosin cDNA directed spermatid-specific expression of the boar proacrosin protein in five transgenic lines (O'Brien *et al.*, 1992).

More extensive studies have examined both 5' and 3' regulatory regions of the mouse *Prm-1* gene in transgenic mice. Initial studies demonstrated that a 2.4-kb restriction fragment containing the 880 bp upstream of the transcription start site directs expression of the transgene specifically in round spermatids (Peschon *et al.*, 1987). Deletion analysis of this construct has shown that sequences between −40 and −465 bp relative to the transcription start site (−132 to −557 from the ATG) are essential for stage-specific transcription (Peschon *et al.*, 1989).

Constructs replacing the coding sequence and 3' UTR of *Prm-1* with the human growth hormone *(GH)* gene are still expressed exclusively in spermatids, indicating that the 5' sequences are sufficient for appropriate developmental expression of *Prm-1* (Braun *et al.*, 1989b; Peschon *et al.*, 1989). GH protein was detected by immunological methods in >90% of sperm in the caput epididymis of hemizygous transgenic mice, suggesting that genotypically distinct spermatids share the *GH* gene product via the intercellular bridges that connect these haploid cells (Braun *et al.*, 1989a).

Comparisons of the expression of chimeric *Prm-1/GH* constructs also have established that sequences in the *Prm-1* 3' UTR are important for translational regulation (Braun *et al.*, 1989b). Fusion genes containing identical *Prm-1* 5' and *GH* coding sequences but distinct 3' UTRs are translated at different stages of spermatogenesis, although they are transcribed coordinately. The transgene containing the 3' UTR of *GH* is translated immediately after transcription, and protein is detected first in the acrosomes of developing spermatids in 25-day-old mice. In contrast, translation of the transgene containing 156 bp of the *Prm-1* 3' UTR is delayed until day 28, when GH protein is detected in the cytoplasm of elongating spermatids. This delayed pattern of translation parallels the synthesis of endogenous *Prm-1*. Transcripts from the *Prm-1* 3' UTR construct also persist later in spermatid development than transcripts from the *GH* 3' UTR construct, suggesting that sequences in this region influence both translation and mRNA stability. Additional deletions in the *GH/Prm-1* 3' UTR chimeric construct indicate that a 62-nucleotide sequence of the 3' UTR is sufficient for *Prm-1*-like translational regulation (Braun, 1990). This 62-bp region contains the AAUAAA polyadenylation signal and a 20-nucleotide sequence just 5' to this signal that is also present in the 3' UTR of *Prm-2*, as well as in human, bovine, and mouse *Prm-1* (Johnson *et al.*, 1988).

B. Genes Expressed in Somatic and Germ Cells

Several transgenic mouse studies of somatic cell genes have found unexpected expression in spermatogenic cells. These findings have provided information useful in understanding regulation of gene expression in somatic and germ cells. For example, a transgene containing 185 bp of the 5' flanking sequence of the mouse metallothionein 1 *(Mt-1)* gene and human ornithine transcarbamoylase *(OTC)* and rat growth hormone *(Gh)* genes is expressed exclusively in male germ cells in seven transgenic lines (Kelley *et al.*, 1988). However, expression occurs in a variety of tissues in addition to testis when a 1.6-kb *Mt-1* promoter sequence is used. Transgenes with either the 185-bp or the 1.6-kb *Mt-1* promoter sequence are expressed in pachytene spermatocytes. High levels of *Mt* mRNA since have been found in mouse pachytene spermatocytes and spermatids. The expression of *Mt-1* and *Mt-2* genes in male germ cells is regulated developmentally and relatively unresponsive to treatments that dramatically increase levels in liver and ovary (De *et al.*, 1991). These results suggest that the 185-bp *Mt-1* 5' sequence contains a spermatogenic cell-specific promoter, whereas upstream sequences act as promoters in other tissues.

The regulation of gene expression by promoter sequences sometimes appears to be influenced by other sequences within the transgene. Although the 185-bp *Mt-1* promoter caused expression of the *OTC* coding region exclusively in male germ cells in the studies just described, this same promoter region used with the *Gh* gene sequence resulted in transgene expression in multiple tissues in a pattern similar to that of the endogenous *Mt-1* gene (Kelley *et al.*, 1988). The *Gh* gene appears to modify the specificity of the 185-bp *Mt-1* germ cell promoter sequence when it is immediately adjacent to it, but insertion of the 1166-bp *OTC* cDNA between the two sequences eliminates this influence.

Unexpected tissue expression also occurs when the SV40 T-antigen coding region is fused to the 5' and 3' flanking sequences of the *Prm-1* gene (Behringer *et al.*, 1988). As is discussed elsewhere (Section III,A), other transgenes containing the 5' *Prm-1* sequence are expressed exclusively in round spermatids (Peschon *et al.*, 1987,1989). Although the *Prm-1*/SV40 transgene is expressed in round spermatids, it is expressed also in heart and temporal bone and results in tumors in these two sites (Behringer *et al.*, 1988). Promoter sequences responsible for tissue-specific expression usually are located 5' to the transcription initiation site, but have been found occasionally in regions within or downstream of genes (e.g., Gillies *et al.*, 1983; Behringer *et al.*, 1987; Lozano and Levine, 1991) and may function independently of position or orientation (Serfling *et al.*, 1985). This result suggests that the structural sequence within a hybrid gene sometimes has significant effects on the action of the regulatory sequences to which it is linked (Behringer *et al.*, 1988). However, although *Prm-1*/SV40 mRNA is present in round spermatids, T antigen was not detected, suggesting that the transgene may have lacked sequence information for translation in spermatids (Behringer *et al.*, 1988).

Other transgenic mouse lines in which SV40 T antigen or other viral sequences have been used as reporters express T antigen in the testis. In most of these studies, the expressing cell type either has not been determined (Bautch *et al.*, 1987) or was found to be somatic (Leder *et al.*, 1986; Choi *et al.*, 1987). However, expression may occur in germ cells in transgenic mice carrying the major histocompatibility complex (MHC) Class I K^b gene enhancer linked to the SV40 T-antigen gene (Reynolds *et al.*, 1988). These mice develop tumors in multiple endocrine tissues and in seminiferous tubules. The SV40 T antigen is present at high levels in the testis, and seminiferous tubules are distended and filled with neoplastic cells with hyperchromatic nuclei. Three founder males were fertile but failed to transmit the transgene, whereas three female founders were fertile and transmitted the transgene to offspring (Reynolds *et al.*, 1988). Although the male founders may have been mosaics that failed to incorporate the transgene into the germ cell line, expression of the transgene also may have affected transmission, as appeared to occur for the herpes simplex virus thymidine kinase (HSV *tk*) gene (Braun *et al.*, 1990).

C. Expressed Transgenes that Affect Fertility

Several lines of transgenic mice have been produced with impaired fertility in males. These studies have involved use of HSV *tk* or interferon *(Ifn)* reporter genes; the products of these genes are likely to be the cause of infertility. This effect is often dominant; transgenic male founders and male heterozygous offspring of female founders are infertile. Alternatively, transgenic founder males and heterozygous offspring males are sometimes fertile but fail to transmit the transgene to their progeny.

Studies of the promoter region of a mouse group 1 major urinary protein *(Mup)* in transgenic mice used HSV *tk* as the reporter gene (Al-Shawi *et al.*, 1988). Three transgenic founder males were infertile, whereas two others were fertile but did not pass the transgene on to the next generation. The fertile males that did not transmit the transgene were presumed to be mosaics. Mosaic mice occur when the transgene is not incorporated into the genome until some time after the first cleavage division. Some male germ cells then may be descendants of early cleavage stage cells that did not incorporate the transgene.

The HSV *tk/Mup* hybrid gene is transmitted to half the offspring in five transgenic lines established from founder females. Transcripts for the hybrid gene are present in testes of males in all five transgenic lines and TK enzyme activity is high in testes of the males in these lines, as well as in two of the three sterile founder males. Although transgenic males have reduced sperm counts and most of their sperm are structurally abnormal, they mate with females and produce copulatory plugs. The cause of male infertility in these mice may be the presence of TK enzyme in the testis, so spermiogenesis is impaired rather than blocked completely. The TK enzyme has been speculated to affect fertility because (1) the viral enzyme has a substrate specificity different from that of cellular enzymes and efficiently phosphorylates deoxycytidine, (2) high levels of phosphorylated deoxynucleosides produced by the enzyme inhibit the normal pathway of pyrimidine phosphate metabolism, or (3) the enzyme is active at times in the cell cycle when the eukaryotic enzyme is repressed (Al-Shawi *et al.*, 1988). Although HSV *tk* was presumed to be expressed in spermatogenic cells, this was not demonstrated.

Male infertility also occurs in transgenic mice carrying the 5' region of the MHC Class II E_α gene as the promoter and HSV *tk* as the reporter gene (Pinkert *et al.*, 1985). Male founder mice are infertile or fail to transmit the transgene to their offspring; male offspring of female founders are also infertile. This observation was examined further using hybrid gene constructs containing different segments of the E_α sequence or transgenes containing the HSV *tk* sequence only. Neither gene is sufficient to cause infertility alone, but enhancers appear to exist in the E_α sequence that increase HSV *tk* expression when they are used together (Braun *et al.*, 1990).

The effect of HSV *tk* expression was analyzed also using the *Prm-1* promoter to target expression specifically to round spermatids. This construct produces the same phenotype as other HSV *tk* constructs: transgenic founder males are either sterile or fail to transmit the transgene, whereas male offspring of founder females are infertile (Braun *et al.*, 1990). HSV *tk* mRNA and TK protein are found at low levels in round spermatids and at higher levels in elongating spermatids. In addition, degeneration is seen in some seminiferous tubules; elongating spermatids are retained late in the seminiferous epithelium. Sperm have chromatin that is incompletely condensed and have abnormally shaped heads and acrosomes. The HSV *tk* gene expressed in spermatids was speculated to cause these effects by (1) metabolizing novel substrates into nucleotide analogs that interfere with nucleotide biosynthesis important for nuclear shaping and condensation or (2) establishing a futile cycle of thymidine phosphorylation and dephosphorylation at the expense of ATP (Braun *et al.*, 1990).

HSV *tk* expression also affects male fertility in the MyK-103 transgenic mouse line. The transgene consists of an *Mt-1* promoter sequence linked to the HSV *tk* coding region (Palmiter *et al.*, 1982). Males and females in the MyK-103 line are fertile, but males do not transmit the transgene (Palmiter *et al.*, 1984). Initially, the transgene was believed to have disrupted expression of a gene required for normal development of sperm or to have activated a gene that should be quiescent (Palmiter *et al.*, 1984; Wilkie and Palmiter, 1987). Subsequent studies suggested that the transgene occasionally is deleted, probably by intrachromosomal recombination, during early phases of germ cell development. The investigators hypothesized that fertile sperm that do not carry the transgene arise from these stem cells (Wilkie *et al.*, 1991). On the other hand, the HSV *tk* transgene is expressed in round and elongating spermatids when the transgene is not deleted (Wilkie *et al.*, 1991).

Thus, MyK-103 male mice might fail to transmit the transgene because sperm that contain the transgene are incapable of fertilizing eggs due to the expression of HSV *tk* during their formation, whereas sperm arising from stem cells that have deleted the transgene can fertilize eggs.

However, how the transgene is transmitted through the female germ cell line in MyK-103 mice is not clear. This event presumably would occur if the HSV *tk* gene either is not expressed during oocyte development or is expressed but does not affect the subsequent ability of eggs to be fertilized. In addition, homologous recombination might be predicted to occur in the female germ cell line as well as in the male. If so, MyK-103 females should sometimes fail to transmit the transgene to their offspring.

Impaired male fertility also has been seen in studies with transgenic mice using mouse *Ifn* as a reporter gene. In one study, an *Mt-1* enhancer–promoter sequence was linked to a hybrid *Ifn-β*/HSV *tk* gene (Iwakura *et al.*, 1988). This construct results in infertile male founders and infertile male offspring of female founders. The *Ifn* mRNA is synthesized in the testes of these mice; substantial levels of IFN protein are detected in testis homogenates. Testis weight is reduced 60% and testes contain abnormal pachytene spermatocytes and many degenerating spermatids. High levels of IFN protein in the testes were proposed to be responsible for male infertility (Iwakura *et al.*, 1988). However, because TK enzyme activity in the testis was elevated as well, other investigators suggested that infertility in these transgenic mice was caused by expression of the HSV *tk* gene (Wilkie *et al.*, 1991).

Infertility is seen also in founder males and male offspring of founder females in transgenic mice with an *Mt* promoter linked to the *Ifn-α*1 coding sequence and the rabbit *β*-globin 3′ region (Hekman *et al.*, 1988). Some male offspring of founder females are fertile at puberty but become infertile by 3 months of age. *Ifn-α*1 mRNA is detected in testes and not other tissues; biologically active IFN protein is present in testis homogenates. The testes are small and seminiferous tubules show vacuolization, progressive degeneration, and diminished spermatid production. Expression of the *Ifn-α*1 gene was concluded to be responsible for these effects (Hekman *et al.*, 1988), supporting the idea that the infertility noted earlier in mice expressing the *Ifn-β*/HSV *tk* hybrid gene in the testis (Iwakura *et al.*, 1988) may have been caused by the presence of both IFN and TK proteins.

Mice containing a human *IFN* cDNA transgene also exhibit male infertility (Gordon *et al.*, 1989). Mating behavior is normal, but testes are small and some seminiferous tubules are empty or contain arrested primary spermatocytes. These studies involved a single transgenic line. Karyotyping identified a translocation associated with the chromosome in which the transgene was integrated. Chromosomal rearrangement and faulty chromosome synapsis caused by the transgene were postulated to be responsible for the male infertility (Gordon *et al.*, 1989). Although the results of Hekman *et al.* (1988)

would suggest that *IFN* expression might have contributed to or been the primary cause of male infertility in the studies of Gordon *et al.* (1989), this possibility apparently has not been tested.

Another transgene that appears to cause male infertility when expressed in germ cells carries the activated N-*ras* oncogene under the transcriptional control of the mouse mammary tumor virus long terminal repeat (MMTV-LTR). Of four founder male mice, one was infertile, one failed to transmit the transgene, and two others were initially fertile and transmitted the transgene, but they and their offspring males became infertile with age (Mangues *et al.*, 1990). The sperm of infertile mice has poor motility, some have detached heads, and they are unable to fertilize eggs *in vitro*. RNase protection assays indicate that N-*ras* transcripts are present in the testes of these transgenic mice (Mangues *et al.*, 1990).

D. Insertional Mutations that Affect Fertility

Transgenes integrating into the genome occasionally cause an insertional mutation that disrupts expression of a gene affecting male fertility. In most cases, the mutations are recessive and observed only after mice are bred to homozygosity. Such insertional mutations may result in pleiotropic phenotypes, with consequences in addition to male infertility, whereas others specifically perturb male fertility. The insertional mutations that disrupt spermatogenesis that have been reported to date apparently produce their effects in postmeiotic cells. Although using the transgene as a tag to clone and identify the mutated gene is possible, this procedure has not yet been reported for an insertional mutation affecting male fertility. Usually, insertional mutations are assumed to affect male fertility by disrupting a gene expressed in germ cells. However, in some cases the effect on germ cells and fertility may be due to disruption of a gene normally expressed in Sertoli cells or other somatic cells of the testis.

Two insertional mutations have been reported that alter male fertility and also affect the central nervous system. One of those is allelic with a spontaneous mutation in mice—Purkinje cell degeneration *(pcd)*. The spontaneous mutation causes loss of cerebellar Purkinje cells, retinal photoreceptor cells, and olfactory bulb cells, as well as producing abnormalities in spermatogenesis (Mullin and La Vail, 1975; Mullen *et al.*, 1976). The insertional mutation is recessive and not all mutant males are sterile, possibly because the transgenic line was established in a hybrid mouse strain (Krulewski *et al.*, 1989). The homozygous infertile males are azoospermic and their testes lack late spermatids (Krulewski *et al.*, 1989), suggesting that the mutation disrupts a gene active during the postmeiotic developmental phase of spermatogenesis.

The other insertional mutation affecting both male fertility and the central nervous system is allelic with the hotfoot *(ho)* mutation causing motor disorders and male infertility (Gordon *et al.*, 1990). Sperm appear to move abnormally and have flagellar defects detected by electron microscopy. Sperm are unable to penetrate the zona pellucida to fertilize eggs *in vitro* until holes are "drilled" in the zona pellucida. Spermiogenesis was speculated to be disrupted by the mutation, leading to abnormal motility or the absence of a surface molecule(s) required for interaction with the zona pellucida (Gordon *et al.*, 1990).

An insertional mutation apparently affecting only spermatogenesis is named "symplastic spermatids" *(sys)* because the intercellular bridges between clones of round spermatids appear to open and allow them to become giant cells (MacGregor *et al.*, 1990; Russell *et al.*, 1991). Spermiogenesis is arrested near the beginning of spermatid elongation, but variability exists between animals; pachytene spermatocytes are sometimes affected as well. Since both Sertoli cells and spermatogenic cells appeared to be disrupted, determining the primary cell type(s) affected by the mutation by histology was not possible (Russell *et al.*, 1991).

Another transgenic line, referred to as AE24, contains an insertional mutation causing infertility in about half the homozygous males (Merlino *et al.*, 1991). The sperm have poor motility and aberrant axonemal structures. Outer doublet microtubules are missing from the middle piece, and the flagellar axoneme disassembles in the epididymis. The outer doublets of the flagellum were suggested to be assembled normally, but with a latent instability that led to later disruption. The transgene contains the human epidermal growth factor receptor *(EGFR)* cDNA. Elevated levels of *EGFR* mRNA are present in the testis; the EGFR protein is seen by immunostaining in elongating spermatids. However, EGFR also is expressed in the testis of other transgenic lines in which male fertility is not affected, strongly suggesting that expression of the transgene is not responsible for the sperm defects (Merlino *et al.*, 1991).

Male infertility also is observed in a line of transgenic mice carrying the hematopoietic cell kinase *(HCK)* oncogene, but the mutant phenotype was completely penetrant only on some genetic backgrounds (Magram and Bishop, 1991). This dominant mutation was named "lacking vigorous sperm" *(Lvs)*. Spermatogenesis appears to progress normally until nuclear condensation in spermatids, but later spermatid nuclei are abnormal. The nucleus is stained more darkly and is short and wide rather than elongated. Investigators noted that insertional mutations caused by transgenes typically result in a loss of function that becomes apparent only in homozygous animals. Possible explanations for the dominant effect seen are that the *Lvs* mutation might (1) result in insufficient amounts of a gene product in haploid cells; (2) truncate or alter the product of a gene, resulting in an anomalous protein that may disrupt assembly of a multimeric protein; (3) activate a gene that is not normally expressed in the testis; or (4) inactivate one allele at a locus that does not express the other allele because of genomic imprinting (Section IV,A), effectively resulting in a null mutation (Magram and Bishop, 1991).

IV. Genomic Imprinting and Gene Methylation

Circumstantial evidence suggests that genomic imprinting occurs during spermatogenesis and that changes in DNA methylation during germ cell development may be involved. In some cases, expression in offspring of different genes or transgenes has been shown to depend on the parent of origin and apparently changes on passage through the germ line (Sections IV,A,B). Further, changes in expression of some imprinted transgenes correlate with changes in methylation of these transgenes (Section IV,C). Finally, DNA methylation usually is reduced when tissue-specific genes are expressed. Changes in DNA methylation appear to be related to the expression of some spermatogenic cell-specific genes (Section IV,D). However, whether DNA methylation is a cause or an effect of genomic imprinting and/or the regulation of spermatogenic cell gene expression has not yet been established.

A. Imprinting of Endogenous Genes

Some autosomal genes function differently depending on whether they come from the male or the female parent. This difference apparently arises during gametogenesis by an epigenetic modification referred to as parental or genomic imprinting. Although the imprinted gene remains modified through subsequent rounds of DNA replication in that individual, the imprint is erased during passage of the gene through the germ line of the opposite sex. The result is that some genes change their pattern of expression depending on whether they are maternally or paternally inherited. The phenomenon of imprinting has been identified through studies of the developmental potential of embryos following nuclear transplantation, the patterns of expression of some endogenous genes and transgenes, and the genetic and molecular analysis of certain mutant mice and heritable human diseases (reviewed by Surani, 1984; Surani *et al.*, 1986,1988,1990; Solter, 1988; Howlett *et al.*, 1989; Reik, 1989; Sapienza *et al.*, 1989; Haig and Graham, 1991).

Genetic studies on the effects of transmission of chromosomal rearrangements have identified portions of certain mouse chromosomes that contain imprinted genes. Robertsonian translocations that result in combinations of maternal duplications and paternal deficiencies of chromosomal regions (and vice versa) are lethal or produce opposite phenotypes, depending on the disomy or nullisomy produced. Portions of chromosomes 2, 6, 7, 8, 11, and 17 show parental origin effects on development (reviewed by Searle and

Beechey, 1985; Cattanach, 1986; Howlett *et al.*, 1989; Searle *et al.*, 1989; Beechey *et al.*, 1990), indicating that parental imprinting during gametogenesis may be a considerably more significant mechanism for regulating gene expression than previously realized, particularly during early embryogenesis.

More recent studies have identified four mouse genes that undergo genomic imprinting: the insulin-like growth factor II *(Igf2)* gene (DeChiara *et al.*, 1990), the insulin-like growth factor II receptor *(Igf2r)* gene (Barlow *et al.*, 1991), the *H19* gene (Bartolomei *et al.*, 1991), and the *Tme* gene (Forejt and Gregorova, 1992). These genes are located on regions of mouse chromosomes previously known to be imprinted: *Igf2* and *H19* are on chromosome 7 (Lalley and Chirgwin, 1984; Bartolomei *et al.*, 1991) and *Igf2r* and *Tme* are on chromosome 17 (Barlow *et al.*, 1991; Forejt and Gregorova, 1992). The *Igf2* gene encodes a bifunctional receptor that binds both IGF-II and mannose 6-phosphate-containing ligands in a cation-independent fashion (Morgan *et al.*, 1987; Lobel *et al.*, 1988). This gene is expressed at low levels in mouse spermatogenic cells and at higher levels in Sertoli cells. The *Igf2r* may play an important role in cell interactions in the seminiferous epithelium (O'Brien *et al.*, 1989).

The *Igf2* gene was found to be imprinted when mice carrying a targeted disruption of the gene were studied (DeChiara *et al.*, 1990,1991). Germ-line transmission of the inactivated *Igf2* gene from male chimeras resulted in heterozygous offspring that were approximately 60% as large as their wild-type littermates (DeChiara *et al.*, 1990). However, when the disrupted gene was transmitted maternally, the heterozygous offspring were phenotypically normal. The function of the intact *Igf2* gene was shown to be switched by passage through the germ line of the opposite sex in successive generations, indicating that only the paternal copy of the intact *Igf2* gene is expressed in embryos, whereas the maternal copy of the gene is silent (DeChiara *et al.*, 1991).

The *Igf2r* gene was found to be imprinted; the copy inherited from the male parent was transcriptionally inactive (Barlow *et al.*, 1991). Initial mapping studies indicated that the gene was closely linked or identical to the locus of the T-associated maternal effect *(Tme)* mutation, a deletion that causes embryonic death at day 15 when inherited from the mother (Barlow *et al.*, 1991). More recent studies indicate that *Igf2r* and *Tme* are separate genes, although imprinting causes the paternal copy of both to be inactive (Forejt and Gregorova, 1992). The two genes show different patterns of imprinting in interspecific hybrids. Other data suggest that the *Imprintor-1 (Imp-1)* gene causes imprinting at the *Tme* locus (Forejt and Gregorova, 1992). Three other genes that map closely to *Igf2r* on chromosome 17 (mitochondrial superoxide dismutase 2, T-complex peptide 1, and plasminogen) are expressed from both chromosomes (Barlow *et al.*, 1991), suggesting that imprinting is specific to the *Igf2r* and *Tme* genes and not the result of a regional modification of chromosome 17.

The *H19* gene encodes an abundant 2.5-kb mRNA present in a variety of tissues in mouse embryos; an equivalent gene is present in humans (Brannon *et al.*, 1990). Although *H19* lacks an open reading frame and apparently is untranslated, its overexpression in transgenic mice is lethal between day 14 and birth (Brunkow and Tilghman, 1991). Using an RNase protection assay that distinguishes between alleles in four subspecies of mice, this gene was shown to be parentally imprinted, with the active copy derived from the mother (Bartolomei *et al.*, 1991). The *H19* and *Igf2* genes map to the distal third of chromosome 7 and are tightly linked genetically, yet are imprinted in opposite directions (Bartolomei *et al.*, 1991). This result provides additional evidence of a relatively high degree of specificity of the parental imprinting process that probably occurs during gametogenesis.

B. Imprinting of Transgenes

Several transgenes display expression patterns that are indicative of parental imprinting. Because transgenes serve as specific and easily identifiable markers of particular sites in the genome, they are potentially useful for determining the mechanism and molecular nature of imprinting (Surani *et al.*, 1988). For example, when only one transgenic line of several prepared with the same construct shows evidence of imprinting, the location of the transgene rather than a sequence element (e.g., an "imprinting box") determines that it is imprinted. However, imprinting of transgenes may occur because they integrate nonrandomly into the genome or are treated as foreign DNA (DeLoia and Solter, 1990). Further, transgene imprinting and endogenous gene imprinting have not been established to be equivalent processes.

One example of transgene imprinting was observed in studies of a transgene containing a major portion of the Rous sarcoma virus (RSV) LTR linked to a c-*myc*-containing fusion gene. Expression of the transgene was found to be determined by parental origin (Swain *et al.*, 1987). The transgene was expressed in the heart if inherited from the male parent, but not expressed if inherited from the female parent. The potential for expression of the transgene in a mesodermal organ was suggested to be due to the RSV LTR component (Overbeek *et al.*, 1986), whereas parental inheritance of expression of the transgene probably was acquired during gametogenesis in the male or shortly after fertilization (Swain *et al.*, 1987).

Imprinting also occurs in acro-dysplasia *(Adp)* transgenic mice that were produced with a transgene containing the *Mt-1* promoter linked to a bovine papilloma virus sequence and human growth hormone releasing factor hybrid reporter gene (DeLoia and Solter, 1990). Half the offspring of the original male founder had deformities of the limbs whereas, in subsequent generations, the female parents carrying the transgene produced normal offspring and male parents produced deformed offspring. The transgene probably caused the *Adp* mutation by insertion into an endogenous gene involved in limb formation (DeLoia and Solter, 1990). Because only deformed mice developed skin papillomas, expression and imprinting of both the endogenous gene and the transgene were suggested to be determined by the parent of origin (DeLoia and Solter, 1990).

C. Methylation of Transgenes

Methylation of some transgenes occurs according to parental origin. Although whether changes in methylation are causes or effects of imprinting is not clear, modification of methylation correlates closely with the imprinting phenomenon. In the RSV LTR promoter and c-*myc* reporter transgenic line discussed earlier (Section IV,B), methylation of the transgene occurs during its passage through the female parent and apparently is eliminated during gametogenesis in the male (Swain *et al.*, 1987). If a male inherits the transgene from a female parent, the transgene is methylated in somatic tissues and is not expressed. However, the transgene is undermethylated in the testes (Swain *et al.*, 1987). In animals inheriting the transgene from the male parent, the transgene is undermethylated and expressed in the heart.

Other studies have reported similar parental effects on changes in methylation of transgenes that apparently occur during spermatogenesis. A quail troponin I transgene is more methylated when inherited from a female parent than when inherited from a male parent (Sapienza *et al.*, 1987). The transgene in testis DNA is undermethylated compared with the transgene DNA from somatic tissues, suggesting a decrease in transgene methylation during spermatogenesis (Sapienza *et al.*, 1987). A change in methylation also was seen for the *CAT17* transgene; the methylation pattern was reversed after germ-line transmission to the opposite sex (Reik *et al.*, 1987). The paternally inherited transgene is undermethylated relative to the maternally inherited transgene. The *CAT17* transgene was suggested to have integrated into a region of the genome subject to changes in methylation as determined by parental origin and to have been modified accordingly (Reik *et al.*, 1987).

An inverse correlation between transgene methylation and expression also was observed for a transgene consisting of the HSV *tk* promoter and a *β-gal* reporter gene (Allen *et al.*, 1990). However, methylation changes with successive generations were seen in both somatic tissues and testis. This transgene segregates into low and high expressing lines. Expression and DNA methylation varied with the genetic background. Expression of the transgene was suggested to be influenced by strain-specific modifier genes that regulate transcription factors (Allen *et al.*, 1990).

D. Methylation of Endogenous Genes

DNA methylation is believed to be one of the mechanisms involved in regulating gene activity. Methylation probably inhibits gene expression by interfering with protein–DNA interactions required for transcription, either indirectly by altering the overall structure of the gene or directly by interfering with binding of transcription factors (reviewed by Cedar, 1988). Studies of methylation usually focus on regions of genes enriched for cytosine nucleotides linked on the 3' side to guanine (CpG islands). The typical housekeeping gene is unmethylated at these sites in both germ cells and somatic cells. However, tissue-specific genes often are highly methylated in sperm and oocytes, remain methylated during and after development in most tissues, but become demethylated in the tissue of expression (Cedar, 1988). Not only does a correlation exist between gene expression and undermethylation, but transfection experiments have shown that methylation represses transcription (Razin and Cedar, 1991). If these rules apply to spermatogenic cell-specific genes, such genes should be methylated in somatic cells and be demethylated during gametogenesis.

Significant changes in methylation occur during gametogenesis. Both total DNA and several individual genes are relatively undermethylated in primordial germ cells (Monk *et al.*, 1987; Driscoll and Migeon, 1990; Kafri *et al.*, 1992), whereas DNA from sperm is highly methylated (e.g., Jagiello *et al.*, 1982; Rahe *et al.*, 1983). Further, several studies have examined the methylation patterns of spermatogenic cell-specific genes during spermatogenesis. One study compared the methylation patterns of three mouse spermatogenic cell-specific genes (*Tp1*, *Prm-1*, and *Prm-2*) expressed in postmeiotic cells with a housekeeping gene (*β*-actin) expressed throughout spermatogenesis. The *Tp1* gene was found to become progressively less methylated during spermatogenesis, whereas the *Prm-1* and *Prm-2* genes become progressively more methylated; the *β*-actin housekeeping gene remains unchanged (Trasler *et al.*, 1990).

Another study compared the methylation patterns of the spermatogenic cell-specific *Ldh-c* gene and its somatic cognate gene *(Ldh-a)* in mouse spermatogenic cells. Although expression of both genes is regulated temporally during spermatogenesis, DNA methylation at the CpG sites monitored does not change markedly for either gene during germ cell development (Alcivar *et al.*, 1991b). Other studies found that the methylation of selected CpG sites in the spermatogenic cell-specific *Pgk-2* gene and the housekeeping apolipoprotein A-1 *(ApoA1)* gene transcribed during spermatogenesis were the same in pachytene spermatocytes and in round spermatids (Ariel *et al.*, 1991; Kafri *et al.*, 1992). Although methylation and demethylation can occur in haploid germ cells during spermatogenesis (Trasler *et al.*, 1990), a general correlation has not yet been established between the methylation status of the specific CpG sites examined and expression of spermatogenic cell-specific genes.

Differences in gene methylation have been detected when spermatogenic cells, somatic cells, and sperm are compared. CpG sites that are demethylated in the *Pgk-2* and *ApoA1* genes in mouse pachytene spermatocytes and round spermatids are methylated in somatic cell and sperm DNA (Ariel *et al.,* 1991). The *Ldh-a* gene is less methylated in adult testis DNA than in spleen DNA, but the methylation pattern for *Ldh-c* does not differ between the two tissues (Alcivar *et al.,* 1991). Although these studies suggest that hypomethylation coincides with expression for some spermatogenic cell-specific genes, this is not the case for others. Studies of additional genes may help clarify these relationships. To date, most studies have used restriction enzyme or polymerase chain reaction (PCR) methylation analysis techniques that allow analysis of only a limited number of potential methylation sites for a particular gene. Therefore, changes in the methylation status in a key structural or regulatory region of a gene might be missed with these methods.

Other promising approaches have been used to study the methylation status of a spermatogenic cell gene: *in vivo* DNase footprinting and DNA mobility shift experiments and *in vitro* DNA methylation and transfection assays. These approaches have provided evidence that expression is regulated by methylation for a rat spermatogenic cell-specific histone *(TH2B)* gene (Choi and Chae, 1991). The *TH2B* gene is hypomethylated in spermatogenic cells, but methylation inhibits its expression in transfected cells and blocks binding of transcription factors *in vitro* (Choi and Chae, 1991), suggesting that DNA methylation may play an important role in repressing the transcription of *TH2B* in somatic tissues.

V. Chromosomal Location of Spermatogenic Cell-Specific Genes

Changes in the structure of chromatin and DNA have been proposed to affect the regulation of tissue-specific gene expression (Elgin, 1990; Herbomel, 1990; Svaren and Chalkley, 1990; Eissenberg and Elgin, 1991). These changes may involve binding of chromosomal proteins, methylation of DNA (discussed in Section IV,D), and modifications associated with DNA replication. Evidence also suggests that such changes in DNA structure can be restricted to specific DNA domains, the boundaries of which are defined by associations with the nuclear matrix (Elgin, 1990; Herbomel, 1990; Eissenberg and Elgin, 1991; Getzenberg *et al.,* 1991). Since these domains can span relatively large segments of DNA and may encompass more than one gene, the grouping of genes in close proximity to one another could allow parallel regulation. Although representative of one level of transcriptional control, structural changes alone probably would not trigger gene expression. These changes are more likely to open DNA structure, including gene enhancers and promoters, for additional regulation by the specific transcription factors mentioned earlier (Section I,A). Such selective access to *cis* elements would allow the same transcription factors to activate different tissue-specific genes, depending on the structural accessibility of these elements to transcription factors (Elgin, 1990; Svaren and Chalkley, 1990; Eissenberg and Elgin, 1991).

In a preliminary attempt to determine whether genes expressed in spermatogenic cells show clustering within the genome and may be regulated in parallel, we have grouped those genes known to be expressed in spermatogenic cells by their chromosome location. Localization of mouse spermatogenic cell-expressed genes shows a dispersal across most of the murine chromosomes (see Table 4). Indeed, 19 of the 21 mouse chromosomes contain loci currently known to be expressed in male germ cells. The concentration of 25 germ cell-expressed genes on chromosome 17, where the t complex is located, was the only obvious grouping observed. Although the localization of 25 spermatogenic cell-expressed genes to chromosome 17 partially may reflect extensive research into the molecular basis of the t haplotype, the presence of such a large number of similarly expressed genes is intriguing and suggests the possibility of coordinated regulation of spermatogenic cell-expressed genes.

VI. Conclusions

Advances are occurring rapidly in identifying the components and processes involved in gene expression during spermatogenesis. The application of modern cell and molecular biology approaches to the study of spermatogenesis indicates that the events leading to the production of the male gamete probably occur under the command of overlapping developmental programs. A key issue is what the intrinsic and extrinsic mechanisms regulating these programs are. Since the proteins responsible for the distinctive structures and functions of spermatogenic cells and sperm have been identified, investigators have realized that many of the genes and transcripts encoding these proteins are unique to male germ cells. Such genes and transcripts are being identified constantly; these efforts are beginning to provide an understanding of spermatogenic cell gene regulation.

Research on the intrinsic regulation of expression of spermatogenic cell genes is still at an early stage. The list of transcription factor present in spermatogenic cells continues to grow and includes some factors that appear to be unique to these cells and many others that are also present in somatic cells. A few genes unique to spermatogenic cells have been shown to have upstream sequences comparable to those in other cells, and bind specific transcription factors. In addition, studies with transgenic mice have demonstrated that regulatory elements necessary and sufficient for spermatogenic cell-specific expression are present a few hundred base pairs upstream of the transcription start site. However, the connection between one or more particular transcription factors and the regulation of expression of a specific spermatogenic cell-specific or stage-specific gene has not been made.

Several studies have identified upstream sequences of spermatogenic cell-specific genes that bind proteins present in nuclei isolated from testes or

TABLE 4   Mouse Chromosome Locations of Spermatogenic Cell Expressed Genes

Chromosome	Protein	Gene	Chromosome	Protein	Gene
1	Transition protein 1	Tp1		t-Complex protein 10b	Tcp-10b
	ATP-dependent RNA helicase	D1Pas1		t-Complex protein 10c	Tcp-10c
				t-Complex protein 11	Tcp-11
2	c-abl Proto-oncogene	Abl		t-Complex testis expressed 1	Tcte-1
3	N-ras Proto-oncogene	N-ras			
	Nerve growth factor β	Ngfb		t-Complex testis expressed 3	Tcte-3
4	c-jun Proto-oncogene	Jun			
	β1,4-Galactosyltransferase	Ggtb		t-Complex testis expressed 1	Tctex-1
	c-mos Proto-oncogene	Mos		t-Complex testis expressed 3	Tctex-3
5	c-kit Proto-oncogene	Kit			
	β-Actin	Actb		t-Complex testis expressed 4	Tctex-4
6	Hox 1.4	Hox-1.4			
	Lactate dehydrogenase B	Ldh-2		t-Complex testis expressed 5	Tctex-5
	γ-Actin	Actg			
7	Lactate dehydrogenase A	Ldh-1		t-Complex testis expressed 6	Tctex-6
	Lactate dehydrogenase C	Ldh-3			
	H-ras Proto-oncogene	Hras-1		t-Complex testis expressed 7	Tctex-7
	Transforming growth factor β	Tgfb-1			
				t-Complex testis expressed 8	Tctex-8
8	jun B Proto-oncogene	Junb			
	jun D Proto-oncogene	Jund		t-Complex testis expressed 9	Tctex-9
	DNA polymerase β	Polb			
9	Pro-enkephalin	Penk		t-Complex testis expressed 10	Tctex-10
	β-Galactosidase	Bgl			
	Cholecystokinin	Cck		t-Complex testis expressed 11	Tctex-11
	Apolipoprotein A-1	Apoa-1			
10	Zinc finger protein, autosomal	Zfa		t-Complex testis expressed 12	Tctex-12
	Insulin-like growth factor 1	Igf-1		c-pim Proto-oncogene	Pim-1
11	Retinoic acid receptor α	Rara		Phosphoglycerate kinase-2	Pgk-2
12	Heat shock protein 86	Hsp86-1		Male enhanced antigen	Mea
	Pro-opiomelanocortin	Pomc-1		Oct-4a transcription factor	Otf-3
	c-fos Proto-oncogene	Fos		Heat shock cognate protein 70t	Hsc70t
15	c-myc Proto-oncogene	Myc			
	int-1 Proto-oncogene	Wnt-1	18	Zinc finger protein 35	Zfp35
	SP1 transcription factor	Sp1-1	19	Pyruvate dehydrogenase E1α subunit	Pdha1
	Preproacrosin	Acr			
	Diaphorase	Dia-1	Y	Zinc finger protein 1, Y-linked	Zfy-1
16	Protamine 1	Prm-1			
	Protamine 2	Prm-2		Zinc finger protein 2, Y-linked	Zfy-2
17	t-Complex protein 1	Tcp-1			
	t-Complex protein 3	Tcp-3		Sex determining region of Chrosome Y	Sry
	t-Complex protein 7	Tcp-7			
	t-Complex protein 10a	Tcp-10a			

[a] The list of germ cell expressed genes was compiled from Hecht (1992) and Wolgemuth and Watrin (1991). Chromosome localizations were obtained from Eppig (1992) and Silver et al. (1992).

seminiferous tubules, but not present in somatic cell nuclei. However, considerable additional work will be necessary to determine whether these proteins are among the transcription factors that regulate expression of spermatogenic cell genes.

Clearly many spermatogenic cell genes also are subject to translational regulation. Further studies are needed to identify the specific regulatory sequences and regulatory proteins that are involved. Also, determining whether common translational regulatory mechanisms are used for spermatogenic cell-specific genes with similar functions of temporal expression patterns will be necessary.

Circumstantial evidence indicates that genomic imprinting of endogenous genes and transgenes occurs during spermatogenesis. The correlations between genomic imprinting and spermatogenesis, changes in DNA methylation and spermatogenesis, and DNA methylation and expression of tissue-specific genes also suggest that changes in imprinting during spermatogenesis involve changes in DNA methylation. However, whether changes in DNA methylation are involved in the regulation of expression of spermatogenic cell-specific genes remains to be determined.

Most of these findings have been made possible by the considerable technological advances in molecular biology. The further development and application of these methods, the use of transgenic mice to study gene regulatory regions, the development of mice with mutations targeted to specific spermatogenic cell genes by homologous recombination in embryonic stem cells, and rapid progress in sequencing the mouse genome should lead to a much better understanding of the regulation of spermatogenic cell-specific genes in the forseeable future. This knowledge should lead to exciting advances in characterizing other cellular and molecular aspects of male reproduction.

Acknowledgments

This work was supported in part by NIH grants HD-26485 to D.A.O. and P30-HD-18968 to the Laboratories for Reproductive Biology, University of North Carolina at Chapel Hill.

References

Adamo, S., Conti, M., Geremia, R., and Monesi, V. (1980). Particulate and soluble adenylate cyclase activities of mouse male germ cells. *Biochem. Biophys. Res. Commun.* **97,** 607–613.

Alcivar, A. A., Hake, L. E., Mali, P., Kaipia, A., Parvinen, M., and Hecht, N. B. (1989). Developmental and differential expression of the ornithine decarboxylase gene in rodent testis. *Biol. Reprod.* **41,** 1133–1142.

Alcivar, A. A., Hake, L. E., Hardy, M. P., and Hecht, N. B. (1990). Increased levels of *jun*B and c-*jun* mRNAs in male germ cells following testicular cell dissociation. Maximal stimulation in prepuberal animals. *J. Biol. Chem.* **265,** 20160–20165.

Alcivar, A. A., Hake, L. E., Kwon, Y. K., and Hecht, N. B. (1991a). *junD* mRNA expression differs from c-*jun* and *jun*B mRNA expression during male germinal cell differentiation. *Mol. Reprod. Devel.* **30,** 187–193.

Alcivar, A. A., Trasler, J. M., Hake, L. E., Salehi-Ashtiani, K., Goldberg, E., and Hecht, N. B. (1991b). DNA methylation and expression of the genes coding for lactate dehydrogenases A and C during rodent spermatogenesis. *Biol. Reprod.* **44,** 527–535.

Allen, N. D., Norris, M. L., and Surani, M. A. (1990). Epigenetic control of transgene expression and imprinting by genotype-specific modifiers. *Cell* **61,** 853–861.

Al-Shawi, R., Burke, J., Jones, C. T., Simons, J. P., and Bishop, J. O. (1988). A *Mup* promoter-thymidine kinase reporter gene shows relaxed tissue-specific expression and confers male sterility upon transgenic mice. *Mol. Cell. Biol.* **8,** 4821–4828.

Anakwe, O. O., Sharma, S., Hardy, D. M., and Gerton, G. L. (1991). Guinea pig proacrosin is synthesized principally by round spermatids and contains O-linked as well as N-linked oligosaccharide side chains. *Mol. Reprod. Devel.* **29,** 172–179.

Anthony, C. T., Kovacs, W. J., and Skinner, M. K. (1989). Analysis of the androgen receptor in isolated testicular cell types with a microassay that uses an affinity ligand. *Endocrinology* **125,** 2628–2635.

Ariel, M., McCarrey, J., and Cedar, H. (1991). Methylation patterns of testis-specific genes. *Proc. Natl. Acad. Sci. U.S.A.* **88,** 2317–2321.

Ashworth, A., Skene, B., Swift, S., and Lovell-Badge, R. (1990). *Zfa* is an expressed retroposon derived from an alternative transcript of the *Zfx* gene. *EMBO J.* **9,** 1529–1534.

Barberis, A., Superti-Furga, G., and Busslinger, M. (1987). Mutually exclusive interaction of the CCAAT-binding factor and of a displacement protein with overlapping sequences of a histone gene promoter. *Cell* **50,** 347–359.

Barlow, D. P., Stoger, R., Herrmann, B. G., Saito, K., and Schweifer, N. (1991). The mouse insulin-like growth factor type-2 receptor is imprinted and closely linked to the Tme locus. *Nature (London)* **349,** 84–87.

Bartolomei, M. S., Zemel, S., and Tilghman, S. M. (1991). Parental imprinting of the mouse H19 gene. *Nature (London)* **351,** 153–155.

Bautch, V. L., Toda, S., Hassell, J. A., and Hanahan, D. (1987). Endothelial cell tumors develop in transgenic mice carrying polyoma virus middle T oncogene. *Cell* **51,** 529–538.

Beechey, C. V., Cattanach, B. M., and Searle, A. G. (1990). Genetic imprinting map. *Mouse Genome* **87,** 64–65.

Behringer, R. R., Hammer, R. E., Brinster, R. L., Palmiter, R. D. and Townes, T. M. (1987). Two 3′ sequences direct adult erythroid-specific expression of human beta-globin genes in transgenic mice. *Proc. Natl. Acad. Sci. U.S.A.* **84,** 7056–7060.

Behringer, R. R., Peschon, J. J., Messing, A., Gartside, C. L., Hauschka, S. D., Palmiter, R. D., and Brinster, R. L. (1988). Heart and bone tumors in transgenic mice. *Proc. Natl. Acad. Sci. U.S.A.* **85,** 2648–2652.

Benbrook, D., and Pfahl, M. (1987). A novel thyroid hormone receptor encoded by a cDNA clone from a human testis library. *Science* **238,** 788–791.

Bernards, R., Schackleford, G. M., Gerber, M. R., Horowitz, J. M., Friend, S. H., Schartl, M., Bogenmann, E., Rapaport, J. M., McGee, T., Dryja, T. P., and Weinberg, R. A. (1989). Structure and expression of the murine retinoblastoma gene and characterization of its encoded protein. *Proc. Natl. Acad. Sci. U.S.A.* **86,** 6474–6478.

Bernstein, P., and Ross, J. (1989). Poly(A), poly(A) binding protein and the regulation of mRNA stability. *Trends Biochem. Sci.* **14,** 373–377.

Bernstein, P., Peltz, S. W., and Ross, J. (1989). The poly(A)–poly(A)-binding protein complex is a major determinant of mRNA stability in vitro. *Mol. Cell. Biol.* **9,** 659–670.

Brannon, C. I., Dees, E. C., Ingram, R. S., and Tilghman, S. M. (1990). The product of the H19 gene may function as an RNA. *Mol. Cell. Biol.* **10,** 28–36.

Braun, R. E. (1990). Temporal translational regulation of the protamine 1 gene during mouse spermatogenesis. *Enzyme* **44,** 120–128.

Braun, R. E., Behringer, R. R., Peschon, J. J., Brinster, R. L., and Palmiter, R. D. (1989a). Genetically haploid spermatids are phenotypically diploid. *Nature (London)* **337**, 373–376.

Braun, R. E., Peschon, J. J., Behringer, R. R., Brinster, R. L., and Palmiter, R. D. (1989b). Protamine 3'-untranslated sequences regulate temporal translational control and subcellular localization of growth hormone in spermatids of transgenic mice. *Genes Dev.* **3**, 793–802.

Braun, R. E., Lo, D., Pinkert, C. A., Widera, G., Flavell, R. A., Palmiter, R. D., and Brinster, R. L. (1990). Infertility in male transgenic mice: Disruption of sperm development by HSV-tk expression in postmeiotic germ cells. *Biol. Reprod.* **43**, 684–693.

Braun, T. (1974). Evidence for multiple, cell specific, distinctive adenylate cyclase systems in rat testis. *Curr. Top. Mol. Endocrinol.* **1**, 243–264.

Brunkow, M. E., and Tilghman, S. M. (1991). Ectopic expression of the H19 gene in mice causes prenatal lethality. *Genes Dev.* **5**, 1092–1101.

Búcan, M., Yang-Feng, T., Colberg-Poley, A. M., Wolgemuth, D. J., Guenet, J. L., Francke, U., and Lehrach, H. (1986). Genetic and cytogenetic localisation of the homeo box containing genes on mouse chromosome 6 and human chromosome 7. *EMBO J.* **5**, 2899–2905.

Bunick, D., Johnson, P. A., Johnson, T. R., and Hecht, N. B. (1990). Transcription of the testis-specific mouse protamine 2 gene in a homologous *in vitro* transcription system. *Proc. Natl. Acad. Sci. U.S.A.* **87**, 891–895.

Carson-Jurica, M. A., Schrader, W. T., and O'Malley, B. W. (1990). Steroid receptor family: Structure and functions. *Endocrinol. Rev.* **11**, 201–220.

Cattanach, B. M. (1986). Parental origin effects in mice. *J. Embryol. Exp. Morphol. (Suppl.)* **97**, 137–150.

Cedar, H. (1988). DNA methylation and gene activity. *Cell* **53**, 3–4.

Chavrier, P., Lemaire, P., Revelant, O., Bravo, R., and Charnay, P. (1988a). Characterization of a mouse multigene family that encodes zinc finger structures. *Mol. Cell. Biol.* **8**, 1319–1326.

Chavrier, P., Zerial, M., Lemaire, P., Almendral, J., Bravo, R., and Charnay, P. (1988b). A gene encoding a protein with zinc fingers is activated during G0/G1 transition in cultured cells. *EMBO J.* **7**, 29–35.

Chavrier, P., Vesque, C., Galliot, B., Vigneron, M., Dollé, P., Duboule, D., and Charnay, P. (1990). The segment-specific gene *Krox-20* encodes a transcription factor with binding sites in the promoter region of the *Hox-1.4* gene. *EMBO J.* **9**, 1209–1218.

Choi, Y.-C., and Chae, C.-B. (1991). DNA hypomethylation and germ cell-specific expression of testis-specific H2B histone gene. *J. Biol. Chem.* **266**, 20504–20511.

Choi, Y., Henrard, D., Lee, I., and Ross, S. R. (1987). The mouse mammary tumor virus long terminal repeat directs expression in epithelial and lymphoid cells to different tissues in transgenic mice. *J. Virol.* **61**, 3013–3019.

Chowdhury, K., Rohdewohld, H., and Gruss, P. (1988). Specific and ubiquitous expression of different Zn finger protein genes in the mouse. *Nucl. Acids Res.* **16**, 9995–10011.

Colberg-Poley, A. M., Voss, S. D., Chowdhury, K., Stewart, C. L., Wagner, E. F., and Gruss, P. (1985). Clustered homeo boxes are differentially expressed during murine development. *Cell* **43**, 39–45.

Cunliffe, V., Koopman, P., McLaren, A., and Trowsdale, J. (1990a). A mouse zinc finger gene which is transiently expressed during spermatogenesis. *EMBO J.* **9**, 197–205.

Cunliffe, V., Williams, S., and Trowsdale, J. (1990b). Genomic analysis of a mouse zinc finger gene, *Zfp-35*, that is up-regulated during spermatogenesis. *Genomics* **8**, 331–339.

Curran, T., and Franza, B. R. (1988). Fos and Jun: The AP-1 connection. *Cell* **55**, 395–397.

De, S. K., Enders, G. C., and Andrews, G. K. (1991). High levels of metallothionein messenger RNAs in male germ cells of the adult mouse. *Mol. Endocrinol.* **5**, 628–636.

DeChiara, T. M., Efstratiadis, A., and Robertson, E. J. (1990). A growth-deficiency phenotype in heterozygous mice carrying an insulin-like growth factor II gene disrupted by targeting. *Nature (London)* **345**, 78–80.

DeChiara, T. M., Robertson, E. J., and Efstratiadis, A. (1991). Parental imprinting of the mouse insulin-like growth factor II gene. *Cell* **64**, 849–859.

DeLoia, J. A., and Solter, D. (1990). A transgene insertional mutation as an imprinted locus in the mouse genome. *Development (Suppl.)*, 73–79.

Denny, P., and Ashworth, A. (1991). A zinc finger protein-encoding gene expressed in the post-meiotic phase of spermatogenesis. *Gene* 106, 221–227.

de Thé, H., Vivanco-Ruiz, M. M., Tiollais, P., Stunnenberg, H., and Dejean, A. (1990). Identification of a retinoic acid responsive element in the retinoic acid receptor-β gene. *Nature (London)* 343, 177–180.

Driscoll, D. J., and Migeon, B. R. (1990). Sex difference in methylation of single-copy genes in human meiotic germ cells: Implications for X chromosome inactivation, parental imprinting, and origin of CpG mutations. *Somatic Cell Mol. Genet.* 16, 267–282.

Duboule, D., Baron, A., Mähl, P., and Galliot, B. (1986). A new homeo-box is present in overlapping cosmid clones which define the mouse *Hox-1* locus. *EMBO J.* 5, 1973–1980.

Dynan, W. S. (1986). Promoters for housekeeping genes. *Trends Genet.* 2, 196–197.

Eddy, E. M., O'Brien, D. A., and Welch, J. E. (1991). Mammalian sperm development in vivo and in vitro. *In* "Elements of Mammalian Fertilization" (P. M. Wassarman, ed.), Vol. I, pp. 1–28. CRC Press, Boca Raton, Florida.

Einck, L., and Bustin, M. (1985). The intracellular distribution and function of the high mobility group chromosomal proteins. *Exp. Cell Res.* 156, 295–310.

Eissenberg, J. C., and Elgin, S. C. R. (1991). Boundary functions in the control of gene expression. *Trends Genet.* 7, 335–340.

Elgin, S. C. R. (1990). Chromatin structure and gene activity. *Curr. Opin. Cell Biol.* 2, 473–445.

Erickson, R. P. (1990). Post-meiotic gene expression. *Trends Genet.* 6, 264–269.

Erickson, R. P., Kramer, J. M., Rittenhouse, J., and Salkeld, A. (1980). Quantitation of mRNAs during mouse spermatogenesis: Protamine-like histone and phosphoglycerate kinase-2 mRNAs increase after meiosis. *Proc. Natl. Acad. Sci. U.S.A.* 77, 6086–6090.

Ernoult-Lange, M., Kress, M., and Hamer, D. (1990). A gene that encodes a protein consisting solely of zinc finger domains is preferentially expressed in transformed mouse cells. *Mol. Cell. Biol.* 10, 418–421.

Escalier, D., Gallo, J.-M., Albert, M., Meduri, G., Bermudez, D., David, G., and Schrevel, J. (1991). Human acrosome biogenesis: Immunodetection of proacrosin in primary spermatocytes and of its partitioning pattern during meiosis. *Development* 113, 779–788.

Eskild, W., Ree, A. H., Levy, F. O., Jahnsen, T., and Hansson, V. (1991). Cellular localization of mRNAs for retinoic acid receptor-α, cellular retinol-binding protein, and cellular retinoic acid-binding protein in rat testis: Evidence for germ cell-specific mRNAs. *Biol. Reprod.* 44, 53–61.

Featherstone, M. S., Baron, A., Gaunt, S. J., Mattei, M.-G., and Duboule, D. (1988). *Hox-5.1* defines a homeobox-containing gene locus on mouse chromosome 2. *Proc. Natl. Acad. Sci. U.S.A.* 85, 4760–4764.

Forejt, J., and Gregorova, S. (1992). Genetic analysis of genomic imprinting: An Imprintor-1 gene controls inactivation of the paternal copy of the mouse Tme locus. *Cell* 70, 443–450.

Foulkes, N. S., Mellström, B., Benusiglio, E., and Sassone-Corsi, P. (1992). Developmental switch of CREM function during spermatogenesis: From antagonist to activator. *Nature (London)* 355, 80–84.

Fujimoto, H., Erickson, R. P., and Toné, S. (1988). Changes in polyadenylation of lactate dehydrogenase-X mRNA during spermatogenesis in mice. *Mol. Reprod. Dev.* 1, 27–34.

Gallie, D. R. (1991). The cap and poly(A) tail function synergistically to regulate mRNA translational efficiency. *Genes Dev.* 5, 2108–2116.

Galliot, B., Dollé, P., Vigneron, M., Featherstone, M. S., Baron, A., and Duboule, D. (1989). The mouse *Hox-1.4* gene: Primary structure, evidence for promoter activity and expression during development. *Development* 107, 343–359.

Garrett, J. E., Collard, M. W., and Douglass, J. O. (1989). Translational control of germ cell-expressed mRNA imposed by alternative splicing: Opioid peptide gene expression in rat testis. *Mol. Cell. Biol.* 9, 4381–4389.

Gebara, M. M., and McCarrey, J. R. (1992). Protein-DNA interactions associated with the onset of testis-specific expression of the mammalian *Pgk-2* gene. *Mol. Cell. Biol.* **12,** 1422–1431.

Getzenberg, R. H., Pienta, K. J., Ward, W. S., and Coffey, D. S. (1991). Nuclear structure and the three-dimensional organization of DNA. *J. Cell. Biochem.* **47,** 289–299.

Gillies, S. D., Morrison, S. L., Oi, V. T., and Tonegawa, S. (1983). A tissue-specific transcription enhancer element is located in the major intron of a rearranged immunoglobulin heavy chain gene. *Cell* **33,** 717–728.

Gold, B., Fujimoto, H., Kramer, J. M., Erickson, R. P., and Hecht, N. B. (1983). Haploid accumulation and translational control of phosphoglycerate kinase-2 messenger RNA during mouse spermatogenesis. *Dev. Biol.* **98,** 392–399.

Goldberg, E., Sberna, D., Wheat, T. E., Urbanski, G. J., and Margoliash, E. (1977). Cytochrome c: Immunofluorescent localization of the testis-sperm form. *Science* **196,** 1010–1012.

Goldsborough, A., Ashworth, A., and Willison, K. (1990). Cloning and sequencing of POU-boxes expressed in mouse testis. *Nucl. Acids Res.* **18,** 1634.

Gonzalez, G. A., and Montminy, M. R. (1989). Cyclic AMP stimulates somatostatin gene transcription by phosphorylation of CREB at serine 133. *Cell* **59,** 675–680.

Gonzalez, G. A., Yamamoto, K. K., Fischer, W. H., Karr, D., Menzel, P., Biggs, W., Vale, W. W., and Montminy, M. R. (1989). A cluster of phosphorylation sites on the cyclic AMP-regulated nuclear factor CREB predicted by its sequence. *Nature (London)* **337,** 749–752.

Goossen, B., Caughman, S. W., Harford, J. B., Klausner, R. D., and Hentze, M. W. (1990). Translational repression by a complex between the iron-responsive element of ferritin mRNA and its specific cytoplasmic binding protein is position-dependent *in vivo*. *EMBO J.* **9,** 4127–4133.

Gordeladze, J. O., and Hansson, V. (1981). Purification and kinetic properties of the soluble Mn^{2+}-dependent adenylyl cyclase of the rat testis. *Mol. Cell. Endocrinol.* **23,** 125–136.

Gordeladze, J. O., Andersen, D., and Hansson, V. (1981). Physicochemical and kinetic properties of the Mn^{2+}-dependent adenylyl cyclase of the human testis. *J. Clin. Endocrinol. Metab.* **53,** 465–471.

Gordon, J. W., Pravtcheva, D., Poorman, P. A., Moses, M. J., Brock, W. A., and Ruddle, F. H. (1989). Association of foreign DNA sequence with male sterility and translocation in a line of transgenic mice. *Somatic Cell Mol. Genet.* **15,** 569–578.

Gordon, J. W., Uehlinger, J., Dayani, N., Talansky, B. E., Gordon, M., Rudomen, G. S., and Neumann, P. E. (1990). Analysis of the hotfoot *(ho)* locus by creation of an insertional mutation in a transgenic mouse. *Dev. Biol.* **137,** 349–358.

Goto, M., Tamura, T., Mikoshiba, K., Masamune, Y., and Nakanishi, Y. (1991). Transcription inhibition of the somatic-type phosphoglycerate kinase 1 gene in vitro by a testis-specific factor that recognizes a sequence similar to the binding site for Ets oncoproteins. *Nucl. Acids Res.* **19,** 3959–3963.

Graham, A., Papalopulu, N., Lorimer, J., McVey, J. H., Tuddenham, E. G., and Krumlauf, R. (1988). Characterization of a murine homeo box gene, *Hox-2.6,* related to the *Drosophila Deformed* gene. *Genes Dev.* **2,** 1424–1438.

Grimes, S. R., Wolfe, S. A., and Koppel, D. A. (1992a). Tissue-specific binding of testis nuclear proteins to a sequence element within the promoter of the testis-specific histone H1t gene. *Arch. Biochem. Biophys.* **296,** 402–409.

Grimes, S. R., Wolfe, S. A., and Koppel, D. A. (1992b). Temporal correlation between the appearance of testis-specific DNA-binding proteins and the onset of transcription of the testis-specific histone H1t gene. *Exp. Cell Res.* **201,** 216–224.

Griswold, M. D., Bishop, P. D., Kim, K. H., Ping, R., Siiteri, J. E., and Morales, C. (1989). Function of vitamin A in normal and synchronized seminiferous tubules. *Ann. N.Y. Acad. Sci.* **564,** 154–172.

Gubbay, J., Koopman, P., Collignon, J., Burgoyne, P., and Lovell-Badge, R. (1990). Normal structure and expression of *Zfy* genes in XY female mice mutant in *Tdy*. *Development* **109,** 647–653.

Haig, D., and Graham, C. (1991). Genomic imprinting and the strange case of the insulin-like growth factor II receptor. *Cell* **64,** 1045–1046.

Hake, L. E., Alcivar, A. A., and Hecht, N. B. (1990). Changes in mRNA length accompany translational regulation of the somatic and testis-specific cytochrome c genes during spermatogenesis in the mouse. *Development* **110,** 249–257.

Harley, V. R., Jackson, D. I., Hextall, P. J., Hawkins, J. R., Berkovitz, G. D., Sockanathan, S., Lovell-Badge, R., and Goodfellow, P. N. (1992). DNA binding activity of recombinant SRY from normal males and XY females. *Science* **255,** 453–456.

Hatzopoulos, A. K., Stoykova, A. S., Erselius, J. R., Goulding, M., Neuman, T., and Gruss, P. (1990). Structure and expression of the mouse Oct2a and Oct2b, two differentially spliced products of the same gene. *Development* **109,** 349–362.

He, X., Treacy, M. N., Simmons, D. M., Ingraham, H. A., Swanson, L. W., and Rosenfeld, M. G. (1989). Expression of a large family of POU-domain regulatory genes in mammalian brain development. *Nature (London)* **340,** 35–41.

Hecht, N. B. (1990). Regulation of 'haploid expressed genes' in male germ cells. *J. Reprod. Fertil.* **88,** 679–693.

Hecht, N. B. (1992). Gene expression during male germ cell development. *In* "Cell and Molecular Biology of the Testis" (C. Desjardins and L. L. Ewing, eds.), pp. 464–503. Oxford University Press, Oxford.

Hekman, A. C. P., Trapman, J., Mulder, A. H., Van Gaalen, J. L. M., and Zwarthoff, E. C. (1988). Interferon expression in the testes of transgenic mice leads to sterility. *J. Biol. Chem.* **263,** 12151–12155.

Herbomel, P. (1990). From gene to chromosome: Organization levels defined by the interplay of transcription and replication in vertebrates. *New Biol.* **2,** 937–945.

Herr, W., Strum, R. A., Clerc, R. G., Corcoran, L. M., Baltimore, D., Sharp, P. A., Ingraham, H. A., Rosenfeld, M. G., Finney, M., Ruvkun, G., and Horvitz, H. R. (1988). The POU domain: A large conserved region in the mammalian *pit-1, oct-1, oct-2,* and *Caenorhabditis elegans unc-86* gene products. *Genes Dev.* **2,** 1513–1516.

Hershey, J. W. B. (1991). Translational control in mammalian cells. *Ann. Rev. Biochem.* **60,** 717–755.

Hiebert, S. W., Chellappan, S. P., Horowitz, J. M., and Nevins, J. R. (1992). The interaction of RB with E2F coincides with an inhibition of the transcriptional activity of E2F. *Genes Dev.* **6,** 177–185.

Höög, C., Schalling, M., Grunder-Brundell, E., and Daneholt, B. (1991). Analysis of a murine male germ cell-specific transcript that encodes a putative zinc finger protein. *Mol. Reprod. Dev.* **30,** 173–181.

Howard, T. E., Shai, S.-Y., Langford, K. G., Martin, B. M., and Bernstein, K. E. (1990). Transcription of testicular angiotensin-converting enzyme (ACE) is initiated within the 12th intron of the somatic ACE gene. *Mol. Cell. Biol.* **10,** 4294–4302.

Howlett, S. K., Reik, W., Barton, S. C., Norris, M. L., and Surani, M. A. (1989). Genomic imprinting in the mouse. *In* "Developmental Biology, A Comprehensive Synthesis. Genomic Adaptability in Somatic Cell Specialization" (M. A. DiBerardino and L. D. Etkin, eds.), Vol. 6, pp. 59–77. Plenum Press, New York.

Huarte, J., Stutz, A., O'Connell, M. A., Gubler, P., Belin, D., Darrow, A. L., Strickland, S., and Vassalli, J.-D. (1992). Transient translational silencing by reversible mRNA deadenylation. *Cell* **69,** 1021–1030.

Huh, N. E., Hwang, I. W., Lim, K., You, K.-H., and Chae, C.-B. (1991). Presence of a bi-directional S phase-specific transcription regulatory element in the promoter shared by testis-specific TH2A and TH2B histone genes. *Nucl. Acids Res.* **19,** 93–98.

Ismail, N., Morales, C., and Clermont, Y. (1990). Role of spermatogonia in the stage-synchronization of the seminiferous epithelium in vitamin-A-deficient rats. *Am. J. Anat.* **188,** 57–63.

Iwakura, Y., Asano, M., Nishimune, Y., and Kawade, Y. (1988). Male sterility of transgenic mice carrying exogenous mouse interferon-beta gene under the control of the metallothionein enhancer-promoter. *EMBO J.* **7,** 3757–3762.

Jackson, R. J., and Standart, N. (1990). Do the poly(A) tail and 3' untranslated region control mRNA translation? *Cell* **62,** 15–24.

Jagiello, G., Tantravahi, U., Fang, J. S., and Erlanger, B. F. (1982). DNA methylation patterns of human pachytene spermatocytes. *Exp. Cell Res.* **141,** 253–259.

Jannini, E. A., Olivieri, M., Francavilla, S., Gulino, A., Ziparo, E., and D'Armiento, M. (1990). Ontogenesis of the nuclear 3,5,3'-triiodothyronine receptor in the rat testis. *Endocrinology* **126,** 2521–2526.

Jantzen, H. M., Admon, A., Bell, S. P., and Tjian, R. (1990). Nucleolar transcription factor hUBF contains a DNA-binding motif with homology to HMG proteins. *Nature (London)* **344,** 830–836.

Johnson, P. F., and McKnight, S. L. (1989). Eukaryotic transcriptional regulatory proteins. *Annu. Rev. Biochem.* **58,** 799–839.

Johnson, P. A., Peschon, J. J., Yelick, P. C., Palmiter, R. D., and Hecht, N. B. (1988). Sequence homologies in the mouse protamine 1 and 2 genes. *Biochim. Biophys. Acta* **950,** 45–53.

Kadonaga, J. T., Carner, K. R., Masiarz, F. R., and Tjian, R. (1987). Isolation of cDNA encoding transcription factor Sp1 and functional analysis of the DNA binding domain. *Cell* **51,** 1079–1090.

Kafri, T., Ariel, M., Brandeis, M., Shemer, R., Urven, L., McCarrey, J., Cedar, H., and Razin, A. (1992). Developmental pattern of gene-specific DNA methylation in the mouse embryo and germ line. *Genes Dev.* **6,** 705–714.

Kashiwabara, S., Arai, Y., Kodaira, K., and Baba, T. (1990). Acrosin biosynthesis in meiotic and postmeiotic spermatogenic cells. *Biochem. Biophys. Res. Commun.* **173,** 240–245.

Kaufmann, S. H., Wright, W. W., Okret, S., Wikstrom, A. C., Gustafsson, J. A., Shaper, N. L., and Shaper, J. H. (1992). Evidence that rodent epididymal sperm contain the M_r approximately 94,000 glucocorticoid receptor but lack the M_r approximately 90,000 heat shock protein. *Endocrinology* **130,** 3074–3084.

Kelley, K. A., Chamberlain, J. W., Nolan, J. A., Horwich, A. L., Kalousek, F., Eisenstadt, J., Herrup, K., and Rosenberg, L. E. (1988). Meiotic expression of human ornithine transcarbamylase in the testis of transgenic mice. *Mol. Cell. Biol.* **8,** 1821–1825.

Kilpatrick, D. L., Howells, R. D., Noe, M., Bailey, L. C., and Udenfriend, S. (1985). Expression of preproenkephalin-like mRNA and its peptide products in mammalian testis and ovary. *Proc. Natl. Acad. Sci. U.S.A.* **82,** 7467–7469.

Kilpatrick, D. L., Borland, K., and Jin, D. F. (1987). Differential expression of opioid peptide genes by testicular germ cells and somatic cells. *Proc. Natl. Acad. Sci. U.S.A.* **84,** 5695–5699.

Kilpatrick, D. L., Zinn, S. A., Fitzgerald, M., Higuchi, H., Sabol, S. L., and Meyerhardt, J. (1990). Transcription of the rat and mouse proenkephalin genes is initiated at distinct sites in spermatogenic and somatic cells. *Mol. Cell. Biol.* **10,** 3717–3726.

Kim, K. H., and Griswold, M. D. (1990). The regulation of retinoic acid receptor mRNA levels during spermatogenesis. *Mol. Endocrinol.* **4,** 1679–1688.

Kim, Y. J., Hwang, I., Tres, L. L., Kierszenbaum, A. L., and Chae, C.-B. (1987). Molecular cloning and differential expression of somatic and testis-specific H2B histone genes during rat spermatogenesis. *Dev. Biol.* **124,** 23–34.

Kleene, K. C. (1989). Poly(A) shortening accompanies the activation of translation of five mRNAs during spermiogenesis in the mouse. *Development* **106,** 367–373.

Kleene, K. C., Distel, R. J., and Hecht, N. B. (1984). Translational regulation and deadenylation of a protamine mRNA during spermiogenesis in the mouse. *Dev. Biol.* **105,** 71–79.

Kleene, K. C., Smith, J., Bozorgzadeh, A., Harris, M., Hahn, L., Karimpour, I., and Gerstel, J. (1990). Sequence and developmental expression of the mRNA encoding the seleno-protein of the sperm mitochondrial capsule in the mouse. *Dev. Biol.* **137,** 395–402.

Koenig, R. J., Lazar, M. A., Hodin, R. A., Brent, G. A., Larsen, P. R., Chin, W. W., and Moore, D. D. (1989). Inhibition of thyroid hormone action by a non-hormone binding c-erbA protein generated by alternative mRNA splicing. *Nature (London)* **337,** 659–661.

Koji, T., Izumi, S., Tanno, M., Moriuchi, T., and Nakane, P. K. (1988). Localization *in situ* of c-*myc* mRNA and c-myc protein in adult mouse testis. *Histochem. J.* **20,** 551–557.

Koopman, P., Gubbay, J., Collignon, J., and Lovell-Badge, R. (1989). *Zfy* gene expression patterns are not compatible with a primary role in mouse sex determination. *Nature (London)* **342,** 940–942.

Koopman, P., Münsterberg, A., Capel, B., Vivian, N., and Lovell-Badge, R. (1990). Expression of a candidate sex-determining gene during mouse testis differentiation. *Nature (London)* **348,** 450–452.

Koopman, P., Ashworth, A., and Lovell-Badge, R. (1991). The ZFY gene family in humans and mice. *Trends Genet.* **7,** 132–136.

Kremer, E. J., and Kistler, W. S. (1992). Analysis of the promoter for the gene encoding the testis-specific histone H1t in a somatic cell line: Evidence for cell-cycle regulation and modulation by distant upstream sequences. *Gene* **110,** 167–173.

Krulewski, T. F., Neumann, P. E., and Gordon, J. W. (1989). Insertional mutation in a transgenic mouse allelic with Purkinje cell degeneration. *Proc. Natl. Acad. Sci. U.S.A.* **86,** 3709–3712.

Kwon, Y. K., and Hecht, N. B. (1991a). Cytoplasmic protein binding to highly conserved sequences in the 3′ untranslated region of mouse protamine 2 mRNA, a translationally regulated transcript of male germ cells. *Proc. Natl. Acad. Sci. U.S.A.* **88,** 3584–3588.

Kwon, Y. K., and Hecht, N. B. (1991b). Cytoplasmic factors from round spermatid bind to the 3′ untranslated region of mouse protamine 2 mRNA and inhibit translation "in vitro." *J. Cell Biol.* **115,** 99a. (abstract)

Kwon, Y. K., Murray, M. T., Nikolajczyk, B. S., and Hecht, N. B. (1992). Mouse male germ cells contain proteins homologous to germ cell-specific RNA- and DNA-binding proteins of Xenopus oocytes. *Mol. Biol. Cell* **3,** 102a. (abstract)

Lalley, P. A., and Chirgwin, J. M. (1984). Mapping of the mouse insulin genes. *Cytogenet. Cell Genet.* **37,** 515.

Landschulz, W. H., Johnson, P. F., and McKnight, S. L. (1988). The leucine zipper: A hypothetical structure common to a new class of DNA binding proteins. *Science* **240,** 1759–1764.

Landsman, D., and Bustin, M. (1991). Assessment of the transcriptional activation potential of the HMG chromosomal proteins. *Mol. Cell. Biol.* **11,** 4483–4489.

Langdon, W. Y., Hyland, C. D., Grumont, R. J., and Morse, H. C., III (1989). The c-*cb1* proto-oncogene is preferentially expressed in thymus and testis tissue and encodes a nuclear protein. *J. Virol.* **63,** 5420–5424.

Langford, K. G., Shai, S.-Y., Howard, T. E., Kovac, M. J., Overbeek, P. A., and Bernstein, K. E. (1991). Transgenic mice demonstrate a testis-specific promoter for angiotensin-converting enzyme. *J. Biol. Chem.* **266,** 15559–15562.

Lazar, M. A., Hodin, R. A., Darling, D. S., and Chin, W. W. (1988). Identification of a rat c-erbA α-related protein which binds deoxyribonucleic acid but does not bind thyroid hormone. *Mol. Endocrinol.* **2,** 893–901.

Lazar, M. A., Hodin, R. A., and Chin, W. W. (1989). Human carboxyl-terminal variant of alpha-type c-erbA inhibits transactivation by thyroid hormone receptors without binding thyroid hormone. *Proc. Natl. Acad. Sci. U.S.A.* **86,** 7771–7774.

Leder, A., Pattengale, P. K., Kuo, A., Stewart, T. A., and Leder, P. (1986). Consequences of widespread deregulation of the c-myc gene in transgenic mice: Multiple neoplasms and normal development. *Cell* **45,** 485–495.

Levine, M., and Hoey, T. (1988). Homeobox proteins as sequence-specific transcription factors. *Cell* **55,** 537–540.

Levy, E. R., and Burgoyne, P. S. (1986). The fate of XO germ cells in the testes of XO/XY and

XO/XY/XYY mouse mosaics: Evidence for a spermatogenesis gene on the mouse Y chromosome. *Cytogenet. Cell. Genet.* **42,** 208–213.

Levy, F. O., Ree, A. H., Eikvar, L., Govindan, M. V., Jahnsen, T., and Hansson, V. (1989). Glucocorticoid receptors and glucocorticoid effects in rat Sertoli cells. *Endocrinology* **124,** 430–436.

Lim, K., and Chae, C.-B. (1992). Presence of a repressor protein for testis-specific H2B (TH2B) histone gene in early stages of spermatogenesis. *J. Biol. Chem.* **267,** 15271–15273.

Lobel, P., Dahms, N. M., and Kornfeld, S. (1988). Cloning and sequence analysis of the cation-independent mannose 6-phosphate receptor. *J. Biol. Chem.* **263,** 2563–2570.

Lönnerberg, P., Parvinen, M., Jahnsen, T., Hansson, V., and Persson, H. (1992). Stage- and cell-specific expression of cyclic adenosine 3′,5′-monophosphate-dependent protein kinases in rat seminiferous epithelium. *Biol. Reprod.* **46,** 1057–1068.

Lozano, G., and Levine, A. J. (1991). Tissue-specific expression of *p53* in transgenic mice is regulated by intron sequences. *Mol. Carcinogen.* **4,** 3–9.

McCarrey, J. R., and Thomas, K. (1987). Human testis-specific PGK gene lacks introns and possesses characteristics of a processed gene. *Nature (London)* **326,** 501–505.

MacGregor, G. R., Russell, L. D., Van Beek, M. E. A. B., Hanten, G. R., Kovac, M. J., Kozak, C. A., Meistrich, M. L., and Overbeek, P. A. (1990). Symplastic spermatids *(sys):* A recessive insertional mutation in mice causing a defect in spermatogenesis. *Proc. Natl. Acad. Sci. U.S.A.* **87,** 5016–5020.

Magram, J., and Bishop, J. M. (1991). Dominant male sterility in mice caused by insertion of a transgene. *Proc. Natl. Acad. Sci. U.S.A.* **88,** 10327–10331.

Mangues, R., Seidman, I., Pellicer, A., and Gordon, J. W. (1990). Tumorigenesis and male sterility in transgenic mice expressing a MMTV/N-*ras* oncogene. *Oncogene* **5,** 1491–1497.

Manzella, J. M., Rychlik, W., Rhoads, R. E., Hershey, J. W. B., and Blackshear, P. J. (1991). Insulin induction of ornithine decarboxylase. Importance of mRNA secondary structure and phosphorylation of eucaryotic initiation factors eIF-4B and eIF-4E. *J. Biol. Chem.* **266,** 2383–2389.

Mardon, G., and Page, D. C. (1989). The sex-determining region of the mouse Y chromosome encodes a protein with a highly acidic domain and 13 zinc fingers. *Cell* **56,** 765–770.

Mardon, G., Luoh, S.-W., Simpson, E. M., Gill, G., Brown, L. G., and Page, D. C. (1990). Mouse Zfx protein is similar to Zfy-2: each contains an acidic activating domain and 13 zinc fingers. *Mol. Cell. Biol.* **10,** 681–688.

Meijer, D., Hermans, A., von Lindern, M., van Agthoven, T., de Klein, A., Mackenbach, P., Grootegoed, A., Talarico, D., Valle, G. D., and Grosveld, G. (1987). Molecular characterization of the testis specific c-*abl* mRNA in mouse. *EMBO J.* **6,** 4041–4048.

Meijlink, F., de Laaf, R., Verrijzer, P., Destree, O., Kroezen, V., Hilkens, J., and Deschamps, J. (1987). A mouse homeobox containing gene on chromosome 11: Sequence and tissue-specific expression. *Nucl. Acids Res.* **15,** 6773–6786.

Meistrich, M. L., Bucci, L. R., Trostle-Weige, P. K., and Brock, W. A. (1985). Histone variants in rat spermatogonia and primary spermatocytes. *Dev. Biol.* **112,** 230–240.

Merlino, G. T., Stahle, C., Jhappan, C., Linton, R., Mahon, K. A., and Willingham, M. C. (1991). Inactivation of a sperm motility gene by insertion of an epidermal growth factor receptor transgene whose product is overexpressed and compartmentalized during spermatogenesis. *Genes Dev.* **5,** 1395–1406.

Mermelstein, F. H., Flores, O., and Reinberg, D. (1989). Initiation of transcription by RNA polymerase II. *Biochim. Biophys. Acta* **1009,** 1–10.

Miller, J., McLachlan, A. D., and Klug, A. (1985). Repetitive zinc-binding domains in the protein transcription factor IIIA from *Xenopus* oocytes. *EMBO J.* **4,** 1609–1614.

Mitchell, P. J., and Tjian, R. (1989). Transcriptional regulation in mammalian cells by sequence-specific DNA binding proteins. *Science* **245,** 371–378.

Monk, M., Boubelik, M., and Lehnert, S. (1987). Temporal and regional changes in DNA methylation in the embryonic, extraembryonic and germ cell lineages during mouse embryo development. *Development* **99**, 371–382.

Morales, C., and Griswold, M. D. (1987). Retinol-induced stage synchronization in seminiferous tubules of the rat. *Endocrinology* **121**, 432–434.

Morgan, D. O., Edman, J. C., Standring, D. N., Fried, V. A., Smith, M. C., Roth, R. A., and Rutter, W. J. (1987). Insulin-like growth factor II receptor as a multifunctional binding protein. *Nature (London)* **329**, 301–307.

Mullen, R. J., and La Vail, M. M. (1975). Two new types of retinal degeneration in cerebellar mutant mice. *Nature (London)* **258**, 528–530.

Mullen, R. J., Eicher, E. M., and Sidman, R. L. (1976). Purkinje cell degeneration, a new neurological mutation in the mouse. *Proc. Natl. Acad. Sci. U.S.A.* **73**, 208–212.

Murre, C., McCaw, P. S., and Baltimore, D. (1989). A new DNA binding and dimerization motif in immunoglobulin enhancer binding, *daughterless*, *MyoD*, and *myc* proteins. *Cell* **56**, 777–783.

Nagamine, C. M., Chan, K., Hake, L. E., and Lau, Y. F. C. (1990). The two candidate testis-determining Y genes (*Zfy-1* and *Zfy-2*) are differentially expressed in fetal and adult mouse tissues. *Genes Dev.* **4**, 63–74.

Nasrin, N., Buggs, C., Kong, X. F., Carnazza, J., Goebl, M., and Alexander-Bridges, M. (1991). DNA-binding properties of the product of the testis-determining gene and a related protein. *Nature (London)* **354**, 317–320.

Nayernia, K., Burkhardt, E., Beimesche, S., Keime, S., and Engel, W. (1992). Germ cell-specific expression of a proacrosin-CAT fusion gene in transgenic mouse testis. *Mol. Reprod. Dev.* **31**, 241–248.

Nelki, D., Dudley, K., Cunningham, P., and Akhavan, M. (1990). Cloning and sequencing of a zinc finger cDNA expressed in mouse testis. *Nucl. Acids Res.* **18**, 3655.

O'Brien, D. A., Gabel, C. A., Rockett, E. L., and Eddy, E. M. (1989). Receptor-mediated endocytosis and differential synthesis of mannose 6-phosphate receptors in isolated spermatogenic and Sertoli cells. *Endocrinology* **125**, 2973–2984.

O'Brien, D. A., Welch, J. E., Goulding, E. H., Taylor, A. A., Jr., Baba, T., Hecht, N. B., and Eddy, E. M. (1992). Expression of boar proacrosin in transgenic mice disrupts spermatogenesis. *Mol. Biol. Cell* **3**, 100a. (abstract)

Odenwald, W. F., Taylor, C. F., Palmer-Hill, F. J., Friedrich, V., Tani, M., and Lazzarini, R. A. (1987). Expression of a homeo domain protein in noncontact-inhibited cultured cells and postmitotic neurons. *Genes Dev.* **1**, 482–496.

Oppenheimer, J. H., Schwartz, H. L., and Surks, M. I. (1974). Tissue differences in the concentration of triiodothyronine nuclear binding sites in the rat: Liver, kidney, pituitary, heart, brain, spleen, and testis. *Endocrinology* **95**, 897–903.

Oppi, C., Shore, S. K., and Reddy, E. P. (1987). Nucleotide sequence of testis-derived c-*abl* cDNAs: Implications for testis-specific transcription and *abl* oncogene activation. *Proc. Natl. Acad. Sci. U.S.A.* **84**, 8200–8204.

Overbeek, P. A., Lai, S.-P., Van Quill, K. R., and Westphal, H. (1986). Tissue-specific expression in transgenic mice of a fused gene containing RSV terminal sequences. *Science* **231**, 1574–1577.

Øyen, O., Myklebust, F., Scott, J. D., Cadd, G. G., McKnight, G. S., Hansson, V., and Jahnsen, T. (1990). Subunits of cyclic adenosine 3′,5′-monophosphate-dependent protein kinase show differential and distinct expression patterns during germ cell differentiation: Alternative polyadenylation in germ cells gives rise to unique smaller-sized mRNA species. *Biol. Reprod.* **43**, 46–54.

Page, D. C., Mosher, R., Simpson, E. M., Fisher, E. M., Mardon, G., Pollack, J., McGillivray, B., de-la-Chapelle, A., and Brown, L. G. (1987). The sex-determining region of the human Y chromosome encodes a finger protein. *Cell* **51**, 1091–1104.

Palmer, M. S., Sinclair, A. H., Berta, P., Ellis, N. A., Goodfellow, P. N., Abbas, N. E., and

Fellous, M. (1989). Genetic evidence that ZFY is not the testis-determining factor. *Nature (London)* **342**, 937–939.

Palmiter, R. D., Brinster, R. L., Hammer, R. E., Trumbauer, M. E., Rosenfeld, M. G., Birnberg, N. C., and Evans, R. M. (1982). Dramatic growth of mice that develop from eggs microinjected with metallothionein-growth hormone fusion genes. *Nature (London)* **333**, 611–615.

Palmiter, R. D., Wilkie, T. M., Chen, H. Y., and Brinster, R. L. (1984). Transmission distortion and mosaicism in an unusual transgenic mouse pedigree. *Cell* **36**, 869–877.

Parisi, M. A., and Clayton, D. A. (1991). Similarity of human mitochondrial transcription factor 1 to high mobility group proteins. *Science* **252**, 965–969.

Peschon, J. J., Behringer, R. R., Brinster, R. L., and Palmiter, R. D. (1987). Spermatid-specific expression of protamine 1 in transgenic mice. *Proc. Natl. Acad. Sci. U.S.A.* **84**, 5316–5319.

Peschon, J. J., Behringer, R. R., Palmiter, R. D., and Brinster, R. L. (1989). Expression of mouse protamine 1 genes in transgenic mice. *Ann. N.Y. Acad. Sci.* **564**, 186–197.

Petkovich, M., Brand, N. J., Krust, A., and Chambon, P. (1987). A human retinoic acid receptor which belongs to the family of nuclear receptors. *Nature (London)* **330**, 444–450.

Pinkert, C., Widera, G., Cowing, C., Heber-Katz, E., Palmiter, R. D., Flavell, R. A., and Brinster, R. L. (1985). Tissue-specific, inducible and functional expression of the Ead MHC class II gene in transgenic mice. *EMBO J.* **4**, 2225–2230.

Ponzetto, C., and Wolgemuth, D. J. (1985). Haploid expression of a unique c-*abl* transcript in the mouse male germ line. *Mol. Cell. Biol.* **5**, 1791–1794.

Ponzetto, C., Wadewitz, A. G., Pendergast, A. M., Witte, O. N., and Wolgemuth, D. J. (1989). P150^{c-abl} is detected in mouse male germ cells by an *in vitro* kinase assay and is associated with stage-specific phosphoproteins in haploid cells. *Oncogene* **4**, 685–690.

Pugh, B. F., and Tjian, R. (1990). Mechanism of transcriptional activation by Sp1: Evidence for coactivators. *Cell* **61**, 1187–1197.

Rahe, B., Erickson, R. P., and Quinto, M. (1983). Methylation of unique sequence DNA during spermatogenesis in mice. *Nucl. Acids Res.* **11**, 7947–7959.

Rao, V. N., Huebner, K., Isobe, M., ar-Rushdi, A., Croce, C. M., and Reddy, E. S. (1989). *elk*, tissue specific *ets*-related genes on chromosomes X and 14 near translocation breakpoints. *Science* **244**, 66–70.

Razin, A., and Cedar, H. (1991). DNA methylation and gene expression. *Microbiol. Rev.* **55**, 451–458.

Reik, W. (1989). Genomic imprinting and genetic disorders in man. *Trends Genet.* **5**, 331–338.

Reik, W., Collick, A., Norris, M. L., Barton, S. C., and Surani, M. A. (1987). Genomic imprinting determines methylation of parental alleles in transgenic mice. *Nature (London)* **328**, 248–251.

Reynolds, R. K., Hoekzema, G. S., Vogel, J., Hinrichs, S. H., and Jay, G. (1988). Multiple endocrine neoplasia induced by the promiscuous expression of a viral oncogene. *Proc. Natl. Acad. Sci. U.S.A.* **85**, 3135–3139.

Richter, J. D. (1991). Translational control during early development. *BioEssays* **13**, 179–183.

Robertson, N. G., Pomponio, R. J., Mutter, G. L., and Morton, C. C. (1991). Testis-specific expression of the human *MYCL2* gene. *Nucl. Acids. Res.* **19**, 3129–3137.

Robinson, M. O., McCarrey, J. R., and Simon, M. I. (1989). Transcriptional regulatory regions of testis-specific *PGK2* defined in transgenic mice. *Proc. Natl. Acad. Sci. U.S.A.* **86**, 8437–8441.

Rogers, M. B., Hosler, B. A., and Gudas, L. J. (1991). Specific expression of a retinoic acid-regulated, zinc-finger gene, *Rex-1*, in preimplantation embryos, trophoblast and spermatocytes. *Development* **113**, 815–824.

Rosner, M. H., Vigano, M. A., Ozato, K., Timmons, P. M., Poirier, F., Rigby, P. W. J., and Staudt, L. M. (1990). A POU-domain transcription factor in early stem cells and germ cells of the mammalian embryo. *Nature (London)* **345**, 686–692.

Rubin, M. R., Toth, L. E., Patel, M. D., D'Eustachio, P., and Nguyen-Huu, M. C. (1986). A mouse homeo box gene is expressed in spermatocytes and embryos. *Science* **233**, 663–667.

Ruppert, S., Cole, T. J., Boshart, M., Schmid, E., and Schütz, G. (1992). Multiple mRNA isoforms of the transcription activator protein CREB: Generation by alternative splicing and specific expression in primary spermatocytes. *EMBO J.* **11**, 1503–1512.

Russell, L. D., Sinha Hikim, A. P., Overbeek, P. A., and MacGregor, G. R. (1991). Testis structure in the *sys (symplastic spermatids)* mouse. *Am. J. Anat.* **192**, 169–182.

Sachs, A. B., and Davis, R. W. (1989). The poly(A) binding protein is required for poly(A) shortening and 60S ribosomal subunit-dependent translation initiation. *Cell* **58**, 857–867.

Saffer, J. D., Jackson, S. P., and Annarella, M. B. (1991). Developmental expression of Sp1 in the mouse. *Mol. Cell. Biol.* **11**, 2189–2199.

Santoro, C., Mermod, N., Andrews, P. C., and Tjian, R. (1988). A family of human CCAAT-box-binding proteins active in transcription and DNA replication: Cloning and expression of multiple cDNAs. *Nature (London)* **334**, 218–224.

Sap, J., Munoz, A., Damm, K., Goldberg, Y., Ghysdael, J., Leutz, A., Beug, H., and Vennstrom, B. (1986). The c-erb-A protein is a high-affinity receptor for thyroid hormone. *Nature (London)* **324**, 635–640.

Sapienza, C., Peterson, A. C., Rossant, J., and Balling, R. (1987). Degree of methylation of transgenes is dependent on gamete of origin. *Nature (London)* **328**, 251–254.

Sapienza, C., Tran, T.-H., Paquette, J., McGowan, R., and Peterson, A. (1989). A methylation mosaic model for mammalian genome imprinting. *In* "Progress in Nucleic Acid Research and Molecular Biology" (M. E. Cohen and K. Moldave, eds.), Vol. 36, pp. 145–157. Academic Press, Orlando, Florida.

Sar, M., Lubahn, D. B., French, F. S., and Wilson, E. M. (1990). Immunohistochemical localization of the androgen receptor in rat and human tissues. *Endocrinology* **127**, 3180–3186.

Sawadogo, M., and Sentenac, A. (1990). RNA polymerase B (II) and general transcription factors. *Ann. Rev. Biochem.* **59**, 711–754.

Schäfer, M., Kuhn, R., Bosse, B., and Schäfer, U. (1990). A conserved element in the leader mediates post-meiotic translation as well as cytoplasmic polyadenylation of a *Drosophila* spermatocyte mRNA. *EMBO J.* **9**, 4519–4525.

Schneider-Gädicke, A., Beer-Romero, P., Brown, L. G., Mardon, G., Luoh, S. W., and Page, D. C. (1989). Putative transcription activator with alternative isoforms encoded by human ZFX gene. *Nature (London)* **343**, 708–711.

Schöler, H. R. (1991). Octamania: The POU factors in murine development. *Trends Genet.* **7**, 323–329.

Schöler, H. R., Hatzopoulos, A. K., Balling, R., Suzuki, N., and Gruss, P. (1989). A family of octamer-specific proteins present during mouse embryogenesis: Evidence for germline-specific expression of an Oct factor. *EMBO J.* **8**, 2543–2550.

Schöler, H. R., Dressler, G. R., Balling, R., Rohdewohld, H., and Gruss, P. (1990). Oct-4: A germline-specific transcription factor mapping to the mouse t-complex. *EMBO J.* **9**, 2185–2195.

Schueler, P. A., Schwartz, H. L., Strait, K. A., Mariash, C. N., and Oppenheimer, J. H. (1990). Binding of 3,5,3'-triiodothyronine (T_3) and its analogs to the *in vitro* translational products of c-erbA protooncogenes: Differences in the affinity of the $\beta\alpha\beta$- and $\beta\beta\beta$-forms for the acetic acid analog and failure of the human testis and kidney $\beta\alpha\beta$-2 products to bind T_3. *Mol. Endocrinol.* **4**, 227–234.

Searle, A. G. and Beechey, C. B. (1985). Noncomplementation phenomena and their bearing on nondisjunctional events. *In* "Aneuploidy" (V. L. Dellarco, P. E. Voytek, and A. Hollaender, eds.), pp. 363–376. Plenum, New York.

Searle, A. G., Peters, J., Lyon, M. F., Hall, J. G., Evans, E. P., Edwards, J. H., and Buckle, V. J. (1989). Chromosome maps of man and mouse. IV. *Ann. Hum. Genet.* **53**, 89–140.

Serfling, E., Jasin, M., and Schaffner, W. (1985). Enhancers and eukaryotic gene transcription. *Trends Genet.* **1**, 224–230.

Sharpe, P. T., Miller, J. R., Evans, E. P., Burtenshaw, M. D., and Gaunt, S. J. (1988). Isolation and expression of a new mouse homeobox gene. *Development* **102**, 397–407.

Shaw, G., and Kamen, R. (1986). A conserved AU sequence from the 3′ untranslated region of GM-CSF mRNA mediates selective mRNA degradation. *Cell* **46**, 659–667.

Siiteri, J. E., Karl, A. F., Linder, C. C., and Griswold, M. D. (1992). Testicular synchrony: Evaluation and analysis of different protocols. *Biol. Reprod.* **46**, 284–289.

Sinclair, A. H., Berta, P., Palmer, M. S., Hawkins, J. R., Griffiths, B. L., Smith, M. J., Foster, J. W., Frischauf, A.-M., Lovell-Badge, R., and Goodfellow, P. N. (1990). A gene from the human sex-determining region encodes a protein with homology to a conserved DNA-binding motif. *Nature (London)* **346**, 240–244.

Singh, J., and Dixon, G. H. (1990). High mobility group proteins 1 and 2 function as general class II transcription factors. *Biochemistry* **29**, 6295–6302.

Solter, D. (1988). Differential imprinting and expression of maternal and paternal genomes. *Ann. Rev. Genet.* **22**, 127–146.

Stewart, T. A., Bellvé, A. R., and Leder, P. (1984). Transcriptional and promoter usage of the *mcy* gene in normal somatic and spermatogenic cells. *Science* **226**, 707–710.

Stewart, T. A., Hecht, N. B., Hollingshead, P. G., Johnson, P. A., Leong, J. A. C., and Pitts, S. L. (1988). Haploid-specific transcription of protamine–*myc* and protoamine–T-antigen fusion genes in transgenic mice. *Mol. Cell. Biol.* **8**, 1748–1755.

Stoykova, A. S., Sterrer, S., Erselius, J. R., Hatzopoulos, A. K., and Gruss, P. (1992). Mini-Oct and Oct-2c: two novel, functionally diverse murine *Oct-2* gene products are differentially expressed in the CNS. *Neuron* **8**, 541–558.

Sturm, R. A., Das, G., and Herr, W. (1988). The ubiquitous octamer-binding protein Oct-1 contains a POU domain with a homeo box subdomain. *Genes Dev.* **2**, 1582–1599.

Sucov, H. M., Murakami, K. K., and Evans, R. M. (1990). Characterization of an autoregulated response element in the mouse retinoic acid receptor type β gene. *Proc. Natl. Acad. Sci. U.S.A.* **87**, 5392–5396.

Surani, M. A. (1984). Differential roles of paternal and maternal genomes during embryogenesis in the mouse. *BioEssays* **1**, 224–227.

Surani, M. A. H., Barton, S. C., and Norris, M. L. (1984). Development of reconstituted mouse eggs suggest imprinting of the genome during gametogenesis. *Nature (London)* **308**, 548–550.

Surani, M. A. H., Reik, W., Norris, M. L., and Barton, S. C. (1986). Influence of germline modifications of homologous chromosomes on mouse development. *J. Embryol. Exp. Morphol. (Suppl.)* **97**, 123–136.

Surani, M. A., Reik, W., and Allen, N. D. (1988). Transgenes as molecular probes for genomic imprinting. *Trends Genet.* **4**, 59–62.

Surani, M. A., Allen, N. D., Barton, S. C., Fundlele, R., Howlett, S. K., Norris, M. L., and Reik, W. (1990). Developmental consequences of imprinting of parental chromosomes by DNA methylation. *Phil. Trans. R. Soc. Lond. B* **326**, 313–327.

Sutcliffe, M. J., and Burgoyne, P. S. (1989). Analysis of the testes of H-Y negative XOSxr[b] mice suggests that the spermatogenesis gene *(Spy)* acts during the differentiation of the A spermatogonia. *Development* **107**, 373–380.

Suzuki, N., Rohdewohld, H., Neuman, T., Gruss, P., and Schöler, H. R. (1990). Oct-6: A POU transcription factor expressed in embryonal stem cells and in the developing brain. *EMBO J.* **9**, 3723–3732.

Svaren, J., and Chakley, R. (1990). The structure and assembly of active chromatin. *Trends Genet.* **6**, 52–56.

Swain, J. L., Stewart, T. A., and Leder, P. (1987). Parental legacy determines methylation and expression of an autosomal transgene: A molecular mechanism for parental imprinting. *Cell* **50**, 719–727.

Tamura, T., Makino, Y., Mikoshiba, K., and Muramatsu, M. (1992). Demonstration of a testis-specific *trans*-acting factor Tet-1 *in vitro* that binds to the promoter of the mouse prot-amine 1 gene. *J. Biol. Chem.* **267**, 4327–4332.

Thach, R. E. (1992). Cap recap: The involvement of eIF-4F in regulating gene expression. *Cell* **68**, 177–180.

Thomas, K., Del Mazo, J., Eversole, P., Bellvé, A., Hiraoka, Y., Li, S. S. L., and Simon, M. (1990). Developmental regulation of expression of the lactate dehydrogenase (LDH) multigene family during mouse spermatogenesis. *Development* **109**, 483–493.

Trasler, J. M., Hake, L. E., Johnson, P. A., Alcivar, A. A., Millette, C. F., and Hecht, N. B. (1990). DNA methylation and demethylation events during meiotic prophase in the mouse testis. *Mol. Cell. Biol.* **10**, 1828–1834.

Tsuruta, J. K., and O'Brien, D. A. (1992). Elevated steady-state levels of c-*fos* mRNA in isolated spermatogenic cells after treatment with fetal bovine serum or phorbol ester. *Mol. Biol. Cell* **3**, 102a. (abstract)

Van der Hoorn, F. A. (1992). Identification of the testis c-mos promoter: Specific activity in a seminiferous tubule-derived extract and binding of a testis-specific nuclear factor. *Onco-gene* **7**, 1093–1099.

Van der Hoorn, F. A., and Tarnasky, H. A. (1992). Factors involved in regulation of the RT7 promoter in a male germ cell-derived *in vitro* transcription system. *Proc. Natl. Acad. Sci. U.S.A.* **89**, 703–707.

Van der Hoorn, F. A., Tarnasky, H. A., and Nordeen, S. K. (1990). A new rat gene RT7 is specifically expressed during spermatogenesis. *Develop. Biol.* **142**, 147–154.

Van Pelt, A. M. M., and De Rooij, D. G. (1990a). The origin of the synchronization of the seminiferous epithelium in vitamin A-deficient rats after vitamin A replacement. *Biol. Reprod.* **42**, 677–682.

Van Pelt, A. M. M., and De Rooij, D. G. (1990b). Synchronization of the seminiferous epithe-lium after vitamin A replacement in vitamin A-deficient mice. *Biol. Reprod.* **43**, 363–367.

Van Pelt, A. M. M., and De Rooij, D. G. (1991). Retinoic acid is able to reinitiate spermatogenesis in vitamin A-deficient rats and high replicate doses support the full development of spermatogenic cells. *Endocrinology* **128**, 697–704.

Van Pelt, A. M. M., Van Den Brink, C. E., De Rooij, D. G., and Van Der Saag, P. T. (1992). Changes in retinoic acid receptor messenger ribonucleic acid levels in the vitamin A-defi-cient rat testis after administration of retinoids. *Endocrinology* **131**, 344–350.

Vassalli, J.-D., Huarte, J., Belin, D., Gubler, P., Vassalli, A., O'Connell, M. L., Parton, L. A., Rickles, R. J., and Strickland, S. (1989). Regulated polyadenylation controls mRNA translation during meiotic maturation of mouse oocytes. *Genes Dev.* **3**, 2163–2171.

Virbasius, J. V., and Scarpulla, R. C. (1988). Structure and expression of rodent genes encoding the testis-specific cytochrome c. Differences in gene structure and evolution between somatic and testicular variants. *J. Biol. Chem.* **263**, 6791–6796.

Waeber, G., Meyer, T. E., LeSieur, M., Hermann, H. L., Gérard, N., and Habener, J. F. (1991). Developmental stage-specific expression of cyclic adenosine 3′,5′-monophosphate re-sponse element-binding protein CREB during spermatogenesis involves alternative exon splicing. *Mol. Endocrinol.* **5**, 1418–1430.

Weinberger, C., Thompson, C. C., Ong, E. S., Lebo, R., Gruol, D. J., and Evans, R. M. (1986). The c-erb-A gene encodes a thyroid hormone receptor. *Nature (London)* **324**, 641–646.

Welch, J. E., Swinnen, J. V., O'Brien, D. A., Eddy, E. M., and Conti, M. (1992). Unique adenosine 3′,5′ cyclic monophosphate phosphodiesterase messenger ribonucleic acids in rat sper-matogenic cells: Evidence for differential gene expression during spermatogenesis. *Biol. Reprod.* **46**, 1027–1033.

West, N. B., and Brenner, R. M. (1990). Estrogen receptor in the ductuli efferentes, epididymis, and testis of rhesus and cynomolgus macaques. *Biol. Reprod.* **42**, 533–538.

Wilkie, T. M., and Palmiter, R. D. (1987). Analysis of the integrant in MyK-103 transgenic mice in which males fail to transmit the integrant. *Mol. Cell. Biol.* **7**, 1646–1655.

Wilkie, T. M., Braun, R. E., Ehrman, W. J., Palmiter, R. D., and Hammer, R. E. (1991). Germ-line intrachromosomal recombination restores fertility in transgenic MyK-103 male mice. *Genes Dev.* **5,** 38–48.

Willison, K., and Ashworth, A. (1987). Mammalian spermatogenic gene expression. *Trends Genet.* **3,** 351–355.

Wingender, E. (1990). Transcription regulating proteins and their recognition sequences. *Crit. Rev. Eukaryot. Gene. Expr.* **1,** 11–48.

Wolfe, S. A., and Grimes, S. R. (1991). Protein-DNA interactions within the rat histone H4t promoter. *J. Biol. Chem.* **266,** 6637–6643.

Wolfes, H., Kogawa, K., Millette, C. F., and Cooper, G. M. (1989). Specific expression of nuclear proto-oncogenes before entry into meiotic prophase of spermatogenesis. *Science* **245,** 740–743.

Wolgemuth, D. J., and Watrin, F. (1991). List of cloned mouse genes with unique expression patterns during spermatogenesis. *Mammalian Genome* **1,** 283–288.

Wolgemuth, D. J., Engelmyer, E., Duggal, R. N., Gizang-Ginsberg, E., Mutter, G. L., Ponzetto, C., Viviano, C., and Zakeri, Z. F. (1986). Isolation of a mouse cDNA conding for a developmentally regulated, testis-specific transcript containing homeo box homology. *EMBO J.* **5,** 1229–1235.

Wolgemuth, D. J., Viviano, C. M., Gizang-Ginsberg, E., Frohman, M. A., Joyner, A. L., and Martin, G. R. (1987). Differential expression of the mouse homeobox-containing gene *Hox-1.4* during male germ cell differentiation and embryonic development. *Proc. Natl. Acad. Sci. U.S.A.* **84,** 5813–5817.

Wolgemuth, D. J., Behringer, R. R., Mostoller, M. P., Brinster, R. L., and Palmiter, R. D. (1989). Transgenic mice overexpressing the mouse homoeobox-containing gene *Hox-1.4* exhibit abnormal gut development. *Nature (London)* **337,** 464–467.

Wolgemuth, D. J., Viviano, C. M., and Watrin, F. (1991). Expression of homeobox genes during spermatogenesis. *Ann. N.Y. Acad. Sci.* **637,** 300–312.

Wright, J. M., and Dixon, G. H. (1988). Induction by torsional stress of an altered DNA conformation 5′ upstream of the gene for a high mobility group protein from trout and specific binding to flanking sequences by the gene product HMG-T. *Biochemistry* **27,** 576–581.

Yelick, P. C., Kwon, Y. K., Flynn, J. F., Borzorgzadeh, A., Kleene, K. C., and Hecht, N. B. (1989). Mouse transition protein 1 is translationally regulated during the postmeiotic stages of spermatogenesis. *Mol. Reprod. Dev.* **1,** 193–200.

Zakeri, Z. F., Ponzetto, C., and Wolgemuth, D. J. (1988). Translational regulation of the novel haploid-specific transcripts for the c-*abl* proto-oncogene and a member of the 70 kDa heat-shock protein gene family in the male germ line. *Dev. Biol.* **125,** 417–422.

7

Molecular Basis of Signaling in Spermatozoa

DANIEL M. HARDY & DAVID L. GARBERS

I. Introduction

The union of sperm and egg to form a cell capable of developing into a new individual must be considered one of the most dramatic events in biology. Successful fusion of the gametes represents the culmination of a complex, multistep process — fertilization — that begins with mating. The duration of this process varies depending on the species; thus, although the moment the two gametes fuse is a discrete point in time, the requisite cellular events preceding it can take from seconds (e.g., sea urchins) to months (e.g., some insects and bats) to complete. This wide range in the duration of fertilization among taxa suggests that the cells involved are receiving instructions. Indeed, more than 80 years have passed since the first report of evidence that factors emanating from the egg affect sperm physiology (Lillie, 1913). In spite of this and other early studies describing effects of egg factors on sperm motility (Gray, 1928; Carter, 1931; Rothschild, 1956) and the sperm acrosome reaction (Dan, 1952,1954), little was known about the chemical nature of these signals until the 1980s.

The past decade of research by several groups has led to the identification of egg-associated factors that regulate motility and induce the acrosome reaction of spermatozoa. Motility is essential both for the journey to the vicinity of the egg and for penetration of its exterior layers. The sperm acrosome reaction, an exocytotic event, exposes enzymes or proteins that facilitate binding and penetration of these egg investments, preparing the fertilizing spermatozoon for fusion with the egg plasma membrane (Yanagimachi, 1981). Progress in understanding the signaling mechanisms that regulate these processes has been most rapid in various mammalian and sea urchin species. Accordingly, we review the current status of research on the regulation of sperm motility and the acrosome reaction in these animals, with particular attention to results published since an earlier general review (Garbers, 1989). For more focused coverage of certain aspects of sperm signaling during fertilization, the reader is referred to reviews of sea urchin egg jelly molecules (Suzuki, 1990), regulation of sperm motility by egg peptides (Domino and Garbers, 1990), species specificity of egg peptides (Suzuki and Yoshino, 1992), and induction of the mammalian sperm acrosome reaction by the egg zona pellucida (Kopf and Gerton, 1990; Storey, 1991).

II. Motility Regulation

A. Sea Urchins

Motility of spermatozoa of several invertebrate species has been studied extensively, largely because the gametes can be obtained relatively easily. Among invertebrates, the motility of echinoderm (Phylum Echinodermata) spermatozoa, especially those of the sea urchins, has been studied the most. When considering results obtained from the different sea urchin species, one should note that they comprise four of the six orders in the Class Echinoidea (Table 1). The other two orders are the sand dollars and the heart urchins. From the standpoint of taxonomic classification, species of sea urchin from different orders are as different from each other as are dogs and rats (Class Mammalia, Orders Carnivora and Rodentia, respectively). Accordingly, one can hypothesize readily that the molecules that mediate fertilization might differ substantially among the various sea urchins. Sperm-activating peptides released from echiniod eggs and the apparent receptors for these peptides do indeed differ among the many echinoid species (discussed in detail in subsequent sections).

1. Egg peptides

In 1981, Hansbrough and Garbers reported the purification of a sperm-activating peptide from acid-solubilized jelly of *Strongylocentrotus pupuratus* eggs. This peptide, designated speract for its sperm activating properties, was purified on the basis of its ability to increase sperm respiration (measured as O_2 consumption) at pH 6.6 (Hansbrough and Garbers, 1981a). The purified peptide stimulated sperm respiration with an EC_{50} of 3×10^{-11} M. Suzuki *et al.* (1981) subsequently reported purification of a sperm activating peptide from acid-solubilized jelly of *Hemicentrotus pulcherrimus* eggs; the *H. pulcherrimus* peptide and *S. purpuratus* speract (Garbers *et al.*, 1982a) proved to

TABLE 1 Egg Peptides of Various Echinoids (Phylum Echinodermata, Class Echinoidea)[a]

Taxonomic designation	Peptide
Subclass Regularia	
Order Echinothurioida	Unknown
Order Diadematoida	SAP-IV
	(GCPWGGAVC)
Order Arbacioida	
Suborder Phymosomatina	Alloresact
	(KLCPGGNCV)
Suborder Arbacina	Resact
	(CVTGAPGCVGGGRL–NH$_2$)
Order Echinoida	Speract
	(GFDLNGGGVG)
Subclass Irregularis	
Order Clypeasteroida	Mosact
	(DSDSAQNLIG)
Order Spatangoida	SAP-V
	(GCEGLFHGMGNC)

[a] Each peptide stimulates respiration of spermatozoa from all species within the order (or suborder in the case of resact and alloresact), but fails to stimulate respiration of spermatozoa from species outside the order.

be identical decapeptides with the amino acid sequence GFDLNGGGVG. Chemically synthesized speract stimulated sperm respiration with the same potency ($EC_{50} = 2.5 \times 10^{-11}$ M) as did speract purified from egg jelly, confirming unequivocally the determination of the primary structure of the peptide (Garbers *et al.*, 1982a).

The purification of speract from both *S. purpuratus* and *H. pulcherrimus* egg jelly, and the observation that speract also activated *Lytechinus pictus* spermatozoa (Hansbrough and Garbers, 1981a), suggested that speract might be a general activator of sea urchin spermatozoa. However, Suzuki and Garbers (1984) reported that speract failed to activate *Arbacia punctulata* spermatozoa under conditions that produce activation of *S. purpuratus* and *L. pictus* spermatozoa. This observation was not due to of the absence of a sperm activating factor in this sea urchin species, since *A. punctulata* egg jelly stimulated respiration of homologous spermatozoa. The sperm activating factor in acid-solubilized *A. punctulata* egg jelly was purified and characterized (Suzuki and Garbers, 1984; Suzuki *et al.*, 1984). This factor, resact (respiration activator), proved to be a tetradecapeptide with the structure CVTGAPGCVGGGRL-NH$_2$ (Suzuki *et al.*, 1984). The cysteine residues in the first and eighth positions of the resact sequence later were found to be oxidized to form an intramolecular disulfide bond in native resact (Yoshino *et al.*, 1991). This observation was somewhat surprising, since chemically synthesized resact stimulated sperm respiration with the same potency as did native resact (Suzuki *et al.*, 1984). Either resact does not need to be disulfide bonded for full activity, or the chemically synthesized peptide was oxidized fortuitously to form the proper structure.

The discovery of significant interspecies differences in the sperm-activating peptides motivated the search for such peptides in the egg jelly of other echinoids, primarily in the laboratory of Suzuki and colleagues at the Noto Marine Laboratory in Japan. Two additional peptides, mosact (DSDSAQNLIG) and alloresact (KLCPGGNCV), were purified from the egg jelly of the sand dollar *Clypeaster japonicus* and the sea urchin *Glyptocidaris crenularis*, respectively (Suzuki *et al.*, 1987, 1988). Mosact stimulated respiration of *C. japonicus* spermatozoa with an EC_{50} of 0.5 nM, and alloresact stimulated respiration of *G. crenularis* spermatozoa with an EC_{50} of 0.5–1.0 nM. Resact also stimulated *G. crenularis* sperm respiration (hence the name "alloresact" for the conspecific peptide), but the EC_{50} was 500 nM. Finally, sperm-activating peptides (SAPs) designated SAP-IV and SAP-V were purified from the primitive sea urchin *Diadema setosum* and from the heart urchin *Brissus agassizii*, respectively (Yoshino *et al.*, 1990, 1992). These peptides also stimulated respiration of the homologous spermatozoa with subnanomolar EC_{50}s. Each of these six peptides—speract, resact, mosact, alloresact, SAP-IV, and SAP-V—represents a different paradigm for sperm-activating factors in echinoids (Table 1), since each is structurally distinct and, with the exception of the marginal cross-reactivity of resact and alloresact noted earlier, each stimulates respiration of spermatozoa from only the same or closely related species (Suzuki and Yoshino, 1992). These data support the hypothesis first proposed by Suzuki *et al.* (1982) that the egg jelly peptides stimulate respiration of spermatozoa of all species from the same taxonomic order (or suborder in the case of resact and alloresact), but not of species from other orders. To date, with dozens of combinations of peptide and spermatozoa tested, no exception to this rule has been reported (Suzuki and Yoshino, 1992).

During the purification of alloresact, five other sperm activating peptides that were similar to alloresact but present in the starting material in smaller amounts were purified also (Suzuki *et al.*, 1988). Earlier studies had provided evidence of multiple forms of speract (Hansbrough and Garbers, 1981a; Suzuki *et al.*, 1981; Garbers *et al.*, 1982a; Shimomura *et al.*, 1986a) and mosact (Suzuki *et al.*, 1987). To date, 69 SAPs other than the six paradigm peptides in Table 1, each a minor variant of the model peptides, have been isolated from the egg jelly of different echinoids (Suzuki and Yoshino, 1992). How did the diversity of these peptides arise? This question was answered, at least partially, by the molecular cloning and sequencing of cDNAs encoding speract (Ramarao *et al.*, 1990). Northern blot analysis indicated that *S. pur-*

puratus ovary produced 2.3-kb and 1.2-kb transcripts encoding speract. Cloned cDNAs corresponding to each of these transcripts encoded four copies of the speract amino acid sequence. In addition to the speract sequences, the 2.3-kb transcript encoded five speract-related sequences and the 1.2-kb transcript encoded six such sequences, all within the same open reading frames, suggesting that large propeptides (e.g., 260 amino acids predicted for the 2.3-kb transcript) are synthesized and then processed to produce the smaller peptides. The existence of multiple speract variants appears, therefore, not to be principally because of individual variation. The predicted speract and speract-related sequences are separated by single lysine residues, suggesting a processing mechanism involving endoproteolysis by a trypsin-like activity followed by removal of the C-terminal lysine by a carboxypeptidase activity (Ramarao *et al.*, 1990). Whether this processing occurs before or after secretion remains unknown.

Analysis of the amino acid sequence deduced from an open reading frame in a cDNA encoding resact revealed that this peptide also may be synthesized as a propeptide (Burks, 1990). The predicted 205-residue propeptide contained the resact sequence only once, and no other similar sequences. This observation is consistent with the fact that no variants of resact similar to the speract, mosact, and alloresact variants have been isolated. At present, no functional explanation has been proposed for the existence of a single SAP in one echinoid suborder and multiple variants of these peptides in individuals from other echinoid orders.

Since reproductive isolation is accepted almost universally as a criterion for whether a given population constitutes a species, the mechanisms that produce reproductive isolation can be examined, as a supplement to morphological analysis, for the purpose of taxonomic classification. The SAPs, because of their species selectivity, may contribute to the reproductive isolation of echinoid species. The fact that resact and alloresact are as different from each other structurally as they are from speract, mosact, and the other SAPs led Suzuki and Yoshino (1992) to propose that the two suborders of the order Arbacioida be designated as two different orders. With this change in taxonomic classification would come strict ordinal specificity of the SAPs.

2. Receptors

Several studies testing the effects of speract and resact analogs for their ability to stimulate respiration revealed that the peptides can tolerate significant modification at their N termini, including additions and deletions of amino acids and derivatization with the Bolton–Hunter reagent, without appreciable loss of potency (Garbers *et al.*, 1982a; Smith and Garbers, 1983; Shimomura and Garbers, 1986). This property greatly facilitated identification of the receptors for the peptides, since it enabled synthesis of radioiodinated, active analogs of the peptides. Using the homobifunctional aminoreactive cross-linker disuccinimidyl suberate, Dangott and Garbers (1984) specifically cross-linked ^{125}I-labeled GGG[Y^2]-speract to a 77,0000 M_r glycoprotein from *S. purpuratus* sperm membranes. Using the same cross-linker, Shimomura *et al.* (1986b) specifically cross-linked ^{125}I-labeled GGGYG-resact to a 160,000 M_r protein from *A. punctulata* spermatozoa. Thus, the apparent speract and resact receptors identified by cross-linking are clearly different.

Dangott *et al.* (1989) cloned a full-length cDNA (2.5 kb; 1596-bp open reading frame) encoding the 77,000 M_r apparent speract receptor. The 532-residue deduced amino acid sequence suggested a 26-residue N-terminal signal sequence, a single transmembrane region, and a 12-amino-acid cytoplasmic extension at the C terminus. The putative 461-residue extracellular domain was relatively cysteine rich (5%) and consisted of four tandem repeats of approximately 115 amino acids that exhibited weak similarity to each other (26% of positions conserved in three or more of the repeated sequences). Comparison of the sequence of the 77,000 M_r apparent speract receptor with sequences in the GenBank and National Biomedical Research Foundation (NBRF) databases initially yielded no significant identities (Dangott *et al.*, 1989). Subsequent comparisons of the cysteine-rich tandem repeats revealed that these sequences in the apparent speract receptor are as much as 48% identical to similar cysteine-rich repeats in the type 1 scavenger receptor from macrophages, the CD5 T-cell surface antigen, and complement factor I (Freeman *et al.*, 1990). The significance of this observation remains unclear.

When Northern blots of *S. purpuratus* and *A. punctulata* testis poly(A)[+] RNA were probed with a 1.5-kb fragment of the apparent speract receptor cDNA, 2.5-kb transcripts were identified in both species (Dangott *et al.*, 1989). Thus, although speract does not stimulate *A. punctulata* spermatozoa and the *A. punctulata* SAP resact does not cross-link to a 77,000 M_r protein, an mRNA transcript homologous to that encoding the 77,000 M_r apparent speract receptor is produced by the *A. punctulata* testis; presumably, the encoded protein is present on *A. punctulata* spermatozoa. However, the cloning of a cDNA encoding a speract receptor-like protein from *A. punctulata* has not yet been reported.

Shimomura *et al.* (1986b) immunoprecipitated the 160,000 M_r protein that cross-linked to resact with a monospecific antibody to *A. punctulata* sperm guanylyl cyclase. This work represented the first demonstration of ligand binding activity and guanylyl cyclase activity within the same polypeptide. Singh *et al.* (1988) subsequently cloned and sequenced a cDNA encoding the *A. punctulata* guanylyl cyclase/apparent resact receptor. The amino acid sequence deduced from the cloned cDNA sequence suggested that the enzyme/receptor possessed a single transmembrane domain that separated an extracellular N-terminal ligand-binding domain and an intracellular C-terminal catalytic domain. A striking feature of the deduced sequence of the intracellular domain was that part of it was similar to conserved sequences of protein tyrosine kinases (Singh *et al.*, 1988). A serine-rich C-terminal 95-residue section also was identified that might represent phosphorylation sites important for regulation of the activity of the enzyme/receptor (discussed in the next section).

Thorpe and Garbers (1989) cloned a cDNA encoding an *S. purpuratus* sperm membrane guanylyl cyclase by homology screening using a 2.3-kb fragment of the *A. punctulata* guanylyl cyclase/resact receptor as a probe. The protein sequence deduced from the *S. purpuratus* cDNA predicted the same membrane topology as the *A. punctulata* enzyme/receptor. The C-terminal 202 amino acids of the deduced sequence were found to be 42% identical to the sequence of a subunit of the soluble heterodimeric guanylyl cyclase of bovine lung (Koesling *et al.*, 1988), implying that the catalytic domain of membrane guanylyl cyclases is the intracellular C-terminal portion of the molecule. This hypothesis was confirmed subsequently by mutagenesis studies of a cDNA encoding the mammalian membrane guanylyl cyclase/atrial natriuretic peptide receptor (Chinkers and Garbers, 1989). Unfortunately, it has not yet been possible to express an active guanylyl cyclase/resact receptor in cultured mammalian cells.

3. Physiological effects

Prior to 1984, one troubling aspect of the study of egg peptides concerned the physiological significance of the actions of these peptides. Seawater is normally alkaline (pH 7.6 – 8.0). However, reproducible effects of egg-derived factors on sperm respiration were not observed routinely until Ohtake (1976) reported that basal sperm respiration rates were low at pH 6.6 – 6.8; respiration could be stimulated consistently at this pH by egg jelly. If basal respiration is high at alkaline pH, what is the normal function of the egg peptides, since both sea water and egg jelly (Holland and Cross, 1983) are alkaline? This important question was answered partially when Suzuki and Garbers (1984) found that egg jelly contained factors that inhibited sperm respiration at alkaline pH and that addition of speract (to *L. pictus* spermatozoa) or resact (to *A. punctulata* spermatozoa) reversed this inhibition. Monensin A (see subsequent text) also overcame the jelly-mediated inhibition of respiration. Thus, one function of the egg peptides may be to maintain sperm respiration and motility when they enter the egg jelly. (Another apparent function of resact is sperm chemotaxis; see subsequent text.) The respiration-inhibiting factor(s) present in the egg jelly has (have) not been purified or characterized further.

Even before the first purification of an SAP, cyclic nucleotides were implicated as second messengers in the signaling mechanism of egg-associated factors. Garbers and Hardman (1975) prepared "egg water" by incubating *S. purpuratus* or *L. pictus* eggs in artificial sea water. Factors in this crude egg-conditioned medium increased sperm adenylyl cyclase activity and decreased guanylyl cyclase activity. Hansbrough *et al.* (1980) showed that a partially purified egg-associated factor, probably speract, stimulated sperm respiration and sperm fatty acid oxidation and increased sperm cAMP and cGMP concentrations more than fourfold in 1 min; these effects were independent of extracellular Ca^{2+}. Incubation with 8-bromo-cGMP but not 8-bromo-GMP, 8-bromo-cAMP, or 8-bromo-AMP also stimulated sperm respiration and fatty acid oxidation. Subsequent experiments revealed that stimulation of sperm motility and respiration by purified speract and by 8-bromo-cGMP required extracellular Na^+ but not extracellular K^+ or Ca^{2+} (Hansbrough and Garbers, 1981b). Monensin A, an ionophore that mediates electroneutral Na^+/H^+ exchange, reproduced these effects, suggesting that Na^+ transport into the cell is part of the mechanism by which speract stimulates spermatozoa. These early studies established that cyclic nucleotides and ion movements both are involved in regulation of sperm respiration.

The requirement of an influx of extracellular Na^+ for speract-induced stimulation of spermatozoa (Hansbrough and Garbers, 1981b) led to the discovery that the peptide activates Na^+/H^+ exchange. Intracellular pH (pH_i) appears to be a primary controller of sperm physiology (see subsequent discussion). Repaske and Garbers (1983) measured a H^+ efflux on stimulation of spermatozoa with speract. Lee (1984a) demonstrated a Na^+-dependent H^+ efflux from isolated sperm tails that was inhibitable by K^+. Further study showed that these ion movements were mediated by a voltage sensitive Na^+/H^+ exchanger (Lee, 1984b). Speract appeared to regulate the activity of this exchanger indirectly by activating sperm K^+ channels, producing K^+ efflux and membrane hyperpolarization, which in turn stimulated Na^+/H^+ exchange (Lee and Garbers, 1986). Activation of sperm K^+ channels by speract, and consequent membrane hyperpolarization, were potentiated by GTP and GTPγS (Lee, 1988). Speract also induced transient increases in intracellular Ca^{2+}, apparently as a consequence of either the increased pH_i or increased intracellular Na^+ concentration effected by the Na^+/H^+ exchanger (Schackmann and Chock, 1986).

Sea urchin sperm adenylyl cyclase may be regulated by Ca^{2+} and calmodulin (Bookbinder et al., 1990). Thus, the increases of cAMP concentrations induced by the SAPs may result from stimulation of adenylyl cyclase by an increase in intracellular Ca^{2+} concentration, which occurs relatively late in the signaling cascade. Although the cAMP produced would be expected to activate cAMP-dependent protein kinase, which could in turn regulate the flagellum as in mammalian spermatozoa (Section II,B,4), the predominant regulator of sea urchin flagellar activity appears to be pH_i since 0.4–0.5 unit increases of pH_i stimulate sperm motility and respiration (Christen et al., 1982). Sea urchin sperm respiration and flagellar activity appear to be very tightly coupled; a majority of the ATP produced is used for flagellation, and the cell's mitochondria do not respire when energy charge is high (Christen et al., 1983). Increased pH_i activates the ATPase activity of flagellar dynein, producing both increased motility and increased respiration (Christen *et al.*, 1983).

The observation that a partially purified egg jelly factor stimulated sea urchin sperm guanylyl cyclase activity (Hansbrough *et al.*, 1980) accelerated the pace of research on this enzyme. Radany *et al.* (1983) purified *S. purpuratus* sperm guanylyl cyclase to apparent homogeneity and found it to be a 135,000–157,000 M_r membrane glycoprotein. Ward and Vacquier (1983) concurrently identified a 160,000 M_r *A. punctulata* sperm membrane phosphoprotein that was dephosphorylated rapidly when the cells were exposed to egg jelly; the M_r of the dephosphorylated polypeptide was 150,000 [determined by polyacrylamide gel electrophoresis in the presence of sodium dodecylsulfate (SDS–PAGE)]. The 160,000 M_r phosphoprotein soon was determined to be guanylyl cyclase (Suzuki *et al.*, 1984; Ward *et al.*, 1985a). Saturating concentrations of purified resact caused complete dephosphorylation of the 160,000 M_r guanylyl cyclase within 5 sec (Suzuki *et al.*, 1984). Monensin A and NH_4Cl, both of which raise intracellular pH, also caused dephosphorylation of guanylyl cyclase (Suzuki *et al.*, 1984). Dephosphorylation decreased the activity of the *A. punctulata* enzyme 38-fold (Ward *et al.*, 1985a). Speract and resact transiently activated guanylyl cyclase in isolated *L. pictus* and *A. punctulata* sperm membranes, respectively, and a subsequent rapid dephosphorylation caused inactivation of the enzymes (Bentley and Garbers, 1986; Bentley *et al.*, 1986a,b). Collectively, these studies established that sea urchin sperm guanylyl cyclase is a phosphoprotein that, on stimulation of the spermatozoon by its cognate egg-derived peptide, is activated transiently, resulting in membrane hyperpolarization and alkalinization of the intracellular space, as described earlier. Dephosphorylation and inactivation of guanylyl cyclase then occurs, possibly by a process requiring increased pH_i.

The molecular events that couple the activation of guanylyl cyclase to changes in ion movements are unknown. At pH 8.0, where basal sperm respiration was high, resact stimulated cGMP accumulation without increasing the respiration rate (Suzuki *et al.*, 1984). Similarly, speract increased cGMP concentrations in cells incubated in high K^+ (100 mM) sea water, but did not increase sperm respiration under these conditions (Harumi *et al.*, 1992). Thus, activation of guanylyl cyclase by speract or resact preceded or was independent of membrane hyperpolarization and subsequent increases of pH_i and respiration. Speract and resact stimulated respiration half maximally at about 100-fold lower concentrations than were required for half maximal stimulation of cGMP accumulation (Hansbrough and Garbers, 1981a; Suzuki *et al.*, 1984). Two resact analogs were synthesized that were 100–1000 times less potent than resact in stimulating cGMP accumulation, but only 2.5–5 times less potent in stimulating sperm respiration (Shimomura and Garbers, 1986). However, no analogs were found to raise cGMP levels without stimulating respiration of spermatozoa. Hence, a very small increase in the intracellular concentration of cGMP induced by low concentrations of the egg-derived peptides may be sufficient to stimulate respiration; the large increases in cGMP at higher peptide concentrations could have some other physiological function. Unfortunately, no sperm molecules regulated by cGMP have been identified, nor has their cGMP concentration dependence been characterized. Stimulation of sperm respiration by low concentrations of speract or resact also could proceed through a signaling mechanism not involving guanylyl cyclase, perhaps mediated by the 77,000 M_r apparent speract receptor or its *A. punctulata* homolog (described in Section II,A,2). However, the early demonstration that 8-bromo-cGMP stimulated sperm respiration (Hansbrough *et al.*, 1980) favors the first hypothesis.

The possibility that speract elicits different physiological responses at different concentrations of the peptide was supported further by work by Babcock *et al.* (1992). Activation of K^+ channels by very low concentrations of speract (<2.5 pM) was identified as an early event in the signaling cascade. This K^+ channel activation produced K^+ efflux and membrane hyperpolarization, as described earlier, without a concomitant increase in pH_i. Thus, the apparent Na^+/H^+ exchange induced by speract appears not to be coupled directly to membrane hyperpolarization. Higher concentrations of speract (>25 pM) increased both K^+ channel activity and pH_i. Concentrations of speract greater than 10 nM caused depolarization that was dependent on extracellular Ca^{2+} (Babcock *et al.*, 1992). Although speract stimulated large increases in cGMP accumulation in *S. purpuratus* spermatozoa at nanomolar concentrations, the dose–response curve was biphasic, with measurable increases in cGMP concentrations induced by 10 pM speract (Hansbrough and Garbers, 1981a). Thus, cGMP may be the second messenger that mediates the diverse physiological effects produced by various concentrations of speract.

In addition to the effects of the egg peptides on respiration and motility, resact is a sperm chemoattractant (Ward *et al.*, 1985b). The chemotactic activity of resact is dependent on extracellular Ca^{2+} (Ward *et al.*, 1985b), although Ca^{2+} is not required for stimulation of respiration and motility (Hansbrough and Garbers, 1981b). To date, none of the other SAPs have been shown to possess chemotactic activity (Garbers, 1989). Although resact cross-links to guanylyl cyclase and speract cross-links to a 77,000 M_r sperm membrane glycoprotein, as described in the previous section, no differences in the effects of these two peptides on activation of guanylyl cyclase activity or on ion movements have been observed (Garbers, 1989). Thus, no explanation is currently available for the difference in apparent chemotactic activity of resact and speract. The effects of the egg peptides on sea urchin sperm physiology are summarized schematically in Fig. 1.

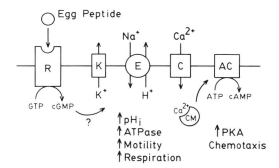

FIGURE 1 Changes in sea urchin sperm physiology induced by the sperm activating peptides. Chemotactic activity has been detected only for resact. Resact appears to activate guanylyl cyclase by binding directly to the putative ligand binding domain of the enzyme, as indicated. Speract may activate guanylyl cyclase indirectly by binding to a 77,000 M_r plasma membrane receptor (see text), although speract also may activate the enzyme directly in a manner similar to resact. Only one of many possible Ca^{2+} channels is shown. R, Receptor; K, K^+ channel; E, Na^+/H^+ exchanger; C, Ca^{2+} channel; CM, calmodulin; AC, adenylyl cyclase; PKA, cAMP-dependent protein kinase activity; ATPase, flagellar dynein ATPase activity.

B. Mammals

Cyclic nucleotides (Garbers *et al.*, 1971, 1973) and Ca^{2+} (Morita and Chang, 1970) have been known for more than 20 years to affect motility and respiration of mammalian spermatozoa. Although increased cyclic nucleotide concentrations (specifically cAMP) consistently stimulate sperm motility, the effects of Ca^{2+} are more variable; both stimulation and inhibition of motility have been documented (reviewed by Garbers and Kopf, 1980; Tash and Means, 1983). Although some of the observed variability in Ca^{2+} effects may be a function of interspecies differences in responsiveness of spermatozoa to this cation (Tash and Means, 1983), epididymal and ejaculated spermatozoa from the same species also have been found to respond differently to Ca^{2+} (Babcock *et al.*, 1979).

1. Adenylyl cyclase and Ca^{2+}

Sperm adenylyl cyclase appears to be distinct from adenylyl cyclases present in other tissues. The most striking difference is the lack of responsiveness of the enzyme to guanine nucleotides or cholera toxin (through activation of guanine nucleotide-binding regulatory proteins; G proteins) or to forskolin (Garbers and Kopf, 1980; Toscano and Gross, 1991). The lack of cholera toxin sensitivity is consistent with evidence that spermatozoa contain pertussis toxin substrates (Gi_α) but not the cholera toxin substrate Gs_α

(Bentley *et al.*, 1986c; Kopf *et al.*, 1986). Hyne and Garbers (1979a) reported that guinea pig sperm adenylyl cyclase was stimulated by Ca^{2+}. This stimulation later was shown to be dependent on the presence of bicarbonate ion (Garbers *et al.*, 1982b). The presence of substantial amounts of calmodulin in both sperm head (Jones *et al.*, 1978,1980) and tail (Feinberg *et al.*, 1981) suggested that this molecule might mediate regulation of sperm adenylyl cyclase by Ca^{2+}. Equine sperm adenylyl cyclase was inhibitable by calmidazolium, a specific calmodulin antagonist. This inhibition was overcome by addition of 50 μM calmodulin from bovine brain, suggesting that at least part of the effect of Ca^{2+} on sperm adenylyl cyclase activity is mediated by calmodulin (Toscano and Gross, 1991).

Okamura *et al.* (1991) reported a 280-fold purification of adenylyl cyclase from porcine epididymal sperm membranes. The specific activity of the purified enzyme was low (1.8 nmol/min/mg protein). The enzyme preparation exhibited a single protein band that migrated with 46,300 M_r and pI 6.9 in two-dimensional isoelectric focusing/SDS–PAGE. In contrast, calmodulin-stimulated adenylyl cyclase purified 15,400-fold from bovine brain membranes had a specific activity of 10,200 nmol/min/mg protein and migrated at an M_r of 120,000 in SDS–PAGE (Smigel, 1986). Molecular cloning of the RNA encoding the brain enzyme revealed a molecular mass of 124,000 g/mol calculated from the deduced sequence of the protein (Krupinski *et al.*, 1989). Either the 46,300 M_r protein of porcine spermatozoa is not adenylyl cyclase or the sperm enzyme has a very much lower turnover number and molecular weight than does calmodulin-stimulated adenylyl cyclase from bovine brain. Calmodulin-stimulated adenylyl cyclase appears to be particularly susceptible to proteolysis by the trypsin-like protease acrosin (Johnson *et al.*, 1985; Yeung *et al.*, 1989; Toscano and Gross, 1991), which is present in large amounts in particulate fractions of mammalian spermatozoa (Hardy *et al.*, 1991; D. M. Hardy and D. L. Garbers, unpublished observations). Thus, the 46,300 M_r protein purified by Okamura *et al.* (1991) may be a proteolytic fragment of sperm adenylyl cyclase that has a much lower turnover number than does the native enzyme.

Analysis of the deduced amino acid sequence of the G protein and calmodulin-stimulated adenylyl cyclase from bovine brain predicted a polypeptide chain (1134 amino acids) that spanned the plasma membrane 12 times and contained two homologous, intracellular, putative catalytic domains (Krupinski *et al.*, 1989). Pitt *et al.* (1992) cloned two cDNAs encoding *Dictyostelium discoideum* adenylyl cyclases, designated ACA and ACG. The deduced sequence (1407 amino acids) for ACA predicted a membrane topology identical to that predicted for the bovine brain enzyme. Interestingly, the deduced sequence (858 amino acids) for ACG predicted a single transmembrane domain and a single intracellular catalytic domain (Pitt *et al.*, 1992), similar to the membrane guanylyl cyclases (Section II,A,2). The predicted molecular mass of the 858-residue ACG polypeptide chain is about 95,000 g/mol (calculated using 110 g/mol for the average molecular mass of an amino acid). Expressed recombinant ACA was responsive to G protein-mediated regulation, but ACG was not. Thus, a G protein-insensitive adenylyl cyclase with a single transmembrane domain/single catalytic domain topology has been identified in a eukaryotic cell.

2. Male tract factors

Mammalian spermatozoa are nonmotile in the male reproductive tract and do not acquire even the capacity for progressive motility until they reach the caudal epididymis. This acquisition of the capacity for motility coincides with the acquisition of capacity for fertilization. The biochemical and physiological changes occuring in the epididymis that render mammalian spermatozoa capable of progressive motility are largely unknown. Because posttesticular spermatozoa no longer synthesize protein, these changes must be posttranslational modifications of some kind, possibly directed by factors released by the different parts of the epididymis. Acott and Hoskins (1978,1981) identified a factor called forward motility protein that was secreted by the epididymis and bound to caput epididymal spermatozoa. Caput epididymal spermatozoa incubated with the protein and cAMP acquired motility similar to that of caudal epididymal spermatozoa (Acott and Hoskins, 1978). Although progress has been made in the identification of other proteins secreted by the epididymis that bind to spermatozoa, the functions of these molecules, including whether they affect the capacity of spermatozoa to acquire motility, remain obscure for the most part.

Caudal epididymal spermatozoa become motile when mixed with accessory gland secretions at ejaculation (Garbers, 1989). Dilution of caudal epididymal spermatozoa into simple media also stimulates motility. Stimulation of epididymal sperm motility by dilution in medium appears to be a consequence of a 0.4-unit increase in intracellular pH that correlates with changes in sperm phosphoprotein patterns (Carr and Acott, 1989). Since dilution of spermatozoa in medium stimulates motility, specific factors from the accessory glands are not essential for activation. Nevertheless, differences in the abilities of epididymal and ejaculated spermatozoa to transport Ca^{2+} (Babcock *et al.*, 1979) and to undergo acrosome reactions in response to the egg zona pellucida (Florman and First, 1988a) clearly demonstrate the importance of the seminal fluid in regulating sperm physiology. The fact that epididymal spermatozoa are capable of motility and are fertile is not evidence that accessory gland secretions lack factors that influence the motility or fertility of these cells.

The observation that bovine epididymal spermatozoa readily transport exogenous Ca^{2+} but ejaculated bovine spermatozoa do not was the first evidence that seminal fluid contained a calcium transport inhibitor (Babcock *et al.*, 1979). This inhibitor, named caltrin, was purified from bovine semen (Rufo *et al.*, 1982), sequenced (Lewis *et al.*, 1985), and found to be a 47-residue peptide containing no disulfide bonds. Purified caltrin bound to the plasma membrane overlying the apical acrosome and the principal piece of the tail of bovine epididymal spermatozoa (San Augustin et al., 1987); the distribution of caltrin binding was identical to that observed on ejaculated spermatozoa. The absence of caltrin binding sites on the plasma membrane overlying the midpiece of the sperm tail is curious, since a majority of Ca^{2+} taken up by mammalian spermatozoa is thought to be destined for mitochondria in the midpiece (Babcock et al., 1978; Breitbart et al., 1985). Purified caltrin inhibited transport of Ca^{2+} into epididymal bovine spermatozoa (Rufo et al., 1982; San Augustin et al., 1987), indicating that caltrin alone could account for the differences in Ca^{2+} transport observed between epididymal and ejaculated spermatozoa. Caltrin appears to inhibit Ca^{2+} import by inhibiting a Na^+/Ca^{2+} exchanger in the sperm plasma membrane (Rufo et al., 1984).

Guinea pig epididymal spermatozoa also imported exogenous Ca^{2+}, but only after a prior incubation (at least 20–30 min) in a Ca^{2+}-free medium containing the metabolic substrates lactate and pyruvate (Coronel and Lardy, 1987). This acquired Ca^{2+} permeability was inhibited by caltrin-like factors present in male guinea pig reproductive tract fluids (Coronel et al., 1988). These factors, designated caltrins I and II, were purified from guinea pig seminal vesicles and their amino acid sequences determined (Coronel et al., 1990). Guinea pig caltrins I and II were found to be 45- and 55-residue glycopeptides, respectively. Neither peptide exhibited significant amino acid sequence identity to bovine caltrin, although caltrin I and bovine caltrin both contained the very short sequence GNRS (Coronel et al., 1990). Future comparisons of the structures and properties of caltrin-like peptides from more mammalian species should help determine whether the apparent diversity of these peptides reflects species-specific differences in function.

Work on caltrin function revealed that its actions on the spermatozoon may change during the course of fertilization. Freshly prepared caltrin inhibits Ca^{2+} import by epididymal spermatozoa, but stored preparations were found to stimulate Ca^{2+} import (San Agustin et al., 1987). This conversion of caltrin from a Ca^{2+} transport inhibitor to a transport enhancer was reproduced by purification by anion exchange chromatography (San Agustin et al., 1987). Addition of anions such as citrate or phosphatidylserine, both of which are present in seminal fluid, to purified enhancer caltrin converted it back to the transport inhibitory form (San Agustin and Lardy, 1990). These observations suggested that the function of caltrin is to prevent premature uptake of Ca^{2+} by the mammalian spermatozoon early in its journey to the site of fertilization. Conversion of caltrin to the enhancer form, possibly by action of female tract factors when the spermatozoon is in the vicinity of the egg, would stimulate Ca^{2+} uptake and thereby potentiate Ca^{2+}-dependent processes such as hyperactivated motility (Section II,B,3) or the acrosome reaction (Section III,B,1). Further study is required to test this intriguing hypothesis.

3. Female tract factors

Because mammals are internal fertilizers, potential sources of factors that might influence sperm motility include not only the male accessory gland secretions but the secretions of the female reproductive tract as well. Freshly ejaculated spermatozoa, although fully motile, cannot fertilize. The quest for conditions that would support fertilization *in vitro* led to the discovery that culturing spermatozoa in the female reproductive tract rendered them capable of fertilizing the egg (Austin, 1951; Chang, 1951). This requisite period in culture was named "capacitation." Although the discovery of capacitation constituted a crucial technical advance in the study of fertilization, our understanding of the essential changes that occur in the spermatozoon during this process is still very limited. Two cellular characteristics that are considered to be diagnostic of the capacitated state are the ability to undergo the acrosome reaction (discussed in Section III) and the manifestation of "hyperactivated" motility, an exaggerated whiplash beating of the sperm tail (Yanagimachi, 1970). Hyperactive motility may improve the ability of the spermatozoon to penetrate the viscous barriers it encounters during fertilization (Suarez *et al.*, 1991).

Although hyperactivated motility can be achieved *in vitro* by incubating epididymal or ejaculated spermatozoa in relatively simple media (Yanagimachi, 1981), factors from the female reproductive tract still may regulate sperm motility. Human follicular fluid appears to stimulate motility of uncapacitated human spermatozoa (Chao *et al.*, 1991). Bradley and Garbers (1983) and Mendoza and Tesarik (1990) reported alteration of bovine and human sperm movement characteristics, respectively, by cumulus oophorus-derived material. However, no factor comparable to the egg peptides of echinoids (i.e., a factor emanating from the egg) has yet been identified in a vertebrate species. Ralt *et al.* (1991) reported evidence that human follicular fluid may contain a sperm chemoattractant. A chamber containing 5000- to 10,000-fold dilutions of follicular fluid accumulated spermatozoa from an adjoining chamber at a higher rate than did a chamber containing medium alone. Although individual samples of follicular fluid varied greatly in potency, the activity of the samples correlated with the fertility of the egg obtained from the same follicle, suggesting that the putative chemoattractive activity correlates with important egg properties, or is itself important. It is important to note that the strict definition of chemotaxis, positive directional movement in a concentration gradient, has not yet been shown for mammalian spermatozoa.

4. Intracellular events

Several enzymes that could regulate motility are present in the mammalian sperm flagellum. Protein kinases identified include casein kinases I and II (Chaudhry *et al.*, 1991a,b), cAMP-dependent protein kinase (Horowitz *et al.*, 1984, 1989; Lieberman *et al.*, 1988), and protein kinase C (Rotem *et al.*,

1990a,b). The possible function of the casein kinases is unknown. Primary functions for cAMP-dependent protein kinase and protein kinase C seem likely.

The consistent stimulatory effect of cAMP on sperm motility strongly implicated cAMP-dependent protein kinase in regulation of sperm motility. Horowitz *et al.* (1984) measured cAMP-dependent protein kinase activity associated with demembranated rat spermatozoa. Addition of cAMP caused release of the kinase activity from the particulate fraction, but the type II regulatory subunit (RII) remained particulate. The RII subunit localized by immunoelectron microscopy to the flagellum of bovine spermatozoa (Lieberman *et al.*, 1988). The flagellar RII subunit later was characterized as the RIIβ subtype (Horowitz *et al.*, 1989). Noland *et al.* (1986,1987) found that RII was a prominent phosphoprotein of bovine spermatozoa. Wasco *et al.* (1989) also identified a calmodulin-stimulated cyclic nucleotide phosphodiesterase activity in rat and bovine sperm flagella. Thus, mammalian sperm flagella contain Ca^{2+}-responsive enzymes that regulate cAMP levels (adenylyl cyclase and cyclic nucleotide phosphodiesterase) as well as a kinase that is responsive to cAMP (type II cAMP-dependent protein kinase) that may function together to regulate flagellar motility.

A possible mediator of the activation of flagellar motility by cAMP-dependent protein kinase has been described. Tash *et al.* (1984,1986) characterized a heat stable activity in an NP-40 extract of canine spermatozoa which, in the presence of added ATP and cAMP, reactivated motility of demembranated canine spermatozoa. Phosphorylation by cAMP-dependent protein kinase of components of the extract was both necessary and sufficient for manifestation of this activity, which was designated axokinin. A partially purified preparation containing axokinin activity was enriched in a 56,000 M_r phosphoprotein that appeared to be the major cAMP-dependent protein kinase substrate in the extract, suggesting that the 56,000 M_r phosphoprotein was axokinin (Tash *et al.*, 1984). However, the major 56,000 M_r soluble phosphoprotein of bovine spermatozoa later was characterized as RII (Paupard *et al.*, 1988). Since axokinin activity was heat stable, RII is unlikely to mediate this activity. Thus, although a soluble activity dependent on phosphorylation by cAMP-dependent protein kinase appears to regulate sperm flagellar motility, the identity of the active principle is still controversial.

Cytosolic pH also appears to be a primary regulator of mammalian sperm motility. Incubating bovine spermatozoa in a pH 7.8–8.2 medium containing 120 mM KCl increased intracellular pH from 6.3 to 6.9 and stimulated motility and respiration (Babcock *et al.*, 1983). This effect was independent of the cAMP-mediated regulation described earlier, since caffeine (a phosphodiesterase inhibitor) did not change cytosolic pH and cAMP-dependent protein kinase activity was not affected by increased cytosolic pH (Babcock *et al.*, 1983). Increased intracellular pH may activate mammalian sperm dynein

ATPase activity, as occurs in sea urchin spermatozoa (discussed in Section II,A,3).

Rotem *et al.* (1990a,b) reported that phorbol-12-myristate-13-acetate, which stimulates protein kinase C, activated motility of previously frozen human spermatozoa. The observed activation of motility did not occur in the presence of protein kinase C inhibitors. Protein kinase C was localized histochemically to the equatorial segment and to the principal piece of the sperm tail (Rotem *et al.*, 1990a,b). These observations imply that a signaling mechanism involving hydrolysis of phosphoinositides by phospholipase C, and activation of protein kinase C by the diacylglycerol produced, may regulate mammalian sperm motility. How this possible mechanism might integrate with the cAMP-mediated mechanism just described is unknown.

III. Acrosome Reaction

A. Sea Urchins

In 1952, Jean C. Dan published results of a series of experiments investigating the effects of "egg-water" on the behavior and morphology of sea urchin spermatozoa. She wrote:

> With phase contrast oil immersion . . . it is also just barely possible to make out a small mass of colorless, non-refringent substance attached to the tips of spermatozoa in egg-water. . . . When such spermatozoa are freely suspended, the substance appears as a flabby, gelatinous "tongue" of uniform diameter (roughly 0.2 μ), which may reach a length of more than 1 μ. . . . That it is exceptionally sticky in normal sea water is attested to by the fact that once this substance has come into contact with any surface, it is never pulled loose even by the most vigorous movements of the spermatozoan *(sic)*.

Dan proposed that the sperm morphology she had described was the result of:

> . . . a response of the acrosome to the chemical stimulation of the dissolved jelly, which is manifested by an almost instantaneous local breakdown of the acrosome membrane so that the acrosome substance is exposed at the tip of the sperm head.

Two years later, Dan used the term "acrosome reaction" for the response that she had earlier characterized (Dan, 1954). From a current perspective, these observations were remarkable for two reasons. First, from a technical standpoint, even with state-of-the-art light microscopes the "tongue" Dan described is very difficult to see. Second, the discovery that egg-derived material induces the acrosome reaction laid the intellectual foundation for four decades of research toward the goal of understanding how animal sperm adhere to and penetrate egg investments.

The "tongue" later was characterized as an actin-containing filament that is constructed within seconds of sperm exposure to egg jelly (Shapiro *et al.*, 1981). The acrosomally derived "sticky substance" is exposed when the acrosome ruptures; it coats the newly formed acrosomal filament, where it is situated uniquely for interaction with the egg vitelline membrane. This substance is composed of the protein bindin, which mediates species-specific adhesion of the sea urchin spermatozoon to the egg (reviewed by Garbers, 1989).

The substance in sea urchin egg jelly that induced the acrosome reaction later was purified and characterized by Segall and Lennarz (1979). The active principle, a fucose sulfate glycoconjugate (FSG), constituted 80% of the mass of the jelly coat. Fucose and sulfate constituted 76–93% of the mass of FSG, with a molar ratio of sulfate to fucose of 1.2 : 1.8 (Segall and Lennarz, 1979; Kopf and Garbers, 1980). As observed with the SAPs, FSG induced acrosome reactions in spermatozoa from organisms in the same taxonomic order (Segall and Lennarz, 1979). Preparations of FSG contained from 4 to 12% or more of covalently bound protein by weight (Segall and Lennarz, 1979; Kopf and Garbers, 1980) and, although a protein-poor portion of FSG induced acrosome reactions, the protein portion may be important as well (Garbers *et al.*, 1983; reviewed by Suzuki, 1990).

Extracellular Ca^{2+} is required for the sea urchin sperm acrosome reaction induced by "egg-water" (Dan, 1954) or by purified FSG (Segall and Lennarz, 1979; Kopf and Garbers, 1980). The Ca^{2+} channel blockers D-600 and verapamil blocked the FSG-induced reaction, indicating that Ca^{2+} transport into the spermaotzoon via Ca^{2+} channels was required (Kopf and Garbers, 1980). Purified FSG also stimulated accumulation of cAMP in intact spermatozoa (Kopf and Garbers, 1980) and in isolated sperm heads (Garbers, 1981). In all instances in which the effects of various treatments were studied — including combinations of ionophores, Ca^{2+} channel blockers, Ca^{2+} concentration changes, and FSG — elevation of intracellular Ca^{2+} concentration always produced cAMP accumulation and induction of the acrosome reaction (Kopf and Garbers, 1980; Garbers, 1981; Schackmann *et al.*, 1984). Maximal elevation of cAMP concentrations preceded maximal acrosome reactions (Garbers, 1981), suggesting that the Ca^{2+}-induced increases in intracellular cAMP control induction of the acrosome reaction. Sea urchin sperm adenylyl cyclase activity bound to a calmodulin affinity column (Bookbinder *et al.*, 1990). Several calmodulin antagonists inhibited adenylyl cyclase activity in a partially purified preparation from sea urchin sperm, and added calmodulin from bovine brain relieved the inhibition (Bookbinder *et al.*, 1990). Hence, sea urchin sperm adenylyl cyclase, like the mammalian sperm enzyme (Section II,B,1), may be regulated by calcium and calmodulin. Although elevated intracellular cAMP concentration may be necessary, it is not sufficient, since added cyclic nucleotides alone did not induce acrosome reactions (Garbers and Kopf, 1980).

What are the minimal requirements, in addition to an increase in intracellular Ca^{2+} concentration, for induction of the sea urchin sperm acrosome reaction by FSG? Conditions that induced acrosome reactions by increasing intracellular Ca^{2+} also caused membrane depolarization and increased pH_i (Kopf and Garbers, 1980; Garbers, 1981; Schackmann et al., 1984). In 1986, Trimmer et al. described a monoclonal antibody to a 210,000 M_r S. purpuratus sperm membrane protein; spermatozoa incubated with the antibody accumulated Ca^{2+} but did not undergo acrosome reactions. The antibody did not cause an increase in pH_i, nor did NH_4Cl-induced increases in pH_i alone cause acrosome reactions. However, the antibody in combination with NH_4Cl induced acrosome reactions (Trimmer et al., 1986). Thus, increased intracellular pH and Ca^{2+} concentration appear to be both necessary and sufficient for induction of the sea urchin sperm acrosome reaction.

How are pH_i and intracellular Ca^{2+} concentrations regulated by FSG? FSG induced a K^+-dependent transient hyperpolarization of the S. purpuratus sperm membrane (Gonzalez-Martinez and Darszon, 1987), possibly mediated by K^+ channels known to be present (Lievano et al., 1986), suggesting that FSG increases pH_i via activation of a hyperpolarization-sensitive Na^+/H^+ exchanger similar to the one that appears to function during signaling by the SAPs (Section II,A,3). Subsequent FSG-induced depolarization of the sperm membrane potential was dependent on extracellular Ca^{2+} concentration, was inhibitable by the Ca^{2+} channel blocker nisoldipine, and was proportional to increases in intracellular Ca^{2+} concentrations (Guerrero and Darszon, 1989a). Calmodulin antagonists inhibited the FSG-induced increase in intracellular Ca^{2+} concentrations (Guerrero and Darszon, 1989a). Guerrero and Darszon (1989b) then characterized two apparent Ca^{2+} channel activities. An immediate Ca^{2+} influx induced by FSG was inhibited by nisoldipine but not by 40 mM extracellular K^+, low Na^+, or tetramethylammonium ion (a K^+ channel blocker). A second putative Ca^{2+} channel activity was activated 5 sec after FSG addition; this activity was inhibited by extracellular K^+, low Na^+, and tetramethylammonium ion. These observations suggested that sequential activation of two different Ca^{2+} channels occurs during induction of the acrosome reaction by FSG (Guerrero and Darszon, 1989b). However, the regulation of these two channels by the putative FSG receptor or by the various ionic events that occur in response to FSG is still poorly understood.

Domino and Garbers (1988) reported that concentrations of inositol 1,4,5-trisphosphate (IP_3) increased up to 10-fold when sea urchin spermatozoa were incubated with FSG. Although these increases were detectable within 30 sec, maximal increases occurred at 2 min, well after the time when most acrosome reactions would have been completed. In addition, the FSG-induced increases in IP_3 concentrations were dependent on extracellular Ca^{2+} and were blocked by the Ca^{2+} channel antagonists verapamil and nifedipine (Domino and Garbers, 1988). Thus, Ca^{2+} influx appeared to activate a phos-

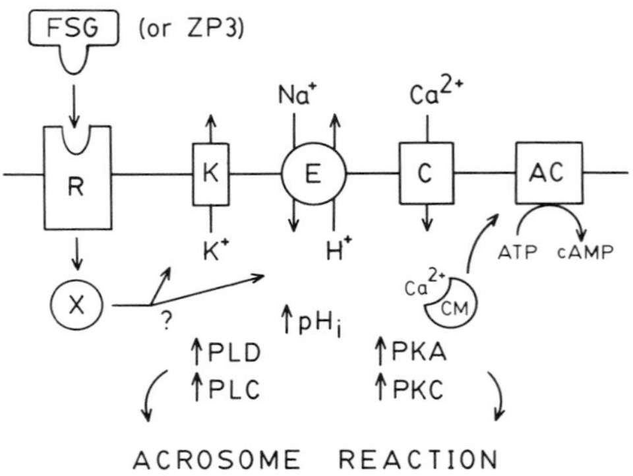

FIGURE 2 Changes in sperm physiology induced by FSG or by ZP3. X designates any activity, for example, a G protein, protein kinase, or receptor-operated ion channel, that could transmit a signal from the receptor to one of the other components of the system. In the mammalian sperm acrosome reaction induced by ZP3, one X component appears to be the heterotrimeric G protein Gi (see text). No X has been identified to date for the FSG-induced acrosome reaction of sea urchin spermatozoa. PLD, Phospholipase D activity. PLC, phospholipase C activity. PKC, protein kinase C activity. R, K, E, C, CM, AC, and PKA designate the same activities as listed in the legend to Figure 1.

pholipase C that hydrolyzed phosphatidyl inositol bisphosphate to produce IP_3 and diacylglycerol. Sperm phosphatidate concentrations also were inreased up to 10-fold 2 min after stimulation by FSG (Domino *et al.*, 1989). The increases in phosphatidate concentrations were dependent on influx of extracellular Ca^{2+}, similar to the IP_3 increases produced by phospholipase C. The phosphatidate increases induced by FSG proved to be a consequence of activation of a phospholipase D activity (Domino *et al.*, 1989) and not a result of diacylglycerol kinase-mediated phosphorylation of the diacylglycerol produced by phospholipase C activity. Phospholipase C and D activities, which are responsive to Ca^{2+} concentrations, may function in various signaling cascades (reviewed by Exton, 1990) but the function of these enzyme activities in the sea urchin sperm acrosome reaction is not known. The physiological effects of FSG on sea urchin spermatozoa are summarized schematically in Fig. 2.

B. Mammals

1. Zona pellucida as ligand

As described in the previous section, the sea urchin acrosome reaction and the origin of the agonist that induced it were discovered simultaneously.

In contrast, the discovery that mammalian spermatozoa also undergo acrosome reactions prior to fertilization was reported by Austin and Bishop in 1958, but a physiological agonist that induced it was not discovered until Florman and Storey (1982) demonstrated that mouse spermatozoa undergo the acrosome reaction at an accelerated rate while attached to the surface of the zona pellucida (ZP) or when stimulated with solubilized ZP. Several research groups subsequently reported stimulation of acrosome reactions by ZP in a number of mammalian species, including hamster (Cherr et al., 1986; Uto et al., 1988), rabbit (O'Rand and Fisher, 1987), human (Cross et al., 1988), bovine (Florman and First, 1988b), and pig (Berger et al., 1989). In addition, ultrastructural studies of sheep (Crozet and Dumont, 1984), goat (Crozet et al., 1987), and rat (Shalgi et al., 1989) fertilization similar to the one first reported for the mouse (Saling et al., 1979) indicate that a majority of these species' spermatazoa attach to the ZP with their acrosomes intact and undergo the acrosome reaction before proceeding further. Thus, by analogy to the mouse system, it is likely that the ZP induces the acrosome reaction of sheep, goat, and rat spermatozoa also.

The ZP first was isolated from pig (Dunbar et al., 1980) and mouse (Bleil and Wassarman, 1980a) oocytes. Physicochemical and functional properties of the ZP (reviewed by Wassarman, 1988) have been characterized best in these two species. Large quantities of the pig ZP can be isolated, which has facilitated chemical characterization of its oligosaccharide complement (Yurewicz et al., 1987; Mori et al., 1991). The mouse ZP is particularly amenable to study because its glycoprotein components can be resolved by one-dimensional SDS–PAGE whereas the pig ZP, as well as the ZP from most other mammals, requires two-dimensional electrophoretic methods to resolve the components. The ZP from both pig and mouse is composed of three heterogeneously glycosylated polypeptides (Bleil and Wassarman, 1980a; Hedrick and Wardrip, 1987); the mouse ZP glycoproteins were designated ZP1, ZP2, and ZP3.

Mouse ZP3 migrates with 83,000 M_r in SDS-PAGE, but the M_r of the ZP3 polypeptide chain is 44,000; the balance of the apparent mass is a result of its extensive glycosylation (Wassarman, 1988). Using binding competition experiments, Bleil and Wassarman (1980b) determined that mouse ZP3 provided the ligand for adhesion of spermatozoa to the ZP. Similarly, hamster ZP3, which migrated with 56,000 M_r in SDS–PAGE, accounted for the adhesion ligand activity of hamster ZP (Moller et al., 1990). The cell adhesion ligand activity of mouse ZP3 was found to reside in its oligosaccharide fraction (Florman et al., 1984), specifically in a 1500–1600 M_r O-linked ZP3 oligosaccharide (Florman and Wassarman, 1985) that contained an essential α-galactosyl residue at its nonreducing terminus (Bleil and Wassarman, 1988).

ZP3 appeared to mediate the acrosome reaction-inducing activity of mouse ZP (Bleil and Wassarman, 1983) and of hamster ZP (Moller et al.,

1990). This activity of mouse ZP3 required not only the oligosaccharide fraction but also the polypeptide chain (Florman *et al.*, 1984). Leyton and Saling (1989a) investigated the basis of this requirement of the ZP3 polypeptide chain for induction of the acrosome reaction. Glycopeptides derived from ZP by pronase digestion failed to induce the acrosome reaction, confirming the earlier work of Florman *et al.* (1984). Addition of monospecific polyclonal anti-ZP3 IgG to spermatozoa with bound ZP glycopeptides caused the spermatozoa to undergo acrosome reactions. Addition of mono-specific polyclonal anti-ZP3 F_{ab} did not induce acrosome reactions; however, subsequent addition of anti-F_{ab} IgG to these cells did induce acrosome reactions (Leyton and Saling, 1989a). These data suggested that the ZP3 polypeptide chain functions to orient ZP3 oligosaccharides so they are able to activate receptors in the sperm plasma membrane, possibly by receptor aggregation.

Molecular cloning of mRNA encoding mouse ZP3 (Ringuette *et al.*, 1986,1988) and human ZP3 (Chamberlin and Dean, 1990) and of genomic DNA containing mouse (Kinloch *et al.*, 1988), hamster (Kinloch *et al.*, 1990), and human (Chamberlin and Dean, 1990) ZP3 genes has been reported. The deduced amino acid sequences of human and hamster ZP3 were 67% and 81% identical, respectively, to the mouse ZP3 amino acid sequence (Chamberlin and Dean, 1990; Kinloch *et al.*, 1990). A partial length cDNA encoding mouse ZP3 hybridized with 1–4 bands in each lane on Southern blots of mouse, rat, rabbit, dog, pig, and human genomic DNA digested with *Bam*HI (Ringuette *et al.*, 1986); a similar probe hybridized on Northern blots with 1.5- to 1.6-kb transcripts in ovarian RNA from mouse, rat, and rabbit (Ringuette *et al.*, 1988). These data clearly showed that the amino acid sequences of ZP3 are conserved substantially among several mammalian species.

Kinloch *et al.* (1991) described the expression of the mouse and hamster ZP3 genes in cultured embryonic carcinoma cells. Cells transfected with the mouse or hamster ZP3 genes secreted ZP3 into the culture medium. The expressed mouse ZP3 inhibited binding of mouse spermatozoa to the ZP of ovulated eggs and induced the acrosome reaction. In contrast, the expressed hamster ZP3 neither inhibited sperm binding to the ZP nor induced the acrosome reaction. The expression of active (adhesion inhibitory and acrosome reaction-inducing) mouse ZP3 in two other cultured mammalian cell lines subsequently was reported by Beebe *et al.* (1992).

As described earlier (Section II,B), intracellular Ca^{2+} and cAMP concentrations are important regulators of mammalian sperm motility. Extracellular Ca^{2+} was shown to be required for spermatozoa to progress to hyperactivated motility or to undergo the acrosome reaction (Yanagimachi and Usui, 1974). Caltrin (Section II,B) bound to both the head and the tail of bovine spermatozoa (San Agustin *et al.*, 1987), suggesting a possible function in regulation of both motility and the acrosome reaction. Hyne and Garbers

(1979b) reported that Ca^{2+} import stimulated adenylyl cyclase activity of guinea pig spermatozoa. The resultant increase in intracellular cAMP concentrations correlated with induction of the acrosome reaction. Increased intracellular pH also is involved in induction of the mammalian sperm acrosome reaction (Endo et al., 1988; Lee and Storey, 1989) and in stimulation of mammalian sperm motility (Babcock et al., 1983). These parallels in the apparent requirements for induction of the sperm acrosome reaction and stimulation of sperm motility complicate the interpretation of experiments aimed at understanding these processes in greater detail. For example, Babcock and Pfeiffer (1987) characterized a voltage-dependent Ca^{2+} channel and a separate voltage-dependent mechanism that increased pH_i, but it was not possible to judge whether these activities function in the induction of the acrosome reaction, or activation of motility, or both. Similarly, stimulation of Ca^{2+} uptake and motility of bovine epididymal spermatozoa by amiloride (Breitbart et al., 1990) could be mediated by channels that normally function in the acrosome reaction. Thus, studying these processes in the context of their stimulation by the relevant physiological agonists is important.

Identification of a physiological agonist (ZP3) that induces the acrosome reaction has enabled dissection of the signaling pathway that regulates this process. Mammalian spermatozoa contain the $Gi_{\alpha 1}$, $Gi_{\alpha 2}$, $Gi_{\alpha 3}$, Gz_α, and G_β subunits of the heterotrimeric G proteins (Bentley et al., 1986c; Kopf et al., 1986; Glassner et al., 1991). A large majority of G_α immunoreactivity, presumably the Gi_α subunits, localizes to the apical acrosomes of mouse and guinea pig spermatozoa. This presumptive Gi_α immunoreactivity disappears coincident with the acrosome reaction (Glassner et al., 1991). In contrast, the Gz_α immunoreactivity is not in the apical acrosome, but instead is confined to the postacrosomal/lateral face (equatorial segment?) of the mouse sperm acrosome. This immunoreactivity persists during early stages of the acrosome reaction, but eventually disappears (Glassner et al., 1991). These studies show that mammalian spermatozoa contain G proteins in the right place for a possible function during the ZP3-induced acrosome reaction.

Two research groups have reported that inactivation of G proteins with pertussis toxin inhibits the induction of the mammalian sperm acrosome reaction by ZP (Endo et al., 1987,1988; Florman et al., 1989; Lee et al., 1992). Pertussis toxin-catalyzed ADP-ribosylation of Gi_α-like proteins of mouse spermatozoa inhibited the first of two resolvable stages of the ZP-induced acrosome reaction (Endo et al., 1987,1988). Similarly, pertussis toxin-catalyzed ADP-ribosylation of G proteins of bovine spermatozoa inhibited the acrosome reaction induced by ZP (Florman et al., 1989). Incubation of human spermatozoa with pertussis toxin also inhibited the ZP-induced acrosome reaction (Lee et al., 1992). Solubilized ZP stimulated high affinity GTPase activity and GTPγS binding of Gi-like protein(s) of mouse sperm membranes (Ward et al., 1992; Wilde et al., 1992). Purified ZP3 also stimulated these activities (Ward et al., 1992). Collectively, these studies showed that activation of G proteins, presumably by a receptor-mediated process, is an early event in the induction of the acrosome reaction by ZP.

Lee and Storey (1989) demonstrated that H^+ and Ca^{2+} permeability increase in the acrosomal region of the sperm head concomitant with the acrosome reaction induced by the ZP. Pertussis toxin inhibited intracellular alkalinization, which normally accompanied the first stage of the mouse sperm acrosome reaction induced by ZP (Endo *et al.*, 1988), and also inhibited increases in intracellular pH and Ca^{2+} concentrations, which occurred during the ZP-induced bovine sperm acrosome reaction (Florman *et al.*, 1989). Thus, a direct or indirect consequence of G protein activation by ZP is regulation of H^+ and Ca^{2+} transport. Florman *et al.* (1992) subsequently demonstrated that a combination of increased pH_i and artificial depolarization of the plasma membrane caused Ca^{2+} entry and induced the acrosome reaction of bovine, ovine, and murine spermatozoa. A Ca^{2+} channel with characteristics of an L-type voltage-dependent channel mediated the Ca^{2+} entry induced either by artificial depolarization and increased pH_i or by the ZP. Incubation of spermatozoa with pertussis toxin inhibited Ca^{2+} entry and acrosome reactions induced by the ZP but did not inhibit acrosome reactions induced by the artificial depolarization/increased pH_i regimen. Thus, induction of the acrosome reaction by ZP appears to proceed via receptor-mediated activation of Gi-like regulatory proteins that directly or indirectly effect membrane depolarization, increased pH_i, and activation of voltage-dependent Ca^{2+} channels.

Two possible consequences of Ca^{2+} influx during the acrosome reaction have been identified. As previously described (Section II,B,1), sperm adenylyl cyclase is stimulated by Ca^{2+}/calmodulin (Hyne and Garbers, 1979a,b; Toscano and Gross, 1991) and cAMP may serve as a second messenger in the acrosome reaction (Hyne and Garbers, 1979b; Mrsny and Meizel, 1980). Bicarbonate was found to be required for activation of sperm adenylyl cyclase by Ca^{2+} (Garbers *et al.*, 1982b) and for induction of the acrosome reaction by the ZP (Lee and Storey, 1986). Thus, one consequence of Ca^{2+} influx stimulated by ZP may be bicarbonate-dependent activation of adenylyl cyclase.

Results of several studies demonstrated Ca^{2+}-induced hydrolysis of phosphoinositides in spermatozoa. Roldan and Harrison (1989,1990,1992) induced Ca^{2+} influx and synchronous acrosome reactions of noncapacitated spermatozoa with the ionophore A23187. Hydrolysis of phosphatidylinositol 4,5-bisphosphate was complete 3 min after ionophore addition, and preceded by several minutes the completion of the acrosome reaction (Roldan and Harrison, 1989). Phosphoinositide hydrolysis required micromolar Ca^{2+} concentrations, but completion of the acrosome reaction required millimolar concentrations (Roldan and Harrison, 1989). Although both diacylglycerol and phosphatidate concentrations increased as a consequence of Ca^{2+}-induced phosphoinositide hydrolysis, the phosphatidate apparently was produced by phosphorylation of diacylglycerol, and exogenous diacylglycerol but not phophatidate accelerated the rate of acrosome reactions (Roldan and Harrison, 1990). Neomycin inhibited ionophore-induced phosphoinositide hydrolysis and acrosome reactions; exogenous diacylglycerol negated the inhibition (Roldan and Harrison, 1992). Thomas and Meizel (1989) reported that progesterone-induced Ca^{2+} influx (Section III,B,3) also stimulated phosphatidylinositol 4,5-bisphosphate hydrolysis in capacitated human spermatozoa. Collectively, these results support the hypothesis that Ca^{2+} influx during the sperm acrosome reaction stimulates a phosphoinositide-specific phospholipase C, and the resultant increase in diacylglycerol concentration contributes in a yet unknown way (possibly by activation of protein kinase C, see subsequent text) to the completion of the acrosome reaction.

Diacylglycerol and Ca^{2+} stimulate all known isozymes of protein kinase C (Nishizuka, 1988). Phorbol esters, which mimic the stimulatory effects of diacylglycerol on protein kinase C (Nishizuka, 1988), accelerated the first stage of the ZP-induced acrosome reaction of mouse spermatozoa (Lee *et al.*, 1987) and stimulated acrosome reactions of bovine spermatozoa (Breitbart *et al.*, 1992). Phorbol esters also caused bicarbonate-dependent increases in cAMP accumulation in hamster spermatozoa (Visconti *et al.*, 1990). Protein kinase C immunoreactivity localized to the apical acrosome and postacrosomal region of bovine spermatozoa (Breitbart *et al.*, 1992), and to the flagellum and equatorial segment of human spermatozoa (Rotem *et al.*, 1990a,b). Protein kinase C-specific activity of human and bovine spermatozoa was reported to be very low in comparison with other cell types (Rotem *et al.*, 1990b; Breitbart *et al.*, 1992). Although the evidence is not overwhelming, a second consequence of Ca^{2+} influx during the acrosome reaction may be activation of protein kinase C by Ca^{2+} and the diacylglycerol produced by phosphoinositide-specific phospholipase C. The changes in sperm physiology induced by ZP3 are summarized in Fig. 2.

2. Candidate receptors

Receptors exhibit at least two activities: specific ligand binding and transduction of a signal, either directly or indirectly, to the intracellular space. Several possible ZP adhesion activities of spermatozoa have been identified, among them an apparent trypsin-like proteolytic activity (Saling, 1981; Benau and Storey, 1987), a galactosyltransferase activity (Lopez *et al.*, 1985; Shur and Neely, 1988; Benau *et al.*, 1990), an α-D-mannosidase activity (Tulsiani *et al.*, 1989; Cornwall *et al.*, 1991), and an antigen defined by the PH-20 monoclonal antibody (Primakoff *et al.*, 1985; Lathrop *et al.*, 1990). Whether any of these apparent ZP adhesion activities also signals is unknown. In fact, the sperm plasma membrane could contain proteins that function in adhesion to the zona pellucida but do not signal. What criteria must be satisfied before one can have confidence that a candidate receptor is the true ZP3 receptor that regulates the acrosome reaction? First, the native receptor should exhibit ligand binding specificity similar to that required for induction of the acrosome reaction. Second, in response to ligand binding, the receptor candidate must produce a change in cell physiology which is required (but not necessarily sufficient) to induce the acrosome reaction. Rigorous satisfaction of these criteria could be achieved either by purifying of the receptor from spermatozoa and reconstituting it into a functional signaling system or by cloning of the mRNA encoding the receptor and then expressing appropriate receptor activity in a cultured cell line. Neither of these goals has yet been met.

Two receptor candidates that appear promising were identified in mouse spermatozoa. Bleil and Wassarman (1990) identified a 56,000 M_r ZP3 binding protein by photoaffinity cross-linking. This binding activity localized to the heads of acrosome-intact spermatozoa, was specific for ZP3, and was saturable. Leyton and Saling (1989b) identified a 95,000 M_r ZP3 binding protein by probing Western blots of sperm proteins with ^{125}I-labeled ZP3. This protein appeared also to contain tyrosine phosphate immunoreactivity that increased during capacitation and as a consequence of stimulation by solubilized ZP.

3. Progesterone as ligand

A second physiological ligand that promotes the mammalian sperm acrosome reaction was identified in follicular fluid. A partially purified fraction of human follicular fluid induced acrosome reactions in capacitated spermatozoa (Suarez et al., 1986). This response was very rapid; 40% of the cells underwent the acrosome reaction 3 min after mixing with follicular fluid (Yudin et al., 1988). The follicular fluid fraction also induced a rapid influx of extracellular Ca^{2+}, which was required for the acrosome reaction (Thomas and Meizel, 1988). The active principle subsequently was purified and identified as progesterone (Osman et al., 1989). Progesterone stimulated Ca^{2+} influx in both noncapacitated and capacitated spermatozoa, possibly by activating a receptor-operated channel (Blackmore et al., 1990). Membrane impermeant conjugates (to bovine serum albumin) of progesterone also induced Ca^{2+} influx (Blackmore et al., 1991; Meizel and Turner, 1991) and the acrosome reaction (Meizel and Turner, 1991), suggesting that the putative receptor is present on the cell surface. Progesterone also induced partial (40%) depolarization of the human sperm plasma membrane (Calzada et al., 1991).

What is the physiological significance of two apparent natural agonists for induction of the acrosome reaction? Unfortunately, this question cannot be answered at this time. During mouse fertilization, induction of the acrosome reaction by the ZP is obligatory, since only acrosome intact mouse spermatozoa attach to and penetrate the ZP (Saling et al., 1979). In contrast, both acrosome intact and acrosome reacted spermatozoa of guinea pig, human, and hamster can attach productively to the ZP (Yanagimachi and Phillips, 1984; Cherr *et al.*, 1986; Myles *et al.*, 1987; Morales *et al.*, 1989). Indeed, guinea pig acrosome-reacted spermatozoa attach more avidly to the ZP than do acrosome intact spermatozoa (Huang *et al.*, 1981; Myles *et al.*, 1987). Hence, progesterone-induced acrosome reactions could be important in some but not all species. However, even in species wherein progesterone-induced acrosome reactions could lead to fertilization, the ZP-induced acrosome reaction might still predominate. Alternatively, progesterone-induced Ca^{2+} influx into noncapacitated spermatozoa could be an important terminal step in capacitation.

IV. Summary

Several generalizations can be made from the material presented in this chapter. First, extracellular signals regulate both motility and the acrosome reaction of sea urchin and mammalian spermatozoa. Second, increases in intracellular pH and Ca^{2+} concentration, and consequent activation of adenylyl cyclase activity, are important intermediate steps in regulation of sperm physiology by each of these extracellular signals (compare Figs. 1, 2). Indeed, concerted activation of overlapping signaling cascades may be the mechanism for the speract-mediated potentiation of FSG-induced acrosome reactions of sea urchin spermatozoa (Yamaguchi *et al.*, 1988). Third, in mammals and in sea urchins, the acrosome reaction is regulated by carbohydrate-rich agonists in the egg investments. Fourth, guanine nucleotides are likely to be key effectors in signaling cascades of sea urchin (activation of guanylyl cyclase by egg peptides) and mammalian (activation of G proteins during the acrosome reaction) spermatozoa. The extent to which the different kinds of extracellular signals that direct fertilization, and the physiological and cellular responses they produce, have been conserved throughout evolution seems quite remarkable considering the taxonomic differences between these two animal classes. These similarities ensure that research on fertilization mechanisms in diverse species will continue to contribute substantially to our understanding of this essential biological process.

References

Acott, T. S., and Hoskins, D. D. (1978). Bovine sperm forward motility protein: Partial purification and characterization. *J. Biol. Chem.* **253**, 6744–6750.

Acott, T. S., and Hoskins, D. D. (1981). Bovine sperm forward motility protein: binding to epididymal spermatozoa. *Biol. Reprod.* **24**, 234–240.

Austin, C. R. (1951). Observations on the penetration of the sperm into the mammalian egg. *Austr. J. Sci. Res.* **4**, 581–596.

Austin, C. R., and Bishop, M. W. H. (1958). Role of the rodent acrosome and perforatorium in fertilization. *Proc. R. Soc. London B* **149**, 241–248.

Babcock, D. F., and Pfeiffer, D. R. (1987). Independent elevation of cytosolic [Ca^{2+}] and pH of mammalian sperm by voltage-dependent and pH-sensitive mechanisms. *J. Biol. Chem.* **262**, 15041–15047.

Babcock, D. F., Stamerjohn, D. M., and Hutchinson, T. (1978). Calcium redistribution in individual cells correlated with ionophore action on motility. *J. Exp. Zool.* **204**, 391–400.

Babcock, D. F., Singh, J. P., and Lardy, H. A. (1979). Alteration of membrane permeability to calcium ions during maturation of bovine spermatozoa. *Dev. Biol.* **69**, 85–93.

Babcock, D. F., Rufo, G. A., and Lardy, H. A. (1983). Potassium-dependent increases in cytosolic pH stimulate metabolism and motility of mammalian sperm. *Proc. Natl. Acad. Sci. U.S.A.* **80**, 1327–1331.

Babcock, D. F., Bosma, M. M., Battaglia, D. E., and Darszon, A. (1992). Early persistent activation of sperm K$^+$ channels by the egg peptide speract. *Proc. Natl. Acad. Sci. U.S.A.* **89**, 6001–6005.

Beebe, S. J., Leyton, L., Burks, D., Ishikawa, M., Fuerst, T., Dean, J., and Saling, P. (1992). Recombinant mouse ZP3 inhibits sperm binding and induces the acrosome reaction. *Dev. Biol.* **151**, 48–54.

Benau, D. A., and Storey, B. T. (1987). Characterization of the mouse sperm plasma membrane zona-binding site sensitive to trypsin inhibitors. *Biol. Reprod.* **36**, 282–292.

Benau, D. A., McGuire, E. J., and Storey, B. T. (1990). Further characterization of the mouse sperm surface zona-binding site with galactosyltransferase activity. *Mol. Reprod. Dev.* **25**, 393–399.

Bentley, J. K., and Garbers, D. L. (1986). Receptor-mediated responses of plasma membranes isolated from *Lytechinus pictus* spermatozoa. *Biol. Reprod.* **35**, 1249–1259.

Bentley, J. K., Shimomura, H., and Garbers, D. L. (1986a). Retention of a functional resact receptor in isolated sperm plasma membranes. *Cell* **45**, 281–288.

Bentley, J. K., Tubb, D. J., and Garbers, D. L. (1986b). Receptor-mediated activation of spermatozoan guanylate cyclase. *J. Biol. Chem.* **261**, 14859–14862.

Bentley, J. K., Garbers, D. L., Domino, S. E., Noland, T. D., and Van Dop, C. (1986c). Spermatozoa contain an guanine nucleotide-binding protein ADP-ribosylated by pertussis toxin. *Biochem. Biophys. Res. Commun.* **138**, 728–734.

Berger, T., Turner, K. O., Meizel, S., and Hedrick, J. L. (1989). Zona pellucida-induced acrosome reaction in boar sperm. *Biol. Reprod.* **40**, 525–530.

Blackmore, P. F., Beebe, S. J., Danforth, D. R., and Alexander, N. (1990). Progesterone and 17α-hydroxyprogesterone. Novel stimulators of calcium influx in human sperm. *J. Biol. Chem.* **265**, 1376–1380.

Blackmore, P. F., Neulen, J., Lattanzio, F., and Beebe, S. J. (1991). Cell surface-binding sites for progesterone mediate calcium uptake in human sperm. *J. Biol. Chem.* **266**, 18655–18659.

Bleil, J. D., and Wassarman, P. M. (1980a). Structure and function of the zona pellucida: Identification and characterization of the proteins of the mouse oocyte's zona pellucida. *Dev. Biol.* **76**, 185–202.

Bleil, J. D., and Wassarman, P. M. (1980b). Mammalian sperm-egg interaction: Identification of a glycoprotein in mouse egg zonae pellucidae possessing receptor activity for sperm. *Cell* **20**, 873–882.

Bleil, J. D., and Wassarman, P. M. (1983). Sperm-egg interaction in the mouse: sequence of events and induction of the acrosome reaction by a zona pellucida glycoprotein. *Dev. Biol.* **95**, 317–324.

Bleil, J. D., and Wassarman, P. M. (1988). Galactose at the nonreducing terminus of O-linked oligosaccharides of mouse egg zona pellucida glycoprotein ZP3 is essential for the glycoprotein's sperm receptor activity. *Proc. Natl. Acad. Sci. U.S.A.* **85**, 6778–6782.

Bleil, J. D., and Wassarman, P. M. (1990). Identification of a ZP3-binding protein on acrosome-intact mouse sperm by photoaffinity crosslinking. *Proc. Natl. Acad. Sci. U.S.A.* **87**, 5563–5567.

Bookbinder, L. H., Moy, G. W., and Vacquier, V. D. (1990). Identification of sea urchin sperm adenylate cyclase. *J. Cell Biol.* **111**, 1859–1866.

Bradley, M. P., and Garbers, D. L. (1983). The stimulation of bovine caudal epididymal sperm forward motility by bovine cumulus-egg complexes *in vitro*. *Biochem. Biophys. Res. Commun.* **115**, 777–787.

Breitbart, H., Rubinstein, S., and Nass-Arden, L. (1985). The role of calcium and Ca^{2+}-ATPase in maintaining motility in ram spermatozoa. *J. Biol. Chem.* **260**, 11548–11553.

Breitbart, H., Cragoe, E. J., and Lardy, H. A. (1990). Stimulation of Ca^{2+} uptake into epididymal bell spermatozoa by analogues of amiloride. *Eur. J. Biochem.* **192**, 529–535.

Breitbart, H., Lax, J., Rotem, R., and Naor, Z. (1992). Role of protein kinase C in the acrosome reaction of mammalian spermatozoa. *Biochem. J.* **281**, 473–476.

Burks, D. J. (1990). The structures and site of synthesis of the precursors for peptides that stimulate spermatozoa. Ph.D. Thesis. Vanderbilt University, Nashville, Tennessee.

Calzada, L., Salazar, E. L., and Macias, H. (1991). Hyperpolarization/depolarization on human spermatozoa. *Arch. Androl.* **26**, 71–78.

Carr, D. W., and Acott, T. S. (1989). Intracellular pH regulates bovine sperm motility and protein phosphorylation. *Biol. Reprod.* **41**, 907–920.

Carter, G. S. (1931). Iodine compounds and fertilization. II. The oxygen consumption of suspensions of sperm of *Echinus esculentus* and *Echinus millaris*. *J. Exp. Biol.* **8**, 176–193.

Chamberlin, M. E., and Dean, J. (1990). Human homolog of the mouse sperm receptor. *Proc. Natl. Acad. Sci. U.S.A.* **87**, 6014–6018.

Chang, M. C. (1951). Fertilizing capacity of spermatozoa deposited into the fallopian tubes. *Nature (London)* **168**, 697–698.

Chao, H. T., Ng, H. T., Kao, S. H., Wei, Y. H., and Hong, C. Y. (1991). Human follicular fluid stimulates the motility of washed human sperm. *Arch. Androl.* **26**, 61–65.

Chaudhry, P. S., Newcomer, P. A., and Casillas, E. R. (1991a). Casein kinase I in bovine sperm: Purification and characterization. *Biochem. Biophys. Res. Commun.* **179**, 592–598.

Chaudhry, P. S., Nanez, R., and Casillas, E. R. (1991b). Purification and characterization of polyamine-stimulated protein kinase (casein kinase II) from bovine spermatozoa. *Arch. Biochem. Biophys.* **288**, 337–342.

Cherr, G. N., Lambert, H., Meizel, S., and Katz, D. F. (1986). *In vitro* studies of the golden hamster sperm acrosome reaction: Completion on the zona pellucida and induction by homologous soluble zonae pellucidae. *Dev. Biol.* **114**, 119–131.

Chinkers, M., and Garbers, D. L. (1989). The protein kinase domain of the ANP receptor is required for signalling. *Science* **245**, 1392–1394.

Christen, R., Schackmann, R. W., and Shapiro, B. M. (1982). Elevation of the intracellular pH activates respiration and motility of sperm of the sea urchin, *Strongylocentrotus purpuratus*. *J. Biol. Chem.* **257**, 14881–14890.

Christen, R., Schackmann, R. W., and Shapiro, B. M. (1983). Metabolism of sea urchin sperm. Interrelationships between intracellular pH, ATPase activity, and mitochondrial respiration. *J. Biol. Chem.* **258**, 5392–5399.

Cornwall, G. A., Tulsiani, D. R. P., and Orgebin-Crist, M.-C. (1991). Inhibition of the mouse sperm surface α-D-mannosidase inhibits sperm-egg binding in vitro. *Biol. Reprod.* **44**, 913–921.

Coronel, C. E., and Lardy, H. A. (1987). Characterization of Ca^{2+} uptake by guinea pig epididymal spermatozoa. *Biol. Reprod.* **37**, 1097–1107.

Coronel, C. E., San Agustin, J., and Lardy, H. A. (1988). Identification and partial characterization of caltrin-like proteins in the reproductive tract of the guinea pig. *Biol. Reprod.* **38**, 713–722.

Coronel, C. E., San Agustin, J., and Lardy, H. A. (1990). Purification and structure of caltrin-like proteins from seminal vesicle of the guinea pig. *J. Biol. Chem.* **265,** 6854–6859.

Cross, N. L., Morales, P., Overstreet, J. W., and Hanson, F. (1988). Induction of acrosome reactions by the human zona pellucida. *Biol. Reprod.* **38,** 235–244.

Crozet, N., and Dumont, M. (1984). The site of the acrosome reaction during *in vivo* penetration of the sheep oocyte. *Gamete Res.* **10,** 97–105.

Crozet, N., Theron, M. C., and Chemineau, P. (1987). Ultrastructure of *in vivo* fertilization in the goat. *Gamete Res.* **18,** 191–199.

Dan, J. C. (1952). Studies on the acrosome. I. Reaction to egg-water and other stimuli. *Biol. Bull.* **103,** 54–66.

Dan, J. C. (1954). Studies on the acrosome. III. Effect of calcium deficiency. *Biol. Bull.* **107,** 335–349.

Dangott, L. J., and Garbers, D. L. (1984). Identification and partial characterization of the receptor for speract. *J. Biol. Chem.* **259,** 13712–13716.

Dangott, L. J., Jordan, J. E., Bellet, R. A., and Garbers, D. L. (1989). Cloning of the mRNA for the protein that crosslinks to the egg peptide speract. *Proc. Natl. Acad. Sci. U.S.A.* **86,** 2128–2132.

Domino, S. E., and Garbers, D. L. (1988). The fucose-sulfate glycoconjugate that induces an acrosome reaction in spermatozoa stimulates inositol 1,4,5-trisphosphate accumulation. *J. Biol. Chem.* **263,** 690–695.

Domino, S. E., and Garbers, D. L. (1990). Mode of action of egg peptides. *In* "Controls of Sperm Motility: Biological and Clinical Aspects" (C. Gagnon, ed.), pp. 91–101. CRC Press, Boca Raton, Florida.

Domino, S. E., Bocckino, S. B., and Garbers, D. L. (1989). Activation of phospholipase D by the fucose-sulfate glycoconjugate that induces an acrosome reaction in spermatozoa. *J. Biol. Chem.* **264,** 9412–9419.

Dunbar, B. S., Wardrip, N. J., and Hedrick, J. L. (1980). Isolation, physicochemical properties, and macromolecular composition of zona pellucida from porcine oocytes. *Biochemistry* **19,** 356–365.

Endo, Y., Lee, M. A., and Kopf, G. S. (1987). Evidence for the role of a guanine nucleotide-binding regulatory protein in the zona pellucida-induced mouse sperm acrosome reaction. *Dev. Biol.* **119,** 210–216.

Endo, Y., Lee, M. A., and Kopf, G. S. (1988). Characterization of an islet-activating protein-sensitive site in mouse sperm that is involved in the zona pellucida-induced acrosome reaction. *Dev. Biol.* **129,** 12–24.

Exton, J. H. (1990). Signaling through phosphatidylcholine breakdown. *J. Biol. Chem.* **265,** 1–4.

Feinberg, J., Weinmann, J., Weinmann, S., Walsh, M. P., Harricane, M. C., Gabrion, J., and Demaille, J. G. (1981). Immunocytochemical and biochemical evidence for the presence of calmodulin in bull sperm flagellum. *Biochim. Biophys. Acta* **673,** 303–311.

Florman, H. M., and First, N. L. (1988a). Regulation of acrosomal exocytosis. II. The zona pellucida-induced acrosome reaction of bovine spermatozoa is controlled by extrinsic positive regulatory elements. *Dev. Biol.* **128,** 464–473.

Florman, H. M., and First, N. L. (1988b). The regulation of acrosomal exocytosis. I. Sperm capacitation is required for the induction of acrosome reactions by the bovine zona pellucida in vitro. *Dev. Biol.* **128,** 453–463.

Florman, H. M., and Storey, B. T. (1982). Mouse gamete interactions: The zona pellucida is the site of the acrosome reaction leading to fertilization in vitro. *Dev. Biol.* **91,** 121–130.

Florman, H. M., and Wassarman, P. M. (1985). O-Linked oligosaccharides of mouse egg ZP3 account for its sperm receptor activity. *Cell* **41,** 313–324.

Florman, H. M., Bechtol, K. B., and Wassarman, P. M. (1984). Enzymatic dissection of the functions of the mouse egg's receptor for sperm. *Dev. Biol.* **106,** 243–255.

Florman, H. M., Tombes, R. M., First, N. L., and Babcock, D. F. (1989). An adhesion-associated agonist from the zona pellucida activates G protein-promoted elevation of internal Ca^{2+} and pH that mediate mammalian sperm acrosomal exocytosis. *Dev. Biol.* **135**, 133–146.

Florman, H. M., Corron, M. E., Kim, T. D.-H., and Babcock, D. F. (1992). Activation of voltage-dependent calcium channels of mammalian sperm is required for zona pellucida-induced acrosomal exocytosis. *Dev. Biol.* **152**, 304–314.

Freeman, M., Ashkenas, J., Rees, D. J. G., Kingsley, D. M., Copeland, N. G., Jenkins, N. A., and Krieger, M. (1990). An ancient, highly conserved family of cysteine-rich protein domains revealed by cloning type I and type II murine macrophage scavenger receptors. *Proc. Natl. Acad. Sci. U.S.A.* **87**, 8810–8814.

Garbers, D. L. (1981). The elevation of cyclic AMP concentrations in flagella-less sea urchin sperm heads. *J. Biol. Chem.* **256**, 620–624.

Garbers, D. L. (1989). Molecular basis of fertilization. *Ann. Rev. Biochem.* **58**, 719–742.

Garbers, D. L., and Hardman, J. G. (1975). Factors released from sea urchin eggs affect cyclic nucleotide metabolism in sperm. *Nature (London)* **257**, 677–678.

Garbers, D. L., and Kopf, G. S. (1980). The regulation of spermatozoa by calcium and cyclic nucleotides. *Adv. Cyclic Nucleotide Res.* **13**, 251–306.

Garbers, D. L., Lust, W. D., First, N. L., and Lardy, H. A. (1971). Effects of hopsphodiesterase inhibitors and cyclic nucleotides on sperm respiration and motility. *Biochemistry* **10**, 1825–1831.

Garbers, D. L., First, N. L., Gorman, S. K., and Lardy, H. A. (1973). The effects of cyclic nucleotide phosphodiesterase inhibitors on ejaculated porcine spermatozoan metabolism. *Biol. Reprod.* **8**, 599–606.

Garbers, D. L., Watkins, H. D., Hansbrough, J. R., Smith, A., and Misono, K. S. (1982a). The amino acid sequence and chemical synthesis of speract and of speract analogues. *J. Biol. Chem.* **257**, 2734–2737.

Garbers, D. L., Tubb, D. J., and Hyne, R. V. (1982b). A requirement of bicarbonate for Ca^{2+}-induced elevations of cyclic AMP in guinea pig spermatozoa. *J. Biol. Chem.* **257**, 8980–8984.

Garbers, D. L., Kopf, G. S., Tubb, D. J., and Olson, G. (1983). Elevation of sperm adenosine 3′,5′-monophosphate concentrations by a fucose-sulfate-rich complex associated with eggs: I. Structural characterization. *Biol. Reprod.* **29**, 1211–1220.

Glassner, M., Jones, J., Kligman, I., Woolkalis, M. J., Gerton, G. L., and Kopf, G. S. (1991). Immunocytochemical and biochemical characterization of guanine nucleotide-binding regulatory proteins in mammalian spermatozoa. *Dev. Biol.* **146**, 438–450.

Gonzalez-Martinez, M., and Darszon, A. (1987). A fast transient hyperpolatization occurs during the sea urchin sperm acrosome reaction induced by egg jelly. *FEBS Lett.* **218**, 247–250.

Gray, J. (1928). The effect of egg secretions on the activity of spermatozoa. *J. Exp. Biol.* **5**, 362–365.

Guerrero, A., and Darszon, A. (1989a). Egg jelly triggers a calcium influx which inactivates and is inhibited by calmodulin antagonists in the sea urchin sperm. *Biochim. Biophys. Acta* **980**, 109–116.

Guerrero, A., and Darszon, A. (1989b). Evidence for the activation of two different Ca^{2+} channels during the egg jelly-induced acrosome reaction of sea urchin sperm. *J. Biol. Chem.* **264**, 19593–19599.

Hansbrough, J. R., and Garbers, D. L. (1981a). Speract: Purification and characterization of a peptide associated with eggs that activates spermatozoa. *J. Biol. Chem.* **256**, 1447–1452.

Hansbrough, J. R., and Garbers, D. L. (1981b). Sodium-dependent activation of sea urchin spermatozoa by speract and monensin. *J. Biol. Chem.* **256**, 2235–2241.

Hansbrough, J. R., Kopf, G. S., and Garbers, D. L. (1980). The stimulation of sperm metabolism by a factor associated with eggs and by 8-bromo-guanosine 3′,5′-monophosphate. *Biochim. Biophys. Acta* **630**, 82–91.

Hardy, D. M., Oda, M. N., Friend, D. S., and Huang, T. T. F. (1991). A mechanism for differential release of acrosomal enzymes during the acrosome reaction. *Biochem. J.* **275,** 759–766.

Harumi, T., Hoshino, K., and Suzuki, N. (1992). Effects of sperm-activating peptide I on *Hemicentrotus pulcherrimus* spermatozoa in high potassium sea water. *Dev. Growth Diff.* **34,** 163–172.

Hedrick, J. L., and Wardrip, N. J. (1987). On the macromolecular composition of the zona pellucida from porcine oocytes. *Dev. Biol.* **121,** 478–488.

Holland, L. Z., and Cross, N. L. (1983). The pH within the jelly coat of sea urchin eggs. *Dev. Biol.* **99,** 258–260.

Horowitz, J. A., Toeg, H., and Orr, G. A. (1984). Characterization and localization of cAMP-dependent protein kinases in rat caudal epididymal sperm. *J. Biol. Chem.* **259,** 832–838.

Horowitz, J. A., Voulalas, P., Wasco, W., MacLeod, J., Paupard M.-C., and Orr, G. A. (1989). Biochemical and immunological characterization of the flagellar-associated regulatory subunit of a type II cyclic adenosine 5'-monophosphate-dependent protein kinase. *Arch. Biochem. Biophys.* **270,** 411–418.

Huang, T. T. F., Fleming, A. D., and Yanagimachi, R. (1981). Only acrosome-reacted spermatozoa can bind to and penetrate zona pellucida: A study using the guinea pig. *J. Exp. Zool.* **217,** 287–290.

Hyne, R. V., and Garbers, D. L. (1979a). Regulation of guinea pig sperm adenylate cyclase by calcium. *Biol. Reprod.* **21,** 1135–1142.

Hyne, R. V., and Garbers, D. L. (1979b). Calcium-dependent increase in adenosine 3',5'-monophosphate and induction of the acrosome reaction in guinea pig spermatozoa. *Proc. Natl. Acad. Sci. U.S.A.* **76,** 5699–5703.

Johnson, R. A., Jakobs, K. H., and Schultz, G. (1985). Extraction of the adenylate cyclase-activating factor of bovine sperm and its identification as a trypsin-like protease. *J. Biol. Chem.* **260,** 114–121.

Jones, H. P., Bradford, M. M., McRorie, R. A., and Cormier, M. J. (1978). High levels of a calcium-dependent modulator protein in spermatozoa and its similarity to brain modulator protein. *Biochem. Biophys. Res. Commun.* **82,** 1264–1272.

Jones, H. P., Lenz, R. W., Palevitz, B. A., and Cormier, M. J. (1980). Calmodulin localization in mammalian spermatozoa. *Proc. Natl. Acad. Sci. U.S.A.* **77,** 2772–2776.

Kinloch, R. A., Roller, R. J., Fimiani, C. M., Wassarman, D. A., and Wassarman, P. M. (1988). Primary structure of the mouse sperm receptor polypeptide determined by genomic cloning. *Proc. Natl. Acad. Sci. U.S.A.* **85,** 6409–6413.

Kinloch, R. A., Ruiz-Seiler, B., and Wassarman, P. M. (1990). Genomic organization and polypeptide primary structure of zona pellucida glycoprotein hZP3, the hamster sperm receptor. *Dev. Biol.* **142,** 414–421.

Kinloch, R. A., Mortillo, S., Stewart, C. L., and Wassarman, P. M. (1991). Embryonal carcinoma cells transfected with ZP3 genes differentially glycosylate similar polypeptides and secrete active mouse sperm receptor. *J. Cell Biol.* **115,** 655–664.

Koesling, D., Herz, J., Gausepohl, H., Niroomand, F., Hinsch, K.-D., Mulsch, A., Bohme, E., Schultz, G., and Frank, R. (1988). The primary structure of the 70 kDa subunit of bovine soluble guanylate cyclase. *FEBS Lett.* **239,** 29–34.

Kopf, G. S., and Garbers, D. L. (1980). Calcium and a fucose-sulfate-rich polymer regulate sperm cyclic nucleotide metabolism and the acrosome reaction. *Biol. Reprod.* **22,** 1118–1126.

Kopf, G. S., and Gerton, G. L. (1990). The mammalian sperm acrosome and the acrosome reaction. *In* "Elements of Mammalian Fertilization. I. Basic Concepts" (P. M. Wassarman, ed.), pp. 153–203. CRC Press, Boca Raton, Florida.

Kopf, G. S., Woolkalis, M. J., and Gerton, G. L. (1986). Evidence for a guanine nucleotide-binding regulatory protein in invertebrate and mammalian sperm. *J. Biol. Chem.* **261,** 7327–7331.

Krupinski, J., Coussen, F., Bakalyar, H. A., Tang, W.-J., Feinstein, P. G., Orth, K., Slaughter, C.,

Reed, R. R., and Gilman, A. G. (1989). Adenylyl cyclase amino acid sequence: possible channel- or transporter-like structure. *Science* **244**, 1558–1564.

Lathrop, W. F., Carmichael, E. P., Myles, D. G., and Primakoff, P. (1990). cDNA cloning reveals the molecular structure of a sperm surface protein, PH-20, involved in sperm-egg adhesion and the wide distribution of its gene among mammals. *J. Cell Biol.* **111**, 2939–2949.

Lee, H. C. (1984a). Sodium and proton transport in flagella isolated from sea urchin spermatozoa. *J. Biol. Chem.* **259**, 4957–4963.

Lee, H. C. (1984b). A membrane potential-sensitive Na^+–H^+ exchange system in flagella isolated from sea urchin spermatozoa. *J. Biol. Chem.* **259**, 15315–15319.

Lee, H. C. (1988). Internal GTP stimulates the speract receptor mediated voltage changes in sea urchin spermatozoa membrane vesicles. *Dev. Biol.* **126**, 91–97.

Lee, H. C., and Garbers, D. L. (1986). Modulation of the voltage-sensitive Na^+/H^+ exchange in sea urchin spermatozoa through membrane potential changes induced by the egg peptide speract. *J. Biol. Chem.* **261**, 16026–16032.

Lee, M. A., and Storey, B. T. (1986). Bicarbonate is essential for fertilization of mouse eggs: Mouse sperm require it to undergo the acrosome reaction. *Biol. Reprod.* **34**, 349–356.

Lee, M. A., and Storey, B. T. (1989). Endpoint of first stage of zona pellucida-induced acrosome reaction in mouse spermatozoa characterized by acrosomal H^+ and Ca^{2+} permeability: Population and single cell kinetics. *Gamete Res.* **24**, 303–326.

Lee, M. A., Kopf, G. S., and Storey, B. T. (1987). Effects of phorbol esters and a diacylglycerol on the mouse sperm acrosome reaction induced by the zona pellucida. *Biol. Reprod.* **36**, 617–627.

Lee, M. A., Check, J. H., and Kopf, G. S. (1992). A guanine nucleotide-binding regulatory protein in human sperm mediates acrosomal exocytosis induced by the human zona pellucida. *Mol. Reprod. Dev.* **31**, 78–86.

Lewis, R. V., San Agustin, J., Kruggel, W., and Lardy, H. A. (1985). The structure of caltrin, the calcium-transport inhibitor of bovine seminal plasma. *Proc. Natl. Acad. Sci. U.S.A.* **82**, 6490–6491.

Leyton, L., and Saling, P. (1989a). Evidence that aggregation of mouse sperm receptors by ZP3 triggers the acrosome reaction. *J. Cell Biol.* **108**, 2163–2168.

Leyton, L., and Saling, P. (1989b). 95 kd sperm proteins bind ZP3 and serve as tyrosine kinase substrates in response to zona binding. *Cell* **57**, 1123–1130.

Lieberman, S. J., Wasco, W., MacLeod, J., Satir, P., and Orr, G. A. (1988). Immunogold localization of the regulatory subunit of a type II cAMP-dependent protein kinase tightly associated with mammalian sperm flagella. *J. Cell Biol.* **107**, 1809–1816.

Lievano, A., Sanchez, J. A., and Darszon, A. (1986). Single channel activity of bilayers derived from sea urchin sperm plasma membranes at the tip of a patch-clamp electrode. *Dev. Biol.* **112**, 253–257.

Lillie, F. R. (1913). Studies on fertilization. V. The behavior of the spermatozoa of *Nereis* and *Arbacia* with special reference to egg-extractives. *J. Exp. Zool.* **14**, 515–574.

Lopez, L. C., Bayna, E. M., Litoff, D., Shaper, N. L., Shaper, J. H., and Shur, B. D. (1985). Receptor function of mouse sperm surface galactosyltransferase during fertilization. *J. Cell Biol.* **101**, 1501–1510.

Meizel, S., and Turner, K. O. (1991). Progesterone acts at the plasma membrane of human sperm. *Mol. Cell. Endocrinol.* **11**, R1–R5.

Mendoza, C., and Tesarik, J. (1990). Effect of follicular fluid on sperm movement characteristics. *Fertil. Steril.* **54**, 1135–1139.

Moller, C. C., Bleil, J. D., Kinloch, R. A., and Wassarman, P. M. (1990). Structural and functional relationships between mouse and hamster zona pellucida glycoproteins. *Dev. Biol.* **137**, 276–286.

Morales, P., Cross, N. L., Overstreet, J. W., and Hanson, F. W. (1989). Acrosome intact and acrosome-reacted human sperm can initiate binding to the zona pellucida. *Dev. Biol.* **133**, 385–392.

Mori, E. Takasaki, S., Hedrick, J. L., Wardrip, N. J., Mori, T., and Kobata, A. (1991). Neutral oligosaccharide structures linked to asparagines of porcine zona pellucida glycoproteins. *Biochemistry* **30**, 2078–2087.

Morita, Z., and Chang, M. C. (1970). The motility and aerobic metabolism of spermatozoa in laboratory animals with special reference to the effects of cold shock and the importance of calcium for the motility of hamster spermatozoa. *Biol. Reprod.* **3**, 169–179.

Mrsny, R. J., and Meizel, S. (1980). Evidence suggesting a role for cyclic nucleotides in acrosome reactions of hamster sperm in vitro. *J. Exp. Zool.* **211**, 153–157.

Myles, D. G., Hyatt, H., and Primakoff, P. (1987). Binding of both acrosome-intact and acrosome-reacted guinea pig sperm to the zona pellucida during *in vitro* fertilization. *Dev. Biol.* **121**, 559–567.

Nishizuka, Y. (1988). The molecular heterogeneity of protein kinase C and its implications for cellular regulation. *Nature (London)* **334**, 661–665.

Noland, T. D., Corbin, J. D., and Garbers, D. L. (1986). Cyclic AMP-dependent protein kinase isozymes of bovine epididymal spermatozoa: Evidence against the existence of an ecto-kinase. *Biol. Reprod.* **34**, 681–689.

Noland, T. D., Abumrad, N. A., Beth, A. H., and Garbers, D. L. (1987). Protein phosphorylation in intact bovine epididymal spermatozoa: Identification of the type II regulatory subunit of cyclic adenosine 3′,5′-monophosphate-dependent protein kinase as an endogenous phosphoprotein. *Biol. Reprod.* **37**, 171–180.

Ohtake, H. (1976). Respiratory behaviour of sea-urchin spermatozoa. I. Effect of pH and egg water on the respiratory rate. *J. Exp. Zool.* **198**, 303–312.

Okamura, N., Tajima, Y., Onoe, S., and Sugita, Y. (1991). Purification of bicarbonate-sensitive sperm adenylylcyclase by 4-acetamido-4′-isothiocyanostilbene-2,2′-disulfonic acid-affinity chromatography. *J. Biol. Chem.* **266**, 17754–17759.

O'Rand, M. G., and Fisher, S. J. (1987). Localization of zona pellucida binding sites on rabbit spermatozoa and induction of the acrosome reaction by solubilized zonae. *Dev. Biol.* **119**, 551–559.

Osman, R. A., Andria, M. L., Jones, A. D., and Meizel, S. (1989). Steroid induced exocytosis: The human sperm acrosome reaction. *Biochem. Biophys. Res. Commun.* **160**, 828–833.

Paupard, M.-C., MacLeod, J., Wasco, W., and Orr, G. A. (1988). Major 56,000-dalton, soluble phosphoprotein present in bovine sperm is the regulatory subunit of a type II cAMP-dependent protein kinase. *J. Cell. Biochem.* **37**, 161–175.

Pitt, G. S., Milona, N., Borleis, J., Lin, K. C., Reed, R. R., and Devreotes, P. N. (1992). Structurally distinct and stage-specific adenyl cyclase genes play different roles in dictyostelium development. *Cell* **69**, 305–315.

Primakoff, P., Hyatt, H., and Myles, D. G. (1985). A role for the migrating sperm surface antigen PH-20 in guinea pig sperm binding to the egg zona pellucida. *J. Cell Biol.* **101**, 2239–2244.

Radany, E. W., Gerzer, R., and Garbers, D. L. (1983). Purification and characterization of particulate guanylate cyclase from sea urchin spermatozoa. *J. Biol. Chem.* **258**, 8346–8351.

Ralt, D., Goldenberg, M., Fetterolf, P., Thompson, D., Dor, J., Mashaich, S., Garbers, D. L., and Eisenbach, M. (1991). Sperm attraction to a factor(s) correlates with human egg fertilizability. *Proc. Natl. Acad. Sci. U.S.A.* **88**, 2840–2844.

Ramarao, C. S., Burks, D. J., and Garbers, D. L. (1990). A single mRNA encodes multiple copies of the egg peptide speract. *Biochemistry* **29**, 3383–3388.

Repaske, D. R., and Garbers, D. L. (1983). A hydrogen ion flux mediates stimulation of respiratory activity by speract in sea urchin spermatozoa. *J. Biol. Chem.* **258**, 6025–6029.

Ringuette, M. J., Sobieski, D. A., Chamow, S. M., and Dean, J. (1986). Oocyte-specific gene expression: Molecular characterization of a cDNA coding for ZP-3, the sperm receptor of the mouse zona pellucida. *Proc. Natl. Acad. Sci. U.S.A.* **83**, 4341–4345.

Ringuette, M. J., Chamberlin, M. E., Baur, A. W., Sobieski, D. A., and Dean, J. (1988). Molecular analysis of cDNA coding for ZP3, a sperm binding protein of the mouse zona pellucida. *Dev. Biol.* **127**, 287–295.

Roldan, E. R. S., and Harrison, R. A. P. (1989). Polyphosphoinositide breakdown and subsequent exocytosis in the Ca^{2+}/ionophore-induced acrosome reaction of mammalian spermatozoa. *Biochem. J.* **259**, 397–406.

Roldan, E. R. S., and Harrison, R. A. P. (1990). Diacylglycerol and phosphatidate production of the exocytosis of the sperm acrosome. *Biochem. Biophys. Res. Commun.* **172**, 8–15.

Roldan, E. R. S., and Harrison, R. A. P. (1992). The role of diacylglycerol in the exocytosis of the sperm acrosome. Studies using diacylglycerol lipase and diacylglycerol kinase inhibitors and exogenous diacylglycerols. *Biochem. J.* **281**, 767–773.

Rotem, R., Paz, G. F., Homonnai, Z. T., Kalina, M., and Naor, Z. (1990a). Protein kinase C is present in human sperm: Possible role in flagellar motility. *Proc. Natl. Acad. Sci. U.S.A.* **87**, 7305–7308.

Rotem, R., Paz, G. F., Homonnai, Z. T., Kalina, M., and Naor, Z. (1990b). Further studies on the involvement of protein kinase C in human sperm flagellar motility. *Endocrinology* **127**, 2571–2577.

Rothschild, N. M. V. (1956). *In* "Fertilization." Methuen, London.

Rufo, G. A., Singh, J. P., Babcock, D. F., and Lardy, H. A. (1982). Purification and characterization of a calcium transport inhibitor protein from bovine seminal plasma. *J. Biol. Chem.* **257**, 4627–4632.

Rufo, G. A., Schoff, P. K., and Lardy, H. A. (1984). Regulation of calcium content in bovine spermatozoa. *J. Biol. Chem.* **259**, 2547–2552.

Saling, P. M. (1981). Involvement of trypsin-like activity in binding of mouse spermatozoa to zonae pellucidae. *Proc. Natl. Acad. Sci. U.S.A.* **78**, 6231–6235.

Saling, P. M., Sowinski, J., and Storey, B. T. (1979). An ultrastructural study of epididymal mouse spermatozoa binding to zonae pellucidae in vitro: Sequential relationship to the acrosome reaction. *J. Exp. Zool.* **209**, 229–238.

San Agustin, J. T., and Lardy, H. A. (1990). Bovine seminal plasma constituents modulate the activity of caltrin, the calcium-transport regulating protein of bovine spermatozoa. *J. Biol. Chem.* **265**, 6860–6867.

San Agustin, J. T., Hughes, P., and Lardy, H. A. (1987). Properties and function of caltrin, the calcium-transport inhibitor of bull seminal plasma. *FASEB J.* **1**, 60–66.

Schackmann, R. W., Christen, R., and Shapiro, B. M. (1984). Measurement of plasma membrane and mitochondrial potentials in sea urchin sperm. *J. Biol. Chem.* **259**, 13914–13922.

Schackmann, R. W., and Chock, P. D. (1986). Alteration of intracellular $[Ca^{2+}]$ in sea urchin sperm by the egg peptide speract. Evidence that increased intracellular Ca^{2+} is coupled to Na^+ entry and increased intracellular pH. *J. Biol. Chem.* **261**, 8719–8728.

Segall, G. K., and Lennarz, W. J. (1979). Chemical characterization of the component of the jelly coat from sea urchin eggs responsible for induction of the acrosome reaction. *Dev. Biol.* **71**, 33–48.

Shalgi, R., Phillips, D. M., and Jones, R. (1989). Status of the rat acrosome during sperm-zona pellucida interactions. *Gamete Res.* **22**, 1–13.

Shapiro, B. M., Schackmann, R. W., and Gabel, C. A. (1981). Molecular approaches to the study of fertilization. *Ann. Rev. Biochem.* **50**, 815–843.

Shimomura, H., and Garbers, D. L. (1986). Differential effects of resact analogues on sperm respiration rates and cyclic nucleotide concentrations. *Biochemistry* **25**, 3405–3410.

Shimomura, H., Suzuki, N., and Garbers, D. L. (1986a). Derivatives of speract are associated with the eggs of *Lytechinus pictus* sea urchins. *Peptide* **7**, 491–495.

Shimomura, H., Dangott, L. J., and Garbers, D. L. (1986b). Covalent coupling of a resact analogue to guanylate cyclase. *J. Biol. Chem.* **261**, 15778–15782.

Shur, B. D., and Neely, C. A. (1988). Plasma membrane association, purification, and partial characterization of mouse sperm β-1,4-galactosyltransferase. *J. Biol. Chem.* **263**, 17706–17714.

Singh, S., Lowe, D. G., Thorpe, D. S., Rodriguez, H., Kuang, W.-J., Dangott, L. J., Chinkers, M., Goeddel, D. V., and Garbers, D. L. (1988). Membrane guanylate cyclase is a cell-surface receptor with homology to protein kinases. *Nature (London)* 334, 708–712.

Smigel, M. D. (1986). Purification of the catalyst of adenylate cyclase. *J. Biol. Chem.* 261, 1976–1982.

Smith, A. C., and Garbers, D. L. (1983). The binding of an ^{125}I-speract analogue to spermatozoa. *In* "Biochemistry of Metabolic Processes" (D. L. F. Lennon, F. W. Stratman, and R. N. Zahlten, eds.), pp. 15–28. Elsevier Biomedical, New York.

Storey, B. T. (1991). Sperm capacitation and the acrosome reaction. *Ann. N.Y. Acad. Sci.* 637, 459–473.

Suarez, S. S., Wolf, D. P., and Meizel, S. (1986). Induction of the acrosome reaction in human spermatozoa by a fraction of human follicular fluid. *Gamete Res.* 14, 107–121.

Suarez, S. S., Katz, D. F., Owen, D. H., Andrew, J. B., and Powell, R. L. (1991). Evidence of the function of hyperactivated motility in sperm. *Biol. Reprod.* 44, 375–381.

Suzuki, N. (1990). Structure and function of sea urchin egg jelly molecules. *Zool. Sci.* 7, 355–370.

Suzuki, N., and Garbers, D. L. (1984). Stimulation of sperm respiration rates by speract and resact at alkaline extracellular pH. *Biol. Reprod.* 30, 1167–1174.

Suzuki, N, and Yoshino, K. (1992). The relationship between amino acid sequences of sperm-activating peptides and the taxonomy of echinoids. *Comp. Biochem. Physiol.* 102B, 679–690.

Suzuki, N., Nomura, K., Ohtake, H., and Isaka, S. (1981). Purification and the primary structure of sperm-activating peptides from the jelly coat of sea urchin eggs. *Biochem. Biophys. Res. Commun.* 99, 1238–1244.

Suzuki, N., Hoshi, M., Nomura, K., and Isaka, S. (1982). Respiratory stimulation of sea urchin spermatozoa by egg extracts, egg jelly extracts and egg jelly peptides from various species of sea urchins: Taxonomical significance. *Comp. Biochem. Physiol.* 72A, 489–495.

Suzuki, N., Shimomura, H., Radany, E. W., Ramarao, C. S., Ward, G. E., Bentley, J. K., and Garbers, D. L. (1984). A peptide associated with eggs causes a mobility shift in a major plasma membrane protein of spermatozoa. *J. Biol. Chem.* 259, 14874–14879.

Suzuki, N., Kurita, M., Toshino, K., Kajiura, H., Nomura, K., and Yamaguchi, M. (1987). Purification and structure of mosact and its derivatives from the egg jelly of the sea urchin *Clypeaster japonicus. Zool. Sci.* 4, 649–656.

Suzuki, N., Yoshino, K., Kurita, M., Nomura, K., and Yamaguchi, M. (1988). A novel group of sperm-activating peptides from the sea urchin *Glyptocidaris crenularis. Comp. Biochem. Physiol.* 90C, 305–311.

Tash, J. S., and Means, A. R. (1983). Cyclic adenosine 3′,5′-monophosphate, calcium and protein phosphorylation in flagellar motility. *Biol. Reprod.* 28, 75–104.

Tash, J. S., Kakar, S. S., and Means, A. R. (1984). Flagellar motility requires the cAMP-dependent phosphorylation of a heat-stable NP-40-soluble 56-kd protein, axokinin. *Cell* 38, 551–559.

Tash, J. S., Hicaka, H., and Means, A. R. (1986). Axokinin phosphorylation by cAMP-dependent protein kinase is sufficient for activation of sperm flagellar motility. *J. Cell Biol.* 103, 649–655.

Thomas, P., and Meizel, S. (1988). An influx of extracellular calcium is required for initiation of the human sperm acrosome reaction induced by human follicular fluid. *Gamete Res.* 20, 397–411.

Thomas, P., and Meizel, S. (1989). Phosphatidylinositol 4,5-bisphosphate hydrolysis in human sperm stimulated with follicular fluid or progesterone is dependent upon Ca^{2+} influx. *Biochem. J.* 264, 539–546.

Thorpe, D. S., and Garbers, D. L. (1989). The membrane form of guanylate cyclase: Homology with a subunit of the cytoplasmic form of the enzyme. *J. Biol. Chem.* 264, 6545–6549.

Toscano, W. A., and Gross, M. K. (1991). Calmodulin-mediated adenylyl cyclase from equine sperm. *Meth. Enzymol.* **195**, 91–110.

Trimmer, J. S., Schackmann, R. W., and Vacquier, V. D. (1986). Monoclonal antibodies increase intracellular Ca^{2+} in sea urchin spermatozoa. *Proc. Natl. Acad. Sci. U.S.A.* **83**, 9055–9059.

Tulsiani, D. R. P., Skudlarek, M. D., and Orgebin-Crist, M.-C. (1989). Novel α-D-mannosidase of rat sperm plasma membranes: Characterization and potential role in sperm-egg interactions. *J. Cell Biol.* **109**, 1257–1267.

Uto, N., Yoshimatsu, N., Lopata, A., and Yanagimachi, R. (1988). Zona-induced acrosome reaction of hamster spermatozoa. *J. Exp. Zool.* **248**, 113–120.

Visconti, P. E., Muschietti, J. P., Flawia, M. M., and Tezon, J. G. (1990). Bicarbonate dependence of cAMP accumulation induced by phorbol esters in hamster spermatozoa. *Biochim. Biophys. Acta* **1054**, 231–236.

Ward, G. E., and Vacquier, V. D. (1983). Dephosphorylation of a major sperm membrane protein is induced by egg jelly during sea urchin fertilization. *Proc. Natl. Acad. Sci. U.S.A.* **80**, 5578–5582.

Ward, C. R., Storey, B. T., and Kopf, G. S. (1992). Activation of a Gi protein in mouse sperm membranes by solubilized proteins of the zona pellucida, the egg's extracellular matrix. *J. Biol. Chem.* **267**, 4061–4067.

Ward, G. E., Garbers, D. L., and Vacquier, V. D. (1985a). Effects of extracellular egg factors on sperm guanylate cyclase. *Science* **227**, 768–770.

Ward, G. E., Brokaw, C. J., Garbers, D. L., and Vacquier, V. D. (1985b). Chemotaxis of *Arbacia punctulata* spermatozoa to resact, a peptide from the egg jelly layer. *J. Cell Biol.* **101**, 2324–2329.

Wasco, W. M., Kincaid, R. L., and Orr, G. A. (1989). Identification and characterization of calmodulin-binding proteins in mammalian sperm flagella. *J. Biol. Chem.* **264**, 5104–5111.

Wassarman, P. M. (1988). Zona pellucida glycoproteins. *Ann. Rev. Biochem.* **57**, 415–442.

Wilde, M. W., Ward, C. R., and Kopf, G. S. (1992). Activation of a G protein in mouse sperm by the zoma pellucida, an egg-associated extracellular matrix. *Mol. Reprod. Dev.* **31**, 297–306.

Yamaguchi, M., Niwa, T., Kurita, M., and Suzuki, N. (1988). The participation of speract in the acrosome reaction of *Hemicentrotus pulcherrimus*. *Dev. Growth Differ.* **30**, 159–167.

Yanagimachi, R. (1970). The movement of golden hamster spermatozoa before and after capacitation. *J. Reprod. Fertil.* **23**, 193–196.

Yanagimachi, R. (1981). Mechanisms of fertilization in mammals. *In* "Fertilization and Embryonic Development In Vitro" (L. Mastroianni and J. D. Biggers, eds.), pp. 81–182. Plenum Press, New York.

Yanagimachi, R., and Phillips, D. M. (1984). The status of acrosomal caps of hamster spermatozoa immediately before fertilization in vivo. *Gamete Res.* **9**, 1–19.

Yanagimachi, R., and Usui, N. (1974). Ca^{++} dependence of sperm acrosome reaction and activation of guinea pig spermatozoa. *Exp. Cell Res.* **89**, 161–174.

Yeung, S.-M. H., Shoshani, I., Stubner, D., and Johnson, R. A. (1989). Ammonium ions enhance proteolytic activation of adenylate cyclase and decrease its sensitivity to inhibition by "P"-site agonists. *Arch. Biochem. Biophys.* **271**, 332–345.

Yoshino, K., Kurita, M., Yamaguchi, M., Nomura, K., Takao, T., Shimonishi, Y., and Suzuki, N. (1990). A species-specific sperm-activating peptide from the jelly of the sea urchin *Diadema setosum*. *Comp. Biochem. Physiol.* **95B**, 423–429.

Yoshino, K., Takao, T., Suhara, M., Kitai, T., Hori, H., Nomura, K., Yamaguchi, M., Shimonishi, Y., and Suzuki, N. (1991). Identification of a novel amino acid, *o*-bromo-L-phenylalanine, in egg-associated peptides that activate spermatozoa. *Biochemistry* **30**, 6203–6209.

Yoshino, K., Takao, T., Shimonishi, Y., and Suzuki, N. (1992). Sperm-activating peptide type-V

(SAP-V), a fifth member of sperm-activating peptide family, purified from the egg-conditioned media of the heart urchin *Brissus agassizii*. *Comp. Biochem Physiol.* **102,** 691–700.

Yudin, A. I., Gottlieb, W., and Meizel, S. (1988). Ultrastructural studies of the early events of the human sperm acrosome reaction as initiated by human follicular fluid. *Gamete Res.* **20,** 11–24.

Yurewicz, E. C., Sacco, A. G., and Subramanian, M. G. (1987). Structural characterization of the Mr = 55,000 antigen (ZP3) of porcine oocyte zona pellucida: Purification and characterization of α- and β-glycoproteins following digestion of lactosaminoglycan with endo-β-galactosidase. *J. Biol. Chem.* **262,** 564–571.

8

Paracrine Mechanisms in Testicular Control

B. JÉGOU & R. M. SHARPE

I. Introduction

In 1952, Roosen-Runge wrote that "the testis is not a mass of independently developing cells." Over 40 years later, the bulk of experimental evidence supports this contention.

II. Germ Cell–Sertoli Cell Cross Talk

The extraordinary structural intricacy of the seminiferous epithelium components is the most obvious evidence of the absolute interdependence of germ cells and Sertoli cells. Germ cell–Sertoli cell dialogue is based on structural elements and chemical factors.

A. Anatomical Devices

The discovery and description of the cellular structural devices within the seminiferous epithelium was decisive in understanding the germ cell–Sertoli cell communication network. The nature and function of these devices have been the subject of extensive reviews (Russell, 1980; Jégou *et al.*, 1992; Jégou, 1993; Pelletier and Byers, 1992). We have classified these mechanisms into three categories: (1) those implicated in cell attachment, movement, and shaping; (2) those that fulfill these functions and are involved in the transfer of molecules and materials from Sertoli cells to germ cells or vice-versa; and (3) those strictly involved in transfer. Desmosome-like and ectoplasmic specializations belong to the first category; spermatogonial processes, tubulobulbar complexes, spermatid processes, and gap junctions belong to the second; and residual bodies belong to the last category (Jégou, 1993).

The different devices interconnecting Sertoli cells and late spermatids are represented in Fig. 1. These devices are considered to be the most sophisticated within the body (Jégou et al., 1992), thus reflecting the complexity and uniqueness of Sertoli cell and spermatid interaction during spermiogenesis.

In contrast to the careful attention paid to the description of the anatomical basis of Sertoli cell–germ cell communication, relatively little attention has been paid to identifying the molecules and the molecular mechanisms involved in the composition and function of the devices. The study of tissues or organs other than the testis has shown that tissue cohesion and restructuring during embryogenesis and in adult vertebrates involve regulation of cell–cell and cell–substrate adhesion mechanisms (Edelman, 1988; Takeichi, 1990; Peyriéras, 1992). Therefore, the major molecular families of cell adhesion molecules—the cell–cell adhesion molecules (CAMs), the cell–substrate adhesion molecules (SAMs), and the cell junction molecules (CJMs)—are most likely to be present in the seminiferous epithelium (Byers et al., 1993; Jégou et al., 1992; Jégou, 1993). The appearance and disappearance of Sertoli–germ cell structural devices is most likely to be coordinated spatio-temporally.

Progress in the identification, the description of cellular distribution, the regulation, and the spatio-temporal interaction of these molecules undoubtedly will lead to a much better understanding of morphogenesis and gene expression within the seminiferous epithelium.

B. Functional Interactions

1. Evidence of Sertoli cell control of spermatogenesis

Sertoli cells are believed to be involved in the synchronous development of germ cells in the stages of the epithelium cycle, although the underlying mechanism is totally obscure. Similarly, that Sertoli cells exert a coordinating influence on the longitudinal direction in the tubule, thus contributing to the maintenance of the wave of spermatogenesis, remains to be established experimentally.

a. Sertoli cell barrier

As reviewed by Ploën and Setchell (1992), the Sertoli cell barrier constitutes a key element of the physiological barriers between the blood and the seminiferous tubule lumen. The Sertoli cell barrier is now clearly established to lie within the specialized Sertoli cell junctional complexes that are located in the basal portion of the seminiferous epithelium. These junctions divide the epithelium into a basal compartment that encompasses premeiotic germ cells and an adluminal compartment that contains meiotic and postmeiotic germ cells (Dym and Fawcett, 1970).

In mammals as well as in birds, reptiles, fish, amphibians, insects, and nematodes, the establishment of inter-Sertoli tight junctions during postnatal life triggers Sertoli cell maturation and is the prerequisite for the normal development of meiosis and spermiogenesis (Jégou, 1992).

The Sertoli cell barrier fulfills the following functions: (1) promotion of the seminiferous epithelium cytoarchitecture; (2) formation of the microenvironment required for meiosis and spermiogenesis; and (3) segregation of the Sertoli cell plasma membrane into distinct domains [e.g., localization of follicle-stimulating hormone (FSH) receptors at the level of the basal portion of the Sertoli cell; Orth and Christensen (1977)]. The Sertoli cell barrier is not impermeable but selective. Its role as an immunological barrier remains to be proven (Ploën and Setchell, 1992).

To prevent any major breach in the Sertoli cell barrier that would perturb the germ cell microenvironment during early spermatocyte migration from the basal to the adluminal compartment at the beginning of the meiotic prophase, a synchronous breakdown and reconstitution of the Sertoli cell barrier must occur (Russell, 1980). FSH probably is involved in the control of Sertoli cell barrier formation (de Kretser and Burger, 1972; Furuya *et al.*, 1980; Janecki *et al.*, 1991). Evidence suggests that testosterone may be involved in its maintenance (Kerr *et al.*, 1993b). However, the exact role of hormones in control of the formation and function of Sertoli cell tight junctions remains obscure. Similarly, how germ cells may interfere with Sertoli cell barrier formation during testicular development and with dismantling and reconstitution during germ cell translocation, from the basal to the adluminal compartment of the tubule in the adult, is still a mystery.

b. Sertoli cell products

Production of seminiferous tubule fluid, proteins, peptides, and steroids by Sertoli cells (Bardin *et al.*, 1988; Griswold, 1988; Jégou, 1992,1993) justifies their title of nurse cells. Tubule fluid is essential for the nutrition of germ cells and the transport of chemical substances from the basal to the apical portion of the seminiferous epithelium. This fluid also is required for the release of spermatozoa and their transport to the epididymis (Jégou, 1992).

Sertoli cells produce some 100 different proteins (Bardin *et al.*, 1988), about one-third of which have been characterized or identified (Jégou 1992, 1993). Most likely, all are involved in Sertoli cell–germ cell across talk as factors controlling cell proliferation, differentiation, and metabolism; transport or binding proteins; proteases; components of the extracellular matrix; energy metabolites; and components of the junctional complexes or of Sertoli cell membranes (Table 1). Establishing this list is easier than establishing the actual involvement of these substances in the spermatogenic process. Most of the Sertoli cell factors identified to date were discovered initially in Sertoli

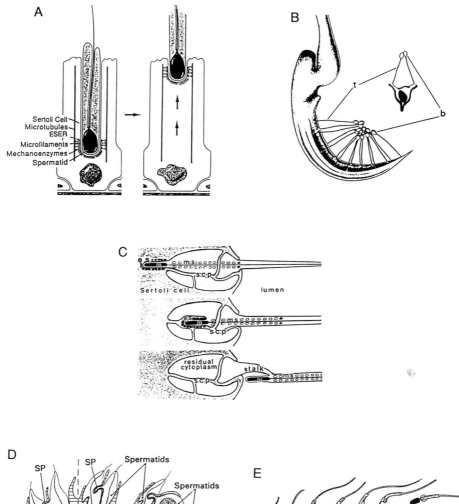

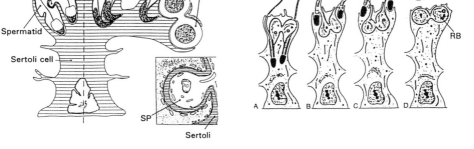

FIGURE 1 Schematic representation of Sertoli cell-late spermatid connection devices. **(A)** *(Left)* Spermatid deeply embedded within a Sertoli cell crypt; *(Right)* Spermatid has moved

cell culture media or in isolated Sertoli cells and have not yet been studied *in vivo* or under conditions allowing an unequivocal extrapolation to physiology. Table 1 gives a selection of the Sertoli cell products identified to date that are likely to be involved in the control of spermatogenesis.

Interestingly, compartmentalization of the seminiferous epithelium allows a bidirectional secretion of Sertoli cell products (Gunsalus and Bardin, 1991). As suggested by Byers and Pelletier (1992), Sertoli cell-secreted proteins that are destined to interact with germ cells in the adluminal compartment and perhaps also in the epididymis are likely to be secreted apically, whereas those controlling early germ cells, myoid cells, and Leydig cells are secreted basally.

2. Evidence of germ cell control of Sertoli cell function

a. Stage-dependent variation in the Sertoli cell

Both structural and functional changes occur in Sertoli cells during the seminiferous epithelium cycle. These changes have been reviewed extensively (Linder *et al.*, 1991; Toppari *et al.*, 1991; Jégou *et al.*, 1992; Parvinen, 1993; Jégou, 1993) and will not be listed here. The Sertoli cells may have an intrinsic rhythm, but the cyclic, morphological, and biochemical variations in Sertoli cells no doubt result from the changing number, nature, and needs of the germ cell complement (Jégou *et al.*, 1992; Jégou, 1993).

---

toward the apex of the Sertoli cell. Ectoplasmic specializations (ES), constituted of actin filaments flanked on their cytoplasmic side by a cistern of endoplasmic reticulum (ESER), line the Sertoli cell crypt. Microtubules occur adjacent to the ESER. The mechanoenzymes, illustrated as a bridge between the ESER and microtubules, are proposed to act as motors to displace the spermatid–ES complex along microtubule tracks toward the lumen of the tubule. ES also may intervene in the shaping of the sperm head. Reprinted with permission from Redenbach and Vogl (1991). (B) Tubulobulbar complexes are watery and organelle-free cytoplasmic evaginations of the spermatid that penetrate into the Sertoli cell cytoplasm and are composed of tubular (t) and bulbous (b) portions. They serve as anchoring devices, permit the elimination of excess spermatid membrane and cytoplasm, and may permit the increase of glycolytic enzymes in spermatids (Jégou, 1993). Reprinted with permission from Russell *et al.* (1990a). (C) Sertoli cell processes (scp) correspond to projections of the Sertoli cell that penetrate the spermatid cytoplasm in the mouse to form a canal complex. Around the spermatid head (n) are ectoplasmic specializations (es). ms, Mitochondrial sheath. The scp may allow transport of molecules between the Sertoli cell and late spermatids, and may retain the cytoplasmic lobe of spermatids during movement to the tubule lumen (Clermont and Morales, 1982; Sakai and Yamashina, 1990; Jégou *et al.*, 1992). Reprinted with permission from Sakai and Yamashina (1990). (D) Spermatid processes (SP) are projections of a spermatid and a Sertoli cell that penetrate a neighboring spermatid. SPs are anchoring devices, but also may transfer materials between cells. The inset shows an enlargement of the different components of spermatid–Sertoli cell processes. Reprinted with permission from Segretain and Decrossas (1991). (E) Residual bodies (RB) are the portions of late spermatid cytoplasm left behind by the mature sperm as they detach from the apex of the Sertoli cell. Reprinted with permission from Fawcett (1975).

TABLE 1 Possible Sertoli Cell Regulators of Germ Cell Divisions, Differentiation, and Metabolism

Sertoli cell agent	Putative action(s) or function(s)	Germ cell target
Factors involved in cell growth/differentiation/metabolism		
Anti-Müllerian hormone (AMH); also known as Müllerian-inhibiting substance (Josso and Picard, 1986; Josso, 1991)	Regression of Müllerian ducts; possible role in preventing meiosis during intra-uterine life is controversial (Josso, 1991)	—
Activin (Grootenhuis *et al.*, 1989)	Stimulates spermatogonial proliferation in Sertoli cell–germ cell cocultures (Woodruff *et al.*, 1992)	Activin-A receptors on spermatogonia, primary spermatocytes (not on leptotene/zygotene), and early spermatids (Woodruff *et al.*, 1992); activin receptor mRNA present in pachytene spermatocytes and early spermatids (not late spermatids) (de Winter *et al.*, 1992)
Inhibin; (Le Gac and de Kretser, 1982; de Jong, 1988; Vaughan *et al.*, 1989)	May inhibit spermatogonial proliferation *in vivo* (Franchimont *et al.*, 1981; Van Dissell-Emiliani *et al.*, 1989)	Inhibin A receptors present on spermatogonia, primary spermatocytes, and early spermatids (Woodruff *et al.*, 1992)
Transforming growth factors β (TGF *β*s) (Avallet *et al.*, 1987; Morera *et al.*, 1987; Skinner and Moses, 1989)	Intervene in the control of germ cell differentiation and divisions.	TGF *β*1 and *β*2 immunoreactivities found at the germ cell level (Watrin *et al.*, 1991; Teerds and Dorrington, 1993); no direct demonstration of receptor for TGF *β* on germ cells, but TGF *β* precursor bears the mannose-6 phosphate (M6P) recognition marker and therefore may be recognized by pachytene spermatocytes and early spermatids via their M6P-receptors (O'Brien *et al.*, 1991)
Insulin growth factor 1 (IGF-1/somatomedin C) (Hall *et al.*, 1983; Benhamed *et al.*, 1987; Chatelain *et al.*, 1987)	Growth regulation and cell metabolism	IGF-1 receptors immunolocalized on secondary spermatocytes and spermatids and, to a lesser extent, on primary spermatocytes (Vannelli *et al.*, 1988)
Interleukin 1α(IL-1α) (Syed *et al.*, 1988b; N. Gérard *et al.*, 1991)	May stimulate spermatogonial divisions (Pöllänen *et al.*, 1989) and meiotic DNA replication (Parvinen *et al.*, 1991)	—
"Steel" Locus Factor/kit ligand (STL/KL) (Zsebo *et al.*, 1990; Rossi *et al.*, 1991; Sorrentino *et al.*, 1991; Tajima *et al.*, 1991)	Product of the *steel* gene; may guide primordial germ cells (PGC) (chemotaxis) toward germinal crests (Keshet *et al.*, 1991); membrane-bound form of KL most likely plays an important role in PGC survival (Besmer, 1991; Dolci *et al.*, 1991; Godin *et al.*, 1991) and has growth factor activity *in vitro* (Tajima *et al.*, 1991); soluble form of KL stimulates proliferation of germ cells in culture (Rossi *et al.*, 1992)	PGC, spermatogonia (in particular type A), and primary spermatocytes possess the KL receptor, which is the transmembrane *c-kit* proto-oncogene, a tyrosine-kinase receptor (Chabot *et al.*, 1988; Geissler *et al.*, 1988; Manova *et al.*, 1990; Sorrentino *et al.*, 1991); survival and/or proliferation of differentiating type A spermatogonia (but not of gonocytes, stem spermatogonia, or spermatocytes) requires *c-kit* in the mouse, as shown *in vivo* using an anti-*c-kit* antibody (Yoshinaga *et al.*, 1991)

3 α-Hydroxy-4-pregnen-20-one (3HP) (Campbell and Wiebe, 1989)	Stimulates pachytene spermatocyte development in vitro and in vivo (Wiebe et al., 1988; Campbell and Wiebe, 1989)	—
Transport/binding proteins		
Transferrin (Skinner and Griswold, 1980; Holmes et al., 1984; Jenkins and Ellison, 1989; Monet-Kuntz et al., 1992)	Transport of iron required for cell division and differentiation	Receptors present on germ cells (Holmes et al., 1983; Steinberger et al., 1984; Sylvester and Griswold, 1984; Vannelli et al., 1986); number of receptors high on spermatogonia, important on spermatocytes, very low on early spermatids, and absent on late spermatids (Segretain et al., 1992)
Androgen-binding protein (ABP) (Hansson et al., 1975a; Jégou, 1992)	May transport androgens to germ cells	Pachytene spermatocytes may possess specific binding sites for ABP (Steinberger et al., 1984); all germ cell classes and spermatozoa can bind ABP (Gérard et al., 1989; A. Gérard et al., 1991)
Retinol-binding protein (RBP) (Huggenvik and Griswold, 1981; Davis and Ong, 1992)	Retinol (Vitamin A) is required for spermatogenesis (Skinner, 1991); RBP may transport retinol to meiotic and postmeiotic germ cells, which may transform it into retinoic acid (Davis and Ong, 1992)	Relatively high levels of cellular retinoic acid-binding protein (CRABP) found in late spermatocytes and spermatids (Huggenvik and Griswold, 1981)
Sulfated glycoprotein 1 (SGP-1/prosaposin) (Sylvester et al., 1984; Collard et al., 1988)	May be a carrier for lipid precursors and specific fatty acids to germ cells (Skinner, 1991)	Binding of SGP-1 to sperm plasma membrane has been shown (Sylvester et al., 1984,1989)
Sulfated glycoprotein 2 (SGP-2/clusterin) (Blaschuk and Fritz, 1984; Cheng et al., 1988)	May solubilize and carry lipids to germ cells and sperm (Sylvester et al., 1984)	Localized over the acrosome and tail of developing spermatids and sperm (Sylvester et al., 1984; Kierzenbaum et al., 1988)
α2-Macroglobulin (α2M) (Cheng et al., 1990)	By binding growth factors, cytokines (Cheng et al., 1992), inhibin, and activin, α2M probably indirectly influences the spermatogenic process	—
γ-Glutamyl transpeptidase (γ-GTP) (Lu and Steinberger, 1977; Schteingart et al., 1988,1989)	May permit the transport of glutathione amino acids to germ cells (Risley and Morse-Gaudio, 1992)	—
Proteases/inhibitors		
Urokinase (predominant) and tissue-type plasminogen activators (uPA and tPA) (Lacroix et al., 1977; Hettle et al., 1986; Vihko et al., 1986,1988; Monet-Kuntz et al., 1991)	May degrade Sertoli–Sertoli and Sertoli–germ cell junctional complexes and therefore intervene in the permanent remodeling of the seminiferous epithelium (Lacroix et al., 1981; Hettle et al., 1986; Vihko et al., 1986)	May be required for the passage of preleptotene spermatocytes through the Sertoli cell tight junctional complexes and for the release of sperm (Lacroix et al., 1981)
Cyclic protein 2/procathepsin L (CP-2) (Wright, 1988; Erickson-Lawrence et al., 1990)	May intervene in sperm release (Wright, 1988)	Procathepsin L bears the M6P recognition marker, which may allow this molecule to interact with pachytene spermatocytes and early spermatids (O'Brien et al., 1991)
Cystatin C (Esnard et al., 1992)	Inhibitor of cathepsin L (see CP-2)	—
Type IV collagenase and other metalloproteinases, (Sang et al., 1990 a,b)	Probably involved in basement membrane and seminiferous epithelium remodeling as well as in sperm release (Sang et al., 1990 a,b)	—

(Continues)

TABLE 1 *(continued)*

Sertoli cell agent	Putative action(s) or function(s)	Germ cell target
Extracellular matrix (ECM) components		
Collagen I and IV, laminin, proteoglycans (Skinner *et al.*, 1985; Skinner and Fritz, 1985)	ECM is necessary for establishment of the columnar shape of Sertoli cells (Byers *et al.*, 1993) and the storage and circulation of growth factors, hormones, and many other factors	—
Energy metabolites		
Lactate and pyruvate (Jutte *et al.*, 1981,1982; Robinson and Fritz, 1981)	Energy substrates crucially required for germ cells (Grootegoed *et al.*, 1989; Risley and Morse-Gaudio, 1992)	Lactate-binding protein is present on germ cell membranes (Nakamura *et al.*, 1991)
Anti-oxidant agent		
Glutathione (Den Boer *et al.*, 1989)	May be transferred to germ cells (Grootegoed *et al.*, 1989; Risley and Morse-Gaudio, 1992)	—
Components of junctional complexes		
Testins (Cheng *et al.*, 1989)	Sertoli–Sertoli and Sertoli–germ cell junctional complex proteins (Zong *et al.*, 1992)	—
Other membrane components		
γ-glutamyl transpeptidase (γ-GTP)	See previous entry	—
Liver regulating protein-like (LRP) (Corlu *et al.*, 1992)	Present in both Sertoli and primary spermatocyte membrane; blockage by an antibody inhibits spermatocyte stimulation of transferrin secretion (Corlu *et al.*, 1992)	LRP-like expression starts on preleptotene, peaks on zygotene, and declines on pachytene spermatocytes; apparently absent from spermatid surface (Corlu *et al.*, 1992)

b. *In vivo* evidence of germ cell control of Sertoli cell structure and function

In mammals, each Sertoli cell is simultaneously in contact with several generations of germ cells, which constitutes the major obstacle to the precise exploration of germ cell control of Sertoli cell activity. The use of various agents (radiation, heat, antimitotic drugs, methoxyacetic acid) that are known to induce the selective loss of a germ cell category or of a restricted number of germ cell categories has demonstrated that Sertoli cell morphology and function are controlled dramatically, positively or negatively, by germ cells (Jégou *et al.*, 1988,1992; Jégou, 1993; Sharpe, 1993). In the adult rat, the presence of late (elongating and elongated) spermatids appears to be an essential prerequisite for normal Sertoli cell structure and function. This fact does not, however, exclude the possibility that other germ cell types (e.g., spermatocytes) are involved in the control of the adult Sertoli cell, as shown by several groups (Vihko *et al.*, 1984; Bartlett *et al.*, 1989; Pineau *et al.*, 1989; Allenby *et al.*, 1991; Kaipia *et al.*, 1991; Jégou *et al.*, 1993). Most, if not all, of the studies performed *in vivo* have used the rat as model. In humans, an inverse and exclusive relationship also has been established between late spermatids and the levels of circulating FSH (Rodriguez-Rigaud *et al.*, 1980), strongly suggesting that, as in the rat, late spermatids control Sertoli cell inhibin production.

Despite the great interest of this experimental approach, to our knowledge, it has only been used once in immature mammals. In the immature rat we have induced experimental cryptorchidism as a means of disrupting spermatogenesis and have demonstrated that depletion primarily of the number of early spermatids, but also of pachytene spermatocytes, is accompanied by a marked reduction in secretion of androgen-binding proteins (ABP) and a rise in FSH levels (Jégou *et al.*, 1984a). This result indicates that these germ cell types are most likely to control Sertoli cell activity during testicular development and contrasts with the situation prevailing in the adult rat, in which Sertoli cell ABP and inhibin levels do not change when early spermatid numbers decrease (Pinon-Lataillade *et al.*, 1988; Velez de la Calle *et al.*, 1988; Pineau *et al.*, 1989; Jégou *et al.*, 1992). This discrepancy has led us to hypothesize that, at each step of testicular development, Sertoli cell activity is supervised essentially by the most advanced generation(s) of germ cells present within the seminiferous epithelium (Jégou *et al.*, 1988; Jégou, 1991).

c. *In vitro* evidence of germ cell control of Sertoli cell function

The development of the techniques of cell separation, cell culture, and coculture has contributed greatly to the demonstration that germ cells regulate Sertoli cell activity. Cocultures have been prepared using Sertoli cells and mixtures of germ cells or fractions enriched in pachytene spermatocytes, early spermatids, or residual bodies/cytoplasts from elongated spermatids (Jégou, 1992,1993).

TABLE 2 Effects of Germ Cells on Sertoli Cells in Coculture

Sertoli cell parameter	Germ cell effects			
	Mixed cells[a]	Pachytene spermatocytes[b]	Early spermatids[b]	Residual bodies[b]
ABP	Stimulation (Le Magueresse et al., 1986; Le Magueresse and Jégou, 1986,1988a)	Stimulation (Galdieri et al., 1984; Le Gac et al., 1984; Le Magueresse et al., 1986; Le Magueresse and Jégou, 1988a,b; Castellon et al., 1989a,b)	Stimulation[c] (Le Gac et al., 1984; Le Magueresse et al., 1986; Le Magueresse and Jégou, 1988a,b)	Weak stimulation (Le Magueresse et al., 1986; Le Magueresse and Jégou, 1988b)
Transferrin	Stimulation (Le Magueresse et al., 1988a; mRNA: Stallard and Griswold, 1990)	Stimulation (Le Magueresse et al., 1988a; Castellon et al., 1989a,b; mRNA: Stallard and Griswold, 1990)	Stimulation (Le Magueresse et al., 1988a)	No effect (Le Magueresse et al., 1988a)
γ GTP		Stimulation (Schteingart et al., 1989)		
Estradiol	Inhibition (Le Magueresse and Jégou, 1986,1988a)	Inhibition (Le Magueresse and Jégou, 1988a,b; Schteingart et al., 1989)	Inhibition (Le Magueresse and Jégou, 1988a,b)	No effect (Le Magueresse and Jégou, 1988b)
Inhibin	Stimulation (Pineau et al., 1990)	Stimulation[d] (Pineau et al., 1990)	Stimulation (protein and mRNA: Pineau et al., 1990)	No effect (Pineau et al., 1990)
SGP2/clusterin	No effect on mRNA (Stallard and Griswold, 1990)	—		
Preproenkephalin		Stimulation (mRNA: Fujisawa et al., 1992)	Stimulation (mRNA: Fujisawa et al., 1992)	
FSH-responsiveness	Inhibition (Le Magueresse and Jégou, 1988a)	Inhibition (Castellon et al., 1989a)		
Interleukin 1α (IL-1α)	—	No effect (Gérard et al., 1992)	No effect (Gérard et al., 1992)	Stimulation (Gérard et al., 1992)
Interleukin 6 (IL-6)	—	No effect (Syed et al., 1993)	No effect (Syed et al., 1993)	Stimulation[e] (Syed et al., 1993)
Protein(s) that stimulate Leydig cell function	—	Inhibition (Onoda et al., 1991a)	—	—
Protein phosphorylation	Stimulation (Welsh and Ireland, 1992)	—	—	—

[a] Mixed cells: crude germ cell preparation. No specific germ cell type exceeded 50% enrichment.

[b] Enriched fractions, enrichment ranging from 70 to 90%.

[c] No effect according to Galdieri et al. (1984).

[d] Inhibition according to Ultee-van Gessel et al. (1986) using Sertoli cells from prenatally irradiated rats and cultured at 37°C (instead of Sertoli cells from normal rats and cultured at 32°C as done by Pineau et al.).

[e] The effect of RB/CES on IL-6 is, in fact, mediated by an autocrine action of IL-1 α (V. Syed, J.P. Stephan, N. Gérard, A. Legrand, M. Parvinen, W. Bardin, and B. Jégou, submitted).

The different effects of germ cells on Sertoli cell parameters in coculture are summarized in Table 2. We have analyzed these *in vitro* results in light of *in vivo* conditions during testicular maturation: at the onset of meiosis, the secretion of ABP, transferrin, and inhibin begins and develops whereas Sertoli cell aromatase activity declines. This behavior has prompted us to suggest that spermatocytes and early spermatids may, in conjunction with FSH and testosterone, play a crucial role in Sertoli cell ontogenesis (Jégou *et al.*, 1988).

Germ cell action in coculture varies according to the germ cell category tested, the Sertoli cell parameter considered, the age of the Sertoli cell donors, and the presence or absence of hormones in Sertoli cell cultures (Table 1; Le Magueresse *et al.*, 1986; Le Magueresse and Jégou, 1988a,b; Castellon *et al.*, 1989a).

d. Mechanisms involved in germ cell control of Sertoli cell function

We have hypothesized that several pathways exist by which germ cells control Sertoli cell structure and function.

i. Morphoregulatory mechanisms and membrane molecules.   As stated earlier, CAMs, SAMs, and CJMs—which are known to influence the formation and disassembly of intercellular or cell–substratum junctions and probably are located on germ cell and Sertoli cell surfaces—probably crucially affect the shape and differentiated status of the Sertoli cell. Progress is needed in this area. Alterations in Sertoli cell shape resulting from the continuous change in the nature and composition of the germ cell complement also may influence Sertoli cell gene transcription directly. This possible route of germ cell action presupposes that changes in Sertoli cell shape could be translated via the cytoskeleton to the nuclear matrix, resulting in a possible physical modulation of transcription (Byers *et al.*, 1993; Jégou *et al.*, 1992).

The presence of germ cells is required for the levels of testins, components of Sertoli cell junctional complexes, to be controlled adequately (Cheng and Bardin, 1987; Cheng *et al.*, 1989; Jégou *et al.*, 1993). Further, an integral plasma membrane molecule, the liver regulating protein (LRP) involved in the regulation of hepatocyte differentiation by cell–cell contact interactions with biliary epithelial cells (LEC), also has been discovered in the membrane of Sertoli cells and preleptotene and early meiotic prophase spermatocytes (Corlu *et al.*, 1992). This LRP is able to regulate Sertoli cell activity *in vitro* (Corlu *et al.*, 1992) and is the first direct evidence that a germ cell membrane component is involved in the control of Sertoli cell function.

ii. Transfer of germ cell materials.   Another important pathway for the germ cell regulation of Sertoli cell activity, at the beginning and at the end of the spermatogenic process, is the transfer of germ cell materials to Sertoli cells via the "spermatogonial processes" (type A and intermediate

spermatogonia; Ulvik, 1983; Kumari and Duraiswami, 1987), the tubulobulbar complexes, and the residual bodies (late spermatids; Jégou et al., 1992).
Residual bodies have been proposed to provide the triggering signal that
regulates the seminiferous cycle because, in several mammalian species including the rat, residual bodies are formed and phagocytosed at stages of the
spermatogenic cycle when marked morphological changes occur in the Sertoli cell, when the first mitosis of A spermatogonia occurs, as well as at the
onset of meiosis and the rapid structural transformation of spermatids (Roosen-Runge, 1952,1962; Lacy, 1960,1962). After a long period during which
this hypothesis was abandoned, we reactivated the concept that residual
bodies may intervene in some way in the local control of spermatogenesis
(see reviews by Jégou, 1991,1992,1993; Jégou et al., 1992) on the following
grounds: late spermatids dramatically influence Sertoli cell function *in vivo*
(see Section II,B,2,b) and, *in vitro*, residual bodies or cytoplasts from elongate
spermatids specifically stimulate the production of interleukin 1 α(IL-1 α)
and interleukin 6 (IL-6) by the Sertoli cell (Table 1). These two major cytokines are involved in a wide range of biological functions including cell
proliferation and differentiation in other tissues (Dinarello, 1988; Van Snick,
1990). We have suggested that residual body RNA may provide a signal(s)
that directly or indirectly influences germ cell activity or that the phagocytosis of these structures by itself, via the activation of Sertoli cell cytokines, may
control some of the key steps in spermatogenesis (Jégou, 1991; Jégou et al.,
1992).

   iii. Germ cell soluble proteins.   Germ cells also can regulate Sertoli cell
activity via the secretion of proteinaceous factors. That germ cells produce
factors that regulate Sertoli cell activity was proposed first by Le Magueresse
and Jégou (1986). Media conditioned by mixed germ cells, spermatocytes,
and spermatids are now well established to stimulate or inhibit several Sertoli
cell parameters *in vitro* (Table 3). The germ cell factor(s) responsible for these
effects are proteins (Le Magueresse and Jégou, 1986,1988b; Djakiew and
Dym, 1988; Onoda and Djakiew, 1990,1991; Stallard and Griswold, 1990;
Fujisawa et al., 1992), the purification of which is in progress. Proteins with a
molecular weight of 20–30 kDa appear to be responsible for the stimulation
of Sertoli cell transferrin (Stallard and Griswold, 1990; Onoda and Djakiew,
1992; Pineau et al., 1991b, 1993a,b) inhibition of SGP2/clusterin (Pineau et
al., 1992a,b), and stimulation of preproenkephalin mRNA (Fujisawa et al.,
1992).

   Two groups of mitogenic factors are produced by germ cells: the nerve
growth factor (NGF, NGF mRNA; Olson et al., 1987; Ayer-Lelievre et al.,
1988; MacGrogan et al., 1991; B. Jégou, N. Gérard and E. Dicou, unpublished
data) or NGF-like proteins (Onoda et al., 1991b) and the basic fibroblast
growth factor (bFGF and bFGF mRNA; Mayerhofer et al., 1991; Lahr et al.,
1992; J. J. Feige, V. Syed, and B. Jégou, unpublished data). Since the low
affinity NGF receptor is expressed in rat Sertoli cells (Persson et al., 1990;

TABLE 3 Effects of Germ Cell Conditioned Media or of Germ Cell Proteins on Sertoli Cell Activity

Sertoli cell parameter	Germ cell effects			
	Mixed cells[a]	Pachytene spermatocytes[b]	Early spermatids[b]	Residual bodies[b]
ABP	Stimulation (Le Magueresse and Jégou, 1986)	Stimulation (Le Magueresse and Jégou, 1988b; Castellón et al., 1989b)	Stimulation (Le Magueresse and Jégou, 1988b; Castellón et al., 1989b)	No effect (Le Magueresse and Jégou, 1988b)
Estradiol	Inhibition (Le Magueresse and Jégou, 1986,1988a)	Inhibition (Le Magueresse and Jégou, 1988b; Schteingart et al., 1989)	Inhibition (Le Magueresse and Jégou, 1988b)	No effect (Le Magueresse and Jégou, 1988b)
Inhibin	—	Stimulation (Pineau et al., 1990)	Stimulation (protein and mRNA: Pineau et al., 1990)	Marginal stimulation (Pineau et al., 1990)
Transferrin	Stimulation (mRNA: Stallard and Griswold, 1990; protein: Pineau et al., 1991b, 1993a)	Stimulation (Djakiew and Dym, 1988; Le Magueresse et al., 1988a; Castellón et al., 1989b; Onoda and Djakiew, 1991; mRNA: Stallard and Griswold, 1990)	Stimulation (Le Magueresse et al., 1988a; Onoda and Djakiew, 1990)	—
γ-GTP	—	Stimulation (Schteingart et al., 1989)	—	—
Transferrin Receptor	No effect (mRNA: Stallard and Griswold, 1990)	—	—	—
Ceruloplasmin	Stimulation (mRNA: Stallard and Griswold, 1990)	Stimulation (Onoda and Djakiew, 1991)	Stimulation (Onoda and Djakiew, 1990)	—
Protein(s) that stimulate Leydig cell function	—	Inhibition (Onoda et al., 1991a)	No effect (Onoda et al., 1991a)	—
α2-macroglobulin	Stimulation (Grima et al., 1992)			
SGP-1	No effect on mRNA (Stallard and Griswold, 1990)	Stimulation (Onoda and Djakiew, 1991)	—	—
SGP-2/clusterin	Inhibition (Pineau et al., 1991b,1993a,b; Grima et al., 1992); No effect on mRNA (Stallard and Griswold, 1990)	Stimulation (Onoda and Djakiew, 1991)	Stimulation (Onoda and Djakiew, 1990); Inhibition (Pineau et al., 1993b)	—
Testins	Inhibition (Grima et al., 1992; Pineau et al., 1993b)	Stimulation (Onoda and Djakiew, 1991)	Stimulation (Onoda and Djakiew, 1990)	—
Preproenkephalin	—	Stimulation (mRNA: Fujisawa et al., 1992)	Stimulation (mRNA: Fujisawa et al., 1992)	—
Overall protein secretion ([35S]methionine incorporation)	—	Stimulation (Djakiew and Dym, 1988; Onoda and Djakiew, 1991)	Stimulation (Onoda and Djakiew, 1990)	—
cAMP production	—	Stimulation (Fujisawa et al., 1992)	Stimulation (Fujisawa et al., 1992)	—

[a] Mixed cells: crude germ cell preparation, no specific germ cell type exceeding 50% enrichment.

[b] Enriched fractions, enrichment ranging from 70 to 90%.

Parvinen *et al.*, 1992) and bFGF is known to affect Sertoli cell activity (Jaillard *et al.*, 1987; Smith *et al.*, 1989), these germ cell growth factors may represent important modulators of Sertoli cell function.

Although the second messenger pathway(s) for germ cell-mediated Sertoli cell modulation has been explored only superficially, indications in the literature are that cAMP does not represent a major pathway (Fujisawa *et al.*, 1992) and that, in contrast, the inositol triphosphate/diacylglycerol pathway may be very important (Welsh and Ireland, 1992).

III. Peritubular–Seminiferous Epithelium Cross Talk

The relationship between the stromal cells — the peritubular myoid cells — that border the seminiferous tubules and Sertoli cells has begun to deliver up some of its secrets (Skinner, 1991; Skinner *et al.*, 1991; Verhoeven, 1992). The importance of the dialogue between these two tubular cell types has been demonstrated in the fetal testis (Paranko *et al.*, 1983), the prepubertal testis (Bresler and Ross, 1972), and the adult testis (Tung and Fritz, 1986a,b).

Peritubular cells insure the structural cohesion and probably the contraction of the seminiferous tubules (Clermont, 1958; Ross and Long, 1966; Kormano and Hovatta, 1972). They also are constituents of the blood–testis barrier (Dym and Fawcett, 1970; Ploën and Setchell, 1992) and, in cooperation with Sertoli cells, produce the extracellular matrix (ECM). The ECM separates Sertoli and peritubular cells and forms the basement membrane of the tubule (Pöllänen *et al.*, 1985; Skinner and Fritz, 1985; Skinner *et al.*, 1985; Hadley and Dym, 1987; Fig. 2). Clearly, the ECM is crucial to cell migration, proliferation, differentiation, polarity, stabilization of phenotypic expression, gene expression, and response to various paracrine and humoral factors including growth factors (Bissell and Barcellos-Hoff, 1987; Engel, 1991; Thiéry and Boyer, 1992). Within the tubules, the ECM is known to promote normal histotypes of both peritubular cells and Sertoli cells, the latter being connected to the basal lamina by hemidesmosomes (Connell, 1974; Russell, 1977). No doubt the ECM is responsible for the promotion and maintenance of cell polarity (Mather *et al.*, 1984; Tung *et al.*, 1984; Hadley *et al.*, 1985; Enders *et al.*, 1986; Anthony and Skinner, 1989), for Sertoli cell migration (Tung and Fritz, 1985,1986a), and for tubule cord formation (Pelliniemi *et al.*, 1984; Hadley *et al.*, 1990). Cell polarity is a prerequisite for normal Sertoli cell function which includes formation of the Sertoli cell barrier and vectorial secretion (Byers *et al.*, 1986; Janecki and Steinberger, 1987; Ailenberg *et al.*, 1988; Ailenberg and Fritz, 1989) and, consequently, for normal spermatogenesis (Hadley *et al.*, 1985; Byers and Pelletier, 1992; Byers *et al.*, 1993). *In vitro*, the ECM has been shown to enhance the Sertoli cell Gs complex of adenyl cyclase and the cyclic AMP response to FSH. This result was interpreted as the demonstration of a role of the ECM in the mainten-

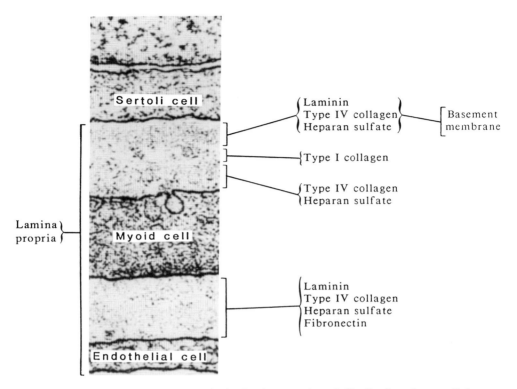

FIGURE 2 Organization of the testicular lamina propria and distribution of extracellular matrix molecules. Reprinted with permission from Hadley and Dym (1987).

ance of the differentiated status of Sertoli cell morphology and function (Dym *et al.*, 1991). This hypothesis is substantiated by the results of another experiment establishing that the ECM also decreases FSH-dependent aromatase activity (Dym *et al.*, 1991). Greater knowledge of the composition of the ECM and of the SAMs that mediate cell–ECM interactions within the tubule undoubtedly would improve understanding of testicular cell–cell interactions.

Sertoli cells and peritubular cells also are known to cooperate metabolically *in vitro* (Hutson, 1983). Peritubular cells have been shown to stimulate the production of Sertoli cell total proteins (Hadley *et al.*, 1985), of ABP (Tung and Fritz, 1980; Hutson and Stocco, 1981), and of transferrin (Holmes *et al.*, 1984; Hadley *et al.*, 1985), and to affect the levels of several Sertoli cell enzymes (Cameron and Snydle, 1985). Several of the peritubular cell effects are mediated *in vitro* by a nonmitogenic factor termed PModS (Skinner, 1991). Interestingly, peritubular cells also have been shown to produce a protease inhibitor that could inactivate Sertoli cell plasminogen activator

(Hettle *et al.*, 1988). This latter protease is supposed to be involved in degradation of the basal lamina breakdown of the Sertoli cell barrier during early spermatocyte cluster translocation, at the onset of meiosis (Lacroix *et al.*, 1981). Other peritubular cell products such as transforming growth factors TGFβ and TGFα may act on Sertoli cells (Skinner, 1991; Skinner *et al.*, 1991; Verhoeven, 1992), although the fact that Sertoli cells also produce these growth regulators blunt this point.

If peritubular cell action on Sertoli cell morphology and activity is relatively well known, how Sertoli cells interact with peritubular cells is much more obscure, primarily because less is known about the biology of the latter. That perturbations in Sertoli cell activity have repercussions for peritubular cell function is suggested strongly by results of experiments showing that disruption of Sertoli cell function, resulting from a depletion in the number of late spermatids, is followed by a dramatic alteration in the peritubular tissue (Pinon-Lataillade *et al.*, 1988) and in peritubular cell morphology (Pinon-Lataillade, 1986). A possible role for Sertoli cell insulin-like growth factor 1 (IGF-1), TGFα, and TGFβ in the migration and growth of peritubular cells has been suggested (Skinner *et al.*, 1991), but this role is extremely difficult to establish experimentally.

A totally unknown aspect of cell–cell communication in the tubules is the relationship that early germ cell types located in the basal compartment of the tubules may establish with the peritubular cells via the ECM or other pathways.

IV. Sertoli Cell–Leydig Cell Cross Talk

The structure of the normal adult seminiferous tubule is completely dependent on testosterone secreted by the neighboring Leydig cells. In view of this dependence, that the number and function of the Leydig cells clearly is modulated by the Sertoli cells is not surprising. Although this interdependence of the Sertoli and Leydig cells is accepted (Sharpe, 1993), we still have only vague ideas about the identity of the secreted factors that mediate these important paracrine interactions. The goal of this section is to highlight the likely physiological significance of the reported interactions between the Sertoli and Leydig cells at different stages during development, to show how these interactions regulate separately the supply and function of the Leydig cells, and to summarize briefly the factors that might be involved in these interactions. The vast majority of these data are derived from studies in the laboratory rat, but probably have general application to most species. Considering the regulation of Leydig cell numbers and their function separately is easiest.

A. Regulation of Leydig Cell Numbers

The adult testis has a radically different requirement for Leydig cells than does the fetal or prepubertal testis, both because of the greater size of the adult testis and because of the requirement for testosterone in high levels. Evidence suggests that the supply of Leydig cells is orchestrated by the Sertoli cells, although luteinizing hormone (LH) and FSH have equally important roles in various aspects of the regulation of Leydig cell numbers (see Fig. 3). The fetal testis contains different Leydig cells (the fetal generation of Leydig cells) than the adult testis. Puberty primarily involves the develop-

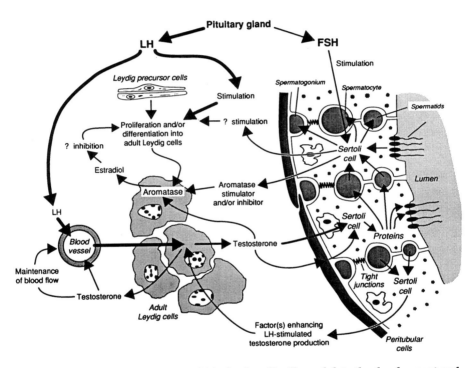

FIGURE 3 Likely major routes by which the Sertoli cells modulate the development and function of the Leydig cells in the adult rat testis. Note that each of the routes by which the Sertoli cell influences the Leydig cells is germ cell modulated, although which germ cell types are involved is not clear. The diagram is probably accurate for a seminiferous tubule at stage VII of the rat spermatogenic cycle, but fewer or different pathways of modulation may exist at other stages. Note that all the paracrine mechanisms indicated either interface with endocrine input from the pituitary gland or are controlled directly by this endocrine input. However, during fetal life and in early puberty, slightly different mechanisms probably operate (see text) in which FSH rather than LH provides the predominate endocrine control.

ment of large numbers of adult-type Leydig cells. The fetal Leydig cells probably persist into adulthood although their function, if any, in the adult is unknown (Kerr and Knell, 1988). Little is known about their paracrine regulation, although the ontogeny, numbers, and function of fetal Leydig cells most likely are regulated via mechanisms analogous to those in the adult, involving both endocrine and paracrine systems (see Christensen, 1975; Gondos, 1977; Kerr et al., 1988).

The development of adult Leydig cells is not completely understood, but a general picture is emerging. The key regulatory factors appear to be FSH and the Sertoli cells. During early puberty, the blood levels of FSH are at their highest. These high FSH levels are responsible for stimulating the initiation and expansion of early spermatogenesis (Means et al., 1976; Russell et al., 1987). During this period, the number of Leydig cells increases progressively (e.g., Hardy et al., 1989), a response that is thought to occur by FSH stimulating the Sertoli cells to produce one or more growth factors that act on Leydig precursor cells to stimulate their multiplication and differentiation into adult-type Leydig cells (e.g., Chen et al., 1976; Odell and Swerdoff, 1976; Kerr and Sharpe, 1985; O'Shaughnessy et al., 1992; reviewed by Sharpe, 1993). The identity of these growth factors is not known, although several potential candidates have been identified, the in vitro evidence pointing to a 30-kDa protein (Ojeifo et al., 1990). In this early pubertal period, the role of LH is probably minimal. This hormone may play a role in the final process of Leydig cell differentiation, as it appears to do in the late pubertal and adult testis (see Fig. 3), but at this stage of development, when testosterone is not needed in high amounts, the requirement is to differentiate Leydig cells without stimulating testosterone production (Sharpe, 1993).

In addition to studies showing the effect of FSH on Leydig cell development, three other key pieces of information point to the Sertoli cell as a central regulator of Leydig cell numbers in the adult testis. The first and the oldest piece of evidence is the demonstration that induction of damage to spermatogenesis (i.e., loss of germ cells) in the rat results in hypertrophy (and perhaps hyperplasia) of the neighboring Leydig cells. This response occurs irrespective of whether the damage is induced by X irradiation (Rich et al., 1979), cryptorchidism (Risbridger et al., 1981a), vitamin A deprivation (Rich et al., 1979), efferent duct ligation (Risbridger et al., 1981b), or chemicals (Roe, 1964; Aoki and Fawcett, 1978), and is restricted to the vicinity of the damage (Aoki and Fawcett, 1978; Risbridger et al., 1981b). As well as pointing to Sertoli cell-derived paracrine factors as the mediators of these changes in Leydig cell size or number, these observations also highlight that the loss of germ cells triggers the changes. In other words, the germ cells modulate Sertoli cell regulation of the Leydig cells (Fig. 3). Other strong evidence supports this effect (see subsequent discussion).

The second piece of evidence for Sertoli cell regulation of Leydig cell numbers in the adult testis derives from studies on the regeneration of Leydig cells in the rat testis after their destruction with ethane dimethane sulfonate (EDS). Commencing at 12–14 days after EDS treatment, Leydig cells regenerate and normal numbers are restored by ~42 days (Sharpe *et al.*, 1990). Although LH (and perhaps FSH) clearly is essential to this regeneration (Molenaar *et al.*, 1986; Teerds *et al.*, 1989b), the rate and the location of Leydig cell regeneration is determined by local factors within the testis, that is, if seminiferous tubules lack germ cells then Leydig cells regenerate more rapidly and in larger numbers in the vicinity of these tubules (Kerr *et al.*, 1986,1987; Sharpe *et al.*, 1990). Presumably, this regeneration involves the same unidentified germ cell-modulated Sertoli cell factors referred to earlier.

The third piece of evidence pinpointing the Sertoli cell as the regulator of final Leydig cell number in the adult testis derives from studies involving experimental manipulation of Sertoli cell number by the induction of neonatal hypothyroidism in the rat (Cooke *et al.*, 1992; van Haaster *et al.*, 1992). These studies show that the more Sertoli cells there are, the larger the size of the adult testis, and that the ratio of Leydig cells to Sertoli cells remains constant, that is, the more Sertoli cells there are, the more Leydig cells there are (see Cooke and Meisami, 1991; Cooke *et al.*, 1991; Kirkby *et al.*, 1992). This result raises the important question of how the testis determines when the normal complement of Leydig cells is achieved. In studies of the regeneration of Leydig cells after EDS treatment, normal intratesticular levels of testosterone are restored when the Leydig cell number is <10% of normal, yet Leydig cell regeneration continues until the normal adult complement of these cells is achieved (Sharpe *et al.*, 1990). Clearly a paracrine signal must exist that negatively regulates Leydig cell proliferation and counterbalances the paracrine stimulatory factors from the Sertoli cells as well as the stimulatory effects of LH. Estradiol is the strongest candidate for this negative regulator (Fig. 3). If estradiol is administered intratesticularly to EDS-treated rats, it is able to prevent the regeneration of Leydig cells completely via local, not endocrine, pathways (Abney and Myers, 1991) and does this by inhibiting the proliferation of Leydig precursor cells (Moore *et al.*, 1992). Since Leydig cells are probably the main source of estradiol in the adult rat testis, this result implies that Leydig cells regulate the numbers of their precursor cells and thus self-regulate their numbers (Fig. 3). The same system is also highly likely to operate in the immature and perhaps also in the fetal and neonatal testis, since administration of estradiol to immature hypophysectomized rats is able to prevent the stimulatory effects of exogenous FSH on Leydig cell numbers and differentiation (Chen *et al.*, 1977; Hsueh *et al.*, 1978). In this respect, an additional point of interest is that, in the fetal and neonatal period, the Sertoli cells rather than the Leydig cells are the source of estradiol, a function that is switched off at 15–21 days of age (e.g., Rosselli and Skinner, 1992). Studies involving the addition of germ cells of various types to Sertoli cells isolated from rats at these ages have demonstrated pronounced inhibition of estradiol production (Table 2). Further, the peritubular cell product PModS also was found to suppress Sertoli cell aromatase activity (Verhoeven, 1992) suggesting that, in the neonatal period as well as in the adult, germ cells — but perhaps also peritubular cells — are able to modulate the ability of the Sertoli cell to regulate the supply of Leydig cells.

A series of studies on the source of regenerating Leydig cells following EDS treatment (Teerds et al., 1988,1989b; Myers and Abney, 1991; Moore et al., 1992) suggests that the sequence of events (and their regulation) is as follows. Destruction of the Leydig cells results, within a few days, in prolif-eration of Leydig precursor cells, probably as a consequence of the removal of inhibitory signals such as estradiol, but perhaps also because of stimula-tory signals from the Sertoli cells because of the elevation of FSH levels (see Fig. 3). Under the actions of LH- and perhaps also FSH-driven Sertoli cell factors, these Leydig precursor cells are stimulated to differentiate into adult Leydig cells. LH also may trigger mitosis of these newly differentiated cells. As more and more Leydig cells develop, the local levels of estradiol increase to a level at which they inhibit further proliferation of Leydig precursor cells and thus stabilize Leydig cell numbers. Physiologically, in the adult rat testis the Sertoli cells secrete an aromatase inhibitor (Boitani et al., 1981; reviewed by Sharpe, 1993) that may represent one mechanism by which the Sertoli cells can modulate the supply of Leydig cells (Fig. 3). Of particular interest is that the secretion of this aromatase inhibitor varies according to the stage of the spermatogenic cycle and is maximal at Stages VII–VIII, which are the androgen-dependent stages (see Sharpe et al., 1992; Kerr et al., 1993a). This local suppression of aromatase probably has little immediate effect on testos-terone levels but will trigger Leydig precursor cell proliferation. The available data suggest that once this proliferation has occurred, the development of Leydig cell precursors into functional adult Leydig cells takes 12–14 days (Teerds et al., 1988,1989a,b; Abney and Myers, 1991). This period is approx-imately equivalent to the duration of the spermatogenic cycle (12.8 days) in the rat (Leblond and Clermont, 1952) and raises the intriguing possibility that the Sertoli cells at Stages VII–VIII insure the presence of adequate adjacent Leydig cells for their passage through Stages VII–VIII in the next spermatogenic cycle. This behavior would represent a remarkable long-term regulatory process, although testing this hypothesis is probably difficult. However, evidence suggests that seminiferous tubules at Stages VII–VIII can regulate the size of neighboring Leydig cells (Bergh, 1982) or their content of smooth endoplasmic reticulum (SER; see subsequent discussion) (Fouquet, 1987).

The data discussed in this section clearly indicate that abundant evidence exists for paracrine regulation of Leydig cell numbers by the Sertoli cells and, indirectly, by the germ cells. With the exception of estradiol, we still have little knowledge of the identity of the growth and differentiation regulators involved in these events, although IGF-1, EGF/TGFα, FGF, PDGF, TGFβ, and IL-1 are strong potential candidates (Bellvé and Zheng, 1989; Skinner, 1991; Sharpe, 1993; see also Gnessi *et al.*, 1992; Khan *et al.*, 1992a; Lin *et al.*, 1992; Teerds and Dorrington, 1993), as are unidentified mitogens in interstitial fluid (Drummond *et al.*, 1988).

B. Modulation of Leydig Cell Function by Sertoli Cells

Considerable evidence suggests that the Sertoli cells exercise some degree of control over Leydig cell steroidogenesis. This evidence derives primarily from two different experimental approaches. First is the demonstration that coculture of Leydig cells with Sertoli cells (SC), seminiferous tubules (ST), or SC- or ST-conditioned medium (CM) is able to modulate Leydig cell testosterone secretion (Table 4). Second is the demonstration that testicular interstitial fluid contains one or more factors that are able to stimulate Leydig cell testosterone secretion.

1. Coculture of Leydig cells with Sertoli cells or seminiferous tubules

Researchers unanimously agree that the coculture of Sertoli cells isolated from immature (pig) or immature or adult (rat) animals enhances basal and LH/human chorionic gonadotropin (hCG)-stimulated testosterone production by the Leydig cells (Table 4). Similar stimulatory effects are demonstrable using SC-CM. In all instances, stimulation of the Sertoli cells with FSH enhances their ability to stimulate Leydig cell testosterone production. This result is of particular interest because of the wealth of evidence, discussed earlier, that FSH is able to stimulate Leydig cell development and differentiation in hypophysectomized immature rats and mice. With the exception of FSH responsiveness, these data probably have relevance to the adult rat (Papadopoulos *et al.*, 1987) and to humans (Papadopoulos, 1991). However, most of these studies have used Sertoli cells isolated from immature animals (Table 4), which clearly have different functions than in the adult (see previous discussion).

Some of the apparent effects of Sertoli cells or SC-CM on Leydig cell testosterone production *in vitro* also may reflect maintenance of the SER in Leydig cells (see Tabone *et al.*, 1984), since the volume of this organelle clearly correlates with the ability of the Leydig cells to secrete testosterone (Ewing and Zirkin, 1983). Moreover, the volume of Leydig cell SER changes quite dramatically *in vivo* when germ cells in the adjacent seminiferous tubules are depleted experimentally (Aoki and Fawcett, 1978; Kerr *et al.*, 1979; Rich *et al.*, 1979). Another observation is that the factors secreted by rat Sertoli cells *in vitro* that enhance testosterone production by the Leydig cells are modulated specifically by pachytene spermatocytes but not by round spermatids (Onoda *et al.*, 1991a). As well as reinforcing the evidence discussed earlier that germ cells play an important role in modulating the ability of the Sertoli cell to regulate the Leydig cells, this evidence also suggests that

Sertoli cells cultured in the absence of germ cells may function very differently from normal adult Sertoli cells, which may explain the rather disparate findings on the effects of isolated seminiferous tubules from adult rats on Leydig cell function (Table 4). Basically, no clear consensus to such findings exists and, although more studies have shown inhibition or no effect, the data are far from convincing. Because other convincing evidence exists that adult rat Sertoli cells can modulate Leydig cell function (see subsequent discussion), the equivocal data on seminiferous tubule–Leydig cell cocultures are likely to reflect problems with this culture system. One problem in cocultures that has been ignored is the ability of the seminiferous tubule to metabolize some of the testosterone produced by the Leydig cells, thus obscuring any stimulatory effects on testosterone production *per se* (Qureshi, 1993).

2. Effects of testicular interstitial fluid

Modulation of Leydig cell structure and function by the Sertoli cells requires first that the modulatory factor is secreted into the interstitial fluid that surrounds the tubules and the Leydig cells. Overwhelming evidence suggests that interstitial fluid from the adult rat contains one or more factors that are able to enhance Leydig cell testosterone production *in vitro*, especially LH/hCG-stimulated testosterone production (Sharpe and Cooper, 1984; Risbridger *et al.*, 1986; Hedger *et al.*, 1990). These effects are dose dependent and far more pronounced than any of the stimulatory effects demonstrated using Sertoli cell–Leydig cell cocultures. Of particular interest are findings that show that induction of seminiferous tubule damage (i.e., germ cell depletion) by a variety of methods results in increased levels of the factor in interstitial fluid (Sharpe and Cooper, 1984; Sharpe *et al.*, 1986a,b; Bartlett and Sharpe, 1987; Ishida *et al.*, 1987; Risbridger *et al.*, 1987a,b; Jansz *et al.*, 1990). Increased secretion of this factor(s) into interstitial fluid also can be induced by testosterone withdrawal (Sharpe *et al.*, 1986b; Risbridger *et al.*, 1987b; Drummond *et al.*, 1988). These situations are remarkably akin to those in which Sertoli cell modulation of Leydig cell number and development are demonstrable (see previous discussion). The factor present in interstitial fluid has not been purified to date, but is probably a partially heat-labile glycoprotein of 57–75 kDa (Jansz *et al.*, 1990; Hedger *et al.*, 1991). The factor appears to modulate steroidogenesis by affecting the side-chain cleavage of pregnenolone (Risbridger *et al.*, 1986). In this respect and others, the factor is similar to a protein termed steroidogenesis-inducing protein (SIP) that has been isolated from human ovarian follicular fluid and which can enhance steroidogenesis by ovarian and adrenal cells as well as by Leydig cells (Khan *et al.*, 1990). SIP is a 60-kDa protein. Of particular interest is the finding that it can stimulate DNA synthesis and induce morphological changes in Leydig cells from immature rats (Khan *et al.*, 1992b). This result raises the possibility that a comparable factor secreted from the Sertoli cells might be responsible for several of the observations described earlier relating to changes in number, structure, and function of the Leydig cells.

C. Mediators of Sertoli Cell – Leydig Cell Interactions

Remarkably, despite the wealth of evidence showing Sertoli cell modulation of the Leydig cells, we still have little knowledge of the identity of the factors that mediate these effects. Estradiol is one such modulator (see earlier discussion) but, even in this instance, beyond 15 – 18 days of age in the rat this steroid clearly is no longer produced in appreciable amounts by the Sertoli cells (e.g., Rosselli and Skinner, 1992) but is produced instead by the Leydig cells themselves (Fig. 3). In all other instances discussed, the mediator(s) has not yet been purified. The approach to this problem taken by many groups has been assessing whether identified factors known to be secreted by Sertoli cells (or for which the mRNA is expressed in Sertoli cells) have any effect on Leydig cells *in vitro;* most such studies have concentrated on the modulation of testosterone production. Indeed, the number of known hormones, releasing hormones, growth factors, cytokines, and so on that have been shown to affect Leydig cell testosterone production *in vitro* is enormous. However, for none of these factors is convincing evidence available that it plays a physiological role in mediating Sertoli – Leydig cell interactions *in vivo* (see Sharpe, 1990,1993). To review these data is beyond the scope of this chapter, so the reader is referred to reviews that have addressed these findings (Risbridger and de Kretser, 1989; Bellvé and Zheng, 1989; Sharpe, 1990,1993; Skinner, 1991; Verhoeven, 1992).

Nevertheless, brief mention is made of the factors for which evidence exists of both Sertoli cell production and effects on the Leydig cells. Prime among these factors are gonadotropin releasing hormone (GnRH) (reviewed by Sharpe, 1993), IL-1 (reviewed by Jégou, 1993; Sharpe, 1993), IGF-1 (reviewed by Sharpe, 1993; see also Lin *et al.*, 1992), and members of the transforming growth factor family such as EGF, TGFα, and TGFβ (see Bellvé and Zheng, 1989; Skinner, 1991). Some or all of these factors may play a role in mediating Sertoli – Leydig cell interactions but, since most of the factors cited are ubiquitous in distribution and action, establishing their definitive roles will prove extremely difficult. The major leads in this area are still likely to come from studies that are able to purify secreted factors from cell-conditioned medium and demonstrate that they have effects similar to those of the unfractionated medium.

V. Conclusions

Paracrine mechanisms are obviously important in all organs since they enable the efficient coordination of the function of the different cell types

TABLE 4 Effects of Isolated Sertoli Cells or Seminiferous Tubules or Their Conditioned Medium on the Function of Isolated Leydig Cells *in Vitro*

Coculture system		Effect on basal and/or LH/hCG-stimulated Testosterone production	Other observations
Sertoli cells or seminiferous tubules	Leydig cells		
Sertoli cells or Sertoli cell CM[a] (immature pig)	Purified (immature pig)	Stimulation (Benahmed *et al.*, 1984; Tabone *et al.*, 1984; Perrard-Sapori *et al.*, 1987)	FSH-stimulated (Benahmed *et al.*, 1984; Tabone *et al.*, 1984; Perrard-Sapori *et al.*, 1987); increase in Leydig cell SER (Tabone *et al.*, 1984)
Sertoli cells or Sertoli cell CM (immature rat)	Purified (immature rat)	Stimulation (Verhoeven and Cailleau, 1985, 1986,1989,1990)	FSH-stimulated (Verhoeven and Cailleau, 1986,1990); mediator is a 30-kDa protein (Verhoeven and Cailleau, 1986)
Sertoli cells or Sertoli cell CM (immature/adult rat)	Purified (adult rat)	Stimulation (Janecki *et al.*, 1985; Papadopoulos *et al.*, 1987; Onoda *et al.*, 1991)	FSH-stimulated (Janecki *et al.*, 1985)
Sertoli cells or Sertoli cell CM (adult human)	Purified (adult human)	Stimulation (Papadopoulos, 1991)	Mediator is a 79-kDa protein (Papadopoulos, 1991)
Seminiferous tubules (adult rat)	Crude (adult rat)	Inhibition (Parvinen *et al.*, 1984; Syed *et al.*, 1985,1988a)	Stage-dependent variation (Parvinen *et al.*, 1984; Syed *et al.*, 1985,1988a)
Seminiferous tubules (ST) or ST CM (adult rat)	Purified (adult rat)	Inhibition (Vihko and Huhtaniemi, 1989; Qureshi, 1993); no effect (Bartlett *et al.*, 1987); stimulation (Parvinen *et al.*, 1984; Syed *et al.*, 1985; Papadopoulos *et al.*, 1986,1987)	Stage-dependent variation (Vihko and Huhtaniemi, 1989; Qureshi, 1993); stage-dependent variation (Parvinen *et al.*, 1984; Syed *et al.*, 1985)
Seminiferous tubules (adult human)	Purified (adult human, rat and mouse)	Stimulation (Verhoeven and Cailleau, 1987)	—

[a] CM, Conditioned medium.

that constitute that organ. The testis probably has a uniquely high requirement for such mechanisms, not only because of the presence of many different cell types with radically different functions but also because of the requirement to coordinate function of these cells in a cyclical and time-dependent manner. Depending on the species, each germ cell takes 8–11 weeks to develop into a spermatozoon. During this period, its functions and requirements change as dramatically as does its morphological appearance. These changing requirements are met by the Sertoli cells. The simplest way in which this can occur is that each germ cell type "informs" the Sertoli cell of its coming needs via paracrine signals. These signals may involve structural or humoral pathways. As this chapter has shown, we are beginning to piece together these key interactions, although they are extremely complex. This situation should not surprise us since spermatogenesis is not just an accumulation of interactions between individual germ cells and their supporting Sertoli cells. This process also involves coordinating the development of each of the different generations of germ cells present at each stage and coordinating this with the development and function of the other cell types in the testis—the peritubular cells, the Leydig cells, and the macrophages and vasculature. Also, all of these events must be coordinated with the development and function of the rest of the body, which is accomplished by a close interaction between the intratesticular paracrine mechanisms and the endocrine drive from the pituitary gland, which is itself under hypothalamic control.

The apparent complexity of the local control mechanisms is therefore not at all surprising and is viewed by many as a daunting barrier to progress in our understanding of the control of spermatogenesis. This barrier is one of the strongest arguments voiced in the support of studying the functions of isolated testicular cells, whether they be Sertoli cells, Leydig cells, peritubular cells, or germ cells, alone or in combination. In so doing, we must remember that removal of the cells from their normal environment must alter their function if paracrine mechanisms are as important as they appear to be. In itself, this effect is not an obstacle, provided that the function or mechanism of the isolated cell under study is evaluated subsequently *in vivo*. Only then can we place that event or mechanism confidently within the overall framework of normal testicular function. Only in this way will we come to understand spermatogenesis completely.

References

Abney, T. O., and Myers, R. B. (1991). 17β-estradiol inhibition of Leydig cell regeneration in the ethane dimethylsulfonate-treated mature rat. *J. Androl.* **12**, 295–304.

Ailenberg, M., and Fritz, I. B. (1989). Influences of follicle-stimulating hormone, proteases, and antiproteases on permeability of the barrier generated by Sertoli cells in a two-chambered assembly. *Endocrinology* **124**, 1399–1407.

Ailenberg, M., Tung, P. S., Pelletier, M., and Fritz, I. B. (1988). Modification of Sertoli cell function in two-chamber assembly by peritubular cells and extracellular matrix. *Endocrinology* **122,** 2604–2612.

Allenby, G., Foster, P. M. D., and Sharpe, R. M. (1991). Evidence that secretion of immunoactive inhibin by seminiferous tubules from the adult rat testis is regulated by specific germ cell types: Correlation between *in vivo* and *in vitro* studies. *Endocrinology* **28,** 467–476.

Anthony, C. T., and Skinner, M. K. (1989). Actions of extracellular matrix on Sertoli cell morphology and function. *Biol. Reprod.* **40,** 691–702.

Aoki, A., and Fawcett, D. W. (1978). Is there a local feedback from the seminiferous tubules affecting activity of the Leydig cell? *Biol. Reprod.* **19,** 144–158.

Avallet, O., Vigier, M., Perrard-Sapori, M. H., and Saez, J. M. (1987). Transforming growth factor β inhibits Leydig cell functions. *Biochem. Biophys. Res. Commun.* **146,** 575–581.

Ayer-Le Lievre, C., Olson, L., Ebendal, T., Hallbook, F., and Persson, H. (1988). Nerve growth factor mRNA and protein in the testis and epididymis of mouse and rat. *Proc. Natl. Acad. Sci. U.S.A.* **85,** 2628–2632.

Bardin, C. W., Cheng, C. Y., Musto, N. A., and Gunsalus, G. L. (1988). The Sertoli cell. *In* "The Physiology of Reproduction" (E. Knobil, J. Neill, *et al.* eds.), Vol. I, pp. 933–974. Raven Press, New York.

Bartlett, J. M. S., and Sharpe, R. M. (1987). Effect of local heating of the testis on the levels in interstitial fluid of a putative paracrine regulator of the Leydig cells and its relationship to changes in Sertoli cell secretory function. *J. Reprod. Fertil.* **80,** 279–287.

Bartlett, J. M. S., Wu, F. C. W., and Sharpe, R. M. (1987). Enhancement of Leydig cell testosterone secretion by isolated seminiferous tubules during co-perifusion *in vitro*: Comparison with static co-culture systems. *Int. J. Androl.* **10,** 603–617.

Bartlett, J. M. S., Weinbauer, G. F., and Nieschlag, E. (1989). Quantitative analysis of germ cell number and relation to intratesticular testosterone following vitamin A-induced synchronization of spermatogenesis in the rat. *Endocrinology* **123,** 403–412.

Bellvé, A. R., and Zheng, W. (1989). Growth factors as autocrine and paracrine modulators of male gonadal functions. *J. Reprod. Fertil.* **85,** 771–793.

Benahmed, M., Tabone, E., Reventos, J., and Saez, J. M. (1984). Role of Sertoli cells in Leydig cell function. *INSERM Coll.* **123,** 363–385.

Benahmed, M., Morera, A. M., Chauvin, M. A., and de Peretti, E. (1987). Somatomedin C/insulin growth factor-I as a possible intratesticular regulator of Leydig cell activity. *Mol. Cell. Endocrinol.* **50,** 69–77.

Bergh, A. (1982). Local differences in Leydig cell morphology in the adult rat testis: Evidence for a local control of Leydig cells by adjacent seminiferous tubules. *Int. J. Androl.* **5,** 325–330.

Besmer, P. (1991). The kit ligand encoded at the murine ligand encoded at the murine *steel* locus: A pleiotropic growth and differentiation factor. *Curr. Opin. Cell Biol.* **3,** 939–946.

Bissell, M. J., and Barcellos-Hoff, M. H. (1987). The influence of extracellular matrix on gene expression: Is structure the message? *J. Cell. Sci. (Suppl.)* **8,** 327–343.

Blaschuk, O., and Fritz, I. B. (1984). Isoelectric forms of clusterin isolated from ram rete testis fluid and from secretions of primary cultures of ram and rat Sertoli cell enriched cultures. *Can. J. Biochem. Cell Biol.* **62,** 456–461.

Boitani, C., Ritzen, E. M., and Parviven, M. (1981). Inhibition of rat Sertoli cell aromatase by factor(s) secreted specifically at spermatogenic stages VII and VIII. *Mol. Cell. Endocrinol.* **23,** 11–22.

Bresler, R. S., and Ross, M. H. (1972). Differentiation of peritubular myoid cells of the testis: Effects of intratesticular implantation of newborn-mouse testis into normal and hypophysectomized adults. *Biol. Reprod.* **6,** 148–159.

Byers, S., and Pelletier, R. M. (1992). Sertoli–Sertoli cell tight junctions and the blood–testis barrier. *In* "Tight Junctions" (M. Cereijido, ed.), pp. 279–304. CRC Press, Boca Raton, Florida.

Byers, S. W., Hadley, M. A., Djakiew, D., and Dym, M. (1986). Growth and characterization of polarized monolayers of epididymal epithelial cells and Sertoli cells in dual environment culture chambers. *J. Androl.* **7**, 59–68.

Byers, S. W., Jégou, B., MacCalman, C., and Blaschuk, O. (1993). The Sertoli cell–cell and cell–substratum adhesion and the collective organization of the testis. *In* "The Sertoli Cell" (L. D. Russell and M. D. Griswold, eds.), pp. 431–446. Cache River Press, Clearwater, Florida.

Cameron, D. F., and Snyddle, E. (1985). Selected enzyme histochemistry of Sertoli cells. 2. Adult rat Sertoli cells in co-culture with peritubular fibroblasts. *Andrologia* **17**, 185–193.

Campbell, S. M. C., and Wiebe, J. P. (1989). Stimulation of spermatocyte development in prepubertal rats by the Sertoli cell steroid, 3α-hydroxy-4-pregnen-20-one. *Biol. Reprod.* **40**, 897–905.

Castellón, E., Janecki, A., and Steinberger, A. (1989a). Influence of germ cells on Sertoli cell secretory activity in direct and indirect co-culture with Sertoli cells from rats of different ages. *Mol. Cell. Endocrinol.* **64**, 169–178.

Castellón, E., Janecki, A., and Steinberger, A. (1989b). Age-dependent Sertoli cell responsiveness to germ cells *in vitro*. *Int. J. Androl.* **12**, 439–450.

Chabot, B., Stephenson, D. A., Chapman, V. M., Besmer, P., and Bernstein, A. (1988). The proto-oncogene c-*kit* encoding a transmembrane tyrosine kinase receptor maps to the mouse W locus. *Nature (London)* **335**, 88–89.

Chatelain, P. G., Naville, D., and Saez, J. M. (1987). Somatomedin-C/insulin-like growth factor I-like material secreted by porcine Sertoli cells *in vitro*. Characterization and regulation. *Biochem. Biophys. Res. Commun.* **146**, 1009–1017.

Chen, Y. -D. I., Payne, A. H., and Kelch, R. P. (1976). FSH stimulation of Leydig cell function in the hypophysectomized immature rat. *Proc. Soc. Exp. Biol. Med.* **153**, 473–475.

Chen, Y.-D. I., Shaw, M. J., and Payne, A. H. (1977). Steroid and FSH action on LH receptors and LH-sensitive testicular responsiveness during sexual maturation of the rat. *Mol. Cell. Endocrinol.* **8**, 291–299.

Cheng, C. Y., and Bardin, C. W. (1987). Identification of two testosterone-responsive testicular proteins in Sertoli cell-enriched culture medium whose secretion is suppressed by cells of the intact seminiferous tubule. *J. Biol. Chem.* **262**, 12768–12779.

Cheng, C. Y., Chen, C. L. C., Feng, Z. M., Marshall, A., and Bardin, C. W. (1988). Rat clusterin isolated from primary Sertoli cell-enriched culture medium is sulfated glycoprotein-2 (SGP-2). *Biochem. Biophys. Res. Commun.* **155**, 398–404.

Cheng, C. Y., Grima, J., Stahler, M. S., Lockshin, R. A., and Bardin, C. W. (1989). Testins are structurally related Sertoli cell proteins whose secretion is tightly coupled to the presence of germ cells. *J. Biol. Chem.* **264**, 21368–21393.

Cheng, C. Y., Grima, J., Stahler, M. S., Gugiolmotti, A., Silverstini, B., and Bardin, C. W. (1990). Sertoli cell synthesizes and secretes a protease inhibitor, α2-macroglobulin. *Biochemistry* **29**, 1063–1068.

Cheng, C. Y., Grima, J., Pineau, C., Stahler, M. S., Jégou, B., and Bardin, C. W. (1992). α-2-Macroglobulin protects the seminiferous epithelium during stress. *In* "Stress and Reproduction" (K. E. Sheppard, J. H. Boublik, and J. W. Funder, eds.), pp. 123–133. Raven Press, New York.

Christensen, A. K. (1975). Leydig cells. *In* "Handbook of Physiology" (D. W. Hamilton and R. O. Greep, eds.), Vol. 5, pp. 57–94. American Physiological Society, Washington, D.C.

Cleland, K. W. (1951). *Aust. J. Sci. Res. B.* **4**, 344–369.

Clermont, Y. (1958). Contractile elements in the limiting membrane of the seminiferous tubules of the rat. *Exp. Cell. Res.* **15**, 438–440.

Clermont, Y., and Morales, C. (1982). Transformation of Sertoli cell processes invading the cytoplasm of elongating spermatids of the rat. *Anat. Rec.* **202**, 32–33A.

Collard, M. W., Sylvester, S. R., Tsuruta, J. K., and Griswold, M. D. (1988). Biosynthesis and molecular cloning of sulfated glycoprotein 1 secreted by rat Sertoli cells: Sequence similarity with the 70-kDA precursor to sulfatide/GM1 activator. *Biochemistry* **27**, 4557–4564.

Connell, C. J. (1974). The Sertoli cell of the sexually mature dog. *Anat. Rec.* **178**, 333.

Cooke, P. S., and Meisami, E. (1991). Early hypothyroidism in rats causes increased adult testis and reproductive organ size but does not change testosterone levels. *Endocrinology* **129**, 237–243.

Cooke, P. S., Hess, R. A., Porcelli, J., and Meisami, E. (1991). Increased sperm production in adult rats after transient neonatal hypothyroidism. *Endocrinology* **129**, 244–248.

Cooke, P. S., Porcelli, J., and Hess, R. A. (1992). Induction of increased testis growth and sperm production in adult rats by administration of the goitrogen propylthiouracil (PTU): The critical period. *Biol. Reprod.* **46**, 146–154.

Corlu, A., Gérard, N., Rissel, M., Kercret, H., Kneip, B., Guguen-Guillouzo, C., and Jégou, B. (1992). Identification of a plasma membrane protein involved in the cell-cell contact-mediated regulation of the Sertoli cell function. 7th European Workshop on Molecular and Cellular Endocrinology of the Testis, Castle Elmau, Germany.

Davis, J. T., and Ong, D. E. (1992). Synthesis and secretion of retinol-binding protein by cultured rat Sertoli cells. *Biol. Reprod.* **47**, 528–533.

De Jong, F. H. (1988). Inhibin. *Physiol. Rev.* **68**, 555–607.

de Kretser, D. M., and Burger, H. G. (1972). Ultrastructural studies of the human Sertoli cell in normal men and males with hypogonadotrophic hypogonadism before and after gonadotrophic treatment. *In* "Gonadotrophins" (B. B. Saxena, C. G. Beling, and H. M. Gandy, eds.), pp. 640–656. Wiley Interscience, New York.

Den Boer, J. P., Machenbach, P., and Grootegoed, J. A. (1989). Gluthatione metabolism in cultured Sertoli cells and spermatogenic cells from hamsters. *J. Reprod. Fertil.* **87**, 391–400.

de Winter, J. P., Themmen, A. P. N., Hooger-Brugge, J. W., Klaij, I. A., Grootegoed, J. A., and de Jong, F. H. (1992). Activin receptor mRNA present in pachytene spermatocytes and early spermatids (not late spermatids). *Mol. Cell. Endocrinol.* **83**, R1–R8.

Dinarello, C. A. (1988). Biology of interleukin-1. *FASEB J.* **2**, 108–115.

Djakiew, D., and Dym, M. (1988). Pachytene spermatocyte proteins influence Sertoli cell function. *Biol. Reprod.* **39**, 1193–1205.

Dolci, S., Williams, D. E., Ernest, M. K., Resnick, J. L., Brannan, C. I., Lock, L. F., Lyman, S. D., Boswell, H. S., and Donovan, P. J. (1991). The membrane-bound form of KL most probably plays an important role in PGC survival. *Nature (London)* **352**, 809–811.

Drummond, A. E., Risbridger, G. P., O'Leary, P. C. O., and de Kretser D. M. (1988). Alterations in mitogenic and steroidogenic activities in rat testicular interstitial fluid after administration of ethane dimethane sulphonate. *J. Reprod. Fert.* **83**, 141–147.

Dym, M., and Fawcett, D. W. (1970). The blood-testis barrier in the rat and the physiological compartmentation of the seminiferous epithelium. *Biol. Reprod.* **3**, 308–326.

Dym, M., Lamsam-Casalotti, S., Jia, M. C., Kleinman, H. K., and Papadopoulos, V. (1991). Basement membrane increases G-protein levels and follicle-stimulating hormone responsiveness of Sertoli cell adenylyl cyclase activity. *Endocrinology* **128**, 1167–1176.

Edelman, G. M. (1988). "Topobiology." Basic Books, New York.

Enders, G. C., Henson, J. H., and Millette, C. F. (1986). Sertoli cell binding to isolated testicular basement membrane. *J. Cell Biol.* **103**, 1109–1119.

Engel, J. (1991). Common structural motifs in proteins of the extracellular matrix. *Curr. Opin. Cell Biol.* **111**, 779–785.

Erickson-Lawrence, M., Zabludoff, S. D., and Wright, W. W. (1990). Cyclic protein-2, a secretory product of rat Sertoli cell, is the proenzyme form of cathepsin-L. *J. Cell Biol.* **111**, 107a.

Esnard, A., Esnard, F., Guillou, F., and Gauthier, F. (1992). Production of the cysteine proteinase inhibitor cystatin C by rat Sertoli cells. *FFBS Lett.* **300**, 131–135.

Ewing, L. L., and Zirkin, B. (1983). Leydig cell structure and steroidogenic function. *Rec. Progr. Horm. Res.* **39**, 599–635.

Fawcett, D. W. (1975). Ultrastructure and function of the Sertoli cell. *In* "Handbook of Physiology" (D. W. Hamilton and R. O. Greep, eds.), pp. 21–55. Williams & Wilkins, Baltimore.

Fouquet, J. P. (1987). Ultrastructural analysis of a local regulation of Leydig cells in the adult monkey *(Macaca fascicularis)* and rat. *J. Reprod. Fertil.* **79,** 49–56.

Franchimont, P., Croze, F., Demoulin, A., Bologne, R., and Hustin, J. (1981). Relationship between azoospermia and oligospermia. *Acta Endorinol.* **98,** 312–320.

Fujisawa, M., Bardin, C. W., and Morris, P. L. (1992). A germ cell factor(s) modulates preproenkephalin gene expression in rat Sertoli cells. *Mol. Cell. Endocrinol.* **84,** 79–88.

Furuya, S., Kumamoto, Y., and Ikegaki, S. (1980). Blood-testis barrier in men with idiopathic hypogonadotropic eunuchoidism and post-pubertal pituitary failure. *Arch. Androl.* **5,** 361–367.

Galdieri, M., Monaco, L., and Stefanini, M. (1984). Secretion of androgen binding protein by Sertoli cells is influenced by contact with germ cells. *J. Androl.* **5,** 409–415.

Geissler, E. N., Ryan, M. A., and Housman, D. E. (1988). The dominant-white spotting *(W)* locus of the mouse encodes the c-*kit* proto-oncogene. *Cell* **55,** 185–192.

Gérard, A., En Nya, A., Egloff, M., Domingo, M., Degrelle, H., and Gérard, H. (1991). Monkey germ cells do internalize the human sex-steroid-binding protein (hSBP). *Ann. N.Y. Acad. Sci.* **637,** 258–276.

Gérard, J. H., En Nya, A., Guéant, J. L., Anwar, W., Fremont, S., and Gérard, A. (1989). Internalisation de l'androgen-binding protein (ABP) par les cellules germinales chez le rat: Étude autohistoradiographique on microscopie électronique. Congrès de Biologie Cellulaire et Biochimie, Orsay, France.

Gérard, N., Syed, V., Bardin, C. W., Genetet, N., and Jégou, B. (1991). Sertoli cells are the site of interleukin-1α synthesis in rat testis. *Mol. Cell. Endocrinol.* **82,** R13–R16.

Gérard, N., Syed, V., and Jégou, B. (1992). Lipopolysaccharide, latex beads and residual bodies are potent activators of Sertoli cell interleukin-1α production. *Biochem. Biophys. Res. Commun.* **185,** 154–161.

Gnessi, L., Emidi, A., Farini, D., Scarpa, S., Modesti, A., Ciampini, T., Silvestroni, L., and Spera, G. (1992). Rat Leydig cells bind platelet-derived growth factor through specific receptors and produce platelet-derived growth factor-like molecules. *Endocrinology* **130,** 2219–2224.

Godin, I., and Wylie, C. C. (1991). TGFβ1 inhibits proliferation and has a chemotropic effect on mouse primordial germ cells in culture. *Development* **113,** 1451–1457.

Godin, I., Doed, R., Cooke, J., Zsebo, K., Dexter, M., and Wylie, C. C. (1991). Effects of the *steel* gene product on mouse primordial germ cells in culture. *Nature (London)* **352,** 807–809.

Gondos, B. (1977). Testicular development. *In* "The Testis" (A. D. Johnson and W. R. Gomes, eds.), Vol. 4, pp. 1–37. Academic Press, New York.

Grima, J., Pineau, C., Bardin, C. W., and Cheng, C. Y. (1992). Rat Sertoli cell clusterin, α2-macroglobulin, and testins: Biosynthesis and differential regulation by germ cells. *Mol. Cell. Endocrinol.* **89,** 127–140.

Griswold, M. D. (1988). Protein secretions of Sertoli cells. *Int. Rev. Cytol.* **110,** 133–156.

Grootegoed, J. A., Den Boer, P. J., and Mackenbach, P. (1989). Sertoli cell–germ cell communication. *Ann. N.Y. Acad. Sci.* **564,** 232–242.

Grootenhuis, A. J., Steebergen, J., Timmerman, M. A., Dorsman, A. N. R. D., Shaaper, W. M. M., Meloen, R. M., and de Jong, F. M. (1989). Activin. *J. Endocrinol.* **122,** 293–301.

Gunsalus, G. L., and Bardin, C. W. (1991). Sertoli–germ cell interactions as determinants of bidirectional secretion of androgen-binding protein. *Ann. N.Y. Acad. Sci.* **637,** 327–339.

Hadley, M. A., and Dym, M. (1987). Immunocytochemistry of extracellular matrix of the lamina propria of the rat testis: Electron microscopy localization. *Biol. Reprod.* **37,** 1283–1289.

Hadley, M. A., Byers, S. W., Suarez-Quian, C. A., Kleinman, H. K., and Dym, M. (1985). Extracellular matrix regulates Sertoli cell differentiation, testicular cord formation and germ cell development *in vitro. J. Cell Biol.* **101,** 1511–1522.

Hadley, M. A., Weeks, B. J., Kleiman, H. K., and Dym, M. (1990). Laminin promotes formation of cord-like structures by Sertoli cells in vitro. *Dev. Biol.* **140,** 318–327.

Hall, K., Ritzén, E. M., Johnsonbaugh, R. E., and Parvinen, M. (1983). Secretion of somatome-din-like compound from Sertoli cells *in vitro*. *In* "Insulin-Like Growth Factors/Somato-medins" (E. M. Spencer, ed.), pp. 611–614. de Gruyter, New York.

Hansson, V., Ritzén, E. M., French, F. S., and Nayfeh, S. N. (1975a). Androgen transport and receptor mechanisms in testis and epididymis. *In* "Handbook of Physiology" (D. W. Hamilton and R. O. Greep, eds.), pp. 173–201. William & Wilkins, Baltimore.

Hardy, M. P., Zirkin, B. R., and Ewing, L. L. (1989). Kinetic studies on the development of the adult population of Leydig cells in the testes of the pubertal rat. *Endocrinology* **12,** 762–770.

Hedger, M. P., Robertson, D. M., de Kretser, D. M., and Risbridger, G. P. (1990). The quantifica-tion of steroidogenesis-stimulating activity in testicular interstitial fluid by an *in vitro* bioassay employing adult rat Leydig cells. *Endocrinology* **127,** 1967–1977.

Hedger, M. P., Leung, A., Robertson, D. M., de Kretser, D. M., and Risbridger, G. P. (1991). Steroidogenesis-stimulating activity in the gonads: Comparison of rat testicular fluid with bovine and human ovarian follicular fluids. *Biol. Reprod.* **44,** 937–944.

Hettle, J. A., Waller, E. K., and Fritz, I. B. (1986). Hormonal stimulation alters the type of plasminogen activatory produced by Sertoli cells. *Biol. Reprod.* **34,** 895–904.

Hettle, J. A., Balekjian, E., Tung, P. S., and Fritz, I. B. (1988). Rat testicular peritubular cells in culture secrete an inhibitor of plasminogen activator activity. *Biol. Reprod.* **38,** 359–371.

Holmes, S. D., Bucci, L. R., Lipshultz, L. I., and Smith, R. G. (1983). Transferrin binds specifi-cally to pachytene spermatocytes. *Endocrinology* **113,** 1916–1918.

Holmes, S. D., Lipshultz, L. I., and Smith, R. G. (1984). Regulation of transferrin secretion by human Sertoli cells cultured in the presence or absence of human peritubular cells. *J. Clin. Endocrinol. Metab.* **59,** 1058–1062.

Hsueh, A. J. W., Dufau, M. L., and Catt, K. J. (1978). Direct inhibitory effect of estrogen on Leydig cell function of hypophysectomized rats. *Endocrinology* **103,** 1096–1102.

Huggenvik, J., and Griswold, M. D. (1981). Retinol-binding protein in rat testicular cells. *J. Reprod. Fertil.* **61,** 403–408.

Hutson, J. C. (1983). Metabolic cooperation between Sertoli cells and peritubular cells in culture. *Endocrinology* **112,** 1375–1381.

Hutson, J. C., and Stocco, D. M. (1981). Peritubular cell influence on the efficiency of androgen-binding protein secretion by Sertoli cells in culture. *Endocrinology* **108,** 1362–1368.

Ishida, H., Risbridger, G. P., and de Kretser, D. M. (1987). Variation in the effect of a nongonad-otropic Leydig cell stimulating factor in testicular interstitial fluid after exposure of the testis to a single episode of heat treatment. *J. Androl.* **8,** 247–252.

Jaillard, C., Chatelain, P. G., and Saez, J. M. (1987). *In vitro* regulation of pig Sertoli cell growth and function: Effects of fibroblast growth factor and somatomedin-C. *Biol. Reprod.* **337,** 665–674.

Janecki, A., and Steinberger, A. (1987). Vectorial secretion of transferrin and androgen binding protein in Sertoli cell cultures: effect of extracellular matrix, peritubular myoid cells and medium composition. *Mol. Cell. Endocrinol.* **52,** 125–135.

Janecki, A., Jakubowiak, A., and Lukaszyk, A. (1985). Stimulatory effect of Sertoli cell secretory products on testosterone secretion by purified Leydig cells in primary culture. *Mol. Cell. Endocrinol.* **42,** 235–243.

Janecki, A., Jakubowiak, A., and Steinberger, A. (1991). Effects of cyclic AMP and phorbol ester on transepithelial electrical resistance of Sertoli cell monolayers in two-compartment culture. *Mol. Cell. Endocrinol.* **82,** 61–69.

Jansz, G. F., Cooke, R. A., and Pomerantz, D. K. (1990). Initial characterization of factors from testicular fluid which alter *in vitro* androgen secretion by normal rat Leydig cells. *J. Androl.* **11,** 131–139.

Jégou, B. (1991). Spermatids are regulators of Sertoli cell function. *Ann. N.Y. Acad. Sci.* **637,** 340–353.

Jégou, B. (1992). The Sertoli cell. *Baillière's Clin. Endocrinol. Metab.* **6,** 273–311.

Jégou, B. (1993). The Sertoli–germ cell communication network in mammals. *Int. Rev. Cytol.* **147**, in press.

Jégou, B., Laws, A. O., and de Kretser, D. M. (1984a). Changes in testicular function induced by short-term exposure of the rat testis to heat: Further evidence for interaction of germ cells, Sertoli cells and Leydig cells. *Int. J. Androl.* **7**, 244–257.

Jégou, B., Peake, R. A., Irby, D. C., and de Krestser, D. M. (1984b). Effects of the induction of experimental cryptorchidism and subsequent orchidopexy on testicular function in immature rats. *Biol. Reprod.* **30**, 179–187.

Jégou, B., Le Magueresse, B., Sourdaine, P., Pineau, C., Velez de la Calle, J. F., Garnier, D. H., Guillou, F., and Boisseau, C. (1988). Germ cell–Sertoli cell interactions in vertebrates. In "Molecular and Cellular Endocrinology of the Testis" (B. Cooke and R. M. Sharpe, eds.), pp. 255–270. Raven Press, New York.

Jégou, B., Syed, V., Sourdaine, P., Byers, S., Gérard, N., Velez de la Calle, J. F., Pineau, C., Garnier, D. H., and Bauché, F. (1992). The dialogue between late spermatids and Sertoli cells in vertebrates: A century of research. In "Spermatogenesis-Fertilization-Contraception. Molecular, Cellular and Endocrine Events in Male Reproduction." (E. Nieschlag and U.-F. Habenicht, eds.), pp. 57–95. Schering Foundation Series. Springer, Berlin.

Jégou, B., Velez de la Calle, J. F., Touzalin, A. M., Bardin, C. W., and Cheng, Y. C. (1993). Germ cell control of testin production is inverse to that of other Sertoli cell products. *Endocrinology* **132**, 2557–2562.

Jenkins, N., and Ellison, J. (1989). Regulation of transferrin secretion in Sertoli cells of the calf testis. *Anim. Reprod. Sci.* **20**, 11–20.

Josso, N. (1991). Anti-Müllerian Hormone. *Balliére's Clin. Endocrinol. Metab.* **5**, 625–654.

Josso, N., and Picard, J. Y. (1986). Anti-Müllerian hormone. *Physiol. Rev.* **66**, 1038–1090.

Jutte, N. H. P. M., Grootegoed, J. A., Rommerts, F. F. G., and van der Molen, H. J. (1981). Exogenous lactate is essential for metabolic activities in isolated rat spermatocytes and spermatids. *J. Reprod. Fertil.* **62**, 399–405.

Jutte, N. H. P. M., Jansen, R., Grootegoed, J. A., Rommerts, F. F. G., Clausen, O. P. F., and Van der Molen, H. J. (1982). Regulation of survival of rat pachytene spermatocytes by lactate supply from Sertoli cells. *J. Reprod. Fertil.* **65**, 431–438.

Kaipia, A., Parvinen, M., Shimasaki, S., Ling, N., and Toppari, J. (1991). Stage-specific cellular regulation of inhibin α-subunit mRNA expression in the rat seminiferous epithelium. *Mol. Cell. Endocrinol.* **82**, 165–173.

Kerr, J. B., and Donachie, K. (1986). Regeneration of Leydig cells in unilaterally cryptorchid rats: Evidence for stimulation by local testicular factors. *Cell. Tissue Res.* **245**, 649–655.

Kerr, J. B., and Knell, C. M. (1988). The fate of fetal Leydig cells during the development of the fetal and postnatal rat testis. *Development* **103**, 535–544.

Kerr, J. B., and Sharpe, R. M. (1985). FSH induction of Leydig cell maturation. *Endocrinology* **116**, 2592–2604.

Kerr, J. B., Rich, K. A., and de Kretser, D. M. (1979). Alterations of the fine structure and androgen secretion of the interstitial cells in the experimentally cryptorchid rat testis. *Biol. Reprod.* **20**, 409–422.

Kerr, J. B., Bartlett, J. M. S., Donachie, K., and Sharpe, R. M. (1987). Origin of regenerating Leydig cells in the testis of the adult rat. An ultrastructural, morphometric and hormonal assay study. *Cell Tissue Res.* **249**, 367–377.

Kerr, J. B., Risbridger, G. P., and Knell, C. M. (1988). Stimulation of interstitial cell growth after selective destruction of foetal Leydig cells in the testis of postnatal rats. *Cell Tissue Res.* **252**, 89–98.

Kerr, J. B., Millar, M., Maddocks, S., and Sharpe, R. M. (1993a). Stage-dependent changes in spermatogenesis and Sertoli cells in relation to the onset of spermatogenic failure following withdrawal and restoration of testosterone. *Anat. Rec.* In press.

Kerr, J. B., Savage, G. N., Millar, M., and Sharpe, R. M. (1993b). Response of the seminiferous epithelium of the rat testis to withdrawal of androgen: Evidence for direct effects upon inter-Sertoli cell tight junctions. *Cell Tissue Res.* In press.

Keshet, E., Lyman, S. D., Williams, D. E., Anderson, D. M., Jenkins, N. A., Copeland, N. G., and Parada, L. F. (1991). Embryonic RNA expression patterns of the c-*kit* receptor and its cognate ligand suggest multiple functional roles in mouse development. *EMBO J.* **10,** 2425–2435.

Khan, S. A., Keck, C., Gudermann, T., and Neischlag, E. (1990). Isolation of a protein from human ovarian follicular fluid which exerts major stimulatory effects on in-vitro steroid production of testicular, ovarian and adrenal cells. *Endocrinology* **126,** 3043–3052.

Khan, S. A., Khan, S. J., and Dorrington, J. (1992a). Interleukin-1 stimulates DNA synthesis in immature rat Leydig cells *in vitro*. *Endocrinology* **131,** 1853–1857.

Khan, S. A., Teerds, K., and Dorrington, J. (1992b). Steroidogenesis-inducing protein promotes DNA synthesis in Leydig cells from immature rats. *Endocrinology* **130,** 599–606.

Kierszenbaum, A. L., Ueda, H., Ping, L., Abdullah, M., and Tres, L. L. (1988). Antibodies to rat Sertoli cell secretory proteins recognize antigenic sites in acrosome and tail of developing spermatids and sperm. *J. Cell Sci.* **91,** 145–153.

Kirkby, J. D., Jetton, A. E., Cooke, P. S., Hess, R. A., Bunick, D., Ackland, J. F., Turek, F. W., and Schwartz, N. B. (1992). Developmental hormonal profiles accompanying the neonatal hypothyroidism-induced increase in adult testicular size and sperm production in the rat. *Endocrinology* **131,** 559–565.

Kormano, M., and Hovatta, O. (1972). Contractility and histochemistry of the myoid cell layer of the rat seminiferous tubules during postnatal development. *Z. Anat. Entwicklungsgesch.* **137,** 239–248.

Kumari, M., and Duraiswami, S. (1987). Ultrastructural observations on Sertoli cell–germ cell interaction. *Cytologia (Tokyo)* **52,** 111–116.

Lacroix, M., Parvinen, M., and Fritz, I. B. (1981). Localization of testicular plasminogen activator in discrete portions (Stages VII and VIII) of the seminiferous tubule. *Biol. Reprod.* **25,** 143–146.

Lacroix, M., Smith, F. E., and Fritz, I. B. (1977). Secretion of plasminogen activator by Sertoli cell-enriched cultures. *Mol. Cell. Endocrinol.* **9,** 227–236.

Lacy, D. (1960). Light and electron microscopy and its use in the study of factors influencing spermatogenesis in the rat. *J. R. Micro. Soc.* **79,** 209–225.

Lacy, D. (1962). Certain aspects of testis structure and function. *Br. Med. Bull.* **18,** 205–208.

Lahr, G., Mayerhöfer, A., Seidi, K., Bucher, S., Grothe, C., Knochel, W., and Gratzi, M. (1992). Basic fibroblast growth factor (bFGF) in rodent testis. Presence of bFGF mRNA and of a 30 kDa bFGF protein in pachytene spermatocytes. *FEBS Lett.* **302,** 43–46.

Leblond, C. P., and Clermont, Y. (1952). Definition of the stages of the cycle of the seminiferous epithelium in the rat. *Ann. N.Y. Acad. Sci.* **55,** 548–573.

Le Gac, F., and de Kretser, D. M. (1982). Inhibin production by Sertoli cell cultures. *Mol. Cell. Endocrinol.* **28,** 487–498.

Le Gac, F., Le Magueresse, B., Loir, M., and Jégou, B. (1984). Influence of enriched pachytene spermatocytes, round spermatids and residual bodies on the Sertoli cell secretory activity *in vitro*. *Proceedings of 5th Anglo-French Meeting of the Society for the Study of Fertility and Société Francaise pour l'Etude de la Fertilité*, Fresnes, December 7–9.

Le Magueresse, B., and Jégou B. (1986). Possible involvement of germ cells in the regulation of oestradiol-17β and ABP secretion by immature rat Sertoli cells (*in vitro* studies). *Biochem. Biophys. Res. Commun.* **14,** 861–869.

Le Magueresse, B., and Jégou, B. (1988a). In-vitro effects of germ cells on the secretory activity of Sertoli cells recovered from rats of different ages. *Endocrinology* **122,** 1672–1680.

Le Magueresse, B., and Jégou, B. (1988b). Paracrine control of immature Sertoli cells by adult germ cells in the rat (an *in vitro* study). *Mol. Cell. Endocrinol.* **58,** 65–72.

Le Magueresse, B., Le Gac, F., Loir, M., and Jégou, B. (1986). Stimulation of rat Sertoli cell secretory activity in vitro by germ cells and residual bodies. *J. Reprod. Fert.* **77,** 489–498.

Le Magueresse, B., Pineau, C., Guillou, F., and Jégou, B. (1988a). Influence of germ cells upon transferrin secretion by rat Sertoli cells in vitro. *J. Endocrinol.* **118,** R13–R16.

Le Magueresse, B., Le Gac, F., Loir, M., and Jégou, B. (1988b). Etude *in vitro* des interactions entre les cellules de Sertoli et les cellules germinales du rat. *In* "A la Recherche du Pouvoir Fécondant du Sperme" (R. Schoysman, ed.), pp. 147–160. Fondazione per gli studi sulla riproduzione umana.

Lin, T., Wang, D., Nagpal, M. L., Chang, W., and Calkins, J. H. (1992). Down-regulation of Leydig cell IGF-1 gene expression by Interleukin-1. *Endocrinology* **130,** 1217–1224.

Linder, C. C., Heckert, L. L., Roberts, K. P., Kim, K. H., and Griswold, M. D. (1991). Expression of Receptors during the Cycle of the Seminiferous Epithelium. *Ann. N.Y. Acad. Sci.* **637,** 313–321.

Lu, C., and Steinberger, A. (1977). Gamma-glutamyl transpeptidase activity in the developing rat testis. Enzyme localization in isolated cell types. *Biol. Reprod.* **17,** 84–88.

MacGrogan, D., Després, G., Romand, R., and Dicou, E. (1991). Expression of the β-nerve growth factor gene in male sex organs of the mouse, rat, and guinea pig. *J. Neuro. Res.* **28,** 567–573.

Magre, S., and Jost, A. (1991). Sertoli cells and testicular differentiation in the rat fetus. *J. Elec. Microsc. Tech.* **19,** 172–188.

Manova, K., Mocka, K., Besmer, P., and Bachvarova, R. F. (1990). Gonadal expression of c-kit encoded at the W locus of the mouse. *Development* **110,** 1057–1069.

Mather, J. P., Wolpe, S. D., Gunsalus, G. L., Bardin, C. W., and Phillips, D. M. (1984). Effect of purified and cell-produced extracellular matrix components on Sertoli cell function. *Ann. N.Y. Acad. Sci.* **438,** 572–575.

Mayerhofer, A., Russell, L. D., Grothe, C., Rudolf, M., and Gratzi, M. (1991). Presence and localization of a 30-kDa basic fibroblast growth factor-like protein in rodent testes. *Endocrinology* **129,** 921–924.

Means, A. R., Fakunding, J. L., Huckins, C., Tindall, D. J., and Vitale, R. (1976). Follicle-stimulating hormone, the Sertoli cell and spermatogenesis. *Rec. Progr. Horm. Res.* **32,** 477–527.

Molenaar, R., de Rooij, D. G., Rommerts, F. F. G., and Van Der Molen, H. J. (1986). Repopulation of Leydig cells in mature rats after selective destruction of the existent Leydig cells with ethylene dimethane sulfonate is dependent on luteinizing hormone and not follicle-stimulating hormone. *Endocrinology* **118,** 2546–2554.

Monet-Kuntz, C., Guillou, F., Fontaine, I., and Combarnous, Y. (1991). Equine follicle-stimulating hormone action in cultured Sertoli cells from rat, sheep and pig. *Acta Endocr. (Copenh.)* **125,** 86–92.

Monet-Kuntz, C., Guillous, F., Fontaine, I., and Combarnous, Y. (1992). Purification of ovine transferrin and study of the hormonal control of its secretion in enriched cultures of ovine Sertoli cells. *J. Reprod. Fertil.* **94,** 189–201.

Moore, A., Findlay, K., and Morris, I. D. (1992). In-vitro DNA synthesis in Leydig and other interstitial cells of the rat testis. *J. Endocr.* **134,** 247–255.

Morera, A. M., Benahmed, M., Cochet, C., Chauvin, M. A., Chambaz, E., and Revol, A. (1987). A TGF β-like peptide is a possible intratesticular modulator of steroidogenesis. *Ann. N.Y. Acad. Sci.* **513,** 494–496.

Myers, R. B., and Abney, T. O. (1991). Interstitial cell proliferation in the testis of the ethylene dimethane sulfonate treated rat. *Steroids* **56,** 91–96.

Nakamura, M., Kan, R., Okinaga, S., and Arai, K. (1991). A lactate-binding protein of rat spermatogenic cell-plasma membranes. *11th North American Testis Workshop*, April 24–27. Montreal, Quebec, Canada. Abstract 115.

O'Brien, D. A., Gabel, C. A., Welch, J. E., and Eddy, E. M. (1991). Mannose 6-phosphate receptors: Potential mediators of germ cell–Sertoli cell interactions. *Ann. N.Y. Acad. Sci.* **637**, 327–339.

Odell, W. D., and Swerdloff, R. S. (1976). Etiologies of sexual maturation: A model system based on the sexually maturing rat. *Rec. Progr. Horm. Res.* **32**, 245–288.

Ojeifo, J. O., Byers, S. W., Papadopoulos, V., and Dym, M. (1990). Sertoli cell-secreted protein(s) stimulates DNA synthesis in purified rat Leydig cells *in vitro. J. Reprod. Fertil.* **90**, 93–108.

Olson, L., Ayer-Lelievre, C., Ebendal, T., and Seiger, A. (1987). Nerve growth factor-like immunoreactivities in rodent salivary glands and testis. *Cell Tissue Res.* **248**, 275–286.

Onoda, M., and Djakiew, D. (1990). Modulation of Sertoli cell secretory function by rat round spermatid protein(s). *Mol. Cell Endocrinol.* **73**, 35–44.

Onoda, M., and Djakiew, D. (1991). Pachytene spermatocyte protein(s) stimulate Sertoli cells grown in bicameral chambers: Dose-dependent secretion of ceruloplasmin, sulfated glycoprotein-1, sulfated glycoprotein-2 and transferrin. *In Vitro Cell Dev. Biol.* **27A**, 215–222.

Onoda, M., and Djakiew, D. (1992). Partial purification of the paracrine factor from round spermatids which stimulates Sertoli cell transferrin secretion. *7th European Workshop on Molecular and Cellular Endocrinology of the Testis*, Castle Elmau, Germany.

Onoda, M., Djakiew, D., and Papadopoulos, V. (1991a). Pachytene spermatocytes regulate the secretion of Sertoli cell protein(s) which stimulate Leydig cell steroidogenesis. *Mol. Cell. Endocrinol.* **77**, 207–216.

Onoda, M., Pflug, B., and Djakiew, D. (1991b). Germ cell mitogenic activity is associated with nerve growth factor-like protein(s). *J. Cell. Physiol.* **149**, 536–543.

Orth, J., and Christensen, A. K. (1977). Localization of ^{125}L-labeled FSH in the testes of hypophysectomized rats by autoradiography at the light and electron microscope levels. *Endocrinology* **101**, 262–278.

O'Shaughnessy, P. J., Bennett, M. K., Scott, I. S., and Charlton H. M. (1992). Effects of FSH on Leydig cell morphology and function in the hypogonadal mouse. *J. Endocr.* **135**, 517–525.

Papodopoulos, V. (1991). Identification and purification of a human Sertoli cell-secreted protein (hSCSP-80) stimulating Leydig cell steroid biosynthesis. *J. Clin. Endocrinol. Metab.* **72**, 1332–1339.

Papadopoulos V., Carreau, S., and Drosdowsky, M. A. (1986). Effects of seminiferous tubule secreted factors (s) on Leydig cell cyclic AMP production in mature rats. *FEBS letts.* **202**, 74–78.

Papadopoulos, V., Kamtchouing, P., Drosdowsky, M. A., Hochereau de Reviers, M. T., and Carreau, S. (1987). Adult rat Sertoli cells secrete a factor or factors which modulate Leydig cell function. *J. Endocrinol.* **114**, 459–467.

Paranko, J., Pelliniemi, L. J., Vaheri, A., Foidart, J. M., and Lakkala-Paranko, T. (1983). Morphogenesis and fibronectin in sexual differentiation of rat embryonic gonads. *Differentiation (Suppl.)* **23**, 72–81.

Paranko, J., Frojdman, K., Grund, S. K., and Pelliniemi, L. J. (1988). Differentiation of epithelial cords in the fetal testis. *In* "Development and Function of the Reproductive Organs" (M. Parvinen, I. Huhtaniemi, and L. J. Pelliniemi, eds.), Vol. VII, pp. 21–28. Ares Serono Symposia, Rome.

Parvinen, M. (1993). Cyclic Functions of Sertoli Cells. *In* "The Sertoli Cell" (L. D. Russell and M. D. Griswold, eds.), pp. 331–347. Cache River Press, Clearwater, Florida.

Parvinen, M., Nikula, H., and Huhtaniemi, I. (1984). Influence of rat seminiferous tubules on Leydig cell testosterone production in vitro. *Mol. Cell. Endocrinol.* **37**, 321–326.

Parvinen, M., Soder, O., Mali, P., Froysa, B., and Ritzen, E. M. (1991). In-vitro stimulation of stage-specific deoxyribonucleic acid synthesis in rat seminiferous tubule segments by interleukin-1α. *Endocrinology* **129**, 1614–1620.

Parvinen, M., Polto-Huikko, M., Soder, O., Schultz, R., Kaipia, A., Mali, P., Toppari, J., Hako-virta, H., Lonnerberg, P., Ritzén, M., Ebendal, T., Olson, L., Hokfelt, T., and Persson, H.

(1992). Expression of β-nerve growth factor and its receptor in rat seminiferous epithelium: Specific function at the onset of meiosis. *J. Cell. Biol.* **117**, 629–641.

Pelletier, R. M., and Byers, S. W. (1992). The blood testis barrier and Sertoli cell junctions: Structural considerations. *Microsc. Res. Tech.* **20**, 3–33.

Pelliniemi, L. J., Paranko, J., Grund, S. K., Frojdman, K., Foidart, J-M., and Lakkala-Parenko, T. (1984). Extracellular matrix in testicular differentiation. *Ann. N.Y. Acad. Sci.* **438**, 405–416.

Perrard-Sapori, M. H., Chatelain, P. C., Rogemond, N., and Saez, J. M. (1987). Modulation of Leydig cell functions by culture with Sertoli cells or with Sertoli cell-conditioned medium: Effect of insulin, somatomedin-C and FSH. *Mol. Cell. Endocrinol.* **50**, 193–201.

Persson, H., Ayer-Le Lievre, C., Soder, O., Villar, M. J., Metsis, M., Olson, L., Ritzen, E. M., and Hokfelt, T. (1990). Expression of β-nerve growth factor receptor mRNA in Sertoli cells downregulated by testosterone. *Science* **247**, 704–707.

Peyriéras, N. (1992). Les mécanismes moléculaires de l'adhérence cellulaire. *Médecine Sci.* **6**, 591–596.

Pineau, C., Velez de la Calle, J. F., Pinon-Lataillade, G., and Jégou, B. (1989). Assessment of testicular function after acute and chronic irradiation: Further evidence for an influence of late spermatids on Sertoli cell function in the adult rat. *Endocrinology* **124**, 2720–2728.

Pineau, C., Sharpe, R. M., Saunders, P. T. K., Gérard, N., and Jégou, B. (1990). Regulation of Sertoli cell inhibin production and of inhibin α-subunit mRNA levels by specific germ cell types. *Mol. Cell. Endocrinol.* **72**, 13–22.

Pineau, C., Le Magueresse, B., Courtens, J. L., and Jégou B. (1991a). Study *in vitro* of the phagocytic function of Sertoli cells in the rat. *Cell Tissue Res.* **264**, 589–598.

Pineau, C., Syed, V., Bardin, C. W., Jégou, B., and Cheng, C. Y. (1991b). Identification and partial purification of germ cell factors that modulate Sertoli cell secretory functions. *Proceedings of the 73rd Annual Meeting of the Endocrine Society,* June 19–22, Washington, D.C. Abstract 510, p. 158.

Pineau, C., Syed, V., Bardin, C. W., Jégou, B., and Cheng, C. Y. (1992). Partial purification of two germ cell factors that modulate Sertoli cell secretory functions. *7th European Workshop on Molecular and Cellular Endocrinology of the Testis,* Castle Elmau, Germany.

Pineau, C., Syed, V., Bardin, C. W., Jégou, B., and Cheng, C. Y. (1993a). Identification and partial purification of a germ cell factor that stimulates transferrin secretion by Sertoli cells. *Rec. Prog. Horm. Res.* **48**, 539–542.

Pineau, C., Syed, V., Bardin, C. W., Jégou, B., and Cheng, C. Y. (1993b). Identification and partial purification of biological factors from germ cell conditioned medium that modulate Sertoli cell secretory functions. *J. Androl.* **1**, 87–98.

Pinon-Lataillade, G. (1986). Effets de différents type d'irradiation dur l'activité testiculaire chez le rat. *These d'Etat Univ. P. M. Curie (Paris)* **VI**, 1–108.

Pinon-Lataillade, G., Velez de la Calle, J. F., Viguier-Martinez, M. C., Garnier, D. H., Folliot, R., Maas, J., and Jégou, B. (1988). Influence of germ cells upon Sertoli cells during continuous low-dose rate γ-irradiation of adult rats. *Mol. Cell. Endocrinol.* **58**, 51–63.

Plöen, L., and Setchell, B. P. (1992). Blood-testis barriers revisited. An homage to Lennat Nicander. *Int. J. Androl.* **15**, 1–4.

Pollänen, P. P., Kallajoki, M., Risteli, L., Risteli, J., and Suominen, J. J. (1985). Laminin and type IV collagen in the human testis. *Int. J. Androl.* **8**, 337–347.

Pollänen, P., Parvinen, M., and Söder, O. (1989). Interleukin-1α stimulation of spermatogonial proliferation *in vivo. Reprod. Fertil. Dev.* **1**, 85–87.

Qureshi, S. J. (1993). Comparative physiology of rat and human Leydig cells. Ph.D. Thesis. University of Edinburgh, Scotland.

Redenbach, D. M., and Vogl, A. W. (1991). Microtubule polarity in Sertoli cells: A model for microtubule-based spermatid transport. *Eur. J. Cell Biol.* **54**, 277–290.

Rich, K. A., and de Kretser, D. M. (1979). Effect of fetal irradiation on testicular receptors and testosterone response to gonadotrophin stimulation in adult rats. *Int. J. Androl.* **2**, 343–352.

Rich, K. A., Kerr, J. B., and de Kretser, D. M. (1979). Evidence for Leydig cell dysfunction in rats with seminiferous tubule damage. *Mol. Cell. Endocrinol.* **13**, 123–135.

Risbridger, G. P., and de Kretser, D. M. (1989). Paracrine regulation of the testis. *In* "The Testis" (H. Burger and D. M. de Kretser, eds.), 2d Ed., pp. 255–268. Raven Press, New York.

Risbridger, G. P., Kerr, J. B., and de Kretser, D. M. (1981a). Evaluation of Leydig cell function and gonadotropin binding in unilateral and bilateral cryptorchidism: Evidence for local control of Leydig cell function by the seminiferous tubule. *Biol. Reprod.* **24**, 534–540.

Risbridger, G. P., Kerr, J. B., Peake, R., Rich, K. A., and de Kretser, D. M. (1981b). The temporal changes in Leydig cell function after the induction of bilateral cryptorchidism. *J. Reprod. Fertil.* **63**, 415–423.

Risbridger, G. P., Kerr, J. B., Peake, R. A., and de Kretser, D. M. (1981c). An assessment of Leydig cell function after bilateral or unilateral efferent duct ligation: Further evidence for local control of Leydig cell function. *Endocrinology* **109**, 1234–1241.

Risbridger, G. P., Jenkin, G., and de Kretser, D. M. (1986). The interaction of hCG, hydroxy-steroids and interstitial fluid on rat Leydig cell steroidogenesis *in vitro*. *J. Reprod. Fertil.* **77**, 239–245.

Risbridger, G. P., Drummond, A. E., Kerr, J. B., and de Kretser, D. M. (1987a). Effect of cryptorchidism on steroidogenesis and mitogenic activities in rat testicular interstitial fluid. *J. Reprod. Fertil.* **81**, 617–624.

Risbridger, G. P., Kerr, J. B., and de Kretser, D. M. (1987b). Influence of the cryptorchid testis on the regeneration of rat Leydig cells after administration of ethane dimethane sulphonate. *J. Endocrinol.* **112**, 197–204.

Risley, M. S., and Morse-Gaudio, M. (1992). Comparative aspects of spermatogenic cell metabolism and Sertoli cell function in *Xenopus laevis* and mammals. *J. Exp. Zool.* **261**, 185–193.

Robinson, R., and Fritz, I. B. (1981). Metabolism of glucose by Sertoli cells in culture. *Biol. Reprod.* **24**, 1032–1041.

Rodriguez-Rigau, L. J., Zukerman, Z., Weiss, D. B., Smith, K. D., and Steinberger, E. (1980). Hormonal control of spermatogenesis in man: comparison with the rat. *In* "Testicular Development, Structure, and Function" (A. Steinberger and E. Steinberger, eds.), pp. 139–140. Raven Press, New York.

Roe, F. J. C. (1964). Cadmium neoplasia: Testicular atrophy and Leydig cell hyperplasia and neoplasia in rats and mice following subcutaneous injection of cadmium salts. *Brit. J. Cancer.* **18**, 674.

Roosen-Runge, E. C. (1952). Kinetics of spermatogenesis in mammals. *Ann. N.Y. Acad. Sci.* **55**, 574–584.

Roosen-Runge, E. C. (1962). The process of spermatogenesis in mammals. *Biol. Rev.* **37**, 343–377.

Ross, M. H., and Long, J. R. (1966). Contractile cells in human seminiferous tubules. *Science* **153**, 1271–1273.

Rosselli, M., and Skinner, M. K. (1992). Developmental regulation of Sertoli cell aromatase activity and plasminogen activator production by hormones, retinoids and the testicular paracrine factor, PModS. *Biol. Reprod.* **46**, 586–594.

Rossi, P., Albanesi, C., Grimaldi, P., and Geremia, R. (1991). Expression of the mRNA for the ligand of c-*kit* in mouse Sertoli cells. *Biochem. Biophys. Res. Commun.* **176**, 910–914.

Rossi, P., Mavrail, G., Albanesi, C., Charleswolth, A., Geremie, R., and Sorrentino, V. (1992). A novel c-*kit* transcript potentially encoding a truncated receptor originates within a Kit gene intron in mouse spermatids. *Dev. Biol.* **152(1)**, 203–207.

Russell, L. D. (1977). Movement of spermatocytes from the basal to adluminal compartment of the rat testis. *Am. J. Anat.* **148**, 313–328.

Russell, L. D. (1980). Sertoli-germ cell interrelations: A review. *Gamete Res.* **3**, 179–202.

Russell, L. D., Alger, L. E., and Nequin, L. G. (1987). Hormonal control of pubertal spermatogenesis. *Endocrinology* **120**, 1615–1632.

Russell, L. D., Ettlin, R. A., Sinha Hikim, A. P., and Clegg, E. D. (1990a). In "Histological and Histopathological Evaluation of the Testis," pp. 1–286. Cache River Press, Clearwater, Florida.

Russell, L. D., Ron, H. P., Sinha, H. I., Schulze, W., Sinha Hikim, A. P., *et al.* (1990b). A comparative study in twelve mammalian species of the volume of selected testis components, emphasizing those related to the Sertoli cell. *Amer. J. Anat.* **188**, 21–30.

Sakai, Y., and Yamashina, S. (1990). Spermiation in the mouse: Contribution of the invading Sertoli cell process to adluminal displacement of the spermatid head. *Dev. Growth Diff.* **32**, 389–395.

Sang, Q. X., Dym, M., and Byers, S. W. (1990a). Secreted metalloproteinases in testicular cell culture. *Biol. Reprod.* **43**, 946–955.

Sang, Q. X., Stetler-Stevenson, W. G., Liotta, L. A., and Byers, S. W. (1990b). Identification of Type IV collagenase in rat testicular cell culture: Influence of peritubular-Sertoli cell interactions. *Biol. Reprod.* **43**, 956–964.

Schteingart, H. F., Cigorraga, S., Leon, M., Moya, S., Pellizzari, E., Chemes, H., and Rivarola, M. A. (1988). Hormonal regulation of rat γ-glutamyl-transpeptidase *in vitro*. *Andrologia.* **20**, 351–359.

Schteingart, H. F., Rivarola, M. A., and Cigorraga, S. B. (1989). Hormonal and paracrine regulation of γ-glutamyl transpeptidase in rat Sertoli cells. *Mol. Cell. Endocrinol.* **67**, 73–80.

Segretain, D., and Decrossas, B. (1991). The Sertoli-spermatid processes in the mouse: new-cellular structure. *Biol. Cell.* **71**, 321–324.

Segretain, D., Egloff, M., Gérard, N., Pineau, C., and Jégou, B. (1992). Receptor-mediated and adsorptive endocytosis by male germ cells of different mammalian species. *Cell Tissue Res.* **268**, 471–478.

Sharpe, R. M. (1990). Intratesticular control of steroidogenesis. *Clin. Endocrinol.* **33**, 878–807.

Sharpe, R. M. (1993). Experimental evidence for Sertoli–germ cell and Sertoli–Leydig cell interactions. *In* "The Sertoli Cell" (L. D. Russell and M. D. Griswold, eds.). Cache River Press, Clearwater, Florida., pp. 391–419.

Sharpe, R. M., and Cooper, I. (1984). Intratesticular secretion of a factor(s) with major stimulatory effects on Leydig cell testosterone secretion in vitro. *Mol. Cell. Endocrinol.* **37**, 159–168.

Sharpe, R. M., Doogan, D. G., and Cooper, I. (1986a). Intratesticular factors and testosterone secretion: The role of luteinizing hormone in relation to changes during puberty and experimental cryptorchidism. *Endocrinology* **119**, 2089–2096.

Sharpe, R. M., Kerr, J. B., Cooper, I., and Bartlett, J. M. S. (1986b). Intratesticular factors and testosterone secretion: The effect of treatment with ethane dimethanesulphonate (EDS) and the induction of seminiferous tubule damage. *Int. J. Androl.* **9**, 285–298.

Sharpe, R. M., Kerr, J. B., Fraser, H. M., and Bartlett, J. M. S. (1986c). Intratesticular factors and testosterone secretion: Effect of treatments that alter the levels of testosterone within the testis. *J. Androl.* **7**, 180–189.

Sharpe, R. M., Maddocks, S., and Kerr, J. B. (1990). Cell-cell interactions in the control of spermatogenesis as studied using Leydig cell destruction and testosterone replacement. *Am. J. Anat.* **188**, 3–20.

Sharpe, R. M., Maddocks, S., Millar, M., Saunders, P. T. K., Kerr, J. B., and McKinnell, C. (1992). Testosterone and spermatogenesis: Identification of stage-dependent, androgen-regulated proteins secreted by adult rat seminiferous tubules. *J. Androl.* **13**, 172–184.

Skinner, M. K. (1991). Cell-cell interactions in the testis. *Endocrinol. Rev.* **12**, 45–77.

Skinner, M. K., and Fritz, I. B. (1985). Structural characterization of proteoglycans produced by testicular peritubular cells and Sertoli cells. *J. Biol. Chem.* **260**, 11874–11883.

Skinner, M. K., and Griswold, M. D. (1980). Sertoli cells synthesize and secrete a transferrin-like protein. *J. Biol. Chem.* **255**, 9523–9525.

Skinner, M. K., and Moses, H. L. (1989). Transforming growth factor β gene expression and action in the seminiferous tubule: Peritubular cell-Sertoli cell interactions. *Mol. Endocrinol.* **3**, 625–634.

Skinner, M. K., Tung, P. S., and Fritz, I. B. (1985). Cooperativity between Sertoli cells and testicular peritubular cells in the production and deposition of extracellular matrix components. *J. Cell. Biol.* **10**, 1941–1947.

Skinner, M. K., Norton, J. N., Mullaney, B. P., Rosselli, M., Whaley, P. D., and Anthony, C. T. (1991). Cell–cell interactions and the regulation of testis function. *Ann. N.Y. Acad. Sci.* **637**, 354–363.

Smith, E. P., Hall, H. S., Monaco, L., French, F. S., Wilson, E. M., and Conti, M. (1989). A rat Sertoli cell factor similar to basic fibroblast growth factor increases c-fos messenger ribonucleic acid in cultured Sertoli cells. *Mol. Endocrinol.* **3**, 954–961.

Sorrentino, V., Giorgi, M., Geremia, R., Besmer, P., and Rossi, P. (1991). Expression of the c-*kit* proto-oncogene in the murine male germ cells. *Oncogene* **6**, 149–151.

Stallard, B. J., and Griswold, M. D. (1990). Germ cell regulation of Sertoli cell transferrin mRNA levels. *Mol. Endocrinol.* **4**, 393–401.

Steinberger, A., Dighe, R. R., and Diaz, J. (1994). Testicular peptides and their endocrine and paracrine functions. *Arch. Biol. Med. Exp.* (Santiago) **17**, 267–271.

Syed, V., Khan, S. A., and Ritzén, E. M. (1985). Stage-specific inhibition of interstitial cell testosterone secretion by rat seminiferous tubules in vitro. *Mol. Cell. Endocrinol.* **40**, 257–264.

Syed, V., Karpe, B., Plöen, L., and Ritzén, E. M. (1986). Regulation of interstitial cell function by seminiferous tubules in intact and cryptorchid rats. *Int. J. Androl.* **9**, 271–284.

Syed, V., Khan, S. A., Lindh, M., and Ritzén, E. M. (1988a). Mechanism of action of the factor(s) secreted by rat seminiferous tubules and inhibiting interstitial cell testosterone production in vitro. *Acta Endocrinol.* **119**, 427–434.

Syed, V., Soder, O., Arver, S., Lindh, M., Khan, S., and Ritzén, E. M. (1988b). Ontogency and cellular origin of an interleukin-1-like factor in the reproductive tract of the male rat. *Int. J. Androl.* **11**, 437–447.

Syed, V., Gérard, N., Kaipia, A., Bardin, C. W., Parvinen, M., and Jégou, B. (1993). Identification, ontogeny, and regulation of an Interleukin-6-like (IL-6) factor in the rat testis. *Endocrinology* **132**, 293–299.

Sylvester, S. R., and Griswold, M. D. (1984). Localization of transferrin and transferrin receptors in rat testis. *Biol. Reprod.* **31**, 195–203.

Sylvester, S. R., Skinner, M. K., and Griswold, M. D. (1984). A sulfated glycoprotein synthesized by Sertoli cells and by epididymal cells is a component of the sperm membrane. *Biol. Reprod.* **31**, 1087–1101.

Sylvester, S. R., Morales, C., Oko, R., and Griswold, M. D. (1989). Sulfated glycoprotein-1 (saposin precursor) in the reproductive tract of the male rat. *Biol. Reprod.* **41**, 941–948.

Tabone, E., Benahmed, M., Reventos, J., and Saez, J. M. (1984). Interactions between immature porcine Leydig and Sertoli cells in vitro. An ultrastructural and biochemical study. *Cell Tissue Res.* **237**, 357–362.

Tajima, Y., Onoue, H., Kitamura, Y., and Nishimune, Y. (1991). Biologically active kit ligand growth factor is produced by mouse Sertoli cells and is defective in Sl^d mutant mice. *Development* **113**, 1031–1035.

Takeichi, M. (1990). Cadherins: A molecular family important in selective cell–cell adhesion. *Ann. Rev. Biochem.* **59**, 237–252.

Teerds, K. J., and Dorrington, J. H. (1993). Localization of transforming growth factor β$_1$ and β$_2$ during testicular development in the rat. *Biol. Reprod.* **48**, 40–45.

Teerds, K. J., De Rooij, D. G., Rommerts, F. F. G., and Wensing, C. J. G. (1988). The regulation of the proliferation and differentiation of rat Leydig cell precursor cells after EDS administration or daily hCG treatment. *J. Androl.* **9**, 343–351.

Teerds, K. J., Closset, J., Rommerts, F. F. G., de Rooij, D. G., Stocco, D. M., Colenbrander, B., Wensing, C. J. G., and Hennen, G. (1989a). Effects of pure FSH and LH preparations on the number and function of Leydig cells in immature hypophysectomized rats. *J. Endocrinol.* **120**, 97–106.

Teerds, K. J., de Rooij, D. G., Rommerts, F. F. G., van den Hurk, R., and Wensing, C. J. G. (1989b). Proliferation and differentiation of possible Leydig cell precursors after destruction of the existing Leydig cells with ethane dimethane sulphonate: The role of LH/human chorionic gonadotrophin. *J. Endocrinol.* **122**, 689–696.

Thiéry, J. P., and Boyer, B. (1992). Les molécules adhésives et la communication cellulaire. *Pour La Science* **179**, 36–43.

Toppari, J., Kangasniemi, M., Kalpla, A., Mali, P., Huhtaniemi, I., and Parvinen, M. (1991). Stage- and cell-specific gene expression and hormone regulation of the seminiferous epithelium. *J. Elec. Microsc. Tech.* **19**, 203–213.

Tung, P. S., and Fritz, I. B. (1980). Interactions of Sertoli cells with myoid cells *in vitro*. *Biol. Reprod.* **23**, 207–217.

Tung, P. S., and Fritz, I. B. (1985). Immunolocalization of clusterin in the ram testis and excurrent ducts. *Biol. Reprod.* **33**, 177–186.

Tung, P. S., and Fritz, I. B. (1986a). Extracellular matrix components and testicular peritubular cells influence the rate and pattern of Sertoli cell migration *in vitro*. *Dev. Biol.* **113**, 119–134.

Tung, P. S., and Fritz, I. B. (1986b). Cell–substratum and cell–cell interactions promote testicular peritubular myoid cell histotypic expression *in vitro*. *Dev. Biol.* **115**, 155–170.

Tung, P. S., Skinner, M. K., and Fritz, J. B. (1984). Fibronectin synthesis is a marker for peritubular cell contaminants in Sertoli-enriched cultures. *Biol. Reprod.* **30**, 199–211.

Ultee-van Gessel, A. M., Leemborg, F. G., de Jong, F. H., and van der Molen, H. J. (1986). In-vitro secretion of inhibin-like activity by Sertoli cells from normal and prenatally irradiated immature rats. *J. Endocrinol.* **109**, 411–418.

Ulvik, N. M. (1983). Selective uptake by the Sertoli cells of cytoplasm from normal spermatogonia in the rat testis. *Int. J. Androl.* **6**, 367–374.

van Dissel-Emiliani, F. M. F., Grootenhuis, A. J., de Jong, F. H., and de Rooij, D. G. (1989). Inhibin reduces spermatogonial numbers in testes of adult mice and chinese hamsters. *Endocrinology* **125**, 1899–1903.

van Haaster, L. H., de Jong, F. H., Docter, R., and de Rooij, D. G. (1992). The effect of hypothyroidism on Sertoli cell proliferation and differentiation and hormone levels during testicular development in the rat. *Endocrinology* **131**, 1574–1576.

Vannelli, B. G., Orlando, C., Barni, T., Natali, A., Serio, M., and Balboni, G. C. (1986). Insulin-like growth factor-1 (IGF-1) and IGF-1 receptor in human testis: An immunohistochemical study. *Fertil. Steril.* **45**, 536–541.

Vannelli, B. G., Barni, T., Orlando, C., Natali, A., Serio, M., and Balboni, G. C. (1988). Insulin-like growth factor-1 (IGF-1) and IGF-1 receptor in human testis: An immunohistochemical study. *Fertil. Steril.* **49**, 666–669.

Van Snick, J. (1990). Interleukin-6: An overview. *Ann. Rev. Immunol.* **8**, 253–278.

Vaughan, J. M., Rivier, J., Corrigan, A. Z., McClintock, R., Campen, C. A., Jolley, D., Voglmayr, J. K., Bardin, C. W., Rivier, C., and Vale, W. (1989). Inhibin. *Meth. Enzymol.* **168**, 588–617.

Velez de la Calle, J. F., Soufir, J. C. L., Chodorge, G., Boisseau, C., Kercret, H., and Jégou, B. (1988). Reproductive effects of the anti-cancer drug procarbazine in male rats at different ages. *J. Reprod. Fertil.* **84**, 51–61.

Verhoeven, G. (1992). Local control systems within the testis. *Baillière's Clin. Endocrinol. Metab.* **6**, 313–333.

Verhoeven, G., and Cailleau, J. (1985). A factor in spent media from Sertoli cell-enriched cultures that stimulates steroidogenesis in Leydig cells. *Mol. Cell. Endocrinol.* **40,** 57–68.

Verhoeven, G., and Cailleau, J. (1986). Specificity and partial purification of a factor in spent media from Sertoli cell-enriched cultures that stimulates steroidogenesis in Leydig cells. *J. Steroid Biochem.* **25,** 393–402.

Verhoeven, G., and Cailleau, J. (1987). A Leydig cell stimulatory factor produced by human testicular tubules. *Mol. Cell. Endocrinol.* **49,** 137–147.

Verhoeven, G., and Cailleau, J. (1989). Tubule–Leydig cell interactions. *In* "Perspectives in Andrology" (M. Serio, eds.), pp. 227–234. Raven Press, New York.

Verhoeven, G., and Cailleau, J. (1990). Influence of coculture with Sertoli cells on steroidogenesis in immature rat Leydig cells. *Mol. Cell. Endocrinol.* **71,** 239–251.

Vihko, K. K., and Huhtaniemi, I. (1989). A rat seminiferous epithelial factor that inhibits Leydig cell cAMP and testosterone production: Mechanism of action, stage-specific secretion and partial characterization. *Mol. Cell. Endocrinol.* **65,** 119–127.

Vihko, K. K., Suominen, J. J. O., and Parvinen, M. (1984). Cellular regulation of plasminogen activator secretion during spermatogenesis. *Biol. Reprod.* **31,** 383–389.

Vihko, K. K., Toppari, J., Saksala, O., Suominen, J. J. O., and Parvinen, M. (1986). Testicular plasminogen activators during postnatal development in the rat. *Acta. Endocrinol. (Copenh.)* **112,** 431–435.

Vihko, K. K., Kristensen, P., Dano, K., and Parvinen, M. (1988). Immunohistochemical localization of urokinase-type plasminogen activator in Sertoli cells and tissue-type plasminogen activator in spermatogenic cells in the rat seminiferous epithelium. *Dev. Biol.* **126,** 150–155.

Watrin, T., Scotto, L., Assoian, R. K., and Wolgemuth, D. J. (1991). Cell lineage specificity of expression of the murine transforming growth factor $\beta3$ and transforming growth factor $\beta1$ genes. *Cell Growth Diff.* **2,** 77–83.

Welsh, M. J., and Ireland, M. E. (1992). The second messenger pathway for germ cell-mediated stimulation of Sertoli cells. *Biochem. Biophys. Res. Commun.* **184,** 217–227.

Wiebe, J. P., Buckingham, K. D., Wood, P. H., and Campbell, S. M. C. (1988). Relative steroidogenic activity of Sertoli and Leydig cells and role of the Sertoli cell steroid 3α-hydroxy-4-pregnen-20-one in spermatogenesis and FSH secretion. *In* "Development and Function of the Reproductive Organs" (M. Parvinen, I. Huhtaniemi, and L. J. Pelliniemi, eds.), pp. 39–53. Ares Serono Symposia, Rome.

Woodruff, T. K., Borre, J., Attie, K. M., Cox, E. T., Rice, G. C., and Mather, J. P. (1992). Stage-specific binding of inhibin and activin to subpopulations of rat germ cells. *Endocrinology* **130,** 871–881.

Wright, W. W. (1988). Germ cell-Sertoli cell interactions: Analysis of the biosynthesis and secretion of cyclic protein-2. *Dev. Biol.* **130,** 45–56.

Yoshinaga, K., Nishikawa, S., Ogawa, M., and Hayashi, S-I. (1991). Role of c-*kit* in mouse spermatogenesis: Identification of spermatogonia as a specific site of c-*kit* expression and function. *Development* **113,** 689–699.

Zong, S. D., Bardin, C. W., Phillips, D., and Cheng, C. Y. (1992). Testins are localized to the junctional complexes of rat Sertoli and epididymal cells. *Biol. Reprod.* **47,** 568–572.

Zsebo, K. M., Williams, D. A., Geissler, E. N., Broudy, V. C., and Martin, F. H. (1990). Stem cell factor is encoded at the *Sl* locus of the mouse and is the ligand for the c-*kit* tyrosine kinase receptor. *Cell* **63,** 213–224.

9
Molecular Biology of Iron Transport in the Testis

STEVEN R. SYLVESTER & MICHAEL D. GRISWOLD

I. Introduction

Iron is a very important molecule in oxidation–reduction reactions that occur in the biosphere: iron is required by all life forms. In eukaryotes, iron is incorporated into heme and nonheme iron proteins that largely occur in the mitochondria. In aqueous media at physiological pH, free iron spontaneously forms precipitates of ferric (Fe^{3+}) hydroxide, which are not biologically useful. Even small amounts of the soluble ferrous ion (Fe^{2+}) are highly toxic because they readily catalyze the peroxidation of membrane lipids (Joshi and Zimmerman, 1988). Neither ferrous nor ferric ions will diffuse passively across lipid bilayers and therefore must be transported actively into cells with the aid of proteins. Thus, systems have evolved for the transport, cellular uptake, and storage of iron in a useful nontoxic state. These functions are the roles of the soluble iron transport protein transferrin, the cell membrane transferrin receptor, and the cytoplasmic storage protein ferritin, respectively (Aisen and Listowsky, 1980).

In the mammalian testis, not only must iron be transported across bilayers into cells but it also must negotiate the blood–testis barrier. A vectorial movement of iron from lymph to germinal cells occurs, where apparently it is stored and incorporated into proteins. During the preleptotene period, spermatocytes are moved inside the blood–testis barrier where they eventually undergo two cell divisions to give rise to four spermatids. The spermatids undergo differentiation, which includes the reorganization of mitochondria so they eventually wrap around the principal piece of the sperm tail. Ejaculation of the sperm results in a net loss of iron from the male.

II. Model of Testicular Iron Transport

A model for iron transport in the rat testis that involved Sertoli cell-secreted transferrin was proposed some years ago (Huggenvik *et al.*, 1984). The original model, based on iron movement in other tissues (Octave *et al.*, 1981), is revised here to reflect recent findings in the testis (Fig. 1). Diferric sero-

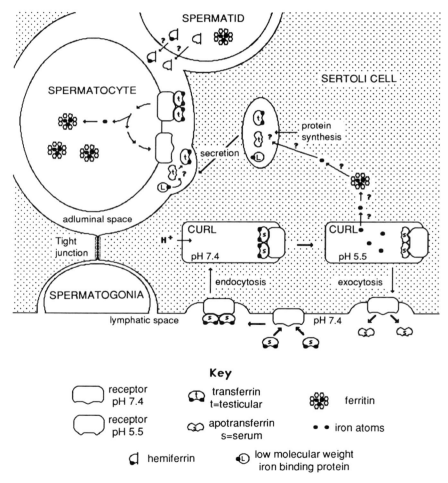

FIGURE 1 Current concepts of iron transport in the testis, as discussed in the text. The question marks represent areas that are currently uncertain.

transferrin, of hepatic origin, has a strong affinity for Sertoli cell basal membrane transferrin receptors at the physiological pH of the lymph of the basal compartment. On binding of two molecules of diferric serotransferrin per receptor, receptor aggregation occurs and the ligand–receptor complex is internalized. In all other cell types studied, occupied transferrin receptors are segregated to a special subcellular compartment called the compartment of uncoupling and recycling of ligand (CURL) (Ciechanover et al., 1983). Acidi-fication of this compartment causes the iron to be released from transferrin but, at the lower pH of 5–6, the receptor affinity is higher for apotransferrin. Finally, the apotransferrin–receptor complex is returned to the cell surface where exposure to the higher pH of lymph causes the receptor affinity to change again, resulting in the release of apotransferrin. Receptor-mediated endocytosis and recycling of serotransferrin by Sertoli cells in vivo and in vitro has been demonstrated (Morales and Clermont, 1986; Wauben-Penris et al., 1986), as has the uptake of iron by these cells (Toebosch et al., 1987a).

Inside the Sertoli cell, or in any other cell type, little is known about the path of iron through compartments of the cytoplasm. However, iron is stored at least temporarily in ferritin present in Sertoli cells (Toebosch et al., 1987b). Testicular transferrin is synthesized by Sertoli cells (Skinner and Griswold, 1980) and secreted with iron into the adluminal space. This step in the model is supported by experiments that show that iron from basally applied human transferrin is transported through rat Sertoli cells and onto rat transferrin, which is secreted apically in bicameral cell cultures (Djakiew et al., 1986). A low molecular mass (less than 30 kDa) iron transport entity has been de-tected in Sertoli cell cultures and is hypothesized to deliver iron to apotrans-ferrin in the adluminal space (Wauben-Penris et al., 1988). The adluminal testicular transferrin is then available to receptors on spermatocytes. This portion of the model is supported by (1) the immunolocalization of trans-ferrin receptors to spermatocytes (Holmes et al., 1983; Sylvester and Gris-wold, 1984), (2) the demonstration of vectorial transport of iron from injected human transferrin through Sertoli cells and into germinal cells in vivo (Mo-rales et al., 1987b), and (3) the demonstration of iron uptake by spermato-cytes (Toebosch et al., 1987a). Most of the iron that enters the germinal cells is incorporated directly into ferritin, the iron storage protein (Toebosch et al., 1987a,b). We have demonstrated by immunohistochemical studies that ferri-tin is abundant in pachytene spermatocytes (S. R. Sylvester and M. D. Griswold, unpublished results). From ferritin, the iron is available for the synthesis of heme and nonheme iron proteins in spermatocytes and sperma-tids. The role that hemiferrin (discussed subsequently) might play in the process of iron transport is totally unknown. The origin of the low molecular weight iron carrier in Sertoli cell cultures (Wauben-Penris et al., 1988) is also unknown. However, this molecule may be hemiferrin, which could have arisen from the few germ cells that routinely contaminate Sertoli cell cul-tures.

III. Transferrin

A. Transferrin Protein Structure

Transferrins are iron binding proteins that belong to a family of glyco-proteins including serotransferrin (often referred to simply as transferrin) in blood, ovotransferrin (conalbumin) in bird egg whites, lactoferrin in mammalian biological fluids, and melanotransferrin in membranes of melanocarcinoma cells (Baker *et al.*, 1987). These proteins share substantial sequence homology, highly conserved cysteine residues, and a similar structure. Further, the C-terminal halves of the proteins are homologous to the N-terminal halves. The X-ray structure of rabbit serotransferrin reveals that each half of the transferrin sequence forms a lobe of approximately 330 amino acids; the two lobes are connected by a short peptide (Bailey *et al.*, 1988). In turn, each lobe is composed of two domains that contribute side chains to ferric ion and accessory anion binding pockets. Thus, each transferrin molecule is capable of binding two iron atoms and two accessory carbonate anions.

B. Transferrin Gene Structure

Apparently, a single gene exists for transferrin (Idzerda *et al.*, 1986) that is approximately 33.5 kb in size (Schaeffer *et al.*, 1987). Hen ovotransferrin and human serotransferrin have 17 exons (Cochet *et al.*, 1979; Schaeffer *et al.*, 1987); the first exon encodes most of the signal peptide. Of the remaining exons, 14 form homologous pairs in the two lobes of the protein structure whereas two encode sequences present only in the C lobe. The transferrins are believed to have originated from a gene duplication event and each half of the molecule evolved further (Williams, 1982).

The promoter region of the human transferrin gene has been mapped by transient expression analysis in hepatic cells (Schaeffer *et al.*, 1989). Liver-specific expression is conferred through a cell type-specific promoter located between -45 and -125 bp. Between -125 and -620 bp are positive and negative *cis*-acting elements, and between -0.6 and -1.0 kb is a negative acting region. A 300-bp enhancer located 3.6 kb upstream of the cap site is active in hepatocytes (Boissiere *et al.*, 1991). Similar analyses in cultured rat Sertoli cells revealed that the enhancer at 3.6 kb is inactive, apparently because of the binding of different nuclear factors (Guillou *et al.*, 1991). Further, in Sertoli cells, a TATA box binding factor is active in the -18 to -34 region; in hepatocytes, an additional element that may be weakly active appears at -425 to -440 bp.

Mice harboring transgene constructs of the human transferrin 5′ flanking region and a reporter gene revealed that -152-bp constructs were expressed poorly in all tissues whereas constructs with sequences to -622 or -1152 were expressed highly in brain and liver (Adrian *et al.*, 1990). Surprisingly, testicular expression remained low for the two longer constructs. In the normal rat, the mRNA content of the liver is reported to be 6500 molecules per cell; in the brain it is 83 molecules per cell, and in the testis it is 114 molecules per cell (Idzerda et al., 1986). Further, homozygous atransferrinemic mice of the Hp strain, which must be kept alive with injections of mouse serum or purified transferrin, are capable of spermatogenesis but have a low fertility rate (Berstein, 1987). All these findings indicate that transferrin synthesis in the testis is unique and complex.

The regions of the transferrin protein lobes that bind to transferrin receptors are unknown. Since exon 2 of transferrin shares similarities with some growth factors, it has been suggested to contribute this property to the mature protein (Bailey et al., 1988). Further, since transferrin exhibits growth promoting properties by interacting with its receptor independent of iron nutrient requirements, and since exon 2 does not encode amino acids that participate directly in iron binding but are exposed to the solvent (as determined by crystal structure), this region may be a good candidate for the receptor binding site. Exon 2 in transferrin encodes the N-terminal peptide of the N-terminal lobe.

C. Regulation of Transferrin Expression

The regulation of transferrin synthesis and secretion by Sertoli cells has been studied by a number of laboratories in relationship to hormones, vitamins, nutrients, growth factors, neighboring cell types, and other factors. Transferrin protein secretion and mRNA levels in rat Sertoli cell cultures were maximal when cultures were treated with a combination of insulin, testosterone, follicle stimulating hormone (FSH), and retinol (Skinner and Griswold, 1982; Huggenvik et al., 1987). Epidermal growth factor (EGF) and plating density were shown to stimulate transferrin protein secretion in another study of rat Sertoli cells in culture, but no effect of FSH was observed (Perez-Infante et al., 1986). Hypophysectomy caused a decrease in transferrin mRNA levels in the testis and was reversed partially by the chronic administration of FSH and testosterone (Hugly et al., 1988). Using a reverse hemolytic plaque assay, basic fibroblast growth factor (bFGF), interleukin 2 (IL-2), interleukin 6 (IL-6), and tumor necrosis factor (TNF) were shown to increase the number of cells secreting transferrin in culture (Boockfor and Schwarz, 1990; Boockfor and Schwarz, 1991). These findings must be tempered, however, with the knowledge that the major secretory product of Sertoli cells, SGP-2, is an inhibitor of complement-mediated cell lysis, which is the basis of the hemolytic plaque assay (Sylvester et al., 1991).

Placing rat Sertoli cells in culture has been reported to induce transferrin mRNA levels dramatically (Lee et al., 1986). Although a relatively low concentration of transferrin mRNA (114 copies per testicular cell) has been reported (Idzerda et al., 1986), note that Sertoli cells are the only testicular cell type with detectable levels of the message (Morales *et al.*, 1987) and make up only 2–5% of the total cell population in adult rat testes. This observation suggests that rat Sertoli cells contain 5000 or more copies of transferrin mRNA, closer to the 6500 copies per cell reported for liver (Idzerda *et al.*, 1986). The increase of transferrin mRNA simply may reflect the enrichment of Sertoli cells on placing them in culture, followed by further enrichment because of the death of the germ cells that contaminate Sertoli cells during the first few days of culture (Karl and Griswold, 1990).

Peritubular cells in coculture with human Sertoli cells caused a 2.4-fold increase in transferrin protein accumulation in the medium, whereas spent medium from peritubular cells was shown to have no effect (Holmes *et al.*, 1984). In another study, transferrin protein production in rat Sertoli cell cultures was stimulated by a factor, P-Mod-S, secreted by rat peritubular cells (Skinner and Fritz, 1985). Finally, germ cell cocultures and protein-aceous factors in germ cell conditioned medium were shown to increase transferrin protein production and transferrin mRNA levels in rat Sertoli cell cultures (Djakiew and Dym, 1988; Le Magueresse *et al.*, 1988; Stallard and Griswold, 1990). If spermatogenesis is allowed to regress in hypophysecto-mized rats and then be reinitiated by testosterone implants, testicular trans-ferrin message increases with germ cell numbers rather than in response to testosterone (Roberts *et al.*, 1991). All these findings suggest a relationship between germ cells and Sertoli cell transferrin synthesis.

Previously, hepatic transferrin gene expression was shown to be related to the concentration of iron. Several findings support the conclusion that the regulation of transferrin observed in the liver is quite different from that observed in the testis. Whereas iron deficiency *in vivo* increased liver trans-ferrin mRNA levels in rats, testicular levels remained unaltered (Idzerda *et al.*, 1986). In the testis, vitamin A deficiency reduced transferrin mRNA steady state levels and could be reversed by retinol administration, whereas no effect was observed on liver transferrin message (Hugly and Griswold, 1987). As noted earlier, differences occur in the regulatory elements control-ling the expression of transferrin in the two tissues. In Sertoli cells, a unique set of transcription factors is believed to bind to elements upstream of the message under the influence of factors other than iron, and thereby confer tissue-specific expression (Guillou *et al.*, 1991).

IV. Transferrin Receptor

A. Transferrin Receptor Protein Structure

Transferrin receptor is a transmembrane glycoprotein composed of two identical disulfide-linked subunits of 90,000 kDa each (Schneider *et al.*, 1982). The structure of the receptor protein has been deduced from biochemical studies and from assumptions based on the amino acid sequence predicted from the cDNA (McClelland *et al.*, 1984; Schneider *et al.*, 1984). Each subunit has a large extracellular domain, a single membrane spanning region, and a small N-terminal cytoplasmic domain. The protein has phosphorylated serine residues and is associated covalently with palmitic acid (Omary and Trowbridge, 1981). Since each subunit has a transferrin binding site, the receptor is capable of binding two transferrin molecules.

B. Transferrin Receptor Gene Structure

The gene for human transferrin receptor is over 31 kb in length, with 19 exons (McClelland *et al.*, 1984). The coding sequence does not exhibit a typical leader sequence found in most transmembrane proteins. Probes specific for transferrin receptor hybridize with a single large transcript of 4.9 kb in RNA isolated from human and rat tissues (Kühn *et al.*, 1984; Roberts and Griswold, 1990). Transferrin receptor mRNA has been found in Sertoli cells and spermatocytes in the rat testis by Northern analysis (Roberts and Griswold, 1990). The promoter of the human transferrin receptor gene is characterized by the presence of a TATA box, GC-rich regions, and cAMP- and phorbol ester-like response elements (CRE and AP-1 sites) (Casey *et al.*, 1988; Beard *et al.*, 1991). The mRNA contains a long 3' untranslated region with five potential hairpin structures known to be iron regulatory elements (IRE) (Casey *et al.*, 1988). All these elements lie within exon 19.

C. Regulation of Transferrin Receptor Expression

Transferrin receptor expression is regulated transcriptionally and translationally. In Hela cells, two nuclear protein factors, TREF1 and TREF2, bind to the response elements in the 5' region and thereby regulate transcription (Roberts *et al.*, 1989). If the IREs in the 3' region of the mRNA are bound by the cytosolic IRE binding protein in the presence of low iron levels, the half-life of the mRNA is prolonged (Casey *et al.*, 1989; Koeller *et al.*, 1989); allowing the cell to synthesize more receptor, thereby capturing more transferrin and ultimately satisfying iron requirements. When cellular iron levels are high, the affinity of the cytosolic IRE binding proteins is reduced and the receptor mRNA is short lived. In rat Sertoli cells in culture, transferrin receptor is regulated similarly by iron levels but not by FSH, retinol, insulin, or testosterone (Roberts and Griswold, 1990). Chronic treatment of hypophysectomized rats with testosterone or FSH resulted in an increase in testicular transferrin receptor mRNA levels compared with hypophysectomized animals. However, the increase paralleled testis weight, which is related to the presence of germ cells. Therefore the higher level of mRNA may have been contributed by the germ cells.

V. Ferritin

A. Ferritin Protein Structure

Ferritin exists as a large spherical structure of 480,000 MW and is composed of 24 protein subunits of two types. Up to 4500 iron atoms can be sequestered inside the core of one ferritin molecule and stored as ferric oxide. The two subunits, ferritin H and ferritin L, are approximately 21 and 19 kDa in the rat and human (Arosio *et al.*, 1978) and occur in different ratios, depending on tissue type and nutritional status (Bomford *et al.*, 1981; Cairo *et al.*, 1991). Ferritin mRNA has been demonstrated in rat testis cDNA libraries (Krawetz *et al.*, 1986) and in testicular mRNA by Northern analysis (Roberts and Griswold, 1990). Ferritin has been immunoprecipitated from isolated rat Sertoli cells and spermatids (Toebosch *et al.*, 1987b). Immunocytochemical studies localize ferritin to testicular macrophages, Sertoli cells, and most germinal cells; high concentrations are apparent in pachytene spermatocytes (S. R. Sylvester and M. D. Griswold, unpublished results).

B. Ferritin Gene Structure

The sequences of the H and L ferritin genes identified to date share approximately 50% similarity in the coding regions. Similarly, the coding sequences are interrupted by three introns in each of the genes (Eisenstein and Munro, 1990). This result has led to the suggestion that the two genes evolved from a common ancestral gene. Although multiple genes or pseudogenes exist, only ferritin subunits of the two types H and L have been found (Theil, 1990a). Probes for the subunits hybridize with transcripts of 0.9 kb in all tissues examined to date.

C. Regulation of Ferritin Expression

Ferritin expression has provided the classic example of a translationally regulated gene (for a review, see Theil, 1990b). The mRNA is stored in the cytoplasm of cells and then rapidly recruited to polysomes in the presence of iron loads. This behavior allows the cell to respond quickly to excess iron and to maintain the metal ion in a nontoxic state. In the 5' untranslated region of the ferritin mRNA is a 28-nucleotide hairpin-forming sequence that is highly conserved between the H and L ferritins. This sequence is the same IRE described earlier that appears five times in the transferrin receptor 3' untranslated region. If the cytosolic binding protein binds to the IRE, as it does at low iron levels, translation of ferritin is blocked and the message is stored in an inactive state. Thus, coordinate regulation of ferritin and transferrin receptor translation occurs through the same IRE binding protein. The transcriptional regulation of ferritin expression has not been studied extensively.

Thyrotropin has been shown to influence the expression of ferritin H subunit in rat thyroid cells (Cox *et al.*, 1988). The increase in H-subunit expression under thyrotropin treatment is believed to be due to a cAMP-mediated increase in transcription (Chazenbalk *et al.*, 1990). Apparent differences of expression exist between the sexes, but the basis of this difference has not been determined (Munro and Linder, 1978).

VI. Hemiferrin

A. Discovery of Hemiferrin

In previous studies that used Northern blots of mRNA from whole testes, we noticed, in addition to the normal 2.4-kb transferrin mRNA, a band at 0.9 kb that strongly hybridized to the transferrin cDNA. This cross-hybridizing band was not present in RNA from Sertoli cells in culture and thus was believed to be contributed by other cell types in the testis. The expression of the transcript was found to be related to the appearance of haploid spermatids in developing testes, and the mRNA was found in isolated round spermatids (Stallard *et al.*, 1991). The 0.9-kb transcript was cloned and sequenced, and found to share substantial sequence similarity with the 3' coding region of rat transferrin. The cDNA sequence exhibited a long open reading frame of 216 amino acids following a potential methionine start site. The predicted amino acid sequence of this open reading frame showed 64% identity and 75% similarity to the C-terminal 216 amino acids of human transferrin. The name hemiferrin was chosen for the mRNA and for the protein that could be made from that transcript. The cysteine residues that are highly conserved in the transferrins among species also are conserved in hemiferrin, suggesting that the cysteines are important in the structure of the protein and implying that the hemiferrin protein may fold up similarly to the C-terminal lobe of transferrin.

B. Evidence that Hemiferrin cDNA Is Translated

Sucrose gradient sedimentation was conducted to separate polysomal mRNA (actively translated) from free mRNA (nontranslated or stored). A large portion of testicular hemiferrin mRNA remained in the upper region of the gradient, indicating that it was not associated with polysomes. However, in the analysis of spermatid mRNA, a significant amount of the hemiferrin mRNA was present in the polysome fraction and was released by ethylenediamine tetraacetic acid (EDTA), a treatment that dissociates ribosomes from mRNA (Stallard *et al.*, 1991). These findings support the conclusion that a portion of the hemiferrin transcript is undergoing translation. A similar observation has been made for at least two other germ cell messages. In one case, the translated protein has been demonstrated (Kleene *et al.*, 1984; Garrett *et al.*, 1989). Some form of translational regulation is likely to exist in spermatids that allows the message to accumulate for a period of time prior to translation at a specific time during differentiation. When round spermatids are isolated from normal testis, all steps of differentiation are represented. Determining those steps in which translation is occurring is difficult.

C. Potential for Hemiferrin Iron Binding

The iron binding sites of transferrin have been studied extensively but, because of incomplete amino acid and nucleic acid sequence availability, the residues involved have been assigned through an accumulation of evidence garnered from a number of species. The amino acid moieties are known for rabbit transferrin from the crystal structure. However, their position in the polypeptide backbone is based on the sequence of human lactoferrin (Bailey *et al.*, 1988). The four residues in the C-terminal lobe that are believed to be involved directly are Asp-412, Tyr-446, Tyr-537, and His-604. Very near the site and potentially assisting in binding are Arg-475 and Lys-553. [The numbering here has been changed from that of Bailey *et al.* (1988) to match the numbering for human transferrin present in databases.] The sequence of hemiferrin lacks three of these six residues. The structure may adjust to position new residues in the pocket or two protein molecules may bind one iron. Although these alternatives must be considered speculative at this time, some experimental support exists for these hypotheses. Two truncated forms of transferrin (less than 18 kDa) are known that are capable of binding iron. Evans and Madden (1984) used partial proteolysis with subtilisin to obtain nonglycosylated fragments of 15 kDa and 17 kDa from the N-terminal region, both of which were capable of binding iron with a stoichiometry of 0.57 g-atoms iron per mole of protein. The sequences of these fragments were not reported. Legrand and co-workers (Legrand *et al.*, 1984) used mild trypsin treatment of human lactoferrin to generate an 18.5-kDa glycopeptide from the N-terminal lobe that retained iron binding capacity with a stoichiometry of 0.42 g-atom iron per mole of protein. Amino acid sequencing of this fragment indicated that the N terminus of the glycopeptide would correspond (by homology arguments) to position 443 in the C-terminal lobe of the human serotransferrin sequence. Although this glycopeptide is smaller than hemiferrin (24,091 calculated MW without glycosylation), it would include the Tyr-446 ligand site and the Arg-475 neighboring residue (again by homology arguments). Thus, hemiferrin might retain iron binding activity or may bind other metal ligands, as do the transferrins.

D. Other Structural Considerations of Hemiferrin

All the transferrins are glycosylated. Serotransferrin and lactoferrin are secreted proteins and melanotransferrin is a membrane protein. The predicted sequence of hemiferrin does not include a typical signal sequence after the putative start of translation, suggesting that hemiferrin might not be secreted. However, alternative explanations are available. Another start site just upstream of that suggested may have been missed because of the cloning process. However, the cloned cDNA of 828 bases is reasonably close to the 0.9-kb mRNA size revealed by electrophoresis. Also, several proteins now are known to be secreted but lack a typical leader sequence. The predicted hemiferrin sequence does contain a potential glycosylation site at Asn-26. The corresponding site is absent in the human serotransferrin sequence. However, rabbit serotransferrin is glycosylated at this site (Bailey *et al.*, 1988). The role that hemiferrin might play in iron transport in the testis is unknown at this time.

VII. Stage-Related Expression of Iron Transport Components

The seminiferous epithelium of rats is highly organized by the regular appearance of specific groups of cell types. These associations have been assigned to 14 distinct stages based on the appearance of 19 steps of spermatid development in association with other cell types (LeBlond and Clermont, 1952). When whole testis is analyzed, all the stages are represented, so biochemical events in specific stages are difficult to discern. We have developed a method of vitamin A depletion and repletion that results in the synchronization of entire testes to a few sequentially related stages of the cycle of the seminiferous epithelium (Morales and Griswold, 1987). By Northern analysis of mRNA isolated from testes synchronized to specific stages of the cycle, one can relate expression to morphological and biochemical events that are known to occur at specific times. Figure 2 shows the levels of mRNA for transferrin, transferrin receptor, ferritin, and hemiferrin that are present in the testis during the cycle of the seminiferous epithelium. Note that transferrin mRNA is detected only in Sertoli cells (Morales *et al.*, 1987a), hemiferrin mRNA is detected only in spermatids (Stallard *et al.*, 1991), transferrin receptor mRNA is highest in Sertoli cells (Roberts and Griswold, 1990) although the protein immunolocalizes predominantly to spermatocytes (Sylvester and Griswold, 1984), and ferritin is likely to be present in all cells but is concentrated in spermatocytes, as revealed by immunocytochemistry (S. R. Sylvester and M. D. Griswold, unpublished results).

Transferrin message is highest in Stages XIII–I during which the two divisions of meiosis occur. Since dividing cells must have occupied transferrin receptors, Sertoli cell transferrin has been suggested to regulate meiotic cell division (Sylvester and Griswold, 1984). Changes in hemiferrin mRNA levels generally parallel those of transferrin. The meaning of this change is unknown because so little is known of hemiferrin at this time. Transferrin receptor is highest when spermatocytes first appear (preleptotene) and re-

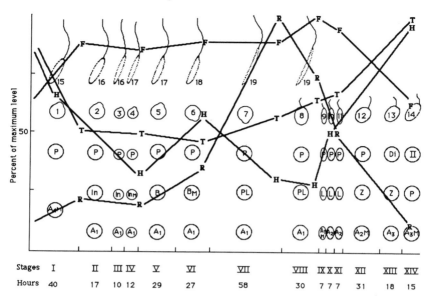

FIGURE 2 Expression of mRNA encoding iron transport-related molecules during the cycle of the seminiferous epithelium. The cycle of the seminiferous epithelium has been drawn to reflect the period (in hours) of each stage (in roman numerals). The cell types diagrammed are spermatogonia types A_1, A_2, A_3, A_4, intermediate (In), and B (B_M indicates mitosis); primary spermatocytes in preleptotene (PL), leptotene (L), zygotene (Z), pachytene (P), and diplotene (Di) stage; secondary spermatocytes (II); and spermatids (numbered 1 through 19). The mRNAs encode ferritin (F), hemiferrin (H), transferrin receptor (R), and transferrin (T).

structuring of the Sertoli cell tight junctions moves the spermatocytes inside the blood–testis barrier. Ferritin levels vary little during the cycle of the seminiferous epithelium, but the peak levels follow closely behind transferrin receptor, as might be predicted given the coordinated translational regulation of the two messages. Minor changes in ferritin message concomitant with greater changes in transferrin receptor mRNA are observed in other tissue types when iron levels are varied (Theil, 1990). Sertoli cell iron levels have been postulated to be low during Stages VII–IX. The increase in transferrin receptor has been proposed to occur largely in Sertoli cells at this time, to load those cells with iron (Griswold, 1991). The presence of translated receptor protein should follow shortly and then dissipate as the higher iron levels cause the release of the cytoplasmic IRE binding protein and destabilize the mRNA. Sertoli cells are then ready to deliver iron with an abundance of transferrin synthesized in Stages XIII and XIV. The transport of iron would cause the lowering of iron levels in the Sertoli cells during Stages I–VI, causing the IRE binding protein to bind transferrin receptor IRE, stabilizing the mRNA and thus completing the cycle.

VIII. Summary

Since the discovery of testicular transferrin, much effort has contributed to the current understanding of iron transport in the testis. However, as can be seen from the information in this chapter, much work remains to be done. In particular, the cytoplasmic movement of iron is not understood, the endocrine and paracrine regulation of iron transport remain a puzzle, and the discovery of hemiferrin has opened a new area of endeavor.

References

Adrian, G. S., Bowman, B. H., Herbert, D. C., Weaker, F. J., Adrian, E. K., Robinson, L. K., Walter, C. A., Eddy, C. A., Riehl, R., Pauerstein, C. J., and Yang, F. (1990). Human transferrin: Expression and iron modulation of chimeric genes in transgenic mice. *J. Biol. Chem.* **265**, 13344–13350.

Aisen, P., and Listowsky, I. (1980). Iron transport and storage proteins. *Ann. Rev. Biochem.* **49**, 357–393.

Arosio, P., Adelman, T. G., and Drysdale, J. W. (1978). On ferritin heterogeneity. *J. Biol. Chem.* **253**, 4451–4458.

Bailey, S., Evans, R. W., Garratt, R. C., Gorinsky, B., Hasnain, S., Horsburgh, C., Jhoti, H., Lindley, P. F., Mydin, A., Sarra, R., and Watson, J. L. (1988). Molecular structure of serum transferrin at 3.3-angstrom resolution. *Biochemistry* **27**, 5804–5812.

Baker, M. E., French, F. S., and Joseph, D. R. (1987). Vitamin K-dependent protein S is similar to rat androgen-binding protein. *Biochem. J.* **243(1)**, 293–6.

Beard, P., Offord, E., Paduwat, N., and Bruggmann (1991). SV40 activates transcription from the transferrin receptor promoter by inducing a factor which binds to the CRE?AP-1 recognition sequence. *Nucl. Acids Res.* **19**, 7117–7123.

Berstein, S. E. (1987). Hedreditary hypotransferrinemia with hemosiderosis, a murine disorder resembling human atransferrinemia. *J. Lab. Clin. Med.* **110**, 690–697.

Boissiere, F., Augé-Gouillou, C., Schaeffer, E., and Zakin, M. M. (1991). The enhancer of the human transferrin gene is organized in two structural and functional domains. *J. Biol. Chem.* **266**, 9822–9828.

Bomford, A., Conlon-Hollingshead, C., and Munro, H. N. (1981). Adaptive responses of rat tissue isoferritins to iron administration. *J. Biol. Chem.* **256**, 948–955.

Boockfor, F. R., and Schwarz, L. K. (1990). Fibroblast growth factor modulates the release of transferrin from cultured Sertoli cells. *Mol. Cell. Endocrinol.* **73(2-3)**, 187–194.

Boockfor, F. R., and Schwarz, L. K. (1991). Effects of interleukin-6, interleukin-2, and tumor necrosis factor a on transferrin release from Sertoli cells in culture. *Endocrinology* **129(1)**, 256–262.

Cairo, G., Rappocciolo, E., Tacchini, L., and Schiaffonati, L. (1991). Expression of the genes for ferritin H and L subunits in rat liver and heart. *Biochem. J.* **275**, 813–816.

Casey, J. L., Di Jeso, B., Rao, K. K., Rouault, T. A., Klausner, R. D., and Harford, J. B. (1988). Deletional analysis of the promoter region of the human transferrin receptor gene. *Nucl. Acids Res.* **16**, 629–646.

Casey, J. L., Koeller, D. M., Ramin, V. C., Klausner, R. D., and Harford, J. B. (1989). Iron regulation of transferrin receptor mRNA levels requires iron-responsive elements and a rapid turnover determinant in the 3′ untranslated region of the mRNA. *EMBO J.* **8**, 3693–3699.

Chazenbalk, G. D., Wadsworth, H. L., and Rapoport, B. (1990). Transcriptional regulation of ferritin H messenger RNA levels in FRTL5 ret thyroid cells by thyrotropin. *J. Biol. Chem.* **265,** 666–670.

Ciechanover, A., Schwartz, A. L., and Lodish, H. F. (1983). Sorting and recycling of cell surface receptors and endocytosed ligands: The asialoglycoprotein and transferrin receptors. *J. Cell. Biochem.* **23,** 107–30.

Cochet, M. F., Gannon, R., Hen, R., Martoeaux, L., Perrin, F., and Chambon, P. (1979). Oranization and sequence studies of the 17-piece chicken conalbumin gene. *Nature (London)* **6,** 567–74.

Cox, F., Gestautas, J., and Rapoport, B. (1988). Molecular cloning of cDNA corresponding to mRNA species with steady state levels in thyroid that are enhanced by thyrotropin: Homology of one of the sequences with ferritin H. *J. Biol. Chem.* **265,** 7060–7067.

Djakiew, D., and Dym, M. (1988). Pachytene spermatocyte proteins influence Sertoli cell function. *Biol. Reprod.* **39,** 1193–1205.

Djakiew, D., Hadley, M. A., Byers, S. W., and Dym, M. (1986). Transferrin-mediated transcellular transport of ^{59}Fe across confluent epithelial sheets of Sertoli cells grown in bicameral cell culture chambers. *J. Androl.* **7(6),** 355–365.

Eisenstein, R. S., and Munro, H. N. (1990). Translational regulation of ferritin synthesis by iron. *Enzyme* **44,** 42–58.

Evans, R. W., and Madden, A. D. (1984). A low molecular weight iron-binding fragment from duck ovotransferrin. *Biochem. Soc. Trans.* **12,** 661–662.

Garrett, J. E., Collard, M. W., and Douglass, J. O. (1989). Translational control of germ-cell expressed mRNA imposed by alternative splicing: opiod peptide gene expression in rat testis. *Mol. Cell. Biol.* **9,** 4381–4389.

Griswold, M. D. (1991). Cyclic functions of Sertoli cells in synchronized testes. *In* "Hormonal Communicating Events in the Testis" (A. Isidori, A. Fabbri, and M. L. Dufau, eds.). Raven Press, New York.

Guillou, F., Zakin, M. M., Part, D., Boissier, F., and Schaeffer, E. (1991). Sertoli cell-specific expression of the human transferrin gene. *J. Biol. Chem.* **266(15),** 9876–9884.

Holmes, S. D., Bucci, L. R., Lipshultz, L. I., and Smith, R. G. (1983). Transferrin binds specifically to pachytene spermatocytes. *Endocrinology* **113,** 1916–1918.

Holmes, S. D., Lipshultz, L. I., and Smith, R. G. (1984). Regulation of transferrin secretion by human Sertoli cells cultured in the presence or absence of human peritubular cells. *J. Clin. Endocrinol. Metab.* **59,** 1058–1062.

Huggenvik, J., Sylvester, S. R., and Griswold, M. D. (1984). Control of transferrin mRNA synthesis in Sertoli cells. *Ann. N.Y. Acad. Sci.* **438,** 1–7.

Huggenvik, J., Idzerda, R. L., Haywood, L., Lee, D. C., McKnight, G. S., and Griswold, M. D. (1987). Transferrin mRNA: molecular cloning and hormonal regulation in rat Sertoli cells. *Endocrinol.* **120,** 332–340.

Hugly, S., and Griswold, M. (1987). Regulation of levels of specific Sertoli cell mRNAs by vitamin A. *Dev. Biol.* **121,** 316–324.

Hugly, S., Roberts, K., and Griswold, M. D. (1988). Transferrin and sulfated glycoprotein-2 messenger ribonucleic acid levels in the testis and isolated Sertoli cells of hypophysectomized rats. *Endocrinology* **122(4),** 1390–1396.

Idzerda, R. L., Huebers, H., Finch, C. A., and McKnight, G. S. (1986). Rat transferrin gene expression: Tissue-specific regulation by iron deficiency. *Proc. Natl. Acad. Sci. U.S.A.* **83,** 3723–3727.

Joshi, J. G., and Zimmerman, A. (1988). Ferritin: An expanded role in metabolic regulation. *Toxicol.* **48,** 21–29.

Karl, A. F., and Griswold, M. D. (1990). Sertoli cells of the testis: Preparation of cell cultures and effects of retinoids. *Meth. Enz.* **190,** 71–75.

Kleene, D. C., Distel, R. J., and Hecht, N. B. (1984). Translational regulation and deadenylation of a protamine mRNA during spermiogenesis in the mouse. *Dev. Biol.* **105,** 71–79.

Koeller, D. M., Casey, J. L., Hentze, M. W., Gerhardt, E. M., Chan, L.-N. C. L., Klausner, R. D., and Harford, J. B. (1989). A cytosolic protein binds to structural elements within the iron regulatory region of transferrin receptor mRNA. *Proc. Natl. Acad. Sci. U.S.A.* **86**, 3574–3578.

Krawetz, S. A., Connor, W., Cannon, R. D., and Dixon, G. H. (1986). A vector–primer–cloner–sequencer plasmid for the construction of cDNA libraries: Evidence for a rat glyceraldehyde-3-phosphate dehydrogenase-like mRNA and a ferritin mRNA within testis. *DNA* **5**, 427–435.

Kühn, L. C., McClelland, A., and Ruddle, F. H. (1984). Gene transfer, expression and molecular cloning of the human transferrin receptor gene. *Cell* **37**, 95–103.

LeBlond, C. P., and Clermont, Y. (1952). Definition of the stages of the cycle of the seminiferous epithelium in the rat. *Ann. N.Y. Acad. Sci.* **55**, 548–573.

Lee, N. T., Chae, C.-B., and Kierszenbaum, A. L. (1986). Contrasting levels of transferrin gene activity in cultured rat Sertoli cells and intact seminiferous tubules. *Proc. Natl. Acad. Sci. U.S.A.* **83**, 8177–8181.

Legrand, D., Mazurier, J., Metz-Boutigue, M., Jolles, J., Jolles, P., Montreuil, J., and Spik, G. (1984). Characterization and localization of an iron binding 18-kDa glycopeptide isolated from the N-terminal half of human lactotransferrin. *Biochim. Biophys. Acta.* **787**, 90–96.

Le Magueresse, B., Pineau, C., Guillou, F., and Jegou, B. (1988). Influence of germ cells upon transferrin secretion by rat Sertoli cells *in vitro. J. Endocrinol.* **118**, R13–R16.

McClelland, A., Kühn, L. C., and Ruddle, F. H. (1984). The human transferrin receptor gene: Genomic organization, and the complete primary structure of the receptor deduced from a cDNA sequence. *Cell* **39**, 267–274.

Morales, C., and Clermont, Y. (1986). Receptor-mediated endocytosis of transferring by Sertoli cells of the rat. *Biol. Reprod.* **35**, 393–405.

Morales, C., and Griswold, M. D. (1987). Retinol induced stage synchronization of the seminiferous tubules of the rat. *Endocrinology* **121**, 432–434.

Morales, C., Hugly, S., and Griswold, M. D. (1987a). Stage-dependent levels of specific mRNA transcripts in Sertoli cells. *Biol. Reprod.* **36**, 1035–1046.

Morales, C., Sylvester, S. R., and Griswold, M. D. (1987b). Transport of iron and transferrin synthesis by the seminiferous epithelium of the rat in vivo. *Biol. Reprod.* **37**, 995–1005.

Munro, H. N., and Linder, M. (1978). Ferritin: Structure, biosynthesis, and the role in iron metabolism. *Physiol. Rev.* **58**, 317–396.

Octave, J. N., Schneider, Y. J., Crichton, R. R., and Trout, A. (1981). Transferrin uptake by cultured rat embryo fibroblasts: the influence of temperature and incubation time, subcellular distribution and short-term kinetic studies. *Eur. J. Biochem.* **115**, 611–18.

Omary, M. B., and Trowbridge, I. S. (1981). Covalent binding of fatty acid to the transferrin receptor in cultured human cells. *J. Biol. Chem.* **256**, 4715–4718.

Perez-Infante, V., Bardin, C. W., Gunsalus, G. L., Musto, N. A., Rich, K. A., and Mather, J. P. (1986). Differential regulation of testicular transferrin and androgen-binding protein secretion in primary cultures of rat Sertoli cells. *Endocrinology* **118(1)**, 383–92.

Roberts, K. P., and Griswold, M. D. (1990). Characterization of rat transferrin receptor cDNA: The regulation of transferrin receptor mRNA in testes and in Sertoli cells in culture. *Mol. Endocrinology* **4(4)**, 531–542.

Roberts, K. P., Awoniyi, C. A., Santulli, R., and Zirkin, B. R. (1991). Regulation of Sertoli cell transferrin and sulfated glycoprotein-2 messenger ribonucleic acid levels during the restoration of spermatogenesis in the adult hypophysectomized rat. *Endocrinology* **129(6)**, 3417–3423.

Roberts, M. R., Miskimins, W. K., and Ruddle, F. H. (1989). Nuclear proteins TREF1 and TREF2 bind to the transcriptional control element of the transferrin receptor gene and appear to be associated as a heterodimer. *Cell Regul.* **1**, 151–164.

Schaeffer, E., Lucero, M. A., Jeltsch, J. M., Py, M. J., Levin, M. C., Chambon, P., Cohen, G. N., and Zakin, M. M. (1987). Complete structure of the human transferrin gene. Comparison with analogous chicken gene and human pseudogene. *Gene* **56**, 109–116.

Schaeffer, E., Boissier, F., Py, M., Cohen, G. N., and Zakin, M. M. (1989). Cell type-specific expression of the human transferrin Gene: Role of promoter, negative and enhancer elements. *J. Biol. Chem.* **264,** 7153–7160.

Schneider, C., Sutherland, R., Newman, R., and Greaves, M. (1982). Structural features of the cell surface receptor for transferrin that is recognized by the monoclonal antibody OKT9. *J. Biol. Chem.* **257,** 8516–8522.

Schneider, C., Owen, M. J., Banville, D., and Williams, J. G. (1984). Primary structure of human transferrin receptor deduced from the mRNA sequence. *Nature (London)* **311,** 675–678.

Skinner, M. K., and Fritz, I. B. (1985). Testicular peritubular cells secrete a protein under androgen control that modulates Sertoli cell functions. *Proc. Natl. Acad. Sci. U.S.A.* **82,** 114–118.

Skinner, M. K., and Griswold, M. D. (1980). Sertoli cells synthesize and secrete transferrin-like protein. *J. Biol. Chem.* **255(20),** 9523–9525.

Skinner, M. K., and Griswold, M. D. (1982). Secretion of testicular transferrin by cultured Sertoli cells is regulated by hormones and retinoids. *Biol. Reprod.* **27,** 211–221.

Stallard, B. J., and Griswold, M. D. (1990). Germ cell regulation of Sertoli cell transferrin mRNA levels. *Mol. Endocrinol.* **4(3),** 393–401.

Stallard, B. J., Collard, M. W., and Griswold, M. D. (1991). A transferrin-like (hemiferrin) mRNA is expressed in the germ cells of rat testis. *Mol. Cell. Biol.* **11(3),** 1448–1453.

Sylvester, S. R., and Griswold, M. D. (1984). Localization of transferrin and transferrin receptors in rat testes. *Biol. Reprod.* **31,** 195–203.

Sylvester, S. R., Morales, C., Oko, R., and Griswold, M. D. (1991). Localization of sulfated glycoprotein-2 (clusterin) on spermatozoa and in the reproductive tract of the male rat. *Biol. Reprod.* **45,** 195–207.

Theil, E. C. (1990a). Ferritin mRNA translation, structure and gene transcription during development of animals and plants. *Enzyme* **44,** 68–82.

Theil, E. C. (1990b). Regulation of ferritin and transferrin receptor mRNAs. *J. Biol. Chem.* **265,** 4771–4774.

Toebosch, A. M. W., Kroos, M. J., and Grootegoed, J. A. (1987a). Transport of transferrin-bound iron into rat Sertoli cells and spermatids. *Int. J. Androl.* **10,** 753–764.

Toebosch, A. M. W., Kroos, M. J., and Grootegoed, M. A. (1987b). Transport of iron into rat Sertoli cells and spermatids. *Ann. N.Y. Acad. Sci.* **513,** 431–433.

Wauben-Penris, P. J. J., Strous, G. J., and van der Donk, H. A. (1986). Transferrin receptors of isolated rat seminiferous tubules bind both rat and human transferrin. *Biol. Reprod.* **35,** 1227–1234.

Wauben-Penris, P. J. J., Veldscholte, J., van der Ende, A., and van der Donk, H. A. (1988). The release of iron by Sertoli cells in culture. *Biol. Reprod.* **38,** 1105–1113.

Williams, J. (1982). The evolution of transferrin. *Trends Biochem. Sci.* **3,** 351–355.

10

Molecular Biology of Testicular Steroid Secretion

PETER F. HALL

This chapter, as the title indicates, addresses the molecular biology of steroid synthesis. The term molecular biology is used commonly in two senses, namely, the study of biology in molecular terms or the study of genes and their expression. The term is used here in both senses, that is, the steroidogenic pathway will be considered as the outcome of molecular events. This discussion includes those specific functions concerned with the regulation of expression of the relevant genes.

The most important testicular steroids are androgens and the most important androgen is testosterone. The testis also secretes androstenedione and dehydroepiandrosterone, but these hormones are relatively weak androgens (Eik-Nes and Hall, 1965). Almost all androgens secreted by the testis are synthesized in Leydig cells; any contribution made by the seminiferous tubules is of minor importance (Eik-Nes and Hall, 1965; Christensen and Mason, 1965). The testis also secretes estrogens, which are made in both Leydig (De Jong *et al.*, 1974) and Sertoli (Dorrington and Armstrong, 1975) cells.

The most important regulatory agent for the synthesis of androgens is the pituitary hormone LH, or luteinizing hormone, which is secreted in pulses (Dierschke *et al.*, 1970). LH acts via a receptor that also responds to the placental gonadotropin that is best known as the human chorionic gonadotropin (hCG). LH and hCG appear to elicit the same responses in their target cells. The relevant receptor may be referred to as the LH/hCG receptor. For the sake of brevity, this protein will be referred to here as the LH receptor.

Molecular Biology of the Male Reproductive System
327

I. Biosynthesis of Testosterone

A. Substrate

Testosterone, like all steroid hormones, is synthesized from cholesterol which is, in turn, made from acetate. This steroidogenic cholesterol is supplied in part by local synthesis from acetate and in part from cholesterol imported from the plasma in lipoproteins. The relative proportions of these two sources vary with cell (e.g., adrenal or Leydig), species, and physiological state. For example, in the tumor cell line MA-10, the cells, when stimulated by hCG, derive more than half their steroidogenic cholesterol from low density lipoprotein (LDL) (Freeman and Ascoli, 1983). On the other hand, rat Leydig cells derive steroidogenic cholesterol from high density lipoprotein (HDL) (Gwynne et al., 1985). Apparently, each cell type must be studied under various physiological conditions. Useful generalizations cannot be made at present other than the fact that the source of the steroid ring system is cholesterol under all conditions, whether it is synthesized or imported.

Steroidogenic cholesterol is stored in conspicuous lipid droplets that form a distinctive feature of steroid-forming cells. The droplets decrease in size and number when Leydig cells are stimulated by LH (Christensen, 1975). Leydig cells contain a cholesterol ester hydrolase that is presumed to mobilize the cholesterol that is stored in ester form in the droplets (Albert et al., 1980). As in other aspects of steroid synthesis, considering what is known about the adrenal cortex (which has been studied in greater detail than Leydig cells) is helpful in at least some aspects of this topic. Clearly cholesterol in LDL is predominantly present as linoleate. Cholesteryl linoleate enters the cell and reaches lysosomes, in which it is de-esterified only to be re-esterified as the oleate (Kovanen et al., 1979). When cholesterol is needed for steroid synthesis the ester is again hydrolyzed to free cholesterol, which is the source of steroid hormones. The ester hydrolase just mentioned is stimulated by hCG (Albert et al., 1980). Details of these events, including the location of the enzyme within the cell, are not clear at present but eventually the steroidogenic cholesterol appears in the inner mitochondrial membrane, where steroidogenesis begins. Adrenal corticotropic hormone (ACTH) is known to promote entry of LDL into adrenal cells (Hall and Nakamura, 1979). LH may have the same role in Leydig cells, although this remains to be demonstrated.

B. Pathway

Four enzymatic reactions are required to convert cholesterol to testosterone.

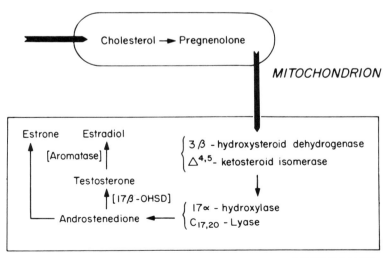

Cholesterol   Testosterone

In this conversion, cholesterol must lose the side chain at C17 and undergo rearrangement of the A and B rings. The side chain is removed in two steps by two cytochromes P450: a six C atom fragment is removed by $P450C_{27}scc$ and a two C atom fragment by $P450C_{21}scc$. Oxidation of the 3β group to a ketone and rearrangement of the double bond are both catalyzed by a single enzyme (dehydrogenase/isomerase). The cleavage of the side chain involves two oxidative steps, the second of which converts the C_{21} substrate to a C_{19} product (androstenedione) and, in doing so, leaves a ketone group at C17. The fourth step involves the reduction of the ketone to the 17β-ol of testosterone.

C. Arrangement in the Cell

The reactions just considered take place in two cellular compartments, mitochondrial and microsomal:

Cholesterol ➔ Pregnenolone

MITOCHONDRION

Estrone   Estradiol

[Aromatase]

3β - hydroxysteroid dehydrogenase

$\Delta^{4,5}$- ketosteroid isomerase

Testosterone

[17β-OHSD]

Androstenedione

17α - hydroxylase

$C_{17,20}$ - Lyase

MICROSOME

The side-chain is first cleaved in mitochondria to give pregnenolone, which leaves the organelle to reach the microsomal compartment where the remaining reactions take place. An additional reaction occurs in this compartment to produce estrogens. This reaction is catalyzed by a cytochrome P450 called aromatase. This enzyme converts androstenedione and testosterone to the corresponding estrogens.

D. Steroidogenic Cytochromes P450

1. Nature and catalytic activity

Cytochromes P450 constitute a large family of heme proteins characterized by noncovalent attachment of the heme to the protein via a thiolate ion between the heme iron and a specific cysteine residue. The heme binding region of the protein is highly conserved in the various members of the family. The following comments, except where indicated, refer to all cytochromes P450, not only those that are steroidogenic.

The conjugated double bond system of the heme is influenced readily by interaction with neighboring charged groups. Such influences are transmitted to the iron, in which the d orbital electrons can exist in two arrangements. When the iron is located in the plane of the ring, the d electrons are all in low energy orbitals called t_{2g} and one is unpaired. This arrangement is called low spin and the iron has a valence of 6 (hexacoordinate) in which the 5th ligand is cysteine (thiolate bond) and the 6th ligand is an $-OH$ group, probably water in most cases (Rein and Ristan, 1978; Ullrich, 1979; White and Coon, 1980; Jsanig et al., 1988). The low spin form is designated 1/2. The second form of the heme iron has all d orbitals unpaired and two electrons in the high energy form called eg. In this case, with loss of the 6th ligand, the iron is pentacoordinate and is displaced slightly from the plane of the heme ring. In either of these forms, the iron can be oxidized or reduced. The short-hand nomenclature of the high spin form is 5/2 (Rein and Ristan, 1978; Griffin et al., 1979; Ullrich, 1979).

When the heme iron of any P450 is reduced, it can bind and activate O_2. The oxygen then can hydroxylate a variety of substrates. This archetypal reaction performed in an atmosphere of $^{18}O_2$ can be written as:

$$R-H + NADPH + H^+ + {}^{18}O_2 \xrightarrow{\text{P450}} R-{}^{18}OH + H_2{}^{18}O + NADP^+$$

The enzyme uses molecular oxygen and, since only one of the two oxygen atoms appears in the product, the reaction is described as monooxygenation and the enzyme is called a monooxygenase. A hydrogen atom in the substrate is replaced by $-OH$, which renders the product more soluble than the substrate. This in turn facilitates the removal of the product from the body, since it is less able to enter cells and, being more soluble in water, can be filtered by the kidney. As a result, cytochromes P450 constitute the most important single defense of the body against xenobiotics (Bresnick, 1980).

The body also uses P450 in biosynthetic pathways in which hydroxylation and other changes take place in the various substrates. The importance of such transformations in the conversion of cholesterol to testosterone is discussed subsequently. Note that P450 can use active oxygen to catalyze variations on the archetypal theme just discussed, for example, N-demethylation or S-demethylation, in connection with xenobiotic metabolism. Finally, the enzymatic activities of various cytochromes are basically similar. Presumably, although the heme-binding regions of P450 are conserved, the different enzymes have evolved by means of genetic reorganization that brings DNA corresponding to the conserved region into juxtaposition with DNA corresponding to different specific substrate binding sites to construct a variety of enzymes capable of catalyzing the same reaction with different substrates and at different sites on the same substrate (Go, 1981). Other domains of the enzymes secure attachment to membranes and interaction with electron carriers.

The prototype P450 reaction reveals two additional features of these enzymes. First, P450 must be reduced by electrons from NADPH. Not shown in the diagram is that transport of electrons requires one or two carrier proteins, which will be discussed later. Second, this typical reaction shows the stoichiometry expected of $1:1:1:1$, that is, one mole of substrate, one mole of oxygen, and one mole of reduced pyridine nucleotide give one mole of product, one mole of oxidized pyridine nucleotide and one mole of water. This stoichiometry can be used as a test for monooxygenation (Mason, 1957).

2. Spectral properties

a. Native spectrum

All heme compounds show a prominent absorbance peak at approximately 420 nm (the Soret peak). The intensity and the exact location of the peak are influenced by the protein moiety. Less conspicuous peaks in cytochromes P450 are to be seen at approximately 570, 540, 410, and 392 nm in the oxidized state.

b. Reduced CO spectrum

Like all heme proteins, cytochromes P450 when reduced bind CO but in this group of proteins the spectral change that results from these two events (reduction followed by CO binding) is marked by a gross shift in the location of the Soret peak towards the red region (420 → 450 nm). This shift gives these proteins their name (pigment 450), and provides the best way to identify them as well as an accurate means of measuring the amount of any P450:

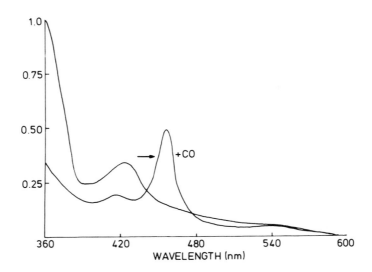

This change, which is attributable to the thiolate attachment of the heme to cysteine, is measured conveniently by difference, that is, absorbance of reduced P450–CO minus that of oxidized P450.

c. P450–Substrate spectrum

The conformational change that results from binding of substrate to any P450 converts the low spin form (substrate-free) to the high spin form with the consequences discussed earlier, that is, displacement of the iron, rearrangement of d orbital electrons, and formation of the pentavalent form. These changes are accompanied by a blue shift of the Soret peak to approximately 390 nm:

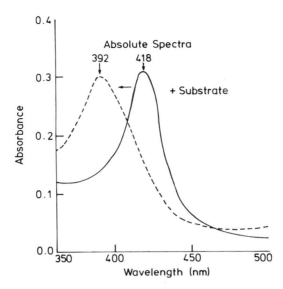

This shift can be measured by difference, that is, enzyme + substrate minus enzyme + solvent:

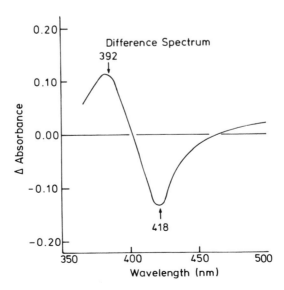

This difference spectrum provides a simple and direct measurement of the proportion of molecules of P450 that are in the bound state, that is, ES. The procedure does not destroy the enzyme and the spectral shift results in a type I spectrum (peak at 390 nm, trough at 420 nm).

d. P450–Product spectrum

In many cytochromes P450, the product of the enzyme activity causes the opposite spectral shift, resulting in a peak at about 420 nm and a trough at about 390 nm (inverse type I spectrum). This shift is believed to result from displacement of bound substrate by the product, which comes to occupy the active site.

e. Type II spectrum

Several nitrogenous compounds bind to cytochromes P450 to produce a spectrum with a peak at 430 nm and a trough between 400 and 420 nm (Griffin et al., 1979). This spectrum resembles but is distinct from the inverse type I spectrum that is seen when these enzymes are bound to their products.

f. Electron paramagnetic resonance

High and low spin forms of P450 can be distinguished by electron paramagnetic resonance (epr), which is sensitive to the paramagnetic properties of the heme iron. This method provides a less empirical approach to the structure of heme proteins than does absorbance spectroscopy.

3. Mechanism of reduction

P450 is reduced by electrons from reduced pyridine nucleotides. Flavoprotein is required for electron transport to P450. In mitochondrial and bacterial P450, an iron–sulfur protein is required also.

a. Mitochondria

Two one-electron transfers occur between the flavoprotein and P450. The reduced (ferrous) P450 activates oxygen and catalyzes the conversion of substrate to product:

The switch from the two-electron carrier (flavoprotein) to the one-electron carrier (iron–sulfur protein) means that two distinct steps are required for this protein to provide P450 with two electrons. The iron-sulfur proteins are named with the suffix -oxin, (adrenodoxin, testodoxin, etc.), according to the source of the protein. The flavoprotein is known as the corresponding reductase, for example, adrenodoxin reductase. The iron–sulfur proteins are members of a general class of the type $Fe_2S_2Cys_4$.

b. Microsomes

The microsomal P450 enzymes are reduced by a reductase called P450 reductase (NADPH–cytochrome P450 reductase); no iron–sulfur protein is involved.

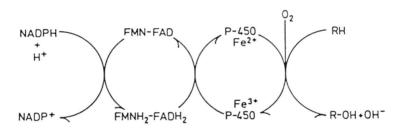

The reductase contains both flavin adenine dinucleotide (FAD) and flavin mononucleotide (FMN). Cytochrome b_5 has been proposed to be involved in the reduction of P450 in microsomes, but the role of this protein is uncertain (Griffin et al., 1979).

4. Catalytic cycle

Since P450 is an oxygenase that is reduced by pyridine nucleotides, to carry out its catalytic function the enzyme must undergo a cycle of changes in which it is reoxidized to continue acting on new molecules of substrate. The cycle can be considered in five steps for descriptive purposes. However, the last of the reactions is, itself, made up of several steps (Gunsalus et al., 1974; Ullrich, 1979; White and Coon, 1980). To illustrate the cycle, P450 can be represented by Fe.

Step 1 involves binding of substrate:

$$Fe^{3+} + RH \longrightarrow Fe^{3+}-RH$$

which leads to a conformational change in the protein. This displaces iron from the plane of the ring system and leads to the formation of high spin iron. As explained earlier, this change causes P450 to show a type I spectral shift. Moreover, binding of substrate promotes the passage of the first electron from adrenodoxin to the enzyme–substrate complex (Step 2):

$$Fe^{3+}-RH + e^- \longrightarrow Fe^{2+}-RH$$

In Step 3, the reduced P450 binds oxygen:

$$Fe^{2+}-RH + O_2 \longrightarrow Fe^{2+}-RH$$
$$|$$
$$O_2$$

This binding is followed by Step 4, in which bound oxygen is activated by the arrival of the second electron from adrenodoxin. One atom of the oxygen is reduced to water:

$$\underset{\overset{|}{O_2}}{Fe^{2+}}-RH + e^- + H^+ \longrightarrow \underset{\overset{|}{O_2^{2-}}}{Fe^{3+}}-RH + OH^-$$

Step 5 is complex and poorly understood. The activated oxygen atom attacks the substrate to produce a hydroxylated product:

$$\underset{\overset{|}{O_2^{2-}}}{Fe^{3+}}-RH \longrightarrow Fe^{3+} + ROH$$

The OH renders the product more highly charged than the substrate. Therefore the product is repelled from the hydrophobic active site of the enzyme. The enzyme is now ready to bind a new molecule of substrate and begin the cycle again:

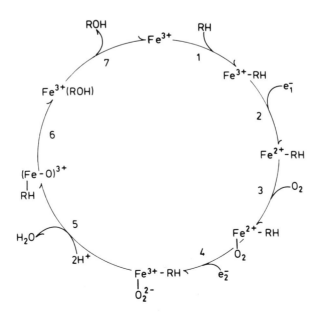

We can now consider the two P450 enzymes involved in the synthesis of androgens.

E. Individual Enzymes

1. C_{27} Side-chain cleavage P450

a. Reaction

The removal of the side chain at C17 of cholesterol takes place in two stages, the first of which is catalyzed by a mitochondrial P450 situated in the inner membrane. The enzyme is supported by two electron carriers — adrenodoxin reductase and adrenodoxin of that organelle (or equivalent proteins in the gonads). This first cleavage converts cholesterol to pregnenolone:

CHOLESTEROL   22-OH CHOLESTEROL   20,22-Di-OH-CHOLESTEROL   PREGNENOLONE

ISOCAPRALDEHYDE

This reaction commonly is referred to as side-chain cleavage unqualified. The reason for the designation C_{27} will become apparent when we consider the second side-chain cleavage reaction.

The first two steps in this reaction represent classical P450 hydroxylation, which prepares the intervening C20–C22 bond for cleavage. Initial doubts that such cleavage is the result of a reaction catalyzed by a cytochrome P450 were dispelled when conversion of the dihydroxycholesterol to pregnenolone was shown to have a typical P450 action spectrum (Hall *et al.*, 1975). The P450–CO complex in all cytochromes P450 is light sensitive and, by definition, shows an absorbance maximum at 450 nm. The last step in the above conversion is inhibited by CO, and inhibition is reversed most effectively by light of wavelength 450 nm because light of this wavelength is best absorbed by the complex (Hall *et al.*, 1975). This is the hallmark of a reaction catalyzed by any P450 enzyme. Moreover, the enzyme was shown to be homogeneous by a number of methods including gel electrophoresis, immunodiffusion, and immunoelectrophoresis (Shikita and Hall, 1973a,b;

Watanuki *et al.*, 1978). No evidence exists for the presence of more than one heme group per molecule (Shikita and Hall, 1973a). Clearly, the three steps in side-chain cleavage are catalyzed by a single P450 protein. Moreover, kinetic and binding studies showed that the enzyme catalyzes these reactions with a single active site (Duque *et al.*, 1978). The occurrence of three successive monooxygenase reactions catalyzed on one substrate by a single active site was without precedent at the time of these reports, but this proposal was confirmed by studies of the stoichiometry of the reaction, which proved to be that of three such reactions (Shikita and Hall, 1974):

$$C_{27}H_{46}O + 3\ O_2 + 3\ NADPH + 3\ H^+ \longrightarrow C_{21}H_{32}O_2 + C_6H_{12}O + 4\ H_2O.$$

Each of the three steps, when examined separately, showed the stoichiometry expected for a single monooxygenation (Shikita and Hall, 1974).

b. Intermediates

When the enzyme was incubated with [^{3}H]cholesterol, the amount of disappearing substrate was found to be approximately equal to the amount of [^{3}H]pregnenolone formed; only trace amounts of the two intermediates could be detected (Hall and Koritz, 1964; Koritz and Hall, 1964a). The existence of the intermediates and the order of the two hydroxylation reactions (22 → 20) finally was settled by the work of Burstein and Gut (1976), in which the hydroxycholesterol intermediates were isolated and characterized and the rates of formation of intermediates and product were determined with large-scale preparations of partly purified enzyme. A detailed study of the binding of substrate, intermediates, and product to the purified enzyme using absorption spectroscopy and epr showed that the enzyme possesses a single binding site for three steroids with the following dissociation constants (Orme-Johnson *et al.*, 1979):

22R-OH cholesterol	4.9 $K_d(nM)$
20R,22R-diOH cholesterol	81.0 $K_d(nM)$
pregnenolone	2900.0 $K_d(nM)$

The value for cholesterol has not been determined because of technical problems related to the insolubility of this substance. Moreover, 22R-OH cholesterol shows a dissociation frequency of 5 sec^{-1}. Clearly the two intermediates are bound tightly to the enzyme and are converted rapidly to pregnenolone, which readily dissociates from the active site. Evidently the intermediates exist almost entirely in enzyme bound form; therefore Burstein and Gut (1976) found it necessary to use such large amounts of enzyme to measure the rates of formation of the intermediates.

Two important features of these reactions should be noticed. First, the catalytic cycle is modified to allow the products of the first two cycles to remain in the active site. The presence of the bound intermediates presumably activates the next cycle. Second, as described earlier, the last step is a typical monooxygenation that also shows a P450 photochemical action spectrum, so this step also is catalyzed by the single P450 active site (Hall *et al.*, 1975). This report was the first of P450 monooxygenation leading to cleavage of a carbon–carbon bond. The detailed mechanism of this step is still not clear.

c. Properties

When pregnenolone is added to homogeneous $P450C_{27}scc$, an inverse type I spectral shift results (see preceding discussion). This shift is the usual response to the product of a reaction catalyzed by P450. In this case, pregnenolone produces product inhibition of enzyme activity (Koritz and Hall, 1964b), which may regulate side-chain cleavage to some degree (Hall and Koritz, 1964; Koritz and Hall, 1964a,b). When pregnenolone is added to mitochondria, it displaces cholesterol previously bound to P450scc and produces an inverse type I shift. The extent of this shift expressed in absorbance units provides a measure of the proportion of molecules of the enzyme that previously were bound to substrate (cholesterol). Molecules of enzyme that are without substrate are always in the low spin state (peak at 420, trough at 380 nm) and therefore do not change on addition of pregnenolone, which produces the low spin form of the enzyme. The inverse type I shift can occur only in those molecules of enzyme that were in the high spin or substrate-bound form (ES) prior to addition of pregnenolone. The extent of the spectral shift is proportional to the number of molecules of enzyme that previously were bound to cholesterol. Such measurements will be referred to in considering the mechanism of action of LH.

In aqueous buffer, $P450C_{27}scc$ aggregates to oligomeric forms (tetramers, octamers, and hexadecamers) (Shikita and Hall, 1973b). When the enzyme is centrifuged through a sucrose gradient containing substrate, NADPH, and electron donors, the only active fractions that produce pregnenolone are those corresponding to the hexadecamer (Takagi *et al.*, 1975). The nature of the active form in the inner mitochondrial membrane is unknown.

2. Side-chain cleavage P450

a. Reaction

The remainder of the side chain at C17 is cleaved to yield C_{19} andro-stenedione and a two C atom fragment:

PROGESTERONE   17α-OH PROGESTERONE   ANDROSTENEDIONE

b. One enzyme

Knowing the mechanism of C_{27} side-chain cleavage, researchers might have predicted that a single P450 would catalyze both steps in the second cleavage reaction by oxygenating the 17 position and cleaving the C17–C20 bond between the two oxygenated carbons. However, because the adrenal cortex synthesizes the 17α-hydroxy C_{21} steroid cortisol, hydroxylation clearly can occur without lyase activity. The idea of a single enzyme was resisted strongly, even after publication of the isolation of a homogeneous cytochrome P450 from porcine testis that was capable of catalyzing both reactions (Nakajin and Hall, 1981a; Nakajin et al., 1981b). The enzyme is microsomal and tightly membrane bound but resembles the C_{27} enzyme sufficiently for the two to be mistaken during purification (Bumpus and Dus, 1982). The evidence for homogeneity rests on immunological, electrophoretic, and amino acid sequence analyses (Nakajin and Hall, 1981a; Nakajiin et al., 1981). This conclusion is supported by isolation of the gene, as discussed subsequently.

When the porcine adrenal enzyme was purified it proved to be virtually identical in physical and kinetic properties to the testicular enzyme and equally capable of catalyzing both hydroxylase and lyase reactions. Clearly the production of cortisol by adrenal cortex and C_{19} androgens by testis cannot be explained by differences between the two enzymes (Nakajin et al., 1981b,1983,1984). On the other hand, microsomes from porcine adrenal express very little lyase activity whereas testicular microsomes not only actively express lyase activity but produce no C_{21} steroids (Nakajin and Hall, 1983). This important example of microsomal regulation is discussed in Section IIb.

c. Properties

The $C_{21}scc$ enzyme acts at crucial points in steroidogenesis. In the Leydig cell, the two reactions catalyzed by this enzyme are essential for the synthesis of androgens and little or no 17α-hydroxy C_{21} steroids are formed. In the zona glomerulosa of the adrenal cortex, $C_{21}scc$ is not expressed because aldosterone does not possess a 17α-hydroxy group, whereas in the zona fasciculata the first step of $C_{21}scc$ (17α hydroxylation) is required without the second (lyase) step for the synthesis of cortisol. The enzyme is subject to complex inhibition. For example, progesterone and pregnenolone produce forward inhibition of lyase activity (Nakajiin and Hall, 1981b,1983). In addition, estrogens cause profound inhibition of lyase activity with the pure enzyme (Onoda and Hall, 1981). This inhibition is competitive and, since the synthesis of estrogens requires lyase activity, may be of physiological significance. (Onoda and Hall, 1981).

The cleavage step has been shown to involve the heme moiety of the enzyme since the conversion of 17α-hydroxypregnenolone to dehydroepiandrosterone and the corresponding conversion of the Δ^4 substrate both show a typical P450 photochemical action spectrum (Nakajin and Hall, 1983). This observation reinforces the analogy between the lyase activities of the two side-chain cleavage enzymes. Once again, it appeared to be important to determine whether a single active site can catalyze these apparently very different reactions of hydroxylation and cleavage of a carbon–carbon bond. One observation pointing to a single active site is that no evidence exists for more than one heme per molecule of enzyme. Since heme is essential for both steps, it seems inevitable that there can be no more than one active site.

Positive evidence for this conclusion came from affinity alkylation of the enzyme using 17α-bromoacetoxyprogesterone. This substrate analog was shown to compete with progesterone for the active site of the enzyme and to be converted to androstenedione. Moreover, the bromine moiety of the analog can combine with the side chains of 9 of the 20 amino acids to produce a derivatized protein and HBr. This interaction can be illustrated for the amino acid cysteine:

$$
\begin{array}{ccc}
\text{17}\alpha\text{-Bromoacetoxy-} & \text{Enzyme} & \text{S-Carboxymethylpeptide} \\
\text{Progestesterone} & &
\end{array}
$$

Structure: 17α-Bromoacetoxyprogesterone (steroid with CH$_3$–C=O, O–C(=O)–CH$_2$–Br) + HS—Cys $\xrightarrow{[-HBr]}$ S-Carboxymethylpeptide (steroid with CH$_3$–C=O, O–C(=O)–CH$_2$–S—Cys)

This reaction results in irreversible inactivation of the enzyme, so that inhibition of enzyme activity gradually changes from competitive to noncompetitive as the reaction proceeds over time. When 17α-bromoacetoxyprogesterone was incubated with C$_{21}$scc, only a single cysteine was attacked. Both activities of the enzyme declined with the same values for $t_{1/2}$. The stoichiometry of inactivation was 1.0 for both activities. Moreover, both substrates (progesterone and 17α-hydroxyprogesterone) protect both enzyme activities (hydroxylation and lyase) against inactivation by the substrate analog (Sweet et al., 1987), suggesting that both enzyme activities take place in the same active site. Inhibition of the two enzyme activities by antibodies and by synthetic competitive inhibitors shows that these inhibitors act at a single active site (Nakajin and Hall, 1981b,1983). Finally, substrate-induced difference spectroscopy also shows that both substrates (17-deoxy and 17-hydroxy) compete for a single active site (Nakajin et al., 1981b).

In spite of the fact that C$_{21}$ and C$_{27}$ side-chain cleavage appear to proceed by the same mechanism, at least two important differences exist between these enzyme activities: (1) the mitochondrial C$_{27}$ system requires three steps to oxygenate both carbons and cleave the intervening bond whereas in the C$_{21}$ system one carbon atom (C20) is already oxygenated in pregnenolone or progesterone, so only two steps are required and (2) the intermediate 17α-hydroxysteroid in the C_{21} system accumulates and can be isolated readily (Nakajin and Hall, 1981b) whereas the hydroxylated intermediates in the C_{27} system do not exist to any measurable extent as free steroids and are detected only by special means, as discussed earlier, in connection with the studies of Burstein and Gut (1976).

3. 3β-Hydroxysteroid dehydrogenase/isomerase

The A and B rings of pregnenolone are converted to the corresponding Δ^4-3-ketosteroid progesterone by a microsomal enzyme:

PREGNENOLONE   PROGESTERONE

The intermediate Δ^5-3-ketosteroid has been detected with the enzyme purified from rat adrenal gland (Ishii-Ohba *et al.*, 1986). The enzyme from human placental microsomes is a tetramer of molecular weight 76,000 (monomer 19,000). This enzyme acts on both C_{21} and C_{19} substrates (pregnenolone and dehydroepiandrosterone) to yield progesterone and androstenedione, respectively (Thomas *et al.*, 1988). These products competitively inhibit enzyme activity. The enzyme occurs in both mitochondria and microsomes in roughly equal proportions. Note that the isomerase activity requires NAD$^+$ or NADH as an allosteric activator, neither activator undergoing net oxidation or reduction in the process (Thomas *et al.*, 1989). The pH optima for dehydrogenase and isomerase activities are 9.8 and 7.5, respectively, suggesting that different amino acid side chains must be involved in the two reactions. Whether or not a single active site or two sites are found remains to be determined (Thomas *et al.*, 1989). Although several steroids influence the activities of the enzyme, only progesterone, which inhibits the enzyme, would be likely to act under physiological conditions (Raimondi *et al.*, 1989). The enzyme has been purified from adrenal tissue as a monomer of MW 46,8000 (Ishii-Ohba *et al.*, 1986), and has been isolated from rat testis (Ishii-Ohba *et al.*, 1986). Dehydrogenase activity is rate limiting for the overall reaction and is irreversible.

4. 17β-Hydroxysteroid dehydrogenase

The microsomal enzyme 17β-hydroxysteroid dehydrogenase (OHSD) catalyzes the conversion of androstenedione to testosterone:

Androstenedione   Testosterone

The reaction catalyzed is reversible and shows the unusual property of product activation, that is, testosterone promotes its own formation from androstenedione and vice versa (Oshima and Ochi, 1973). This dehydrogenase has been purified from placenta and was shown to be composed of two (possibly identical) subunits of molecular weight 35,000. The enzyme possesses two active sites, one for reductase activity and the other for oxidase activity (Oshima and Ochi, 1973; Samuels *et al.*, 1975). Therefore, the enzyme *in situ* is likely to be influenced greatly by ambient levels of substrate, product, and pyridine nucleotide. A separate enzyme has been proposed to catalyze the analogous conversion of estrone to estradiol in ovarian tissue (Luu-Thé *et al.*, 1989b; Tremblay *et al.*, 1989). Also, an additional enzyme may act as a ketoreductase in Leydig cells (Bogovich and Payne, 1980) and a different enzyme may act in the seminiferous tubule than in Leydig cells (Murano and Payne, 1976). This complex enzyme has been reviewed in detail (Ohba *et al.*, 1982).

II. Steroidogenic Membranes

A. Inner Mitochondrial Membrane

A membrane environment is likely to exert a profound influence on the activity of proteins (including enzymes). In the case of steroidogenic cytochromes P450, the membrane anchors the proteins in an appropriate environment within the cell, facilitates transfer of electrons by providing a medium insulated from a water phase, and offers a lipophilic region in which steroid substrates can dissolve. The mitochondrial membrane has been studied in the adrenal cortex by biochemical methods and by electron microscopy. P450scc is found in the inner mitochondrial membrane on the matrix, or inner, side of that membrane (Churchill and Kimura, 1979; Mitani *et al.*, 1982). Immunoelectron microscopy with gold particles conjugated to second antibody showed that all mitochondria within the cortex contain P450scc and P45011β, which are distributed randomly throughout the inner membrane of all mitochondria in the adrenal cortex (Geuze *et al.*, 1987).

Studies of any membrane-bound enzyme in aqueous buffers may not offer an adequate environment for detailed investigation of the activities of such enzymes. To provide an environment resembling the mitochondrial membrane, P450scc was incorporated into lipid vesicles prepared from equimolar mixtures of phosphatidylcholine and phosphatidylethanolamine, with and without cholesterol (Hall *et al.*, 1979a). In this environment, the P450 is converted from the low spin to the high spin form. Moreover, the two electron carriers (adrenodoxin and its reductase) also can be bound to the liposome membrane so that on addition of NADPH side-chain cleavage takes place (Hall *et al.*, 1979a). This system was studied in greater detail by Seybert *et al.* (1978,1979). These workers showed that P450scc in one vesicle does not act on cholesterol in another vesicle and that the active site of the enzyme is located in the hydrophobic part of the membrane, whereas the binding domain for adrenodoxin lies on the surface of the vesicle facing the external aqueous phase (Seybert *et al.*, 1978,1979). These workers also showed that oxidized adrenodoxin binds to reduced reductase. After transfer of one electron to adrenodoxin, this molecule dissociates from the reductase and binds to P450. Binding of adrenodoxin to P450 is facilitated by the presence of cholesterol in the active site of the P450 (Lambeth *et al.*, 1979,1980). In this way, adrenodoxin is said to shuttle between the reductase and P450. The effect of substrate (cholesterol) on binding of adrenodoxin to P450 is consistent with the well-known stimulation by substrate binding on the passage of the first electron to P450 (see previous discussion). The shuttle mechanism also explains the observation that the molar ratios of these proteins in the membrane (at least in the adrenal) are as follows: reductase: adrenodoxin: P450 equals 1:3:4 (Honokoglu and Honokoglu, 1986). Evidently, a ternary complex is not formed and one molecule of reductase can serve numerous molecules of adrenodoxin (Seybert *et al.*, 1978). After donating the first electron to P450, oxidized adrenodoxin dissociates from the enzyme to allow a second molecule of reduced adrenodoxin to provide the second electron to P450. These findings are likely to apply to the Leydig cell also, but this has not been established.

B. Smooth Endoplasmic Reticulum

The membrane of the smooth endoplasmic reticulum also influences the steroidogenic enzymes. Three main questions about this influence have been addressed to date.

1. Sequence of reactions

Since the product of C_{21} side-chain cleavage is a 17-ketosteroid (androstenedione or dehydroepiandrosterone), the last step in the synthesis of testosterone is catalyzed by 17β-hydroxysteroid dehydrogenase. Bearing in mind that both C_{21} side-chain cleavage and dehydrogenase/isomerase reactions are each catalyzed by a single protein, clearly two possible pathways exist from pregnenolone to androstenedione, depending on which of these enzymes acts first:

The "upper" pathway is referred to as the Δ^4 or progesterone pathway, in contrast to the Δ^5 or dehydroepiandrostenedione (DHEA) pathway. These alternatives are not chosen at random. For example, the rat uses the Δ^4 pathway largely if not exclusively (Samuels, 1960). Pig (Ruokenen and Vikko, 1974), rabbit, and dog (Hall *et al.*, 1964) use the Δ^5 pathway to varying degrees. These differences could result from two possible causes: differential specificity of the two enzymes for one or both substrates and arrangement of the enzymes in the membrane. For example, if the C_{21} side-chain cleavage enzyme binds progesterone in preference to pregnenolone and the dehydrogenase/isomerase binds pregnenolone in preference to progesterone, the Δ^4 pathway would be preferred. Alternatively, the enzymes might be arranged so one or the other enzyme is located close to a port through which pregnenolone enters the microsomal membrane. The question is important in understanding the organization of microsomal proteins, but currently the only evidence available shows that porcine C_{21} side-chain cleavage enzyme strongly prefers pregnenolone to progesterone, in keeping with the use of the Δ^5 pathway in that species (Nakajiin and Hall, 1981b; Nakajiin *et al.*, 1981b).

2. Localization of enzymes in the membrane

The endoplasmic reticulum (ER) in the cell possesses an outer or cytoplasmic surface and an inner surface. When the cell and its membranes are broken, fragments of endoplasmic reticulum round up as vesicles that may be right-side-out, as in the cell, or inside-out. Proteins may be associated with the membrane by powerful hydrophobic forces that hold them in the interior of the membrane or by a mixture of hydrophobic and hydrophilic forces that causes the protein to be anchored in the membrane protruding on one or the other surface. When cells are ruptured under standard conditions, in any given cell the vesicles generally prefer, in a reproducible manner, one configuration of the endoplasmic reticulum or the other. Although smooth ER is often assumed to give rise always to right-side-out vesicles, this conclusion must be established with the aid of antibodies in any particular case. Antibodies can interact only with proteins that are exposed on the outer surfaces of vesicles. The antibody can be used on sectioned or permeabilized cells (endoplasmic reticulum intact) to compare the location of the protein(s) in the two situations. Then proteolytic enzymes and phospholipase enzymes can be used to explore the structure of the membrane. Most proteolytic enzymes cannot penetrate a lipid bilayer and therefore can attack only exposed parts of membrane proteins. Methods are available for determining whether or not vesicles are "leaky" (Hall *et al.*, 1979a). If fragments of a membrane protein are released by proteolytic enzymes, that protein must be positioned on the outer side of the vesicle, which may or may not correspond to the outer side of the endoplasmic reticulum *in situ*. If proteolysis only occurs when the enzyme is present at the time the cell is disrupted, the protein is presumed to face the inner surface of the vesicle; the enzyme is excluded from the vesicle by the lipid boundary of the membrane, which it cannot penetrate unless this membrane is ruptured during homogenization of the cell. With phospholipase enzymes, phospholipids are partly removed from the membrane; measurement of enzyme activity before and after treatment with phospholipase will reveal whether or not the enzyme requires phospholipids for full activity. Finally, in hepatic microsomes, P450 has been shown to be located on the inside of microsomes (Welton and Aust, 1974). Incubation of rat testicular microsomes with phospholipase and trypsin (separately) by Samuels *et al.* 1975) was shown to exert the following effects:

	Pregnenolone binding	3β-OHSD[a]	Hydroxylase/lyase activity
Phospholipase	↓	↓↓↓	No change
Trypsin	↓↓↓	↓↓↓	↓↓↓

[a] 3β-OHSD, 3β-hydroxysteroid dehydrogenase.

These results were interpreted to mean that a pregnenolone-binding protein in the membrane faces the cytoplasm, as does the dehydrogenase/isomerase. Since the C_{21}scc is assumed to be on the inside face of the bilayer (by analogy with hepatic P450), inhibition of this activity by trypsin is believed to come about through proteolysis of P450 reductase, which would inhibit electron transport and hence enzyme activity of the P450 (Samuels *et al.*, 1975). Apparently the dehydrogenase/isomerase requires intact phospholipids. The same investigators showed that dehydrogenase/isomerase acts on substrate that approaches the active site from the water phase (i.e., the cytoplasm) whereas P450 acts on intramembrane substrate. We can summarize these findings in a highly diagrammatic form as follows:

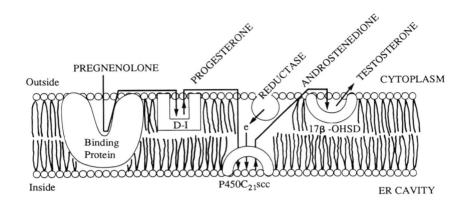

3. Influence of microsomal environment

Comparison of the behavior of homogeneous enzymes in an aqueous phase with that of those *in situ* in microsomes has begun to reveal important features of the influence of the membrane environment. Matsumoto and Samuels (1969) showed that, when testicular microsomes are incubated with [³H]progesterone and 17α-hydroxy-[¹⁴C]progesterone, the [³H] : [¹⁴C] ratio in androstenedione is higher than that in the original mixture of substrates, suggesting that progesterone has greater access to the active site of the C_{21} side-chain cleavage P450 than does the 17α-hydroxysteroid and that this latter steroid is not free to exchange with the exogenous 17α-hydroxyprogesterone (Matsumoto and Samuels, 1969). Moreover, the partition ratio of progesterone from water to microsomes is greater than that of 17α-hydroxyprogesterone (Matsumoto and Samuels, 1969; Samuels and Matsumoto, 1974). In contrast, both steroids show equal affinities for the pure enzyme (Nakajiin *et al.*, 1981b). 17α-Hydroxyprogesterone generated by the enzyme from progesterone exchanges freely with exogenous 17α-hydroxyprogesterone added to aqueous buffer (Nakajin and Hall, 1981b). The difference between equal access of the two substrates to the pure enzyme and the restriction of free exchange in microsomes must result from the microsomal environment. Possibly, protein components of the microsome are important in regulating access to and from the enzyme. Attempts to reconstitute microsomes from purified components extracted from these structures are likely to elucidate the functions of individual microsomal components.

Another such comparison proves instructive. Whereas microsomes from the porcine adrenal gland almost completely fail to convert progesterone to androstenedione, producing 17α-hydroxy C_{21} steroids instead, testicular microsomes produce abundant androstenedione and testosterone. In contrast the pure $C_{21}scc$ P450 enzymes from the two tissues are identical and, as expected, catalyze both activities; that is, both produce 17α-hydroxyprogesterone and androstenedione from progesterone (Nakajin and Hall, 1981a; Nakajin et al., 1983,1984). Moreover, the two enzymes behave identically when incorporated into liposomes prepared from the phospholipids extracted from the microsomes of the two tissues (Yanagibashi and Hall, 1987). Clearly differences in microsomal lipids cannot explain the different activities of the microsomes from the two tissues. Addition of P450 reductase to the pure enzymes or to microsomes from either tissue increases lyase activity relative to hydroxylase activity. On the other hand, antibody to reductase inhibits lyase activity of microsomes relative to hydroxylation (Yanagibashi and Hall, 1987). Cytochrome b_5, which also transports electrons, exerts a similar effect (Onoda and Hall, 1982). Although the rate of reduction of P450 generally is not believed to be rate limiting for enzyme activity, these findings suggest that increase in the rate of reduction of the $C_{21}scc$ enzyme may limit the extent of the second enzyme activity (i.e., lyase). For lyase activity, a second turn of the P450 catalytic cycle is required, hence a second input of electrons from reductase. Enhanced reduction of the enzyme would be expected to increase the rate or the extent of lyase activity, hence the production of androgens at the expense of 17α-hydroxy C_{21} steroids.

A question that remains unanswered for the $C_{21}scc$ system concerns the possibility of a complex between P450 reductase and one or more molecules of P450. In the liver, spectroscopic and sedimentation studies failed to show evidence for clusters of P450 associated with a single molecule of reductase (Dean and Gray, 1983), whereas studies of enzyme activity as a function of temperature suggested that molecules of P450 may cluster around one or a few molecules of reductase (Peterson et al., 1976). The much lower concentrations of P450 in steroidogenic (as opposed to hepatic) microsomes may produce a different microsomal organization from that of the liver. This question should be examined in these organelles.

III. Regulation of Testicular Steroidogenesis

A. Introduction

In hypophysectomized mammals, testicular androgen secretion continues at a low rate of perhaps 10% of normal levels. This defect can be restored to normal by administration of a single pure protein (LH or CG). Although normal plasma concentrations of testosterone can be achieved in this manner, episodic secretion and responses to a variety of stimuli do not return to those seen in the intact animal (Dufau et al., 1983). Other important autocrine and paracrine influences that are necessary for normal secretion of androgens are discussed in Chapter 6.

LH has long been known to bind specifically to the LH receptor. This binding results in the production of cAMP, which in turn acts as a second messenger for LH by stimulating the synthesis of testosterone and other androgens. The events leading up to the synthesis of cAMP are discussed in Chapter 6 also. Regulation of the synthesis of testosterone by Leydig cells, for the purposes of this chapter, therefore reduces to the action of cAMP on the synthetic pathway. The possible involvement of other second messengers however, also is considered in this chapter. Most of the studies described here address the production of testosterone; other minor androgens will not be considered. Finally, as before, we must refer to investigations performed in the adrenal cortex for the relevant information they provide on the activities of the Leydig cell, although these findings eventually must be tested in this cell before one can conclude that such extrapolation is justified.

B. Site of Action of cAMP in the Steroidogenic Pathway

Early studies on the site of action of LH and cAMP were performed with [^{3}H]cholesterol of high specific activity incubated with slices of tissue with and without LH or cAMP. The exogenous substrate was assumed to mix with endogenous steroidogenic cholesterol and degradation of [^{3}H]cholesterol to small molecule steroid precursors was presumed to be negligible. Under these conditions, LH caused considerable stimulation of the production of [^{3}H]testosterone compared to that of slices incubated without LH (Hall, 1966). With [^{3}H]pregnenolone as substrate, conversion to [^{3}H]testosterone was greater than with [^{3}H]cholesterol; LH was without effect on this conversion (Hall, 1966). Moreover, 20R-hydroxycholesterol was converted rapidly to androgens, so the first hydroxylation of the cholesterol side chain is the slow step in the synthesis of androgens (Hall and Young, 1968).

In the meantime, cycloheximide was shown to inhibit the steroidogenic response of adrenal cells to ACTH and to act at a point in the steroidogenic pathway before side-chain cleavage, suggesting that ACTH stimulates some step(s) before this reaction. Since the step before side-chain cleavage must involve transport of cholesterol from lipid depots in the cytoplasm to the inner mitochondrial membrane, this transport was proposed to be stimulated by the trophic hormone (Garren et al., 1965). This idea was confirmed when side-chain cleavage by isolated mitochondria was shown to be the same whether these organelles were prepared from cells previously incubated with or without ACTH (Nakamura et al., 1980). Evidently, the conversion of cholesterol to pregnenolone is more rapid than the transport of this substrate to mitochondria, as was shown directly by incubating cells with and without ACTH and with and without aminoglutethimide to inhibit side-chain cleavage. With ACTH (without aminoglutethimide) no increase in inner membrane cholesterol was observed with time, but with ACTH plus aminoglutethimide cholesterol accumulated in that membrane, thereby giving the first available assay for intracellular transport of cholesterol and establishing directly that ACTH promotes transport of cholesterol to the inner membrane from the cytoplasm (Nakamura *et al.*, 1980).

When preparations of rat Leydig cells became available (Dufau and Catt, 1975), these ideas were tested with Leydig cells and LH, confirming the findings made with adrenal cells (Hall *et al.*, 1979b). Note that interpretation of these mitochondrial studies calls for consideration of the state of the mitochondria used. Without appropriate precautions, these organelles quickly become anaerobic, and since side-chain cleavage requires oxygen (Shikita and Hall, 1974), anaerobiosis acts (like aminoglutethimide) to inhibit side-chain cleavage. With ACTH, such mitochondria accumulate cholesterol as they do in the presence of the chemical inhibitor (Bell and Harding, 1974). When the organelles are incubated, oxygen is admitted to the system and a burst of synthesis of pregnenolone consumes the accumulated cholesterol. The greater synthesis with ACTH is not the result of stimulation of the enzyme itself but the result of increased transport of cholesterol to the mitochondria during anaerobiosis (Hall, 1985).

LH stimulation of transport of cholesterol to mitochondria now seems clear. Studies in luteal cells show that LH may have an additional effect on the side-chain cleavage system itself because swollen mitochondria from superovulated rats treated with LH showed greater side-chain cleavage than those from untreated rats (Mori and Marsh, 1982). Assuming that swelling of mitochondria does no more than permit free entry of cholesterol to the interior of the mitochondrion, these studies suggest that transport of cholesterol may not be the only site of regulation, at least in luteal cells. Unfortunately, swelling may exert numerous effects on mitochondrial function and the hormonal regimen required to superovulate rats makes these experiments difficult to interpret. Certainly numerous investigators have found no effect of LH administered *in vivo* on mitochondria from testes or ovary when side-chain cleavage is measured *in vitro*. Moreover, neither LH nor cAMP added directly to mitochondria alters the rate of side-chain cleavage. These conclusions are reached on the basis of the unpublished observations of the author and of other investigators. The situation appears to be the same in the adrenal cortex. Nevertheless, the studies of Mori and Marsh (1982) present an unanswered challenge, since cycloheximide does not inhibit the response to LH seen in swollen mitochondria, suggesting that LH exerts some effect on side-chain cleavage that is independent of protein synthesis and not directly involved in accelerating transport of cholesterol.

Subsequently, two groups showed that, in adrenal cells, ACTH in the presence of cycloheximide causes cholesterol to accumulate in the outer mitochondrial membrane. Presumably transport to the inner membrane was inhibited (Ohno *et al.*, 1983; Privale *et al.*, 1983). Evidently ACTH increases transport of cholesterol both to and within mitochondria; the second of these responses requires protein synthesis whereas the first does not. These effects are considered in greater detail in a subsequent section. In the meantime, LH seems likely to increase side-chain cleavage activity under certain conditions, in addition to promoting transport of cholesterol (Marsh, 1976). However, the effect on transport would appear to be more important than a possible effect on side-chain cleavage under conditions of acute stimulation.

One further response is necessary to prepare cholesterol for its role as substrate for steroid synthesis, namely, the conversion of cholesterol ester to free cholesterol. For this conversion, cell type and experimental conditions are of prime importance. In rat and bovine adrenal cortex, cAMP-dependent protein kinase (i.e. protein kinase A, PKA) clearly phosphorylates cholesterol esterase and thereby increases the activity of this enzyme. Hence, available levels of free cholesterol which acts as a source of steroid hormones increase (Trzeciack and Boyd, 1973,1974). cAMP and, hence, LH appear to exert the same effect in Leydig cells (Moyle and Ramachandron, 1973; Moyle *et al.*, 1973). On the other hand, the effect of cAMP on esterase activity is less certain in *corpus luteum* (Marsh, 1976). Bovine corpus luteum contains cholesterol largely as the free steroid with little ester, so the response of the esterase to LH is less obvious (Marsh, 1976). Some of these differences arise from variations in the degree to which a given cell type uses lipoprotein as a source of steroidogenic cholesterol. Such cholesterol is transported into the cell by a receptor mechanism, de-esterified in lysosomes, and re-esterified as oleate (Brown and Goldstein, 1976). This cholesterol ester is stored in lipid droplets. In adrenal cells, this receptor uptake of LDL is stimulated by ACTH (Hall and Nakamura, 1979). The other source of steroidogenic cholesterol is synthesis *de novo* from acetate in the steroidogenic cell. In this case, storage of cholesteryl oleate is presumably less important than under conditions in which cholesteryl ester is transported from plasma into the cell. Note that the amount of cholesterol stored in droplets is increased considerably when steroidogenic cells are grown in culture (P. F. Hall, unpublished results). Presumably, LH activates cholesterol lipase by phosphorylating the enzyme via cAMP and PKA. This action is important when stores of cholesterol ester must be mobilized to provide steroidogenic cholesterol.

C. Mechanism of Action of cAMP

1. Role of protein kinase A

The role of PKA in the response of steroidogenic cells to trophic stimulation has been approached indirectly by numerous studies that were designed to determine whether cAMP is an obligatory intermediate in this stimulation. If cAMP can be shown not to be obligatory for the action of LH, then presumably PKA cannot be obligatory for the action of the gonadotropin.

The most direct evidence on this point derives from adrenal tumor (Y-1) cells, in which a number of mutations of PKA were studied by Schimmer and co-workers (1984). The capacity of such mutants to respond to ACTH was shown to be closely associated with the presence of a functional PKA. Mutants with defective kinase enzyme showed a defective response to ACTH. These studies strongly suggest that PKA and, hence, cAMP are obligatory for the steroidogenic response to ACTH under the conditions tested. A strong case also can be made for Ca^{2+} as a second messenger for LH and ACTH under different conditions. PKA seems likely to be an essential component in the steroidogenic responses to these trophic hormones under most circumstances. The significance of possible alternatives to cAMP are considered in subsequent sections. Note that these remarks apply to acute steroidogenic responses to LH. We later consider slower responses in which gene expression is involved and in which cAMP appears to act on gene transcription. Such effects of the cyclic nucleotide may occur without the intervention of PKA.

2. Is cAMP the only second messenger for LH?

Absolutes are notoriously difficult to prove and the obligatory role of cAMP in the response to LH is no exception. The fact that low concentrations of LH (Dufau et al., 1978), epidermal growth factor (EGF; Verhoeven and Cailleau, 1986), and LH releasing hormone (LHRH; Cooke and Sullivan, 1985) are capable of stimulating steroidogenesis without detectable increase in the cellular concentration of cAMP does not mean that this cyclic nucleotide is not involved in the responses to these agents, although these observations are compatible with that idea. In this connection, four points must be remembered.

1. It is most unlikely that any cell contains a single homogeneous pool of cAMP; direct evidence suggests multiple pools in Leydig cells (Moger, 1991).

2. Exogenous cAMP cannot be assumed to be distributed in the cell in exact correspondence with the distribution of the endogenous nucleotide. The mere fact that investigators find it necessary to use the dibutyryl analog to allow the nucleotide to enter the cell means that an analog of very different solubility from that of the endogenous nucleotide enters the cell to be distributed and metabolized, yet, to the best of present knowledge, the responses to the nucleotide and its analog are the same.

3. Comparisons of thresholds of and sensitivities to LH, cholera toxin, and cAMP are difficult to interpret. For example, cholera toxin may not have access to the same pool of a given species of G protein that is stimulated by LH. This objection gains greater force when we consider the various forms of G protein α subunits present in a given cell (Gilman, 1987). Such comparisons between two or more stimulating agents that enter a pathway at different points are uncertain at best. Consider a hypothetical cell with a single

pool of cAMP. When a specific stimulus is applied to such a cell, every cAMP-dependent process in the cell, whether relevant to the applied stimulus or not, will be stimulated to the same degree. Could such a cell show the amazing flexibility we have come to expect from all living matter?

4. Finally, the idea that, if our methods for measuring cAMP become sufficiently sensitive, we will be able to uncover differences in total cellular cAMP that result from increases in one pool of this nucleotide (no matter how small that pool may be relative to total cellular cAMP) places too much confidence in brute force. In the end, the investigator may be faced with the problem of discerning small differences between large numbers, which usually turns out to be self-defeating.

If these approaches to the negative case prove unrewarding, we should turn to the positive, that is, to evidence for the existence of a specific second messenger. To date, the best success has been obtained with Ca^{2+}. Ca^{2+} may stimulate the synthesis of androgens in Leydig cells by a mechanism that involves Cl^- channels (Duchatelle and Joffre, 1987; Choi and Cooke, 1990). The situation in adrenal cells provides an interesting contrast. Bovine fasciculata cells respond to exogenous Ca^{2+} with a considerable increase in production of cortisol. This response involves voltage-dependent Ca^{2+} channels (Yanagibashi et al., 1989). In contrast, rat fasciculata cells do not respond to exogenous Ca^{2+} (Yanagibashi et al., 1989). Evidently considering Ca^{2+} as a second messenger species may be significant. Learning whether bovine adrenal acquires some advantage or must accept some disadvantage by responding to two agents, as opposed to the rat adrenal, which does not respond to one of these stimuli, will be important.

An interesting approach to this problem in rat Leydig cells showed that, at low concentrations of LH, where correlation between steroid synthesis and total cellular cAMP is poor, an inhibitor of cAMP [(Rp)-cAMPS] decreases the steroidogenic response to the gonadotropin, implying that cAMP is involved in this response (Pereira et al., 1987). If cAMP is not the only second messenger for LH, it may be a necessary element in the responses to the trophic hormone. Moreover, Ca^{2+} and cAMP may both be necessary for a maximal response to LH (Pereira et al., 1987). Clearly more information is required before we can interpret these findings.

3. Action of cAMP on gene transcription

a. Introduction

Molecular biology is rapidly transforming current understanding of every cellular activity. In a given branch of biology, the first approach with molecular biology often involves determining whether or not increased activity of a certain protein results from additional protein, that is, synthesis of the protein, and whether this synthesis also requires new mRNA. These questions often are answered first using inhibitors such as cycloheximide and actinomycin D for proteins and RNA, respectively. More direct evidence can be obtained by labeling newly synthesized proteins with [^{35}S]methionine and immunoprecipitating with a specific antibody to determine the amount of a particular protein present in stimulated and unstimulated cells. Increased production of mRNA can be demonstrated by Northern blotting with a labeled DNA probe. Frequently these methods do little more than confirm results obtained with the inhibitors, although the details of the response (e.g., time, specificity, and extent) can be determined more readily and directly with these two highly specific methods. Incidentally, increase in the amount of specific mRNA can result from increased synthesis or decreased breakdown of mRNA (greater stability). These possibilities can be distinguished by nuclear run-on assay, which measures the rate of production of specific mRNA by nuclei.

The next level of molecular biology involves preparation of DNA that is complementary to mRNA (cDNA), which allows rapid determination of the sequence of mRNA, hence the inferred sequence of the corresponding protein(s). The cDNA also reveals DNA that is transcribed but not translated. This, however, is usually of little interest in relation to regulation of synthesis of the protein concerned. In addition, cDNA makes possible the demonstration of the presence of extensions of a peptide, to include such signals as those required for the passage of the protein to the appropriate cellular compartment, for example, mitochondria.

Further application of molecular biology requires isolation of the relevant gene, which reveals the size and structure of the gene as well as the number, size, and nucleotide sequences of introns and exons in conjunction with nucleotide sequences of flanking DNA that are not transcribed (5' upstream and 3' downstream). This untranscribed DNA (especially the 5' upstream) is often responsible for the regulation of transcription of the gene concerned. At this stage, the gene can be assigned to a specific chromosome and the number of genes for a particular protein, including genes that are not expressed (pseudogenes), can be determined.

For most purposes, the regulation of expression or transcription of the gene proves to be the most important application of molecular biology to the study of regulation of cellular functions. The effects of various substances that regulate cellular activity (called messengers and signals) directly or indirectly influence regulatory DNA, usually DNA that is adjacent to the 5' end of the gene. This influence involves the binding of specific proteins to specific sequences of DNA. These two components are referred to as factors (proteins) and elements (DNA), respectively. The usual approach is to construct a vector composed of regulatory DNA from the gene under consideration attached to another gene referred to as a reporter gene. When this construct is transfected into host cells, the production of the protein corresponding to the reporter gene can be used as a quantitative measure of gene expression. The particular reporter gene is chosen for a number of reasons, including absence of expression in host cells (e.g., use of a bacterial gene in eukaryotic cells) and ease with which the protein can be measured. The current favorite is the bacterial gene chloramphenicol acetyltransferase (CAT). Vectors containing the luciferase gene are gaining popularity (Wood, 1991). The underlying assumption is that, if the regulatory DNA is capable of directing expression of the reporter gene, it can regulate expression of any gene and the effects of various messengers and signals in the chosen host cell will be the same for the transfected constructs as for the same DNA *in situ* within the chromosome of the cells from which the gene was isolated originally. Clearly this may be assuming a great deal; interpretation of such results should be cautious.

Great importance attaches to the choice of the host cell. Unfortunately, a prejudice exists that transformed cells are more appropriate than primary cultures of the cell in which regulation is under examination. This serious misunderstanding arises from alleged difficulties with primary cultures. These difficulties often are rationalized on the grounds that primary cultures of cells are not homogeneous, a view attributable to the fact that many molecular biologists are not well versed in cell biology. Without doubt, mutant cells are potentially misleading. Primary cultures, despite the inevitable difficulties encountered, are preferred. We will have occasion to show how different responses in various host cells can be.

b. Regulation of genes for individual enzymes

i. Cytochrome P450 scc

a. Induction by LH and cAMP.   Studies measuring amounts of P450scc by immunoprecipitation show that LH (hCG) and cAMP promote synthesis of the enzyme in rat Leydig cells (Anderson and Mendelson, 1985) and in Leydig cells from immature pigs (Mason *et al.*, 1984). Similar observations have been made in ovarian cells (Goldring *et al.*, 1987; Voutilainen and Miller, 1987). The question of whether new protein is required to permit this response illustrates some of the problems mentioned in the preceding section. Cycloheximide inhibits the response of the transcription of the gene for P450scc to cAMP in bovine adrenocortical cells (John *et al.*, 1986) and in JEG-3 choriocarcinoma (placental) cells (Picado-Leonard and Miller, 1987) but not in human granulosa cells (Golos *et al.*, 1988) or in mouse Leydig cells (Mellon and Vaisse, 1989). In mutant Leydig cells (MA-10), cycloheximide causes great decreases in mRNA of P450scc but cAMP restores the amount of mRNA to that seen in untreated cells (Mellon and Vaisse, 1989). If the action of cycloheximide is specific for protein synthesis, we must assume that regulation of this gene is different in normal adrenal and normal Leydig cells. If protein synthesis is required for enhanced transcription of a given gene, considering that cAMP promotes synthesis of a nuclear protein that binds to regulatory DNA would be reasonable. If protein synthesis is not required, cAMP is likely to promote phosphorylation of an existing protein, thereby altering its function. Such a radical difference between adrenal cells and Leydig cells is unexpected and not in keeping with the general similarity in other aspects of the regulation of steroid synthesis between the two cell types.

b. Gene. The gene for P450scc is called *CYP11A1* and has been isolated as cDNA from bovine (John *et al.*, 1984; Matteson *et al.*, 1986), pig (Mulheron, 1989), and mouse (Schimmer *et al.*, 1990) adrenal. The human genomic DNA has been isolated (Morohashi *et al.*, 1987) and shown to have 9 exons and 8 introns coding for a sequence of 521 amino acids with 72% homology to the bovine gene product. An extension peptide of 39 amino acids has been identified as the signal required for admission of the enzyme to mitochondria. Rat genomic DNA for *CYP11A1* also has been isolated (Oonk *et al.*, 1990). In humans, only one gene exists for this enzyme and is located on chromosome 15 (Chung *et al.*, 1986).

c. Translocation into mitochondria. The requirement for an extension peptide for admission of the bovine peptide to mitochondria shows tissue specificity. The precursor protein (i.e., with the extension peptide) is admitted to mitochondria from adrenal cortex but not to mitochondria from myocardium, whereas the opposite is true for the enzyme transcarbamylase (Matocha and Waterman, 1984). The bovine enzyme with as few as 19 (of 35) amino acids in the extension peptide can enter mitochondria. The presence of some basic amino acids in this peptide is required for binding and processing (Kumamoto *et al.*, 1987,1989). The mature protein enters the mitochondrion without any signal peptide if chelating compounds such as O-phenanthroline are present or if the system is kept at low temperature (20°C). Translocation and proteolytic removal of the signal peptide can be separated because, under these two conditions, the protein is not subject to proteolysis once admitted to the mitochondrion unless the chelating agent is removed and metal ions are restored or the temperature is raised, as the case may be (Ogishima *et al.*, 1985; Ou *et al.*, 1986).

d. Regulation of expression. Regulation of basal transcription of *CYP11A1* was reported to require DNA near base pair -44 (extending to -56 in the 5′ direction) in Y-1 cells (Morohashi *et al.*, 1987) whereas another group (Chung *et al.*, 1989) reported that basal transcription of this gene in the same cells requires sequences between -145 and -76. Evidently, Y-1 cells do not provide a reproducible system for studies of expression of *CYP11A1*. Moreover, unsynchronized transformed cells do not constitute a uniform population. In contrast, 5′ upstream DNA from the mouse gene from -1500 permits, in Y-1 cells, basal transcription that decreases gradually with deletions to -219. The meaning of such a gradual decline in expression is not clear. Certainly these investigators were not able to delineate a specific sequence of DNA required for basal activity of expression of the mouse gene (Rice *et al.*, 1990). With the upstream sequence of the bovine gene, DNA between −183 and −83 permits basal transcription of a globin gene with a minimal promoter. Results were similar with a CAT expression vector. These various sequences that permit basal expression with the 5′ upstream sequences (human, mouse, and bovine) show no homology to each other. Clearly several different sequences can stimulate transcription of CYP11A1 above the basal level.

Current views on the regulation of transcription suggest that such upstream sequences act to enhance the activity of the common 5′ sequences that occur within the first 50 bp of the start of transcription. These common sequences, collectively referred to as promoters, include a TATA box at approximately −30 and a variety of other conserved sequences. These common elements are believed to permit binding of a number of proteins (transcription factors) as well as RNA polymerase in a complex array that positions the polymerase to begin transcription at the correct base pair (usually referred to as +1). The promoter alone may permit some (basal) level of transcription. Enhancers are believed to provide specific stimulation for the transcription of each gene above that produced by the promoter alone (Johnson and McKnight, 1989). The fact that different elements of DNA can exert such stimulation of expression on the same gene is disturbing and suggests that expression vectors may lack some forms of regulation that are exercised on the same DNA in the chromosome. The first requirement for the clarification of this issue is the use of relevant cells cultured under defined conditions instead of convenient but potentially misleading mutant cells. Using constructs in which the physiological promoter is included and the test DNA ends as close to +1 as possible will also be helpful.

Since cAMP stimulates transcription of CYP11A1, examining upstream sequences for elements that are necessary for such stimulation using expression vectors in which 5′ upstream DNA from the gene is attached to the CAT, luciferase, or another appropriate gene is logical. The results of such studies are confusing. Upstream DNA from human CYP11A1 shows three elements necessary for stimulation of expression by cAMP in Y-1 and bovine adrenal cells (Inoue et al., 1988). In many genes transcribed under the influence of cAMP, a responsive sequence (CGTCA) called CRE (cAMP-responsive element) is found. In the human CYP11A1, the three elements referred to earlier, although different from each other, include a sequence of four of these five bases. These elements may be important for the action of cAMP and, hence, LH on transcription of the human P450scc gene.

In the bovine gene, the sequence from −183 to −83 contains the most important region necessary for a response to cAMP. This sequence contains no element that resembles a CRE (Ahlgren et al., 1990). In the mouse gene, two sequences appear to be important for the response to cAMP in Y-1 cells: −424 to −327 and −219 to −77. Neither of these sequences shows homology to those described for the human or bovine genes. Interestingly, two sequences in the mouse gene at −70 and −40 resemble sequences in the upstream DNA of the 21-hydroxylase gene, sequences that are both involved in the regulation of expression of that gene (Rice *et al.,* 1990). When the effect of -1500 to $+28$ on expression of growth hormone by an expression vector was compared in Y-1 adrenal and MA-10 Leydig cells, expression was observed to be much lower in MA-10 cells than in Y-1 cells. (Rice *et al.,* 1990), serving as a reminder that the host cell is important in the regulation of transcription and suggesting that transcription in MA-10 cells (at least for *CYP11A1*) may require different regulatory factors and elements than are required in normal cells.

The upshot of this seems to be that, although a given gene may be highly conserved between species, transcription of the gene in question may respond to an enhancer sequence that bears no resemblance, in either location relative to start site $(+1)$ or nucleotide sequence, to the corresponding enhancer in other species. At first it is puzzling that, whereas enhancers show these differences, the promoters they stimulate, if not always very similar, are at least basically composed of common elements found in many genes and in many species and located at similar distances from $+1$. Further, if the enhancer must reach the promoter by some mechanism involving looping of DNA in which the loop is kept in place by protein–protein interactions between proteins bound to the distant enhancer and those associated with the proximal promoters, the differences in DNA sequences of enhancers are all the more puzzling. The proteins that bind to different enhancers must bind to the proteins associated with promoters, but these proteins appear to be basically similar in various species and tissues. Some of these problems may be resolved when we know more about the factors or proteins involved.

e. Nuclear proteins. The binding of nuclear proteins to sequences of DNA that appear to be important in regulation of gene expression usually is studied by two methods, gel retardation and DNase footprinting. In the former method, free DNA (no protein) moves faster in agarose gels than the same DNA bound to protein. DNaseI footprinting is based on the fact that DNA sequences to which protein is bound are protected from cleavage by DNaseI. The specificity of these interactions must be established carefully by competition with other DNA. In the human *CYP11A1* gene, nuclear proteins have been found that specifically bind to CRE-like elements mentioned earlier (Inoue *et al.,* 1988). Y-1 cells contain nuclear proteins that bind specifically to upstream sequences of the mouse gene. These sequences of upstream DNA include two motifs also seen in the 21-hydroxylase gene and an AP-1 motif at -120 (Rice *et al.,* 1990). The relevance of these proteins to the regulation of expression of the mouse gene remains to be determined.

One of the most surprising differences between species is in the effect of cycloheximide on expression of *CYP11A1,* as discussed earlier. The fact that this substance does not inhibit the cAMP response in normal Leydig cells (Mellon and Vaisse, 1989) but does so in normal adrenal cells (John *et al.,*

1986) suggests that the cyclic nucleotide can promote synthesis of one or more nuclear proteins in the adrenal but acts in a completely different way in Leydig cells, in which it presumably stimulates transcription of *CYP11A1* by phosphorylating an existing protein(s). More serious is the observation that cycloheximide inhibits stimulation by cAMP of the expression of *CYP11A1* in Y-1 cells but does not inhibit stimulation by this cyclic nucleotide of the expression of a reporter gene that includes 5' upstream DNA from bovine *CYP11A1* transfected into the same cells or into BAC cells (Ahlgren *et al.*, 1990). The response of the construct to cAMP appears to be quite different from that of the endogenous gene. If the construct is thought to report on the expression of the normal gene, it is not reporting accurately in this case. This result suggests that the chromosomal environment may be important in gene expression in ways that are not yet apparent, which casts doubt on current interpretation of expression studies.

ii. Cytochrome P450 C$_{21}$scc

a. Induction by LH.   In rat Leydig cells, LH and cAMP increase amounts of cytochrome P450 C$_{21}$scc (17α) mRNA (Nishihara *et al.*, 1988). In adrenocortical cells, ACTH induces synthesis of the enzyme (McCarthy and Waterman, 1983). This induction accounts for the increase in the ratio of cortisol to corticosterone produced by prolonged treatment with ACTH.

b. Gene.   The gene for the hydroxylase/lyase enzyme is called *CYP17*, was isolated from human cDNA, and was shown to be the same gene in testis and adrenal (Chung *et al.*, 1987). The sequence of bases in the gene showed significant similarity to steroid 21-hydroxylase (Picado-Leonard and Miller, 1987). The gene also was isolated from bovine cDNA and was shown to belong to a new family within the P450 superfamily (Zuber *et al.*, 1986). ACTH was found to increase the concentration of P45017α mRNA and, hence, the amount of enzyme in adrenal cells. Fevold *et al.* (1989) isolated the gene from rat testicular cDNA. This gene shows homology with the human (69%) and bovine (64%) genes. The human gene has been assigned to chromosome 10 (Matteson *et al.*, 1986) and is expressed in adrenal and testis (Chung *et al.*, 1987). The bovine gene is composed of 8 exons with considerable homology to the human gene in both the coding and 5' upstream regions (Bhasker *et al.*, 1989). The rat (Nason *et al.*, 1992) and porcine genes show similar exon–intron structures.

c. Insertion in microsomal membrane.   Microsomal cytochromes P450 insert into the microsomal membrane by the usual mechanism of a hydrophobic N-terminal tail that serves as a signal and as an anchor to fix the protein in the membrane (Walter and Blobel, 1981). By expressing the *CYP17* gene in COS-1 monkey kidney cells, Clark and Waterman (1991) showed that, when the *CYP17* gene lacked the sequence for amino acids 2–17, the resulting protein was 2.5 times less successful than the full-length molecule in binding to the microsomal membrane. Moreover, the enzyme molecules that did succeed in binding were inactive and did not show the characteristic CO difference spectrum. These observations suggest that the hydrophobic tail is not a passive anchor but is necessary for correct folding of the molecule to an active conformation.

   d. Regulation of expression.   When expression vectors were used to identify regulatory regions 5' upstream of bovine *CYP17*, the response to cAMP was the first consideration. Two elements were found that can each respond to cAMP when the gene is transfected into Y-1 mouse adrenal cells and into bovine adrenocortical cells, with subsequent increased expression of reporter genes (Lund *et al.*, 1990). These elements are situated at -243 to -225 and -80 to -40. Future studies will be necessary to determine the significance of these findings.

   Neither element resembles the other and neither resembles the well known CRE through which cAMP stimulates transcription of other genes. Again, a similar paradox to that discussed earlier for *CYP11A1* is seen. Whereas cAMP produces stimulation of expression of *CYP17* (measured by immunoprecipitation and by Northern blotting) that is slow in onset and is inhibited by cycloheximide when tested in bovine adrenocortical cells (Zuber *et al.*, 1986), the analogous response in Y-1 cells transfected with the expression vector (-243 to -225) OV is relatively rapid in onset and is not inhibited by cycloheximide. Again, this result suggests that the chromosomal environment is important in the regulation of expression of the gene. Note that interpretation is further complicated because *CYP17* normally is not expressed in Y-1 cells. Clearly the expression vector does not provide an entirely reliable system for studying events in the normal cell.

   e. Nuclear proteins.   A start has been made in determining the nature and number of nuclear proteins responsible for regulation of expression of *CYP17*. A protein of MW 47,000 binds specifically to DNA -243 to -225 of the bovine gene. Part of this sequence resembles the classical CRE, so the 47-kDa protein may play a role in the action of cAMP on transcription of the gene. Moreover, a nuclear protein of the same molecular weight from bovine adrenocortical cells binds specifically to a synthetic CRE (Zanger *et al.*, 1991).

   iii. 3β-Hydroxysteroid dehydrogenase/isomerase.   A full-length cDNA clone of the human placental enzyme has been reported (Zhao *et al.*, 1989). The gene has been assigned to chromosome 17 (Luu-Thé *et al.*, 1989a). This gene has been expressed in COS-1 cells and the resulting protein catalyzes both dehydrogenase and isomerase activities, showing beyond doubt that the conversion of the Δ^5-3β-ol to the Δ^4-3-one is catalyzed by one enzyme (Zhao *et al.*, 1989). A start has been made on investigating regulation of expression of 3β-hydroxysteroid dehydrogenase/isomerase (3β-HSDI). Northern blotting showed that cAMP is necessary for maximal expression but basal expression is high. Both cortisol and testosterone inhibit expression of the gene (Payne and Shah, 1991).

iv. **17β-Hydroxysteroid dehydrogenase.** Two genes in tandem corresponding to 17β-hydroxysteroid dehydrogenase (17β-HSD) have been isolated and are referred to as I and II (Luu-Thé *et al.*, 1990). These genes have been assigned to chromosome 17 (Tremblay *et al.*, 1989). The 17β-HSDII gene is the form found exclusively expressed in adrenal and gonads. This gene is probably responsible for the usual form of 3β-HSD deficiency (Lachance *et al.*, 1991). The gene gives rise to the estrogenic enzyme (Tremlay *et al.*, 1989). Gene I may correspond to the second or androgenic 17β-HSD enzyme (Luu-Thé *et al.*, 1989b). The human type I gene has been assigned to chromosome 1 (Lachance *et al.*, 1990) and is found to be highly expressed in the human placenta. The first studies on regulation of expression of gene II show that hCG increases the levels of the corresponding mRNA in porcine granulosa cells (Chedrese *et al.*, 1990).

c. Mechanism of action of cAMP on gene expression

These studies do not approach the question of how cAMP influences transcription of genes. Investigations with nonsteroidogenic cells also have failed to determine whether or not PKA mediates this action of cAMP or whether phosphorylation of proteins is involved. Note that the responses of gene transcription to cAMP are much slower than the acute response of steroid synthesis.

4. Role of proteins in the response to cAMP

a. Introduction

In 1962, steroidogenic response to LH was reported to be inhibited by puromycin (Hall and Eik-Nes, 1962). Several technical advances, including methods for the preparation of Leydig cells, were required to demonstrate stimulation by LH of the incorporation of amino acids into total Leydig cell protein (Irby and Hall, 1971). Actinomycin D inhibits the response to LH (Mendelson *et al.*, 1975). Evidently, new mRNA is required to permit a response to LH. Details of the inhibitory action of cycloheximide on the response to ACTH also suggested that the newly synthesized proteins are short lived. The question raised by these studies is whether the new proteins are necessary to permit the response to the hormone or whether specific proteins discharge various steps in the response, that is, could continued synthesis of new protein be essential for normal cellular function so new proteins are not specifically concerned with the action of LH? When separation of cellular proteins by gel electrophoresis became widely used, specific proteins could be studied. A search was made for those proteins synthesized under the influence of LH. Eventually researchers hoped to find proteins capable of specific functions related to the action of LH. Relevant proteins were expected to show short half-lives. This search is not yet completed, although interesting proteins have been isolated.

b. Specific proteins

i. 21K and 33K. Janszen *et al.* 1977) reported the occurrence of two specific proteins synthesized in rat Leydig cells under the influence of LH. Neither protein shows a short half-life and the influence of LH on these proteins may be indirect. The functions of the two proteins are unknown.

ii. 30K. Epstein and Orme-Johnson (1991) studied the properties of two mitochondrial proteins in rat adrenal cells. These proteins are made in the cytoplasm under the influence of ACTH and are both of short half-life. 37K appears to be processed in mitochondria to 32K, which in turn gives rise to a mitochondrial phosphoprotein of 30K. Although the precursors (37K and 32K) are labile, 30K decays more slowly. The increased synthesis of 37K produced by ACTH is inhibited by cycloheximide. Similar or identical proteins showing the same properties have been found in Leydig cells (Orme-Johnson, 1991) and bovine corpus luteum (Pon and Johnson, 1986). The characteristics of these proteins have been reviewed in detail (Orme-Johnson, 1990). These features of the proteins give them strong claims to important roles in the mechanism of trophic stimulation of steroidogenesis: they are conserved, labile, found in mitochondria, are phosphorylated, and accumulate in small amounts following stimulation by LH or dibutyryl 3',5'-cAMP. Elucidation of the function of the 30-kDa protein should provide important information about the actions of LH and ACTH.

iii. Sterol carrier protein 2. Work in liver revealed that synthesis of cholesterol takes place with intermediates bound to protein (Ritter and Dempsy, 1971; Scallen *et al.*, 1971). Such a protein, called SCP-2 (sterol carrier protein), was also isolated from steroidogenic tissues. Among these tissues, most work was performed on adrenal cortex, but Leydig cells are likely to contain the same or a similar protein (Chanderbahn *et al.*, 1982). SCP-2 is believed to promote the movement of cholesterol from lipid droplets into mitochondria and from outer to inner mitochondrial membrane (Chanderbahn *et al.*, 1982; Vahouny *et al.*, 1983). These changes result in increased synthesis of pregnenolone when SCP-2 is added to isolated mitochondria. SCP-2 does not cause the spectral shift described earlier that demonstrates increased binding of cholesterol by P450 (Vahouny *et al.*, 1983). Therefore, how SCP-2 stimulates production of pregnenolone if the effects on cholesterol transport do not result in greater loading of the enzyme with substrate is not clear. Mitochondrial SCP-2 is found in the inner membrane but more than 50% of the total adrenal cell SCP-2 is found outside that organelle. The concentrations of SCP-2 in the inner membrane show a diurnal rhythm consistent with changes in adrenal steroid production (Conneely et al., 1984). Moreover, these concentrations of SCP-2 in the inner membrane correlate under various experimental conditions with those of cholesterol (Conneely *et al.*, 1984). Although ACTH stimulates synthesis of SCP-2, the response is slow and is more nearly consistent with a trophic rather than an acute regulatory function in steroid synthesis (Trizeciak *et al.*, 1987).

iv. Sterol activating peptide.

An interesting protein of MW 2200 was isolated from rat adrenal cell cytosol by acid extraction (Pedersen and Brownie, 1983). This protein stimulates production of pregnenolone by mitochondria from rat adrenal when the organelles are prepared from hypophysectomized rats treated with cycloheximide and ACTH. This elaborate preparation is required to load the mitochondria with cholesterol (cycloheximide and ACTH) and to provide a low background activity of side-chain cleavage (hypophysectomy) (Pedersen, 1984). A similar but slightly different protein was isolated from Leydig tumor cells (Pedersen and Brownie, 1987). A peptide that closely resembles sterol activating peptide (SAP) has been found in Leydig and ovarian cells also (Pedersen, 1987). These proteins are believed to be closely related and show the following properties:

1. They are increased in amount by ACTH or LH. This effect is blocked by cycloheximide.
2. They give spectral evidence of increasing the loading of P450 with cholesterol (substrate-induced difference spectrum).
3. They do not bind cholesterol.
4. The amino acid sequence of the proteins closely resembles the C-terminal sequence of a heat-shock protein (glucose regulated protein GRP78).
5. Although GRP78 is widely distributed in various tissues, only the steroidogenic tissues appear to produce more SAP under the influence of cAMP.

Investigators propose that a larger precursor protein (possibly GRP78) is cleaved under the influence of ACTH (LH) or cAMP to give SAP. The cleavage may occur during translation in view of the inhibitory effect of cycloheximide on the synthesis of SAP (Mertz and Pedersen, 1989). Clearly the role of this protein in the response to LH deserves further investigation, especially in relation to its turnover and its action in the inner mitochondrial membrane.

v. Endozepine.

Endozepine already was well known as a form of endogenous valium, possibly the counterpart of endogenous morphine (endorphin), when a protein of ~8200 MW was isolated from bovine adrenal cortex (Yanagibashi *et al.*, 1989). This 8.2-kDa protein was isolated as the result of a search for a protein that could discharge the critical actions of ACTH on the transport of cholesterol to and within mitochondria (Yanagibashi *et al.*, 1988). Endozepine was found to promote movement of cholesterol from outer to inner mitochondrial membrane and loads P450scc with substrate (Yanagibashi *et al.*, 1988). This last action of endozepine is of interest because most enzymes do not require loading with substrate. In the case of P450scc, loading appears to be a limiting factor in the activity of this enzyme. How endozepine accomplishes this loading is uncertain, but the protein exerts this effect on adrenal inner mitochondrial membrane when it is added to this membrane *in vitro* (Brown and Hall, 1991). Recall that binding of P450scc promotes reduction of the enzyme (see previous discussion). When endozepine is added to an enzyme system reconstituted from pure P450scc and electron carriers, the rate of reduction of P450 is increased, but the protein does not have this effect unless substrate (cholesterol) is present (Brown and Hall, 1991). Presumably endozepine loads the enzyme with substrate, promoting reduction of P450 and accelerating side-chain cleavage of cholesterol (Brown and Hall, 1991). Since the synthesis of endozepine in adrenal cells is accelerated by ACTH (Yanagibashi *et al.*, 1988) this protein may be one of the newly synthesized proteins that mediate the steroidogenic action of ACTH. Endozepine has been isolated from bovine Leydig cells, so these observations are likely to apply to the action of LH on Leydig cells.

vi. **Miscellaneous proteins.** Stocco and Sodeman (1991) reported the synthesis of a 30-kDa protein under the influence of hCG and cAMP in MA-10 cells. This protein is formed as the result of proteolytic processing of larger proteins (37 and 32 kDa) in mitochondria. This 30-kDa protein may be important in intramitochondrial transport of cholesterol. Other proteins that may be involved in the response to ACTH have been reviewed elsewhere (Pedersen, 1984).

vii. **Involvement of GTP.** Clearly each of these proteins has some claim to a role in the response of adrenal (and presumably Leydig and ovarian) cells to trophic stimulation. Whether one is a prime mover or whether the full response requires all the proteins remains to be seen. Meanwhile, the interesting observation has been made that GTP added to mitochondria promotes conversion of cholesterol to pregnenolone (Xu *et al.*, 1989). GTP must be hydrolyzed to produce this response, and the effect is specific for this nucleotide to the extent that this has been tested. GTP therefore becomes another candidate for an effect on cholesterol transport, although this has not been examined in Leydig cells. No doubt this action of GTP requires a specific protein.

c. **Posttranslational modification of proteins**

Phosphorylation has been the most intensely studied modification of proteins by LH, no doubt because of the primary role of cAMP as second messenger. To date, no clear relationships have been established between the phosphorylation of specific proteins and the events required to accelerate steroid synthesis. Cooke *et al.* (1977) showed that LH promotes phosphorylation of three Leydig proteins (14 kDa, 57 kDa, and 78 kDa). The extent of phosphorylation of these proteins shows some correlation with the steroidogenic response to LH. In the ovary, attempts have been made to associate the action of LH with phosphorylation of mitochondrial proteins because of the possibility that such changes may be related to regulation of the rate-limiting side-chain cleavage step. LH increases phosphorylation of two mitochondrial proteins in ovarian follicles but no functional significance for these changes was demonstrated (Neymark et al., 1984). Inaba and Wiest (1985) showed that disruption of the mitochondrial membranes of rat luteal cells by Ca^{2+} allowed PKA to accelerate side-chain cleavage, possibly by increasing access of cholesterol to the enzyme. That cAMP and, hence, PKA increase access of substrate to P450scc is in keeping with ideas expressed earlier in connection with the site of action of LH.

The question of phosphorylation has been approached by examining the action of protein kinase C in Leydig cells. Phorbol ester stimulates phosphorylation of Leydig cell proteins (Lin, 1985; Moger, 1985). This response may depend on extracellular Ca^{2+} since it is blocked by nifedipine, an inhibitor of Ca^{2+} channels (Lin, 1985). In keeping with a role for protein kinase C in the regulation of steroid synthesis in Leydig cells, 4β-phorbol-12-myristate-13-acetate stimulates synthesis of testosterone, but nontumorigenic phorbol esters are without effect (Moger, 1985). Interestingly, although phorbol esters stimulate steroid synthesis in adrenal cells, this action is not inhibited by cycloheximide, suggesting that protein kinase C is not on the direct pathway from ACTH to increased steroidogenesis (Widmaier and Hall, 1985). This kinase may serve an ancillary role in regulating adrenal steroidogenesis. More information is necessary before the Leydig cell system can be understood.

The cytoskeleton plays an important role in the regulation of steroid synthesis and specifically in the regulation of cholesterol transport. Significantly, both protein kinase C (Papadopoulos and Hall, 1989) and the Ca^{2+}-dependent phosphatase calcineurin (Papadopoulos et al., 1990a) are found tightly attached to the cytoskeletons of adrenal cells. Both these proteins are capable of causing rapid and extensive changes in the state of phosphorylation of cytoskeletal proteins. Note also that protein kinase C added to the cytoskeletons of adrenal cells causes rounding of these structures (Papadopoulos and Hall, 1989). This finding suggests a possible role for this kinase in cholesterol transport.

5. Role of phospholipids in response to cAMP

The phosphoinositide cycle is now well known as part of the response of cells to numerous stimuli. The cycle involves cleavage of phosphatidylinositol 4,5-bisphosphate to give the two important products inositol triphosphate (IP_3), which is a Ca^{2+} ionophore, and diacylglycerol (DG), which activates protein kinase C (Berridge et al., 1983; Burgess et al., 1984). Although the regulation of the cycle is complex and tetra- and pentaphosphates of inositol may be involved, two major consequences of the cycle are increased levels of activated protein kinase C and increased concentrations of intracellular Ca^{2+}. One group has proposed that the cycle is important in the response of Leydig cells to LH (Lowitt et al., 1982; Farese, 1984). According to this hypothesis, the cycle acts primarily not to activate protein kinase C but to provide phosphatidylphosphoinositides that act directly on mitochondria to accelerate side-chain cleavage (Farese, 1984). These studies were based on measurements of the mass of unlabeled phosphoinositides, the small amounts of which are close to the threshold of the available methods. These studies also led to the idea that, unlike a number of stimulating agents that affect the phosphoinositol cycle by promoting hydrolysis of phosphoinositides catalyzed by phospholipase C, the trophic hormones LH and ACTH, and perhaps other agents, act on the cycle by increasing synthesis of phosphatidic acid (Farese, 1983). Further studies are needed to clarify the mechanism of action of the phosphoinositides on the side-chain cleavage of cholesterol. Analogous studies were reported in adrenal cortex with the same result for ACTH (Farese *et al.*, 1980,1983). The classical phosphoinositide pathway leading to increased intracellular Ca^{2+} does not appear to operate in mouse adrenal tumor cells or in bovine fasciculata cells (Iida *et al.*, 1986). Clearly more evidence is needed to demonstrate the special role of phosphoinositides in the stimulation of side-chain cleavage in Leydig cells. At present no evidence supports the involvement of the classical phosphoinositide cycle via protein kinase C in response to LH. Protein kinase C may be involved in the regulation of steroid synthesis in Leydig cells, as mentioned earlier. In that case, the activation of Leydig cell protein kinase C may take place by another mechanism.

6. Role of Ca^{2+} in response to LH

Earlier, Ca^{2+} was suggested to be the prime candidate for a second messenger for LH in addition to cAMP. Ca^{2+} is involved in many cellular activities in two principal ways. First, Ca^{2+} itself exerts a permissive effect, that is, the ion must be present but does not exert a primary and specific influence in its own right. Second, Ca^{2+} can act through specific binding proteins of which calmodulin is the best known. In the case of steroid synthesis, the extent of the permissive effects of Ca^{2+} is not known. For example, Ca^{2+} must be present for ACTH to stimulate adrenal cells (Birmingham *et al.*, 1953). The same may be true for LH. However, in view of the ubiquitous nature of Ca^{2+}, this is hardly surprising.

Study of the more specific effects of Ca^{2+} in Leydig cells has begun. Trifluoperazine inhibits the steroidogenic action of LH (Hall *et al.*, 1981). This inhibition is exerted at the step(s) involved in transport of cholesterol to mitochondria. When calmodulin is injected into Leydig cells by way of liposomes, the transport of cholesterol to mitochondria and the subsequent synthesis of testosterone are accelerated (Hall *et al.*, 1981). These studies were interpreted to mean that LH promotes redistribution of calmodulin to those regions of the cell in which transport of cholesterol occurs. Such redistribution has been reported in pituitary cells under the influence of gonadotropin releasing hormone (Gn RH) (Conn *et al.*, 1981). In keeping with this idea, when Ca^{2+} and calmodulin are injected into Leydig cells from separate liposomes, stimulation of the rate of synthesis of androgens was intermediate between that seen with calmodulin alone and that with calmodulin saturated with Ca^{2+}. In addition, in adrenal cells, calmodulin appears to be responsible for the coupling of occupied ACTH receptors to the heterotrimeric G protein Gs in the plasma membrane (Papadopoulos et al., 1990b). Moreover, an important mechanism for the entry of Ca^{2+} into bovine adrenal cells is through voltage-dependent Ca^{2+} channels. Ca^{2+} entering via these channels stimulates steroid synthesis without ACTH or an increase in cellular cAMP. Rat adrenal cells show no evidence of such channels (Yanagibashi et al., 1990). Determining whether the same is true of Leydig cells from the two species will be important.

7. Role of the cytoskeleton in response to cAMP

a. Introduction

The cytoskeleton is the most recent cellular component to come under intensive study. This structure provides the cell with shape and polarity and divides the cytoplasm into loosely organized compartments. The three major elements in the cytoskeleton — microtubules, microfilaments, and intermediate filaments — offer surfaces to which water and other cytosolic molecules bind to facilitate productive encounters between components of various pathways (Ottoway and Mowbray, 1977; Hall, 1982,1984; Clegg, 1982). Microfilaments contain actin, which has the potential to trigger myosin ATPase with consequent shortening. Shortening in turn is capable of promoting intracellular movement and, in some cases, movement of the whole cell. Microfilaments bind proteins in the plasma membrane and limit the mobility of these molecules in the plane of the plasma membrane (Tank et al., 1982). Intermediate filaments are less well understood but provide structural stability and contact between the nucleus and the rest of the cell (Fey et al., 1984). These filaments show dynamic activities under a variety of conditions (Steinert and Liem, 1990). Microtubules can shorten by treadmilling, can promote changes in cell shape, and are important in mitosis and secretion (Hall, 1982,1984).

b. Microfilaments

The first hint that the cytoskeleton may be involved in steroid synthesis came with the discovery that ACTH stimulates the transport of cholesterol to mitochondria (Garren et al., 1965; Hall et al., 1979b; Crivello and Jefcoate, 1978). In testing the possibility that this transport involves microfilaments, acceleration of cholesterol transport by ACTH and cAMP was found to be inhibited by cytochalasin under conditions in which cytochalasin was acting by specific inhibition of microfilaments (Mrotek and Hall, 1974,1977). This conclusion was verified by injecting antiactin into adrenal cells, which also inhibited the increased transport of cholesterol produced by ACTH (Hall *et al.*, 1979b). Later this response was demonstrated to be prevented by another inhibitor of actin (DNaseI); also the intracellular concentration of DNaseI required to cause half maximal inhibition of the response to ACTH was determined (Osawa *et al.*, 1984). These findings were confirmed with LH in Leydig cells (Hall *et al.*, 1979c). Although how actin accelerates the transport of cholesterol to mitochondria remains unclear, ACTH and cAMP cause intense rounding of adrenal cells (Yasumara, 1968) and, when cells are grown on plastic surfaces coated with poly(HEMA), they are caused to grow in rounded form. [Poly (HEMA) covers charges on the surface of plastic dishes and limits the intensity of attachment of the cells to the dishes.] This rounding is associated with accelerated steroid synthesis and transport of cholesterol to mitochondria in the absence of ACTH and cAMP (Betz and Hall, 1987). These observations suggest that the shape of the cell may influence transport of cholesterol and that contraction associated with actin (and presumably myosin) may be involved.

c. Intermediate filaments

The intermediate filament system was regarded until recently to be the Cinderella of the cytoskeleton, since it was believed to behave as a static scaffolding for the cell. More recent studies suggest that these filaments are capable of rapid turnover and reorganization (Steinert and Liem, 1990). Despite this new insight into the dynamic nature of intermediate filaments, very little is known about the functional significance of these changes. Work in this laboratory with cultured adrenal cells has revealed the surprising fact that lipid droplets are bound tightly to intermediate filaments. One can remove all cell components except intermediate filaments and bound lipid droplets (Almahbobi and Hall, 1990; Almahbobi *et al.*, 1991). A greater surprise came when mitochondria also were found to be attached to inter-mediate filaments (Almahbobi *et al.*, 1992b). Since steroidogenic cholesterol is stored in lipid droplets and is transported to mitochondria, these findings raise tantalizing possibilities to explain the mechanism of this targeted trans-port process. Clearly it is tempting, if perhaps premature, to think of a possible role of intermediate filaments as guidewires for the transport of lipid droplets. These findings with intermediate filaments in adrenal cells all have been demonstrated in rat Leydig cells (Almahbobi *et al.*, 1993). How actin microfilaments fit into this picture is unclear. However, work in fibroblasts shows that an ATP-dependent contractile process can collapse intermediate filaments and cause the cell to round up (Tint *et al.*, 1991). Perhaps actin acts in a similar fashion in steroidogenic cells by causing collapse of intermediate filaments and allowing the attached droplets and mitochondria to interact, with consequent transfer of cholesterol from droplets to mitochondria.

IV. Summary and Conclusions

The synthesis of androgens by Leydig cells and the mechanisms by which this complex process is regulated are far from clear, despite many exciting discoveries. Each new discovery opens new vistas, but increases the complexity of the problem. Researchers who hope for an incisive answer to the riddle of how LH acts, that will stand beside the brilliant elucidation of the genetic code, can be forgiven for expressing some sense of despair.

Testosterone is synthesized form cholesterol by four enzymatic reactions, each of which is complicated in its own way. Two of these enzymes are cytochromes P450 ($C_{27}scc$ and $C_{21}scc$, also called 17α). P450 is a monooxygenase that uses atmospheric oxygen as a substrate. This enzyme is reduced by electron carriers and, in consequence, activates oxygen that is used to attack bound steroid substrates. These two enzymes remove the side chain of cholesterol in two oxidative steps. The two remaining enzymes are both dehydrogenases that employ pyridine nucleotides. One of the two also catalyzes isomerization in the A and B rings. P450scc is located in the inner mitochondrial membrane; the remaining three enzymes are found in the microsomal membrane. In both cases, the membrane and the other proteins that it contains influence the course of the reactions catalyzed. We have some limited insight into the important influence of the membrane environment on the reactions catalyzed.

The principal but not the only source of regulation of the synthesis of androgens consists of pulses of LH from the adenohypophysis. cAMP mediates the stimulating influence of LH on the synthesis of androgens. Among other agents that may share this regulatory function, Ca^{2+} has the greatest claim on such a role. However, cAMP is likely always to be involved in the action of LH (to the extent that anyone can say "always" in this field).

cAMP stimulates the synthesis of androgens, at least in the case of rapid responses, by increasing the production of pregnenolone by P450scc in mitochondria. Clearly the cyclic nucleotide does this by accelerating the transport of cholesterol from lipid depots in the cytosol to the enzyme, which lies buried in the inner membrane of mitochondria. cAMP may or may not also exert an effect on the enzyme itself. Transport of cholesterol involves movement to that organelle and movement within the mitochondrion from outer to inner membrane. This second step requires the synthesis of new protein. The first step involves microfilaments and intermediate filaments. A recent discovery in adrenal and Leydig cells is that lipid droplets that store steroidogenic cholesterol and mitochondria in which the synthesis of pregnenolone occurs are both tethered to intermediate filaments. In fibroblasts, intermediate filaments can be collapsed by ATP-dependent contraction of actomyosin. Perhaps this latter mechanism occurs in Leydig cells and facilitates the movement of cholesterol to the inner mitochondrial membrane.

To increase synthesis of androgens under the influence of cAMP, Leydig cells employ a number of their biological resources. These cells make new proteins that are not the steroidogenic enzymes themselves. Some of the new proteins are numerous and short lived. These proteins have not all been catalogued and in no case are their functions completely clear. A carrier protein for the insoluble cholesterol, a homolog of a heat-shock protein, and the "endogenous valium" called endozepine are among these proteins. In addition, proteins of no known function are synthesized rapidly under the influence of cAMP under conditions that speak strongly for their importance in the response to cAMP. Some of these new proteins, and others that are already present in the resting cell, undergo phosphorylation. Changes in the cellular content of phospholipids also occur; possible effects of these changes include a direct influence on the mitochondrial step of pregnenolone synthesis.

LH and cAMP also influence transcription of genes for the steroidogenic enzymes. These responses are striking but occur in a slower time frame than the effects discussed already. Whether this function is one of maintenance that secures appropriate levels of these vital proteins or whether the Leydig cell regulates gene expression for more leisurely changes demanded by long-term circadian or seasonal changes must still be determined. Whereas all the effects of cAMP discussed so far appear to result from the action of PKA, the effects of the nucleotide on gene expression may or may not be entirely dependent on the kinase.

If we think of all this information gathered from the work of many students, fellows, and other investigators as a jigsaw puzzle under construction, the sense of despair mentioned at the outset returns, for we are still unable to see clear outlines of the final structure of the puzzle. Insertion of each new piece brings us ever closer to the solution or solutions of the riddle of the action of LH and, in doing so, extends and reinforces contemporary understanding of cells in general, not only of steroidogenic cells. A brilliant flash of enlightenment is not likely to resolve the whole issue. We must continue patiently to identify the new fragments of the puzzle as they slowly emerge.

References

Ahlgren, R., Simpson, R. R., Waterman, M. R., and Lund, J. (1990). Characterization of the promoter/regulatory region of bovine CYP11AI(P450scc) gene. *J. Biol. Chem.* **265,** 3313–3319.

Albert, D. H., Ascoli, M., Puett, D., and Coniglio, J. G. (1980). Lipid composition and gonadotropin-mediated lipid metabolism of the M5480 murine Leydig cell tumor. *J. Lipid Res.* **21,** 862–865.

Almahbobi, G., and Hall, P. F. (1990). The role of intermediate filaments in adrenal steroidogenesis. *J. Cell Sci.* **97,** 679–687.

Almahbobi, G., Williams, L. J., and Hall, P. F. (1992a). Attachment of steroidogenic lipid droplets to intermediate filaments in adrenal cells. *J. Cell Sci.* **101**, 383–393.

Almahbobi, G., Williams, L. J., and Hall, P. F. (1992b). Attachment of mitochondria to intermediate filaments in adrenal cells. *Exp. Cell Res.* **200**, 361–369.

Almahbobi, G., Williams, L. J., Han, X. G., and Hall, P. F. (1993c). Lipid droplets and mitochondria are bound to intermediate filaments in Leydig cells. *J. Reprod. Fertil.* **98**, 209–217.

Anderson, C. M., and Mendelson, C. R. (1985). Regulation of steroidogenesis in rat Leydig cells in culture. *Arch. Biochem. Biophys.* **238**, 378–387.

Ang, D., Liberek, K., Skowysna, D., Zylicz, M., and Georgopoulos, C. (1991). Biological role and regulation of the universally conserved heat shock proteins. *J. Biol. Chem.* **266**, 24, 233–24, 236.

Besman, M. J., Yanagibashi, K., Lee, T. D., Kawamura, M., Hall, P. F., and Shively, J. E. (1989). Identification of des(Gly-Ile)-endozepine as an effector of corticotropin-dependent adrenal steroidogenesis: Stimulation of cholesterol delivery is mediated by the peripheral benzodiazepine receptor. *Proc. Natl. Acad. Sci. U.S.A.* **86**, 4897–4901.

Berridge, M. J., Dawson, R. M. C., Downes, C. P. P., Heslop, J. P., and Irvine, R. F. (1983). Changes in levels of inositol phosphates after agonist-dependent hydrolysis of membrane phosphoinositide. *Biochem. J.* **212**, 473–479.

Bell, J. J., and Harding, B. (1974). The acute action of ACTH on adrenal steroidogenesis. *Biochim. Biophys. Acta* **348**, 285–292.

Betz, G., and Hall, P. G. (1987). Steroidogenesis in adrenal tumor cells: Influence of cell shape. *Endocrinology* **120**, 2547–2554.

Bhasker, C. R., Adler, B. S., Dee, A., John, M. E., Kagimoto, M., Zuber, M. X., Algren, R., Wang, Z., Simpson, E. R., and Waterman, M. R. (1989). Structural characterization of the bovine CYP17 gene. *Arch. Biochem. Biophys.* **271**, 479–487.

Birmingham, M. K., Elliot, F. H., and Valere, P. H. L. (1953). The need for the presence of Ca^{2+} for the stimulation *in vitro* of rat adrenal glands by ACTH. *Endocrinology* **53**, 687–693.

Boggaram, V., Zuber, M. X., and Waterman, M. R. (1984). Turnover of newly synthesized cytochromes P450scc and P45011β and adrenodoxin in bovine adrenocortical cells in monolayer culture: Effect of adrenocorticotropin. *Arch. Biochem. Biophys.* **231**, 518–523.

Bogovitch, K., and Payne, A. H. (1980). Purification of rat testicular microsomal 17β-ketosteroid reductase. *J. Biol. Chem.* **255**, 5552–5558.

Bresnick, E. (1980). The molecular biology of the induction of the hepatic mixed function oxidase. *In* "Hepatic Cytochrome P450 Monooxygenase System" (J. B. Schenkman and D. Kupfer, eds.), pp. 191–224. Pergamon Press, New York.

Brown, A. S., and Hall, P. F. (1991). Stimulation by endozepine of the side-chain cleavage of cholesterol in a reconstituted enzyme system. *Biochem. Biophys. Res. Commun.* **180**, 609–614.

Brown, A. S., Papadopoulos, V., and Hall, P. F. (1992a). The steroidogenic effects of endozepine are independent of hormonal stimulation in hormone tumour cells. *Mol. Cell. Endocrinol.* **83**, 1–9.

Brown, A. S., Hall, P. F., and Yanagibashi, K. (1992b). Regulation of the side-chain cleavage of cholesterol by endozepine. *In* "Fidia Research Foundation Symposium Series" (E. Costa and S. Paul, eds.), Vol. 8, pp. 161–164. Thieme Medical, New York.

Brown, M. S., and Goldstein, J. L. (1976). Receptor-mediated control of cholesterol metabolism. *Science* **191**, 150–154.

Bumpus, J. A., and Dus, K. M. (1982). Bovine adrenocortical microsomal hemeproteins P450 17α and P450 C_{21}. *J. Biol. Chem.* **257**, 12696–12703.

Burgess, G. M., Godfrey, P. P., McKinney, J. S., Berridge, M. J., Irvine, R. R., and Putney, J. W. (1984). The second messenger linking receptor activation to internal Ca release in liver. *Nature (London)* **309**, 63–65.

Burstein, S., and Gut, M. (1976). Intermediates in the conversion of cholesterol to pregnolone. *Steroids* **38**, 115–129.

Chanderbahn, R., Noland, B. J., Scallen, T. J., and Vahouny, G. V. (1982). Sterol carrier protein 2: Delivery of cholesterol from adrenal lipid droplets to mitochondria for pregnenolone synthesis. *J. Biol. Chem.* **257**, 8028–8935.

Chedrese, P. J., Luu-The, V., Labrie, F., Juris, A. V., and Murphy, B. D. (1990). Evidence for the regulation of 3β-hydroxysteroid dehydrogenase mRNA by hCG. *Endocrinology* **126**, 2228–2230.

Choi, M. S. K., and Cooke, B. A. (1990). Evidence for two independent pathways in the stimulation of steroidogenesis by LH. *FEBS Lett* **261**, 402–404.

Christensen, A. K. (1975). Leydig Cells. *In* "Handbook of Physiology" (R. O. Greep and E. B. Astwood, eds.), Vol. V, pp. 21–47. American Physiological Society, Washington, D.C.

Christensen, A. K., and Mason, N. R. (1965). Comparative ability of seminiferous tubules and interstitial tissue of rat testes to synthesize androgens from progesterone-4-^{14}C *in vitro*. *Endocrinology* **76**, 646–650.

Chung, B., Matteson, K. J., Voutilainen, R., Mohandas, T. K., and Miller, W. L. (1986). Human cholesterol side chain cleavage enzyme, P450scc: cDNA cloning, assignment of the gene to chromosome 15 and expression in the placenta. *Proc. Natl. Acad. Sci. U.S.A.* **83**, 8962–8966.

Chung, B., Picardo-Leonard, J., Haniu, M., Bienkowski, M., Hall, P. F., Shively, J. E., and Miller, W. L. (1987). Cytochrome P450C17 cloning of human adrenal and testis cDNAs indicates the same gene is expressed in both tissues. *Proc. Natl. Acad. Sci. USA* **84**, 407–411.

Chung, B., Hu, M., Lai, C., and Lin, C. (1989). The 5' region of the P450IIAI (P450scc) gene contains a basal promoter and an adrenal specific activating domain. *Biochm. Biophys. Res. Commun.* **160**, 276–284.

Churchill, P. F., and Kimura, T. (1979). Topological studies of cytochromes P450scc and P450IIβ in bovine adrenocortical inner mitochondrial membrane. *J. Biol. Chem.* **254**, 10443–10448.

Clark, B. J., and Waterman, M. R. (1991). The hydrophobic amino terminal sequence of bovine 17α-hydroxylase is required for the expression of a functional hemoprotein in COSI cells. *J. Biol. Chem.* **266**, 5898–5604.

Clegg, J. S. (1982). Interrelationships between water and cell metabolism in Artemia cysts. *Cold Spring Harbor Symp. Quant. Biol.* **46**, 23–38.

Conn, P. M., Chafouleas, J. G., Rogers, D., and Means, A. R. (1981). Gonadotropin releasing hormone stimulates calmodulin redistribution in rat pituitary. *Nature (London)* **292**, 264–266.

Conneely, O. M., Headon, D. R., Olson, E. D., Ungar, F., and Dempsey, M. E. (1984). Intramito-chondrial movement of adrenal sterol carrier protein with cholesterol in response to corticotropin. *Proc. Natl. Acad. Sci. U.S.A.* **81**, 2970–2976.

Cooke, B. A., and Sullivan, M. H. F. (1985). The mechanisms of LHRH agonist action in gonadal tissues. *Mol. Cell. Endocrinol.* **41**, 115–122.

Cooke, B. A., Lindh, M. L., and Janzen, F. H. A. (1977). Effect of lutropin on phosphorylation of endogenous proteins in testis Leydig cells. *Biochem. J.* **168**, 43–48.

Crivello, J. F., and Jefcoate, C. R. (1978). Mechanism of corticotripin action in rat adrenal cells. I: Effects of inhibitors of protein synthesis and microfilament formation on corticosterone synthesis. *Biochim. Biophys. Acta* **542**, 315–322.

Dean, W. L., and Gray, R. D. (1983). Relationship between state of aggregation and catalytic activity of P450 LM$_2$ and P450 reductase. *J. Biol. Chem.* **257**, 14679–14686.

De Jong, F. H., Hey, A. H., and Van der Molen, H. J. (1974). Estradiol and testosterone in rat testis tissue: Localization and production *in vitro*. *J. Endocrinol.* **60**, 409–416.

Dierschke, D. J., Battacharya, A. N., Atkinson, L. E., and Knobil, E. (1970). Circhosal oscillation of plasma LH levels in the ovarietomized rhesus monkey. *Endocrinology* **87**, 850–854.

Dorrington, J. H., and Armstrong, D. T. (1975). FSH stimulates estadiol-17β synthesis in cultured Sertoli cells. *Proc. Natl. Acad. Sci. U.S.A.* **72**, 2677–2681.

Duchatelle, P., and Joffrey, M. (1987). Ca^{2+}-dependent chloride and potassium currents in rat Leydig cells. *FEBS Lett.* **217,** 335–339.

Dufau, M. L., and Catt, K. J. (1975). Gonadotropic stimulation of interstitial cell functions of the rat testis in vitro. *In Meth. Enzymol.* **39,** 252–257.

Dufau, M. L., Horner, K. A., Hayashi, K., Tsuruhara, T., Conn, P. M., and Catt, K. J. (1978). Actions of choleragen and gonadotropin in isolated Leydig cells. *J. Biol. Chem.* **253,** 3721–3729.

Dufau, M. L., Veldhuis, J., Fraioli, F., Johnson, M. H., and Catt, K. J. (1983). Mode of bioactive LH secretion in man. *J. Clin. Endocrinol. Metab.* **57,** 993–1003.

Duque, C., Morisaki, M., Ikekawa, N., and Shikita, M. (1978). The enzyme activity of bovine adrenocortical cytochrome P450 producing pregnenolone from cholesterol. *Biochem. Biophys. Res. Commun.* **82,** 179–185.

Eik-Nes, K. B., and Hall, P. F. (1965). Secretion of steroid hormones in vivo. *Vit. Horm.* **23,** 153–181.

Epstein, L. F., and Orme-Johnson, N. R. (1991). Regulation of steroid hormone biosynthesis. Identification of precursors of a phosphoprotein targeted to the mitochondrion in stimulated rat adrenal cortex cells. *J. Biol. Chem.* **266,** 19,739–19,745.

Farese, R. V. (1983). The phosphatidate-phophoinositide cycle: An intracellular messenger system in the action of hormones and neurotransmitters. *Metabolism* **32,** 628–640.

Farese, R. V. (1984). Phospholipids as intermediaries in hormone action. *Mol. Cell. Endocrinol.* **35,** 1–24.

Farese, R. V., Sabir, M. A., Vendor, S. L., and Larson, R. E. (1980). Are phospoinositides the cycloheximide sensitive actions of ACTH and cyclic AMP? *J. Biol. Chem.* **255,** 5728–5734.

Farese, R. V., Sabir, M. A., Larson, R. E., and Trudeau, W. III. (1983). Further observations on the increases in phospholipids after stimulation by ACTH, cyclic AMP and insulin. *Cell Calcium* **4,** 195–203.

Fevold, H. R., Lorence, M. C., McCarthy, J. L., Trant, J. M., Kagimoto, M., Waterman, M. R., and Mason, J. I. (1989). Rat P45017α from testis: Characterization of a full-length cDNA. *Mol. Endocrinol.* **3,** 968–975.

Fey, E. G., Wan, K. M., and Penman, S. (1984). Epithelial cytoskeletal framework and nuclear matrix-intermediate filament scaffold: Three dimensional organization and protein composition. *J. Cell Biol.* **98,** 1973–1984.

Freeman, D. A., and Ascoli, M. (1983). The LDL pathway of cultured Leydig tumor cells. *Biochim. Biophys. Acta* **754,** 72–76.

Garren, L. D., Ney, R. H., and Davis, W. W. (1965). Studies on the role of protein synthesis in the regulation of corticosterone production by ACTH *in vivo. Proc. Natl. Acad. Sci. U.S.A.* **53,** 1443–1447.

Geuzse, H. J., Slot, J. W., Yanagibashi, K., McCracken, J. A., and Hall, P. F. (1987). Immunoelectron microscopy of cytochromes P450 in porcine adrenal cortex: Two enzymes (11β-hydroxylase and side-chain cleavage) are co-localized in the same mitochondria. *Histochemistry* **86,** 551–559.

Gilamn, A. G. (1987). G proteins: Transducers of receptor-generated signals. *Ann. Rev. Biochem.* **56,** 615–649.

Go, M. (1981). Correlation of DNA exonic regions with protein structural units in haemoglobin. *Nature (London)* **291,** 90–92.

Goldring, N. B., Durica, J. M., Lifka, J., Hedin, L., Ratoosh, S. L., Miller, W. L., Orly, J., and Richards, J. S. (1987). Cholesterol side-chain cleavage P450 messenger ribonucleic acid: Evidence for hormonal regulation in rat ovarian follicles and constitutive expression in corpora lutea. *Endocrinology* **120,** 1942–1950.

Golos, T. G., Miller, W. L., and Stauss, J. F., III (1988). Human chorionic gonadotropin and 8-bomo-cyclic adenosine monophosphate promote an acute increase in cytochrome P450scc and adrenodoxin messenger RNAs in cultured human granulosa cells by cycloheximide-insensitive mechanism. *J. Clin. Invest.* **80,** 869–879.

Griffin, B. W., Petersen, J. A., and Estabrook, R. W. (1979). Cytochrome P450: Biophysical properties and catalytic activity. *In* "The Porphyrins" (D. Dolphin, ed.), Vol. VII, pp. 333–375. Academic Press, New York.

Gunsalus, I. C., Meeks, J. R., Lipscomb, J. D., De Brunner, P., and Munck, E. (1974). Bacterial monooxygenase—the P450 cytochrome system. *In* "Molecular Mechanisms of Oxygen Activation" (O. Hayaishi, ed.), pp. 559–572. Academic Press, New York.

Gwynne, J. T., Hess, B., Hughes, T., Rountree, R., and Mahaggee, D. (1985). The role of high density lipoproteins in adrenal steroidogenesis. *Endocrinol. Res.* **10**, 411–429.

Hall, P. F. (1966). On the stimulation of testicular steroidogenesis in the rabbit by interstitial cell-stimulating hormone. *Endocrinology* **78**, 690–697.

Hall, P. F. (1982). The role of the cytoskeleton in endocrine function. *In* "Cellular Regulation of Secretion and Release" (M. P. Conn, ed.), pp. 195–221. Academic Press, New York.

Hall, P. F. (1984). The role of the cytoskeleton in the responses of target cells to hormones. *In* "Regulation of Target Cell Responsiveness" (K. W. McKerns, A. Aakvaag, and V. Hanson, eds.), Vol. 1, pp. 205–227. Plenum, New York.

Hall, P. F. (1985). Trophic stimulation of steroidogenesis: In search of the elusive trigger. *Rec. Prog. Horm. Res.* **41**, 1–39.

Hall, P. F., and Eik-Nes, K. B. (1962). The action of gonadotropic hormones upon rabbit testis *in vitro. Biochim. Biophys. Acta* **63**, 411–419.

Hall, P. F., and Kortiz, S. B. (1964). Inhibition of the biosynthesis of pregnenolone by 20α-hydroxycholesterol. *Biochim. Biophys. Acta* **93**, 441–445.

Hall, P. F., and Nakamura, M. (1979). The influence of adrenocorticotropin on transport of a cholesteryl linoleate–low density lipoprotein complex into adrenal tumor cells. *J. Biol. Chem.* **254**, 12547–12554.

Hall, P. F., and Young, D. G. (1968). Site of action of trophic hormones upon the biosynthetic pathways to steroid hormones. *Endocrinology* **82**, 559–565.

Hall, P. F., Sozer, C. C., and Eik-Nes, K. B. (1964). Formation of dehydroepiandrosterone during *in vivo* and *in vitro* biosynthesis of testosterone by testicular tissue. *Endocrinology* **74**, 35–43.

Hall, P. F., Lee Lewes, J., and Lipson, E. D. (1975). The role of mitochondrial cytochrome P450 from bovine adrenal cortex in side chain cleavage of 20S, 22R-dihydroxycholesterol. *J. Biol. Chem.* **250**, 2283–2290.

Hall, P. F., Watanuki, M., and Hamkalo, B. A. (1979a). Adrenocortical cytochrome P450 side chain cleavage: Preparation of membrane bound side chain cleavage system from purified components. *J. Biol. Chem.* **254**, 547–553.

Hall, P. F., Charppponnier, C., Nakamura, M., and Gabbiani, G. (1979b). The role of microfilaments in the response of adrenal tumor cells to adrenocorticotropic hormone. *J. Biol. Chem.* **254**, 9080–9084.

Hall, P. F., Charponnier, C., Nakamura, M., and Gabbiani, G. (1979c). The role of microfilaments in the response of Leydig cells to luteinizing hormone. *J. Steroid Biochem.* **11**, 1361–1369.

Hall, P. F., Osawa, S., and Mrotek, J. J. (1981). Influence of calmodulin on steroid synthesis in Leydig cells from rat testis. *Endocrinology* **109**, 1677–1684.

Hanukoglu, I., and Honukoglu, Z. (1986). Stoichiometry of mitochondrial cytochromes P450, adrenodoxin and adrenodoxin reductase in adrenal cortex and corpus luteum. *Eur. J. Biochem.* **157**, 27–31.

Iida, S., Widmaier, E., and Hall, P. F. (1986). The phosphoinositide-Ca^{2+} hypothesis does not apply to the steroidogenic action of ACTH. *Biochem. J.* **236**, 53–61.

Inaba, T., and Wiest, W. G. (1985). Protein kinase stimulation of steroidogenesis in rat luteal cell mitochondria. *Endocrinology* **117**, 315–322.

Inoue, H., Watanabe, N., Hiyashi, Y., and Fujii-Kuriyama, Y. (1986). Purification and characterization of rat adrenal 3β-hydroxysteroid dehydrogenase with steroid 5-ene-4-ene-isomerase. *J. Steroid Biochem.* **24**, 753–760.

Inoue, H., Hiyashi, Y., Morohashi, K., and Fujii-Kuriyama, Y. (1988). The 5′-flanking region of the human P450scc gene shows responsiveness to cAMP-dependent regulation in a transient gene-expression system of Y-1 adrenal tumor cells. *Eur. J. Biochem.* **171,** 435–440.

Irby, D. C., and Hall, P. F., (1971). Stimulation by ICSH of protein biosynthesis in isolated Leydig cells from hypophysectomized rats. *Endocrinology* **89,** 1367–1374.

Ishii-Ohba, H., Inano, H., and Tamaoki, B. (1986). Purification and properties of testicular 3β-hydroxysteroid dehydrogenase and 5-ene-4-ene-isomerase. *J. Steroid Biochem.* **25,** 555–560.

Janszen, F. H. A., Cooke, B. A., and Van der Molen, H. J. (1977). Specific protein synthesis in isolated rat Leydig cells. Influence of LH and cycloheximide. *Biochem. J.* **162,** 341–346.

John, M. E., John, M. C., Ashley, P., MacDonald, R. J., Simpson, E. R., and Waterman, M. R. (1984). Identification and characterization of cDNA clones specific for cholesterol side-chain cleavage cytochrome P450. *Proc. Natl. Acad. Sci. U.S.A.* **81,** 5628–5632.

John, M. E., John, M. C., Boggaram, V., Simpson, E. R., and Waterman, M. R. (1986). Transcriptional regulation of steroid hydroxylase genes by corticotropin. *Proc. Natl. Acad. Sci. U.S.A.* **83,** 4715–4719.

Johnson, F. F., and McKnight, S. L. (1989). Eukaryotic transcriptional regulatory proteins. *Ann. Rev. Biochem.* **58,** 799–839.

Jsanig, G. R., Heaft, R., Rabe, H., Nakower, A., and Ruckpaul, K. (1988). Comparative studies on the accessibility and functional importance of tyrosine residues in cytochrome P450 isozymes. *Biomed. Biochim. Acta* **47,** 565–579.

Kovanen, P. T., Faust, J. R., Brown, M. S., and Goldstein, J. L. (1979). Low density lipoprotein receptors in bovine adrenal cortex. Receptor-mediated uptake of LDL for steroid synthesis. *Endocrinology* **104,** 599–609.

Koritz, S. B., and Hall, P. F. (1964a). Feedback inhibition by pregnenolone: A possible mechanism. *Biochim. Biophys. Acta* **92,** 215–218.

Koritz, S. B., and Hall, P. F. (1964b). End-product inhibition of the conversion of cholesterol to pregnenolone in an adrenal extract. *Biochemistry* **3,** 1298–1304.

Kumamoto, T., Morohashi, K., Ito, A., and Omura, T. (1987). Site directed mutagenesis of basic amino acid residues in the extension of peptide P450scc precursor: Effects on the import of the precursor into mitochondria. *J. Biochem.* **102,** 833–838.

Kumamoto, T., Ito, A., and Omura, T. (1989). Critical region in the extension peptide for the import of cytochrome P450(scc) precursor into mitochondria. *J. Biochem.* **105,** 72–78.

Labrie, F., Luu-Thé, V., Labrie, C., Bérubé, D., Conet, J., Zaho, H., and Simard, J. (1989). Characterization of two mRNA species encoding human estradiol 17β-dehydrogenase and assignment to chromosome 17. *J. Steroid Biochem.* **34,** 189–197.

Lachance, Y., Luu-Thé, V., Labrie, C., Simard, J., Dumont, M., de Launoit, Y., Guerin, S., Leblanc, G., and Labrie, F. (1990). Characterization of human 3β-hydroxysteroid dehydrogenase/Δ⁵, Δ⁴-isomerase gene. *J. Biol. Chem.* **265,** 20,469–20,475.

Lachance, Y., Luu-Thé, V., Verreault, H., Dumont, M., Rhe'aume, E., Leblanc, G., and Labrie, F. (1991). Structure of the human type II 3β-hydroxysteroid dehydrogenase/Δ⁵-Δ⁴ isomerase gene. *DNA Cell Biol.* **10,** 701–711.

Lambeth, J. D., Seybert, D. W., and Kamin, H. (1979). Ionic effects on adrenal steroidogenic electron transport. *J. Biol. Chem.* **254,** 7,255–7,261.

Lambeth, J. D., Seybert, D. W., and Kamin, H. (1980). Phospholipid vesicle-reconstituted cytochrome P450 scc. *J. Biol. Chem.* **255,** 138–145.

Leblanc, G., and Labrie, F. (1990). Structure of two in tandem human 17β-hydroxysteroid dehydrogenase genes. *Mol. Endocrinol.* **4,** 268–275.

Lin, T. (1985). The role of Ca²⁺/phospholipid-dependent protein kinase in Leydig cell steroidogenesis. *Endocrinology* **117,** 119–126.

Lorence, M. C., Murry, B. H., Trant, J. M., and Mason, J. I. (1990). Human 3β-hydroxysteroid/Δ⁵-Δ⁴ isomerase from placenta: expression in non-steroidogenic cells. *Endocrinology* **126,** 2493–2498.

Lowitt, S., Farese, R. V., Sabir, M. A., and Root, A. W. (1982). Rat Leydig cell phospholipid content is increased by LH and 8-bromo-cyclic AMP. *Endocrinology* **111,** 1415–1422.

Lund, J., Ahlgren, R., Wu, D., Kagimoto, M., Simpson, E. R., and Waterman, M. R. (1990). Transcriptional regulation of the bovine CYP17 (P-45017α) gene. *J. Biol. Chem.* **265**, 3304–3312.

Luu-Thé, V., Lachance, Y., Labrie, C., Leblanc, G., Thomas, J. L., Stricker, R. C., and Labrie, F. (1989a). Full length cDNA structure and deduced amino acid sequence of human 3β-hydroxy-5-ene steroid dehydrogenase. *Mol. Endocrinol.* **3**, 1310–1313.

Luu-Thé, V., Labrie, C. Zhao, F., Conet, J., Lachance, Y., Simard, J., Leblanc, G., Côté, J., Bérubé, D., Gagne, R., and Labrie, F. (1989b). Characterization of cDNAs for human estradiol 17β-dehydrogenase. *Mol. Endocrinol.* **3**, 1301–1309.

Luu-Thé, V., Labrie, C., Simard, J., Lachance, Y., Zhao, H. F., Conet, J., Leblanc, G., and Labrie, F. (1990). Structure of two in tandem human 17β-hydroxysteroid dehydrogenase genes. *Molecular Endocrinology* **4**, 268–275.

McCarthy, J. L., and Waterman, M. R. (1983). Co-induction of 17α-hydroxylase and -17,20-lyase activities in primary cultures of bovine adrenocortical cells in response to ACTH treatment. *J. Steroid Biochem.* **29**, 307–312.

Marsh, J. M. (1976). The role of cyclic AMP in gonadal steroidogenesis. *Biol. Reprod.* **14**, 30–55.

Mason, H. S. (1957). Mechanisms of oxygen metabolism. *Science* **125**, 1185–1189.

Mason, J. I., MacDonald, A. A., and Laptook, A. (1984). The activity and biosynthesis of cholesterol side-chain cleavage enzyme in cultured immature pig testis cells. *Biochim. Biophys. Acta* **795**, 504–512.

Matocha, M. F., and Waterman, M. R. (1984). Discriminatory processing of the precursor forms of cytochrome P-450scc and adrenodoxin by adrenocortical and heart mitochondria. *J. Biol. Chem.* **259**, 8672–8678.

Matsumoto, K., and Samuels, L. T. (1969). Influence of steroid distribution between microsomes and soluble fraction on steroid metabolism by microsomal enzymes. *Endocrinology* **85**, 402–409.

Matteson, K. L., Picado-Leonard, J., Chung, B., Mohandas, T. K., and Miller, W. L. (1986). Assignment of the gene for adrenal P-450C17 (steroid 17α-hydroxylase/17,20-lyase) to human chromosome 10. *J. Clin. Endocrinol. Metab.* **63**, 798–791.

Mellon, S. H., and Vaisse, C. (1989). cAMP regulates P-450scc gene expression by a cycloheximide insensitive mechanism in cultured mouse Leydig MA-10 cells. *Proc. Natl. Acad. Sci.* **86**, 7775–7779.

Mendelson, O., Dufau, M., and Katt, K. (1975). Dependence of gonadotropin-induced steroidogenesis upon RNA and protein synthesis in the interstitial cells of the rat testis. *Biochem. Biophys. Acta* **411**, 222–230.

Mertz, L. M., and Pedersen, R. C. (1989). The kinetics of steroidogenesis activator polypeptide in the rat adrenal cortex. *J. Biol. Chem.* **264**, 15274–15279.

Mitani, F., Shimizu, T., Ueno, R., Ishimura, Y., Komatsu, N., and Watanabe, K. (1982). Cytochrome P-45011β and Cytochrome P-450scc in adrenal cortex. *J. Histochem. Cytochem.* **30**, 1066–1076.

Moger, W. H. (1985). Stimulation and inhibition of Leydig cell steroidogenesis by the phorbol ester 12-O-tetradecanoylphorbol-13-acetate. *Life Sci.* **37**, 869–874.

Moger, W. H. (1991). Evidence for compartmentalization of cAMP-dependent protein kinases in rat Leydig cells. *Endocrinology* **128**, 1414–1418.

Mori, M., and Marsh, J. M. (1982). The site of LH stimulation of steroidogenesis in mitochondria of the rat corpus luteum. *J. Biol. Chem.* **257**, 6178–6185.

Morohashi, K., Sogawa, K., Omura, T., and Fujii-Kuriyama, Y. (1987). Gene structure of human cytochrome P-450scc cholesterol desmolase. *J. Biochem.* **101**, 879–887.

Moyle, W. R., and Ramachandran, J. (1973). Effect of LH on steroidogenesis and cyclic AMP accumulation in rat Leydig cell preparations and mouse tumor Leydig cells. *Endocrinology* **93**, 127–134.

Moyle, W. R., Kong, Y. C., and Ramachandran, J. (1973). Steroidogenesis and cyclic AMP accumulation in rat adrenal cells. *J. Biol. Chem.* **248**, 2409–2417.

Mrotek, J. J., and Hall, P. F. (1974). The influence of cytochalasin B on the response of adrenal tumor cells to ACTH and cyclic AMP. *Biochem. Biophys. Res. Commun.* **4**, 891–896.

Mrotek, J. J., and Hall, P. F. (1977). Response of adrenal tumor cells to adrenocorticotropin: Site of inhibition of cytochalasin B. *Biochemistry* **16**, 3177–3181.

Mulheron, G. W., Stone, R. T., Miller, W. L., and Wise, T. (1989). Nucleotide sequence of cytochrome P-450 cholesterol side-chain cleavage cDNA isolated from porcine testis. *Nucl. Acids Res.* **17**, 1773–1780.

Murano, E. P., and Payne, A. H. (1976). Distinct testicular 17β-ketosteroid reductases, one in interstitial tissue and one in seminiferous tubules. *Biochim. Biophys. Acta* **450**, 89–95.

Nakajin, S., and Hall, P. F. (1981a). Microsomal cytochrome P-450 from neonatal pig testis: Purification and properties of a C_{21} steroid side-chain cleavage (17α-hydroxylase and $C_{17,20}$-lyase). *J. Biol. Chem.* **256**, 3871–3878.

Nakajin, S., and Hall, P. F. (1981b). Side-chain cleavage of C_{21} steroids to C_{19} steroids by testicular microsomal cytochrome P-450: 17α-hydroxy-C_{21} steroids as obligatory intermediates. *J. Steroid Biochem.* **14**, 1249–1255.

Nakajin, S., and Hall, P. F. (1983). Side-chain cleavage of C_{21} steroids by testicular microsomal cytochrome P-450 (17α-hydroxylase/lyase): Involvement of heme. *J. Steroid. Biochem.* **19**, 1345–1348.

Nakajin, S., Shively, J., Yuan, P. M., and Hall, P. F. (1981a). Microsomal cytochrome P-450 from neonatal pig testis: Two enzymatic activities (17α-hydroxylase and $C_{17,20}$-lyase) associated with one protein. *Biochemistry* **20**, 4037–4045.

Nakajin, S., Hall, P. F., and Onoda, M. (1981b). Testicular microsomal cytochrome P-450 for C_{21} steroid side-chain cleavage: Spectral and binding studies. *J. Biol. Chem.* **256**, 6134–6141.

Nakajin, S., Shinoda, M., and Hall, P. F. (1983). Purification and properties of 17α-hydroxylase from microsomes of pig adrenal: A second C_{21} side-chain cleavage system. *Biochem. Biophys. Res. Commun.* **111**, 512–516.

Nakajin, S., Shinoda, M., Hanui, M., Shively, J. E., and Hall, P. F. (1984). The C_{21} steroid side-chain cleavage enzyme from porcine adrenal microsomes: Purification and characterization of the 17α-hydroxylase-$C_{17,20}$-lase cytochrome P450. *J. Biol. Chem.* **259**, 3971–3978.

Nakamura, M., Watanuki, M., Tilley, B. E., and Hall, P. F. (1980). Effect of adrenocorticotropin on intracellular cholesterol transport. *J. Endocrinol.* **84**, 179–188.

Nason, T. F., Han, X-G., and Hall, P. F. (1992). Cyclic AMP regulates expression of the rat gene for steroid 17α-hydroxylase-$C_{17,20}$-lase (CYP17) in rat Leydig cells. *Biochim. Biophys. Acta* **1171**, 73–80.

Neymark, M. A., Bieszczad, R. R., and Dimino, M. J. (1984). Phosphorylation of mitochondrial proteins in isolated porcine follicles after treatment with LH. *Endocrinology* **114**, 588–595.

Nishihara, M., Winters, C. A., Buzko, E., Waterman, M. R., and Dufau, M. L. (1988). Hormonal regulation of rat Leydig cell cytochrome P45017α mRNA levels. *Biochem. Biophys. Res. Commun.* **154**, 151–158.

Ogishima, T., Okada, Y., and Omura, T. (1985). Import and processing of the precursor of cytochrome P450(scc) by bovine adrenal cortex mitochondria. *J. Biochem.* **98**, 781–791.

Ohba, H., Inano, H., and Tamaoki, B. (1982). Kinetic mechanism of porcine testicular 17β-hydroxysteroid dehydrogenase. *J. Steroid Biochem.* **17**, 381–389.

Ohno, Y., Yanagibashi, K., Yonezawa, Y., Ishiwatari, S., and Matsuba, M. (1983). Effect of ACTH, cycloheximide and aminoglutethimide on the content of cholesterol in the outer and inner mitochondrial membrane of rat adrenal cortex. *Endocrinol. Jpn.* **30**, 335–344.

Onoda, M., and Hall, P. F. (1981). Inhibition of testicular microsomal cytochrome P450 (17α-hydroxylase-$C_{17,20}$-lase) by estrogens. *Endocrinology* **109**, 763–767.

Onoda, M., and Hall, P. F. (1982). Cytochrome b_5 stimulates purified testicular microsomal cytochrome P450 (C_{21} side chain cleavage). *Biochem. Biophys. Res. Commun.* **108**, 454–458.

Onoda, M., Haniu, M., Yanagibashi, K., Sweet, F., Shively, J. E., and Hall, P. F. (1987). Affinity alkylation of the active site of C_{21} side-chain cleavage cyptochrome P450: A unique cysteine residue alkylated by 17- (bromoacetoxy) progeterone. *Biochemisty* **26,** 57–662.

Oonk, R. B., Parker, K. L., Bibson, J. L., and Richards, J. S. (1990). Rat cholesterol side-chain cleavage cytochrome P450 (P450scc) gene. *J. Biol. Chem.* **265,** 22,392–22,401.

Orme-Johnson, N. R. (1991). Distinctive properties of adrenal cortex mitochondria. *Biochim. Biophys. Acta* **1020,** 213–231.

Orme-Johnson, N. R., Light, D. R., White-Stevens, R. W., and Orme-Johnson, W. H. (1979). Steroid-binding properties of beef adrenal cortical cytochrome P450 which catalyzes conversion of cholesterol to pregnenolone. J. Biol. Chem. **254,** 2103–2109.

Osawa, S., Betz, G., and Hall, P. F. (1984). The role of actin in the responses of adrenal cells to ACTH and cyclic AMP: Inhibition by DNaseI. *J. Cell Biol.* **99,** 1335–1342.

Oshima, H., and Ochi, A. I. (1973). On testicular 17β-hydroxysteroid oxidoreductase: Product activation. *Biochim. Biophys. Acta* **306,** 227–235.

Ottaway, J. H., and Mowbray, J. (1977). The role of compartmentation in the regulation of glycolysis. *Curr. Topics Cell Regul.* **12,** 108–149.

Ou, W., Ito, A., Morohashi, K., Fujii-Kuriyama, Y., and Omura, T. (1986). Processing-independent in vitro translocation of cytochrome P450(scc) precursor across mitochondrial membranes. *J. Biochem.* **100,** 1287–1296.

Papadopoulos, V., and Hall, P. F. (1989). Isolation and characterization of protein kinase C from Y-1 adrenal cell cytoskeleton. *J. Cell Biol.* **108,** 553–567.

Papadopoulos, V., Brown, A. S., and Hall, P. F. (1990a). Calcium-calmodulin-dependent phosphorylation of cytoskeletal proteins from adrenal cells. *Mol. Cell. Endocrinol.* **74,** 109–123.

Papadopoulos, V., Widmaier, E. P., and Hall, P. F. (1990b). The role of calmodulin in the responses to adrenocorticotropin of plasma membranes from adrenal cells. *Endocrinology* **126,** 2465–2473.

Payne, A. H., and Shah, L. (1991). Multiple mechanisms for regulation of 3β-hydroxysteroid dehydrogenase/isomerase, 17α-hydroxylase/lyase and side-chain cleavage P450 mRNA levels in primary cultures in primary cultures of mouse Leydig cells. *Endocrinology* **129,** 1429–1435.

Pedersen, R. C. (1984). Polypeptide activators of cholesterol side-chain cleavage. *Endocr. Res.* **10,** 533–561.

Pedersen, R. C. (1987). Steroidogenesis activator polypeptide (SAP) in rat ovary and testis. *J. Steroid Biochem.* **27,** 731–735.

Pedersen, R. C., and Brownie, A. C. (1983). Cholesterol side-chain cleavage in the rat adrenal cortex: Isolation of a cycloheximide sensitive activator protein. *Proc. Natl. Acad. Sci. U.S.A.* **80,** 1882–1886.

Pedersen, R. C., and Brownie, A. C. (1987). Steroidogenesis-activator polypeptide isolated from a Leydig cell tumor. *Science* **236,** 188–190.

Pereira, M. E., Segaloff, D. L., Ascoli, M., and Eckstein, F. (1987). Inhibition of gonadotropin-activated steroidogenesis in cultured Leydig tumor cells by the Rp diastereoisomer of adenosine-3′,5′-cyclic phosphorothioate. *J. Biol. Chem.* **262,** 6093–6100.

Peterson, J. A., Ebel, R. E., O'Keefe, D. H., Matsubara, T., and Estrabrook, R. W. (1976). Temperature dependence of cytochrome P450 reduction. *J. Biol. Chem.* **251,** 4010–4016.

Picado-Leonard, J., and Miller, W. L. (1987). Cloning and sequence of the human gene for P450C17 (17α-hydroxylase/17.20 lyase). *DNA* **6,** 439–448.

Picado-Leonard, J., Voutilainen, R., Kao, L., Chung, B., Strauss, J. F. III, and Miller, W. L. (1988). Human adrenodoxin: Cloning of three cDNAs and cycloheximide enhancement in JEG-3 cells. *J. Biol. Chem.* **263,** 3240–3244.

Pon, L. A., and Orme-Johnson, N. R. (1986). Acute stimulation of steroidogenesis in corpus luteum and adrenal cortex by peptide hormones. *J. Biol. Chem.* **261,** 6,594–6,599.

Pon, L. A., Epstein, L. F., and Orme-Johnson, N. R. (1986). Protein synthesis requirement for acute ACTH stimulation of adrenal corticosteroidogenesis. *Endocr. Res.* **12**, 429–446.

Privalle, C. T., Crivello, J. F., and Jefcoate, C. R. (1983). Regulation of intramitochondrial cholesterol side-chain cleavage P450 in rat adrenal gland. *Proc. Natl. Acad. Sci. U.S.A.* **80**, 702–706.

Raimondi, S. G., Oliver, N. S., Pabrito, L. C., and Flury, A. (1989). Regulation of the 3β-hydroxysteroid dehydrogenase activity in tissue fragments and microsomes from human term placenta. *J. Steroid Biochem.* **32**, 413–420.

Rein, H., and Ristan, O. (1978). The importance of the high spin-low spin equilibrium existing in the cytochrome P450 for the enzyme mechanism. *Pharmazie* **33**, 325–338.

Rice, D. A., Kirkman, M. S., Aitken, L. D., Mouw, A. R., Schimmer, B. P., and Parker, K. L. (1990). Analysis of the promoter region of the gene encoding mouse cholesterol side-chain cleavage enzyme. *J. Biol. Chem.* **265**, 11713–11720.

Ritter, M. C., and Dempsey, M. E. (1971). Specificity and role in cholesterol biosynthesis of a squalene and sterol carrier protein. *J. Biol. Chem.* **246**, 1536–1539.

Ruokenen, A., and Vihko, R. (1974). Concentrations of unconjugated and sulfated neutral sterols in boar testis. *J. Steroid Biochem.* **5**, 33–39.

Samuels, L. T. (1960). Metabolism of steroid hormones; *In* "Metabolic Pathways" (D. M. Greenberg, ed.), 2d Ed., Vol. I, pp. 431–474. Academic Press, New York.

Samuels, L. T., and Matsumoto, K. (1974). Localization of enzymes involved in testosterone biosynthesis in mouse testis. *Endocrinology* **94**, 55–61.

Samuels, L. T., Bussman, L., Matsumoto, K., and Huseby, R. A. (1975). Organization of androgen biosynthesis in the testis. *J. Steroid Biochem.* **6**, 291–301.

Scallen, T. J., Schuster, M. W., and Dhar, A. K. (1971). Evidence for a non-catalytic carrier protein in cholesterol biosynthesis. *J. Biol. Chem.* **246**, 224–230.

Schimmer, B. P., Tsao, J., Collie, G., Wong, M., and Schulz, P. (1984). Analysis of the mutation to forskolin-resistance in Y-1 adrenocortical tumor cells. *Endocr. Res.* **10**, 365–386.

Seybert, D. W., Lambeth, J. D., and Kamin, H. (1978). The participation of a second molecule of adrenodoxin in cytochrome P-450 catalyzed 11β-hydroxylation. *J. Biol. Chem.* **253**, 8355–8361.

Seybert, D. W., Lancaster, J. R., Lambert, J. D., and Kamin, H. (1979). Participation of the membrane in the side-chain cleavage of cholesterol. *J. Biol. Chem.* **254**, 12088–12093.

Shikita, M., and Hall, P. F. (1973a). Cytochrome P-450 from bovine adrenocortical mitochondria: An enzyme for the side-chain cleavage of cholesterol. I. Purification and properties. *J. Biol. Chem.* **248**, 5598–5606.

Shikita, M., and Hall, P. F. (1973b). Cytochrome P-450 from bovine adrenocortical mitochondria: An enzyme for the side-chain cleavage of cholesterol. II. Subunit structure. *J. Biol. Chem.* **248**, 5605–5610.

Shikita, M., and Hall, P. F. (1974). The stoichiometry of the conversion of cholesterol and hydroxycholesterols to pregnenolone (3β-hydroxypregn-5-en-20-one) catalyzed by adrenal cytochrome P-450. *Proc. Natl. Acad. Sci. U.S.A.* **71**, 1441–1446.

Simpson, E. R., Jefcoate, C. R., Brownie, A. C., and Boyd, G. S. (1972). Effect of ether anaesthesia stress on side-chain cleavage in rat adrenal mitochondria. *Eur. J. Biochem.* **28**, 442–446.

Steinert, P. M., and Liem, R. K. H. (1990). Intermediate filament dynamics. *Cell* **60**, 521–523.

Stocco, D. M., and Sodeman, T. C. (1991). The 30 kDa mitochondrial protein induced by hormone stimulation in MA-10 mouse Leydig tumor cells from larger precursors. *J. Biol. Chem.* **266**, 19731–19738.

Tagaki, Y., Shikita, M., and Hall, P. F. (1975). The active form of cytochrome P-450 from bovine adrenocortical mitochondria. *J. Biol. Chem.* **250**, 845–851.

Tank, D. W., Wu, E. S., and Webb, W. W. (1982). Enhanced molecular diffusibility in muscle membrane blebs: release of lateral constraints. *J. Cell Biol.* **92**, 207–212.

Thomas, J. L., Berko, E. A., Faustino, A., Myers, R. P., and Strickler, R. C. (1988). Human placental 3β-hydroxy-5-ene-steroid dehydrogenase and steroid 5-4-ene-isomerase. *J. Steroid Biochem.* **31**, 785–793.

Thomas, J. L., Myers, R. P., and Strickler, R. C. (1989). Human placental 3β-hydroxy-5-ene-steroid dehydrogenase and steroid 5-4-ene-isomerase. Purification and kinetic profiles. *J. Steroid Biochem.* **33**, 209–217.

Tint, L. S., Hollenbeck, P. J., Berkhovsky, A. B., Surgucheva, L. G., and Bershadsky, A. D. (1991). Evidence that intermediate filament reorganization is induced by ATP-dependent contraction of the actomyosin cortex in permeabilized fibroblasts. *J. Cell Sci.* **98**, 375–384.

Tremblay, Y., Ringler, G. E., Morel, Y., Mohandras, T. K., Labrie, F., Strauss, J. F., and Miller, W. L. (1989). Regulation of the gene for estrogenic 17-ketosteroid reductase lying on chromosome 17cen-q25. *J. Biol. Chem.* **264**, 20458–20462.

Trzeciak, W. H., and Boyd, G. S. (1973). The effect of stress on cholesterol content and cholesteryl esterase activity in rat adrenal cortex. *Eur. J. Biochem.* **37**, 327–333.

Trzeciak, W. H., and Boyd, G. S. (1974). Activation of cholesteryl esterase in bovine adrenal cortex. *Eur. J. Biochem.* **46**, 201–207.

Trzeciak, W. H., Simpson, E. R., Scallen, T. J., Vahouny, G. V., and Waterman, M. R. (1987). Studies on the synthesis of SCP2 in rat adrenocortical cells. *J. Biol. Chem.* **262**, 3713–3717.

Ullrich, V. (1979). Cytochrome P-450 and biological hydroxylation reactions. *Topics Curr. Chem.* **83**, 68–115.

Vahouny, G. V., Chanderbhan, R., Noland, B. J., Irwin, D., Dennis, P., Lambeth, J. D., and Scallen, T. J. (1983). Sterol carrier proteins. Identification of adrenal SCP and site of action for mitochondrial cholesterol utilization. *J. Biol. Chem.* **258**, 11731–11737.

Verhoeven, G., and Cailleau, J. (1986). Stimulatory effects of epidermal growth factor on steroidogenesis in Leydig cells. *Mol. Cell. Endocrinol.* **47**, 99–106.

Voutilainen, R., and Miller, W. L. (1987). Coordinate tropic hormone regulation of mRNAs for insulin-like growth factor II and the cholesterol side-chain cleavage enzyme P-450scc in human steroidogenic tissues. *Proc. Natl. Acad. Sci. U.S.A.* **84**, 1590–1594.

Walter, P., and Blobel, G. (1981). Translocation of proteins across the endoplasmic reticulum. III Signal recognition protein causes signal sequence-dependent arrest. *J. Cell Biol.* **91**, 557–561.

Watanuki, M., Granger, G. A., and Hall, P. F. (1978). Cytochrome P-450 from bovine adrenocortical mitochondria: Immunochemical properties and purity. *J. Biol. Chem.* **253**, 2927–2934.

Welton, A. F., and Aust, S. D. (1974). The effects of 3-methylcholanthrene on the structure of the rat liver endoplasmic reticulum. *Biochim. Biophys. Acta* **373**, 197–205.

White, R. E., and Coon, M. J. (1980). Oxygen activation by cytochrome P-450. *Ann. Rev. Biochem.* **49**, 315–349.

Widmaier, E. P., and Hall, P. F. (1985). Protein kinase C in adrenal cells: Possible role in regulation of steroid synthesis. *Mol. Cell. Endocrinol.* **43**, 181–190.

Wood, K. V. (1991). Recent advances and prospects for use of beetle luciferases as genetic reporters. *In* "Bioluminescence and Chemiluminescence: Current Status" (P. Staney and L. Kricka, eds.), pp. 174–221. John Wiley & Sons, Chichester, England.

Xu, X., Xu, T., Robertson, D. G., and Lambeth, J. D. (1989). GTP stimulates pregnenolone generation in isolated rat adrenal mitochondria. *J. Biol. Chem.* **264**, 17674–17680.

Yanagibashi, K., and Hall, P. F. (1987). Role of electron transport in the regulation of lyase activity of C_{21} side-chain cleavage P-450 from porcine adrenal and testicular microsomes. *J. Biol. Chem.* **261**, 8429–8433.

Yanagibashi, K., Ohno, Y., Kawamura, M., and Hall, P. F. (1988). The regulation of intracellular transport of cholesterol in bovine adrenal cells. Purification of a novel protein. *Endocrinology* **123**, 2075–2082.

Yanagibashi, K., Papadopoulos, B., Masaki, E., Iwaki, T., Kawamura, M., and Hall, P. F. (1989). Forskolin activates voltage-dependent Ca^{2+} channels in bovine but not in rat fasciculata cells. *Endocrinology* **124**, 2,383–2,391.

Yanagibashi, K., Kawamura, M., and Hall, P. F. (1990). Voltage-dependent Ca^{2+} channels are involved in regulation of steroid synthesis by bovine but not rat fasciculata cells. *Endocrinology* **126**, 311–318.

Yasumara, Y. (1968). Shape change in cultured adrenal cells induced by ACTH. *Am. Zool.* **8**, 285–290.

Zanger, U. M., Lund, J., Simpson, E. R., and Waterman, M. R. (1991). Activation of transcription in cell-free extracts by a novel cAMP-responsive sequence from the bovine CYP17 gene. *J. Biol. Chem.* **266**, 11417–11420.

Zhao, H., Simard, J., Labrie, C., Breton, N., Rhéaume, E., Luu-Thé, U., and Labrie, F. (1989). Molecular cloning, cDNA structure and predicted amino acid sequence of bovine 3β-hydroxy-5-ene steroid dehydrogenase/Δ^5-Δ^4 isomerase. *FEBS Lett.* **259**, 153–157.

Zuber, M. X., John, M. E., Okamura, T., Simpson, E., and Waterman, M. R. (1986). Bovine adrenocortical P-450 17α. Regulation of gene expression by ACTH. *J. Biol. Chem.* **261**, 2475–2482.

11

Hormonal Control Mechanisms of Leydig Cells

ILPO HUHTANIEMI

I. Introduction

The function of the Leydig cells is producing testosterone for the support of spermatogenesis and for extratesticular actions of androgens (e.g., secondary male sex characteristics, anabolic effects, feedback regulation of gonadotropins). Sufficient androgen synthesis is possible only through tropic stimulation by the gonadotropin luteinizing hormone (LH). Although LH is the main stimulator of steroidogenesis, several other blood-borne endocrine and intratesticular paracrine, autocrine, and intracrine factors function as modulators of the LH action, often having little influence on their own. This chapter is not a comprehensive review of all aspects of the control of Leydig cell function. Instead, some of the recent developments in our understanding of the key elements of Leydig cell endocrine regulation are reviewed. For the sake of clarity, most of the information presented in this chapter is based on the most extensively studied species, the rat, although some facets in Leydig cell regulation and function are species specific. The numerous para- and autocrine connections in Leydig cell functions, as well as the recent developments in the molecular aspects of testicular steroidogenesis, are discussed in other chapters. For more general overviews on Leydig cell regulation and function, other reviews are recommended (de Kretser and Kerr, 1988; Dufau, 1988; Hall, 1988; Rommerts and van der Molen, 1989; Huhtaniemi and Pelliniemi, 1992).

Molecular Biology of the Male Reproductive System
383

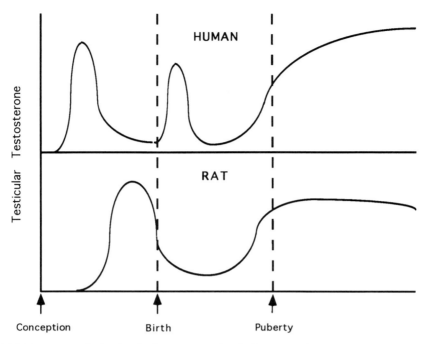

FIGURE 1 Intratesticular level of testosterone in the human and rat male throughout life. The fetal–neonatal peaks of testosterone are caused in both species by activity of the fetal Leydig cell populations. The adult Leydig cell populations produce testosterone from puberty onward.

II. Ontogeny of Tropic Regulation of Leydig Cells

In most mammalian species, two growth phases of Leydig cells are observed (de Kretser and Kerr, 1988; Huhtaniemi and Pelliniemi, 1992) (Fig. 1). The first, the fetal population, appears in fetal life and disappears gradually in the neonatal period. The second, the adult population, appears at puberty and persists for the adult life. Despite some differences in their regulation and function, both Leydig cell populations actively produce androgens in response to LH stimulation. LH from fetal pituitary or chorionic gonadotropin from the placenta is present in fetal circulation in sufficient concentrations to stimulate the fetal Leydig cells. The pubertal activation of Leydig cell function parallels the pubertal increase of gonadotropin secretion (see Huhtaniemi and Warren, 1990).

A. Acquisition of LH Responsiveness

One of the most conspicuous features of the Leydig cell is the dependence of its steroidogenesis on LH. The mechanism of action of LH, and how the LH receptor is maintained in the adult Leydig cell, have been studied extensively (for references, see the reviews just listed). However, how LH dependency is acquired during ontogeny is not very well understood. Although LH itself plays a pivotal role in the recruitment and differentiation of the Leydig cells in postnatal life, the first steps in the onset of Leydig cell differentiation in the embryonic gonad from undifferentiated mesonephric mesenchymal cells (de Kretser and Kerr, 1988; Wartenberg, 1989; Huhtaniemi and Pelliniemi, 1992) are assumed to occur independently of gonadotropins and androgens (Jost, 1970). The machinery to produce androgens and the expression of LH receptors occur in the rat testis in close temporal proximity with morphological differentiation (Niemi and Ikonen, 1961; Warren et al., 1984; Rabinovici and Jaffe, 1990; Huhtaniemi and Pelliniemi, 1992). Some still unknown factor(s) may turn on the initial differentiation of Leydig cells, inducing in these cells the molecular mechanisms responsible for steroidogenesis and LH responsiveness. Thereafter, once LH appears in the circulation, the Leydig cells acquire their morphological characteristics and the quantitatively important androgen production starts.

Apparently the morphologically undifferentiated precursor Leydig cells possess some of the functional features of mature Leydig cells, including the steroidogenic enzymes and responsiveness to LH. The clinical evidence for this statement is that, in prepubertal boys in whom no differentiated or morphologically identifiable Leydig cells are present (Wartenberg, 1989), the level of spermatic venous blood testosterone still exceeds the peripheral concentration (Forti et al., 1981). The presence of LH receptors in the prepubertal testis is implicated by the fact that human chorionic gonadotropin (hCG) is able to stimulate their testosterone secretion (Rivarola et al., 1970). The undifferentiated mesenchymal precursors for human Leydig cells also produce testosterone in response to hCG in culture (Chemes et al., 1992). Hence, the functional differentiation of the Leydig cells must begin before they are differentiated morphologically. Likewise, when Leydig cells are destroyed by treatment with ethylene dimethyl sulfonate (EDS), the subsequent reappearance of Leydig cells is an LH-dependent phenomenon (Teerds et al., 1988). This event is analogous to processes occurring in the fetal testis at the onset of LH-dependent Leydig cell differentiation and during pubertal maturation. We do not know whether similar mechanisms regulate Leydig cell differentiation during the fetal and adult growth phases, although such a possibility definitely exists.

The finding we made on the ontogeny of rat ovarian LH receptor also may be relevant to the development of LH responsiveness in Leydig cells. Functional LH receptor is found in the ovary on day 7 post partum (Sokka and Huhtaniemi, 1990) but, when the appearance of LH receptor mRNA was studied (Sokka et al., 1992), a truncated version of the message encoding the extracellular domain of the receptor was detected in the ovary as early as day 17 of fetal life. Concomitantly with the postnatal appearance of the LH action was a switch in the alternative splicing pattern of LH receptor mRNA and appearance of an mRNA species corresponding to the full-length receptor protein. Hence, the LH receptor gene may be expressed constitutively in the developing gonad, but the onset of its function is regulated through alternative splicing of its mRNA, not at transcription.

B. Induction of Steroidogenesis with Respect to LH Responsiveness

Current evidence suggests that the development of the steroidogenic machinery and the onset of LH responsiveness are two independent phenomena. These events are the first phases of functional differentiation of Leydig cells. Whether the factor(s) inducing these responses are intratesticular (paracrine/autocrine) or extratesticular (endocrine) is not clear. An intratesticular inducer (or constitutive onset?) of steroidogenesis is likely since, in cultured rabbit ovaries and testes, the appearance of the steroidogenic machinery seems to occur in defined medium at the appropriate time without the influence of gonadotropins or other serum factors (Wilson et al., 1981). The fact that, despite concomitant appearance of steroidogenesis, the LH receptors appear in the rat ovary more than 10 days later than in testes (day 7 post partum vs. day 15.5 in utero) (Warren et al., 1984; Sokka and Huhtaniemi, 1990) also suggests that different factors are responsible for the induction of steroidogenesis and LH receptors. Is the sex difference caused by early induction of LH receptor translation in the testis or by suppression of this event in the ovary?

Although steroidogenesis seems to start simultaneously in the fetal ovary and testis, some quantitative sex differences between the steroidogenic enzymes are evident: 3β-hydroxysteroid dehydrogenase is higher in the testis and aromatase is higher in the ovary (Wilson et al., 1981). The activities of the other steroidogenic enzymes in the male and female embryonic gonad are similar. However, because of the absence of LH responsiveness, fetal ovarian steroidogenesis remains quantitatively very low. The anti-Müllerian hormone, a product of fetal Sertoli cells, may be responsible for the absence of aromatase in fetal Leydig cells (Vigier et al., 1989).

C. Possible Inducers of Leydig Cell Differentiation

The role of prolactin (Prl) in the induction and maintenance of adult Leydig cell LH receptors is clear (Bartke et al., 1980; Catt et al., 1980). Also, although not verified, placental Prl could play a role in the fetal onset of Leydig cell LH responsiveness, depending on the timing of appearance of Prl receptors in Leydig cells, which has not been studied. Follicle stimulating hormone (FSH) as the inducer of fetal testicular LH receptor is unlikely, although its indirect effects on Leydig cell maturation through paracrine Sertoli cell factors are very plausible (Odell and Swerdloff, 1976; Kerr and

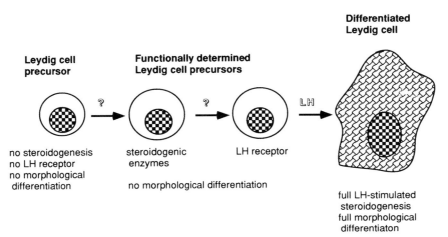

FIGURE 2   Developmental stages of fetal Leydig cells from precursor to full differentiation. Apparently, the steroidogenic machinery and LH receptor are acquired either constitutively or through induction of intratesticular effectors that remain unknown. The last step of differentiation, that is, acquisition of the morphological and functional (active steroidogenesis) characteristics of Leydig cells, is caused by tropic action of LH.

Sharpe, 1985; Closset and Hennen, 1988). However, both in the human and in the rat, the secretion of FSH from the fetal pituitary starts much later than the onset of testicular LH action (Aubert *et al.*, 1985; Huhtaniemi and Pelliniemi, 1992). The only other factor reported to date that mimics gonadotropin action before the onset of gonadotropin responsiveness is the vasoactive intestinal peptide (VIP), which stimulates cAMP production and aromatase in neonatal rat ovary before the appearance of FSH response (George and Ojeda, 1987). Whether this result implies that peptidergic inputs via neuronal connections play a role in the onset of gonadotropin responsiveness still remains an intriguing possibility. Hardy *et al.* (1990) found that the maturation of Leydig cell precursors *in vitro* was induced by a combination of LH and dihydrotestosterone, but not by either hormone alone. Whether this kind of mechanism could be functional in the initial steps of fetal Leydig cell differentiation remains obscure. Finally, the role of intratesticular growth factors in fetal testicular differentiation have been discussed (Rabinovici and Jaffe, 1990).

In conclusion, a multistep process of Leydig cell differentiation during ontogeny can be proposed (Fig. 2). The initial stage is the undifferentiated mesenchymal cell with only the potential to differentiate into a Leydig cell. The first step of differentiation involves the acquisition of the steroidogenic machinery, an event that occurs simultaneously in the ovary and testis. The factor that induces this step is not known, but is evidently intrinsic to the gonad. The expression of the LH receptor gene may start simultaneously or even earlier (constitutive expression?). The next step is the appearance of functional LH receptor, which probably requires changes in the alternative splicing pattern of its mRNA. This step is most likely to occur independently, and after the onset of steroidogenesis, since it clearly is delayed in the ovary despite concomitant onset of steroidogenesis in both sexes. Up to this stage, no marked changes occur in the morphology of Leydig cells. This type of Leydig cell is observed, for example, in prepubertal or posthypophysectomy testes, which are devoid of mature Leydig cells yet are responsive to LH. The final step of Leydig cell maturation is distinctively LH dependent and involves morphological differentiation and quantitative increase in steroidogenesis.

III. Molecular Aspects of LH Receptor Function

A. Structure of the LH Receptor Gene

The cDNA for the LH receptor was cloned and sequenced (Loosfelt et al., 1989; McFarland et al., 1989). The structure of the LH receptor gene is known also (Koo et al., 1991; Tsai-Morris et al., 1991; Huhtaniemi et al., 1992). The 93-Da receptor belongs to the family of the G protein-associated seven-times transmembrane receptors, with an unusually long extracellular domain with leucine-rich repeats that constitutes almost half the molecule (Fig. 3). This domain distinguishes the LH receptor from the other members of this family (e.g., the β-adrenergic receptor), but makes it similar to the FSH and thyroid stimulating hormone (TSH) receptors (Sprengel et al., 1990; Vassart et al., 1991). The gene of this protein is long, over 70 kb, and consists of 11 exons and 10 introns (Koo et al., 1991; Tsai-Morris et al., 1991). The extracellular domain of the receptor is encoded by the first 10 exons; the last large exon (11) encodes the transmembrane and intracellular domains (Fig. 3).

The promoter structure of the LH receptor gene reveals only few canonical recognition sequences for known transcription factors (several Sp1 and AP-1 sites) (Tsai-Morris et al., 1991; Huhtaniemi et al., 1992; Wang et al., 1992), a finding that still leaves the deduction of the regulatory factors of this receptor uncompleted. Likewise, no TATA or CCAAT elements are found at appropriate distances from the transcription initiation sites of the gene. Multiple transcription initiation sites are detected on primer extension analysis (Tsai-Morris et al., 1991; Huhtaniemi et al., 1992; Wang et al., 1992). The promoter structure resembles those of constitutively expressed household genes, which supports our finding on LH receptor gene expression in fetal rat ovary (see previous text). The details of the LH receptor gene and protein structure have been discussed in several reviews (Metsikkö et al., 1990; Segaloff et al., 1990; Hsueh and LaPolt, 1992). Therefore, only the special aspects of the expression of this gene in the testis are reviewed in this section.

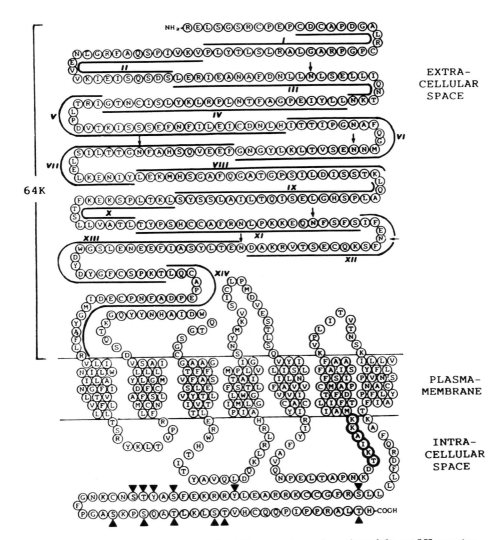

FIGURE 3 The primary structure and putative membrane insertion of the rat LH receptor. The protein belongs to the family of G protein-associated receptors with seven plasma membrane-spanning regions. Unlike most other members of this gene family, the LH receptor has an unusually long extracellular domain (64 K), composing almost half the molecule. The extracellular domain possesses the main ligand binding site(s). The sites for *N*-linked glycosylation in the extracellular domain are marked by arrows. The roman numerals indicate the repeats related to proteins belonging to the leucine-rich glycoprotein family. The boldfaced circles in the third intracellular loop indicate conserved amino acids putatively involved in G protein coupling. The triangles indicate putative phosphorylation sites of the intracellular domain. Reprinted with permission from Metsikkö *et al.* (1990).

B. Multiple LH Receptor mRNA Species

Northern blot analyses of LH receptor mRNA revealed a very complex transcription pattern of this gene (Metsikkö *et al.*, 1990; Segaloff *et al.*, 1990; Hsueh and LaPolt, 1992). Several mRNA species are present in Leydig cells from testes of different ages and after different hormonal treatments. The transcripts vary in length from 1 to over 7 kb, whereas a transcript of 2.6 kb in length encodes the full-length receptor protein. Differences in polyadenylation and alternative or incomplete splicing of the receptor gene are likely explanations for this phenomenon. Also differences are seen in the intensities of the different mRNAs in the ovary, the testis, and various gonadal cell lines that express the LH receptor (Loosfelt *et al.*, 1989; Segaloff *et al.*, 1990; Wang *et al.*, 1991a; Hsueh and LaPolt, 1992). One of the shorter mRNAs (1.2 or 1.8 kb, depending on the tissue source) encodes a truncated receptor protein that corresponds to the extracellular domain (Xie *et al.*, 1990; LaPolt *et al.*, 1991) and has raised considerable interest. This protein could be secreted into the extracellular space, where it could function as an LH antagonist (see subsequent discussion). The physiological correlates of the complicated mRNA structure are still completely obscure.

C. Regulation of LH Receptor Gene Expression

Several studies have been carried out to relate the steady-state levels of LH receptor mRNA to the physiological regulation of this receptor. Although most of the studies on mRNA levels have yielded results in accordance with earlier receptor binding measurements, some new findings have emerged also. The receptor down-regulation (LaPolt *et al.*, 1991; Wang *et al.*, 1991a) and its absence in the fetal–neonatal Leydig cells (Pakarinen *et al.*, 1990) is partially the result of changes at the level of transcription. LaPolt *et al.* (1991) observed that ligand-induced down-regulation of the LH receptor included loss of all other LH receptor mRNA species, except for a 1.8-kb transcript missing portions of the transmembrane domain. In the ovary, in contrast, all mRNA species disappeared on similar treatment. In the MA10 murine Leydig cell tumor model, all LH receptor mRNA species were regulated coordinately. Most of the down-regulation was caused by receptor internalization and degradation, since the mRNA loss was quantitatively minor compared with the total loss of receptor sites (Wang *et al.*, 1991a). The down-regulation of mRNA was shown further to be a cAMP-dependent phenomenon. The decreased steady-state levels of LH receptor mRNA during ligand-induced down-regulation indicate that decreased LH receptor gene transcription plays a role in this phenomenon. However, whether the reason for this change is suppressed transcription or mRNA stability remains to be shown. Events such as tissue-specific posttranscriptional regulation of alternative or incomplete mRNA splicing, alternative start sites of transcription, and RNA stability may be involved (Tsai-Morris *et al.*, 1991; Huhtaniemi *et al.*, 1992; Wang *et al.*, 1992). Also, heterologous down-regulation of LH receptors by epidermal growth factor (EGF) and phorbol ester induces suppression of steady-state levels of the cognate mRNA (Wang *et al.*, 1991b).

D. Structure – Function Relationships of the LH Receptor

Apparently ligand binding resides in the extracellular domain of the LH receptor molecule. However, other information suggests that the transmembrane region contains a ligand binding site in addition to the structures needed for G_s protein activation (Ji and Ji, 1991). Some data (Rodriguez *et al.*, 1992) indicate that both LH-stimulated cAMP production and ligand internalization are enhanced if the distal part of the cytoplasmic tail of the receptor is removed, and that ligand-induced uncoupling (desensitization) also is prevented by removal of the cytoplasmic C terminus (Sanchez-Yagüe *et al.*, 1992).

Mutated LH receptor containing only the extracellular domain, when transfected into kidney 293 cells, binds hCG with high affinity that is comparable to that of full-length receptor (Xie *et al.*, 1990). However, the function of such a protein, possibly formed from some of the truncated LH receptor mRNA species (Loosfelt *et al.*, 1989; Segaloff *et al.*, 1990, see previous discussion) is unclear, since the mutated receptor was not found to be secreted but trapped inside the cells. This matter is still controversial since, in another study by Tsai-Morris *et al.* (1990), a similar truncated form of the receptor was found to be secreted. West and Cooke (1992) have shown that antisense oligonucleotides to the coding region of LH receptor inhibit release of the LH binding proteins from MA10 cells. Whether this protein represents a translation product of the truncated mRNA species or shedding of the extracellular domain of full-length receptors from the plasma membrane still remains unknown. The question is intriguing, since demonstration of secretion of LH receptor species with only the ligand binding activity would bring the Leydig cell response to LH stimulation into new perspective. The truncated receptor could represent an LH binder that could inhibit LH action at the functional receptor site; in this way, the Leydig cell could modulate its response to LH. Whether this hypothesis has any relevance to the known binders of LH in the serum and the testicular interstitial fluid (see subsequent discussion) also remains unknown. The regulation/structure/function relationships of the LH receptor have just started being addressed. A great deal of new information is expected to emerge in the near future.

IV. Signal Transduction Systems Involved in LH Action in Leydig Cells

A. cAMP

That the mechanism of LH action employs cyclic AMP (cAMP) as the second messenger is considered dogma. LH binds to its receptor and activates the G_s protein, which results in activation of adenylyl cyclase and finally in conversion of ATP to cAMP. cAMP then activates protein kinase A, which leads through specific protein phosphorylations to stimulation of Leydig cells, including steroidogenesis by these cells. The details of steroidogenesis are described in Chapter 10. The classical paradigm has remained valid, since ample experimental evidence suggests that LH increases cAMP production of the Leydig cells, phosphodiesterase inhibitors potentiate the LH action, and cAMP analogs can mimic LH action on steroidogenesis. Also the Gi protein-mediated inhibition of cAMP production and steroidogenesis is functional in Leydig cells (Platts et al., 1988), but the ligand activating this pathway still remains unknown.

Some details of the cAMP-mediated signal transduction system have remained controversial. One of these aspects has been the different stoichiometry of cAMP generation and steroidogenic stimulation, that is, maximal steroidogenesis is stimulated by LH concentrations that do not evoke measurable elevation in intracellular cAMP. Although the role of cAMP as the second messenger of LH is not challenged, increasing evidence suggests that this second messenger is not able to explain all LH actions. In fact, the phenomenon that one receptor employs multiple signal transduction systems may be more general (Thompson, 1992). Increasing evidence exists for cross talk between the different signal transduction systems. LH action may be modulated by other endocrine/paracrine/autocrine factors through these alternative mechanisms. Finally, the quantitative importance of the different regulatory systems seems to differ in the different Leydig cell models used. The classical mechanism of LH action by generation of cAMP has been discussed extensively in numerous earlier reviews (see Dufau, 1988; Hall, 1988). Therefore, we summarize only the recently accumulated new information on the signal transduction mechanisms involved in LH action and its modulation.

Doubts have been cast on the obligatory role of cAMP in LH stimulated steroidogenesis in Leydig cells (Themmen et al., 1985; Cooke, 1990). The main argument has been that testosterone synthesis is stimulated maximally by LH concentrations that result in less than 1% saturation of the receptors and in no detectable increase of intracellular cAMP levels. The low occupancy of receptors has been explained by the spare receptor concept (Catt and Dufau, 1973), which provides a mechanism for insurance of continued Leydig cell responsiveness to LH stimulation under physiological conditions through constant supply of unoccupied receptors to transmit the LH signal. The lack of cAMP response has been explained by compartmentalization of cAMP within the target cell and increased cAMP concentration only in a limited intracellular subcompartment that cannot be detected if total cellular cAMP is monitored. To date, only limited evidence for such intracellular compartmentalization of the second messengers has existed. Another piece of evidence against the role of cAMP is the fact that deglycosylated LH has increased affinity for the LH receptor, evokes no cAMP production, but may stimulate steroidogenesis in some cell systems (Sairam, 1990).

Evidence is also available for the obligatory role of cAMP in stimulation of Leydig cell steroidogenesis. The discrepancy between cAMP and steroidogenic response does not exist if protein kinase (PK)-bound cAMP is measured instead of total cellular cAMP (Dufau *et al.*, 1977). If PKA is inhibited competitively in cultured mouse MA 10 Leydig cells with the Rp diastereoisomer of cAMP, steroidogenesis is inhibited at hCG concentrations that do not increase cAMP levels (Pereira *et al.*, 1987). Moger (1991) made intriguing observations suggesting compartmentalization of Leydig cell PKA. He showed that exogenous cAMP synergized with both PKA type I and type II in stimulation of steroidogenesis, whereas the LH-stimulated endogenous cAMP was able to synergize only with type I selective cAMP analogs. A close proximity of adenylyl cyclase and PKA type I was suggested to allow low LH stimulation to increase cAMP locally. This increase is not measurable when total cellular cAMP is monitored. Additional evidence for compartmentalization of endogenous cAMP during gonadotropin stimulation was obtained by patch clamp stimulation tests of individual Leydig cells (Podesta *et al.*, 1991). The controversy of the missing cAMP response at LH levels that stimulate steroidogenesis has been resolved by data on compartmentalization of cAMP. However, as indicated in the next section, the possibility remains that LH also triggers other signal transduction systems and that LH action is under modulatory or permissive influences of other intracellular messengers. Further, the other messenger systems, bypassing LH, also may exert their independent effects on Leydig cell steroidogenesis.

B. Calcium

Another candidate for second messenger of LH action is intracellular free calcium. Ca^{2+} acts in Leydig cells at least as a permissive agent, since maximal testosterone production cannot be achieved in calcium-free medium (Janszen *et al.*, 1976). Injection of Ca^{2+}–calmodulin complexes into Leydig cells in liposomes stimulates steroidogenesis (Hall *et al.*, 1981). Likewise, gonadotropin releasing hormone (GnRH), which increases intracellular calcium but does not affect cAMP, can stimulate rat Leydig cell steroidogenesis (Sharpe, 1984). Studies with calmodulin inhibitors suggest that Ca^{2+}–calmodulin can stimulate steroidogenesis in the absence of elevated cAMP, whereas cAMP production also may be Ca^{2+}–calmodulin dependent (Cooke, 1990). Whether this observation means that Ca^{2+}–calmodulin affects steroidogenesis via amplification of the cAMP response or independently has not been clarified. Gudermann *et al.* (1992a,b) showed that the murine LH receptor expressed in L cells and in *Xenopus* oocytes was able to activate phospholipase C in addition to adenylyl cyclase, resulting in inositol phospholipid breakdown and liberation of Ca^{2+} from intracellular stores.

These findings provide strong evidence that the LH receptor is able to activate two signal transduction mechanisms, that using cAMP and that using the phosphoinositide breakdown products inositol trisphosphate (IP_3) and diacylglycerol (DG). IP_3 results in increased cytosolic free calcium and DG in activation of protein kinase C (see subsequent discussion). Whether all LH effects on intracellular free Ca^{2+} are mediated by activation of phospholipase C is uncertain. cAMP also can increase cytosolic Ca^{2+}, which in this case originates from the extracellular medium through opening of cell membrane calcium channels (Sullivan and Cooke, 1986).

Considering the observations on positive Ca^{2+}–calmodulin effects on cAMP and steroid synthesis, and the positive effect of cAMP on cytosolic free Ca^{2+}, we could postulate that the two signal transduction systems act in Leydig cells in a synergistic fashion, providing an amplification mechanism of LH action. Some of the current discrepancies in the existing literature may be the result of differential stoichiometry of the two signal transduction systems that mediate LH action in Leydig cells. Note that species differences may exist in the signal transduction systems of LH action (e.g., rat vs. mouse). The cytoplasmic domains of the rat and mouse LH receptor display structural differences that imply differences in their signal transduction mechanisms (Gudermann *et al.*, 1992a).

C. Phosphoinositides and Protein Kinase C

Presence of the calcium and phospholipid-activated protein kinase C (PKC) has been demonstrated in Leydig cells (Kimura *et al.*, 1984; Nikula *et al.*, 1987; Pelosin *et al.*, 1991). Of the endogenous activators of this enzyme, DG is released from phosphatidylinositol bisphosphate (PIP_2) after ligand-induced activation of phospholipase C. The other cleavage product of PIP_2, IP_3, stimulates release of the other PKC activator, free calcium, from intracellular stores (Nishizuka, 1986). Phorbol esters [e.g., 12-O-tetradecanoyl-phorbol 13-acetate (TPA)] generally are used as pharmacological probes to activate PKC. Activation of PKC by phorbol esters results in stimulation of basal testosterone production but inhibition of LH-stimulated cAMP and testosterone production in Leydig cells (Mukhopadhyay and Schumacher, 1985; Themmen *et al.*, 1986; Nikula *et al.*, 1987). The former effect may be a direct effect on steroidogenesis that bypasses cAMP or a cAMP-mediated effect through inhibition of the Gi protein, which is another documented effect of TPA in Leydig cells (Platts *et al.*, 1988; Nikula and Huhtaniemi, 1989). The inhibitory effect involves inhibition of LH receptor–G_s protein coupling in the same manner as seen on ligand-induced desensitization to LH action (Rebois and Patel, 1985; Lopez-Ruiz *et al.*, 1992). In addition, some steps beyond cAMP formation seem to be involved in the PKC-mediated inhibition of steroidogenesis (Nikula and Huhtaniemi, 1989).

The inhibitory effect of PKC on Leydig cell LH response is similar, but not identical, to the LH/hCG-induced homologous (Rebois and Patel, 1985; Inoue and Rebois, 1989) and EGF-induced heterologous (Pereira *et al.*, 1988) desensitization. PKC is able to desensitize the LH response, but LH-induced desensitization may not involve activation of PKC. Instead, this event is mediated by some components that are associated stably with the cell membrane, probably an additional gonadotropin-sensitive protein kinase that phosphorylates the LH receptor (Inoue and Rebois, 1989). Although shown in rat granulosa and bovine luteal cells (Davis *et al.*, 1986a,b) and in heterologous cells transfected with the murine LH receptor gene (Gudermann *et al.*, 1992a,b), no direct evidence exists for LH/hCG-stimulated activation of PKC in Leydig cells (Inoue and Rebois, 1989; Cooke, 1990). Neither the catalytic nor the regulatory GTP binding protein seems to be affected by desensitization (Inoue and Rebois, 1989).

LH may activate PKC via an alternative pathway, through activation of phospholipase A_2, directly or via cAMP, which then leads to release of arachidonic acid. Additional metabolism through the lipoxygenase pathway may be involved in stimulation of Leydig cell steroidogenesis (Dix *et al.*, 1984; Cooke *et al.*, 1991), presumably through activation of PKC (Nishizuka, 1988). In fact, the short-term inhibitory action of arachidonic acid on LH-stimulated testosterone production was shown to be mediated through PKC activation (Lopez-Ruiz *et al.*, 1992). The acute effect of PKC on steroidogenesis is therefore inhibitory, but when PKC is down-regulated the effect through alternative mechanisms is stimulatory. These results also suggest that steroidogenesis is normally under tonic inhibitory control by PKC. The locus of PKC action is still unknown, but the protein could affect intracellular Ca^{2+} concentration or phosphorylate some proteins involved in steroidogenesis.

Clearly PKC is present in Leydig cells. Its activation has a multitude of effects on Leydig cell functions, including stimulation of basal steroidogenesis, desensitization of LH action at the LH receptor site, and, in addition to cAMP formation, inhibition of the Gi protein function. Collectively, the evidence for a physiological role of the PKC-mediated events in Leydig cells is still very confusing. Perhaps LH action itself involves PKC activation but other hormones can, via PKC activation, modulate LH action or have direct effects on Leydig cell functions. The physiological effectors using this pathway are still largely unknown.

D. Cl⁻ Channels

In addition to calcium, Cl^- channels also have been demonstrated to participate in LH action in adult Leydig cells (Duchatelle and Joffre, 1990). LH/hCG has little effect on K^+ conductance in Leydig cells, but the outward current of Cl^- is increased. Whether the calcium-activated Cl^- channels are involved in LH action is still controversial (Duchatelle and Joffre, 1990;

Gudermann *et al.*, 1992b). This effect is mimicked by intracellular cAMP, which supports the possibility that the Cl⁻ channels play a role in LH action. Submaximally LH-stimulated steroidogenesis was increased in Cl⁻ free medium, and a chloride channel blocker had an opposite effect at low but not high LH levels (Choi and Cooke, 1990). Investigators concluded from these data that the action of LH at low but not high cAMP levels depends on Cl⁻ channels. How the LH-stimulated efflux of Cl⁻ ions is related to stimulation of steroidogenesis still remains obscure.

E. cGMP and Tyrosine Kinase

Finally, an additional signal transduction system that functions in the Leydig cell employs cGMP as the second messenger. In this case, the cytoplasmic domain of the receptor acts as a guanylyl cyclase, converting GTP to cGMP. The ligand involved in this action in the testis is most likely to be atrial natriuretic peptide (Mukhopadhyay *et al.*, 1986). cGMP has been shown to act synergistically with cAMP generating systems, presumably at the activation of the cAMP-dependent protein kinase (Hipkin and Moger, 1991). Ample evidence also exists for tyrosine-specific kinase activities in Leydig cells, but since these signal transduction mechanisms are related closely to growth factor and oncogene actions, they are discussed in connection with testicular paracrine regulation.

F. Mechanisms Involved in Ligand-Induced Refractoriness to LH Action

In addition to the signal transduction systems employed in gonadotropin-induced stimulation of Leydig cell function, additional mechanisms are involved in the refractoriness after strong or prolonged LH stimulation. These events involve sequestration of the LH receptors from plasma membrane and loss of binding (homologous down-regulation), concomitant changes in some other membrane receptors (e.g., lactogen receptors, heterologous down-regulation), uncoupling of the LH receptor from adenylyl cyclase, increased cAMP degradation due to phosphodiesterase activation, and loss of activity of some key enzymes in androgen formation (steroidogenic lesions). (For further details, see Dufau, 1988.)

The existing *in vivo* and *in vitro* evidence suggests that a loss of coupling occurs between the LH receptor and the G_s protein, whereas the interaction between the G_s α subunit and the adenylyl cyclase catalytic unit remains intact (Cooke *et al.*, 1990). Hence, only LH-stimulated cAMP production, not that evoked by cholera toxin or fluoride, is impaired. Therefore, the LH receptor itself is likely to be the target of alteration leading to desensitization. The hormone-induced desensitization can be mimicked by TPA, although whether LH activates PKC or another kinase functioning in an analogous fashion is not known (Inoue and Rebois, 1989). However, this alternative is possible since the intracellular domain of the receptor possesses several phosphorylation sites of PKC (Loosfelt *et al.*, 1989). Basal cAMP production during desensitization is increased, which can be explained by the fact that the G_i protein function also is inhibited by PKC (Platts *et al.*, 1988), resulting in alleviation of the tonic suppression of adenylyl cyclase throughout this mechanism. Direct evidence for LH effects on G_i protein activity is still missing.

Note that Leydig cell desensitization is not functional in fetal Leydig cells (Huhtaniemi *et al.*, 1984; Huhtaniemi and Pelliniemi, 1992). The reasons for this situation remain unclear, but that the fetal Leydig cells lack functional G_i protein (Warren, 1989) is of special interest. Whether this condition is involved in the lack of refractoriness of the fetal Leydig cells after LH stimulation is not known. Since LH receptor phosphorylation through PKC may be involved, the LH receptor isotype expressed in the fetal Leydig cell may not have this domain of the receptor protein.

In conclusion, although the role of cAMP seems clear as the second messenger in LH action at higher hormone concentrations, increasing evidence suggests that other signal transduction systems also may be involved, especially at lower LH levels. These systems include calcium mobilization from intra- and extracellular sources, Cl^- channel function, release of arachidonic acid and its further metabolism through the lipoxygenase pathway, and activation of protein kinase C. These alternative signal transduction systems may be physiologically important, since *in vivo* the LH stimulation of Leydig cells is normally far lower than that maximally evoked *in vitro*. However, the findings that low LH levels do not use cAMP are disputed by some findings of the obligatory role of cAMP. Dissecting out which alternative signal transduction systems really play a role in the regulation of Leydig cell function and which act as modulators that are ineffective alone but alter the magnitude of the primary response to LH is difficult. The issue is complicated by the fact that a variety of *in vitro* systems have been used to generate the data available. The different cell and species models seem to have qualitative and quantitative differences in the various signal transduction systems involved. A schematic presentation of the pleiotypic response of Leydig cells to LH stimulation is depicted in Fig. 4.

V. Effects of Hormones Other than LH on Leydig Cells

That LH is the most important hormonal regulator of Leydig cell function still remains dogma. However, many divergent paracrine and autocrine effects on Leydig cell function have been discovered. The biochemical evidence for such paracrine regulatory mechanisms within the testis is indisputable, but their physiological role is still undetermined. What are the physiological

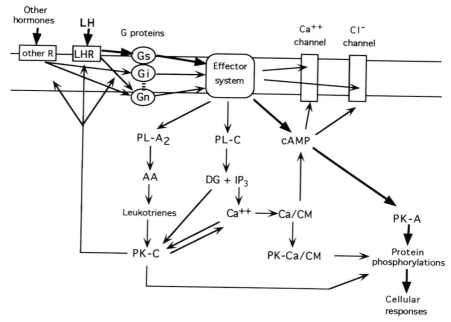

FIGURE 4 Different signal transduction systems involved in LH action in Leydig cells. The scheme emphasizes the involvement of several second messenger systems in LH action and the modulation of LH action by other ligands using a variety of signal transduction systems. The classical pathway of LH action through cAMP is depicted with boldfaced arrows. Some details of the scheme are controversial; all details of the scheme may not be functional in all Leydig cell models. Similarly, some of the effects (stimulation or inhibition) under physiological conditions are still obscure and may be dependent on the magnitude and time relations of the stimulus. Abbreviations: R, receptor; G_s, stimulatory guanine nucleotide binding protein; G_i, inhibitory guanine nucleotide binding protein; G_n, other guanine nucleotide binding proteins; PL, phospholipase; cAMP, cyclic adenosine 3′,5′-monophosphate; AA, arachidonic acid; DG, diacyl glycerol; IP_3, inositol triphosphate; CM, calmodulin; PK, protein kinase.

or pathophysiological conditions under which changes in the paracrine factors interfere with the overall function of the testis? These regulators are discussed in more depth in another section of this book. In this chapter we assess the role of some other blood-borne hormones with likely significance in the physiological regulation of Leydig cell function. The decision of whether a factor is paracrine or endocrine within the testis is increasingly difficult, since a number of classically extragonadal hormones have been shown to be produced by the testis also. Moreover, keep in mind that testicular innervation may play a role in the regulation of Leydig cell function. (For further details, see Mayerhofer *et al.*, 1989.)

A. Prolactin

One hormone with clearly established Leydig cell effects in several species (e.g., rodents) is Prl. In the rodent, Prl receptors are present in the Leydig cells (Aragona *et al.*, 1977; Charreau *et al.*, 1977; Costlow and McGuire, 1977; Catt *et al.*, 1980). Treatment of hypophysectomized rats with Prl significantly augments the testosterone response to LH (Hafiez *et al.*, 1972). Likewise, Prl amplifies *in vitro* the rat Leydig cell steroidogenic response to LH stimulation at low concentration, but inhibits further metabolism of progesterone at high levels (Welsh *et al.*, 1986). LH stimulation induces heterologous down-regulation of Leydig cell Prl receptors in adult but not in neonatal animals. High Prl stimulation can suppress LH receptors (Catt *et al.*, 1980; Huhtaniemi *et al.*, 1984). The mechanisms of these heterologous regulatory events still remain unclear.

The role of Prl in testicular function is especially clear in seasonally breeding hamsters, in which short days are associated with low Prl levels and testicular involution. This effect can be reversed with Prl treatment (Bartke *et al.*, 1980). Direct Prl effects on the human testis are still controversial. Both positive and negative findings on Prl receptors in human Leydig cells exist (Wahlström *et al.*, 1983; Bouhdiba *et al.*, 1989). Hyperprolactinemia has been related to testicular hypofunction (Clayton, 1986) but, in the absence of solid evidence for Prl receptors in the human testis, these effects must be interpreted as indirect, due to suppression of gonadotropin secretion.

B. Growth Hormone

Another anterior pituitary hormone that may have direct effects on Leydig cell functions is growth hormone (GH). Ample evidence supports this concept, both for humans and for experimental animals, since GH deficiency or resistance is associated with delayed puberty and poor Leydig cell response to LH/hCG (Chatelain *et al.*, 1991). The issue of direct GH effects on the Leydig cell is complicated by the fact that many of these studies have been carried out using human GH, which is a potent lactogen in the rat (Posner *et al.*, 1974). Likewise, many GH-deficient animal models also are deficient of Prl (van Buul-Offers, 1983). In favor of a role of GH are the findings that Prl and GH effects on Leydig cells are not identical. Rat GH, but not rat Prl, up-regulated insulin-like growth factor 1 (IGF-1) receptors in purified Leydig cells (Lin *et al.*, 1988). Prl increased testicular LH receptor content without affecting steroidogenesis, whereas GH effects were the reverse (Zipf *et al.*, 1978). Many, although probably not all, of the GH effects on the testis are mediated by IGF-1 (Chatelain *et al.*, 1991). Although GH receptor immunoreactivity has been demonstrated in the rat testis (Lobie *et al.*, 1990), the GH receptor mRNA level in the testis has been shown to be either low or nonexistent (Mathews *et al.*, 1989; Tiong and Herington, 1991). Further studies are needed to resolve this controversy. In conclusion, GH of pituitary origin appears to play a role in Leydig cell function, especially during pubertal maturation. Many, although not all, of the GH actions are mediated through paracrine actions of IGF-1, whose formation in the testis is stimulated by GH.

C. Insulin

Insulin receptors are found on Leydig cells (Abelé *et al.*, 1986; Lin *et al.*, 1986; Rigaudière *et al.*, 1988). Insulin and LH reciprocally up-regulate their receptors (Charreau *et al.*, 1978; Abelé *et al.*, 1986; Bernier *et al.*, 1986), and insulin augments basal and LH-stimulated steroidogenesis of Leydig cells (Adashi *et al.*, 1982; Lin *et al.*, 1988; Rigaudière *et al.*, 1988). Insulin at physiological (i.e., nanomolar) concentrations appears to act through its own receptors, whereas at pharmacological (micromolar) concentrations its action is mediated through insulin and IGF-1 receptors, the latter of which are also present in Leydig cells (Bernier *et al.*, 1986; Lin *et al.*, 1986; Rigaudière *et al.*, 1988). The effects of insulin and IGF-1 on Leydig cells seem to be synergistic through a common saturable mechanism (Rigaudière *et al.*, 1988). Their effect requires protein synthesis and is seen at the LH receptor site and at steps beyond cAMP formation (Lin *et al.*, 1986).

D. Posterior Pituitary Hormones

Of the posterior pituitary hormones, vasopressin (AVP) (Kasson *et al.*, 1986; Tahri-Joutei and Pointis, 1988) is a possible candidate for a blood-borne hormone with effects on Leydig cell steroidogenesis. Data on Leydig cell effects of oxytocin (OT) are controversial (Sharpe and Cooper, 1987; Tahri-Joutei and Pointis, 1988,1989). Receptors for both hormones are present in Leydig cells (Meidan and Hsueh, 1985). AVP may be involved in the testis in acute stimulation and more prolonged inhibition of Leydig cell androgen biosynthesis. The mechanism of action of these hormones involves activation of protein kinase C. Since this signal transduction mechanism is also functional in the Leydig cell (see previous discussion), the physiological role of the hormone is feasible (Vinggaard and Hansen, 1991). Ample bio- and histochemical evidence exists for presence of AVP and OT in the testis (Tahri-Joutei and Pointis, 1988). The expression of genes encoding these hormones also has been demonstrated in the testis, but the level of expression was considered too low to be physiologically important (Foo *et al.*, 1991). Hence, although effects of circulating posterior pituitary hormones on the testis are possible, whether their main action is paracrine or endocrine is still unclear.

E. Atrial Natriuretic Peptide

One additional hormone with a possible endocrine role in Leydig cell regulation is the atrial natriuretic peptide (ANP). Its role in reproductive hormone regulation, including modulation of gonadotropin and Prl secretion, has been emphasized (Samson, 1992). Rat and mouse Leydig cells possess guanylyl cyclase-coupled receptors for ANP; at least in the latter species, ANP stimulates testosterone production (Mukhopadhyay *et al.*, 1986,1989). Although circulating ANP probably can reach the testis in physiologically significant levels, this hormone also may be mainly a paracrine factor. The testicular presence of ANP precursor material and the expression of the ANP gene supports a paracrine role for this hormone (Vollmar *et al.*, 1990).

F. Steroid Hormones

Leydig cell function also is modulated by steroid hormones including androgens, estrogens, and glucocorticoids. Estrogens and androgens are formed by Leydig cells. Therefore their effects can be considered paracrine or autocrine. In contrast, glucocorticoids are of adrenal origin, and therefore contribute to the endocrine regulation of Leydig cells. High systemic glucocorticoid levels, during physical and mental stress and in Cushing's syndrome, suppress testicular androgen production (for references, see Phillips *et al.*, 1989). Glucocorticoids have direct inhibitory effects on Leydig cell steroidogenesis (Saez *et al.*, 1977; Bambino and Hsueh, 1981). The effect is, at least partly, the result of inhibition of cholesterol side-chain cleavage cytochrome P450 (Hales and Payne, 1989). A functional link is suggested by the presence of glucocorticoid receptors in the testicular interstitial cells (Evain *et al.*, 1976; Ortlip *et al.*, 1981). An interesting addition to this system is brought by the discovery of testicular 11β-dehydrogenase activity, which is thought to form a protective mechanism for Leydig cell androgen formation, as a neutralizer of corticosteroid action through their conversion to biologically inactive 11-keto forms (Phillips *et al.*, 1989). Hence, protecting the testis from exposure to high levels of biologically active glucocorticoids is apparently physiologically desirable. Interestingly, the 11β-hydroxylase activity seems to be present only in the adult population of Leydig cells (Phillips *et al.*, 1989). Whether the fetal cells are immune to glucocorticoid action or are protected by other mechanisms is still obscure.

G. Circulating LH Inhibitors

Another factor (or group of factors) that is of potential importance in the endocrine/humoral control of Leydig cell function consists of the different circulating modulators of LH action. The presence of such factors has been observed in *in vitro* bioassay of LH, in which the response of Leydig cell testosterone production to LH is always lower in the presence of peripheral serum or plasma than in the presence of assay buffer (Dufau *et al.*, 1976; Rajalakshmi *et al.*, 1979; Marrama *et al.*, 1983; Ding and Huhtaniemi, 1989). The nature of such inhibiting factors has not been revealed. They may represent LH binding proteins, antibodies in the serum (Rajaniemi and Vanha-Perttula, 1972; Pala *et al.*, 1988), or even fragments of the LH receptor (Tsai-Morris *et al.*, 1990). The effects of such factors can be caused by inhibition of LH action at the receptor site or by interference with the signal transduction system.

The potential small molecular weight LH inhibitors in serum include prostaglandins (Thomas *et al.*, 1978), steroid hormones (Bambino and Hsueh, 1981; Daehlin *et al.*, 1985), fatty acids (Meikle *et al.*, 1989), and a variety of peptides (Magoffin and Erickson, 1982; Welsh and Hsueh, 1982; Adashi *et al.*, 1984; Morera *et al.*, 1988; Lin *et al.*, 1989; Guo *et al.*, 1990). The identification of these factors has not yet been carried out beyond demonstration of a heterogeneous group of chromatographic fractions of peripheral serum with a wide range of molecular sizes (Melsert and Rommerts, 1987; Ding and Huhtaniemi, 1989; Papadopoulos *et al.*, 1990). In addition to the serum, inhibitors and stimulators of LH action also are present in testicular extracts (see Sharpe, 1984,1990). As do the LH inhibitors in peripheral circulation, these molecules suppress the steroidogenic response of Leydig cells to LH stimulation without much effect on basal production. Further characterization of peripheral serum factors with rather dramatic modulatory action on LH effect is a useful strategy to elucidate the endocrine regulation of Leydig cell function.

VI. Conclusions

The purpose of this chapter was to discuss the role of LH in the development and function of Leydig cells. Obviously the initial steps of Leydig cell differentiation occur without influence of LH, but the full steroidogenic activity and morphological characteristics are attained only through LH action. Despite the large body of information on alternative endocrine and paracrine regulators and modulators of Leydig cell function, the key role of LH in the regulation of the Leydig cell function remains unquestioned. The complexity of the mRNA of the LH receptor was revealed by the cloning of the LH receptor gene. New information on structure–function relationships of the LH receptor molecule has started to emerge. With respect to the mechanism of LH action, cAMP seems to hold its key position as the second messenger, although increasing evidence suggests that some LH actions may be mediated by other signal transduction systems and that effectors, acting through other signal transduction mechanisms, likewise modulate the cAMP-mediated LH action. In addition to LH, several other blood-borne hormones may function as physiologically important regulators of Leydig cell function.

References

Abelé, V., Pelletier, G., and Tremblay, R. R. (1986). Radioautographic localization and regulation of the insulin receptors in rat testis. *J. Receptor Res.* **6**, 461–473.

Adashi, E. Y., Fabics, C., and Hsueh, A. J. W. (1982). Insulin augmentation of testosterone production in a primary culture of rat testicular cells. *Biol. Reprod.* **26**, 270–280.

Adashi, E. Y., Tucker, E. M., and Hsueh, A. J. W. (1984). Direct inhibition of rat testicular steroidogenesis by neurohypophysial hormones. Divergent effects on androgen and progestin biosynthesis. *J. Biol. Chem.* **259**, 5440–5446.

Aragona, C., Bohnet, H. G., and Friesen, H. G. (1977). Localization of prolactin binding in prostate and testis: the role of serum prolactin concentration on the testicular LH receptor. *Acta Endocrinol. (Copenhagen)* **84**, 402–409.

Aubert, M. L., Begeot, M., Winiger, B. P., Morel, G., Sizonenko, P. C., and Dubois, P. M. (1985). Ontogeny of hypothalamic luteinizing hormone-releasing hormone (GnRH) and pituitary GnRH receptors in fetal and neonatal rats. *Endocrinology* **116**, 1565–1576.

Bambino, T. H., and Hsueh, A. J. W. (1981). Direct inhibitory effects of glucocorticoids upon testicular luteinizing hormone receptor and steroidogenesis in vivo and in vitro. *Endocrinology* **108**, 2142–2148.

Bartke, A., Goldman, B. D., Klemcke, H. G., Bex, F. J., and Amador, A. G. (1980). Effects of photoperiod on pituitary and testicular function in seasonally breeding species. In "Functional Correlates of Hormonal Receptors in Reproduction" (V. B. Mahesh, T. G. Muldoon, B. B. Saxena, and W. A. Sadler, eds.), pp. 171–186. Elsevier/North-Holland, New York.

Bernier, M., Chatelain, P., Mather, J. P., and Saez, J. M. (1986). Regulation of gonadotropin receptor, gonadotropin responsiveness, and cell multiplication by somatomedin-C and insulin in cultured pig Leydig cells. *J. Cell. Physiol.* **129**, 257–263.

Bouhdiba, M., Leroy-Martin, B., Peyrat, J. P., Saint Pol, P., Djiane, J. and Leonardelli, J. (1989). Immunohistochemical detection of prolactin and its receptors in human testis. *Andrologia* **21**, 223–228.

Catt, K. J., and Dufau, M. L. (1973). Spare gonadotrophin receptors in the rat testis. *Nature New Biol.* **244**, 219–221.

Catt, K. J., Harwood, J. P., Clayton, R. N., Davies, T. F., Chan, V., Katikineni, M., Nozu, K., and Dufau, M. L. (1980). Regulation of peptide hormone receptors and gonadal steroidogenesis. *Rec. Prog. Horm. Res.* **36**, 557–622.

Charreau, E. H., Attramadal, A., Torjesen, P. A., Purvis, K., Calandra, R., and Hansson, V. (1977). Prolactin binding in rat testis: specific receptors in interstitial cells. *Mol. Cell. Endocrinol.* **6**, 303–307.

Charreau, E. H., Calvo, J. C., Tesone, M., de Souza Valle, L. B., and Baranao, J. L. (1978). Insulin regulation of Leydig cell luteinizing hormone receptors. *J. Biol. Chem.* **253**, 2504–2506.

Chatelain, P. G., Sanchez, P., and Saez, J. M. (1991). Growth hormone and insulin-like growth factor I increase testicular luteinizing hormone receptors and steroidogenic responsiveness of growth hormone deficient dwarf mice. *Endocrinology* **128**, 1857–1862.

Chemes, H., Cigorraga, S., Bergadá, C., Schteingart, H., Rey, R., and Pellizzari, E. (1992). Isolation of human Leydig cell mesenchymal precursors from patients with the androgen insensitivity syndrome: Testosterone production and response to human chorionic gonadotropin stimulation in culture. *Biol. Reprod.* **46**, 793–801.

Choi, M. S. K., and Cooke, B. A. (1990). Evidence for two independent pathways in the stimulation of steroidogenesis by luteinizing hormone involving chloride channels and cyclic AMP. *FEBS Lett.* **261**, 402–404.

Clayton, R. N. (1986). Reproductive disorders. In "Neuroendocrinology" (S. L. Lightman and B. J. Everitt, eds.), pp. 563–587. Blackwell, Oxford.

Closset, J., and Hennen, G. (1988). Biopotency of highly purified porcine FSH and human LH on gonadal function. *J. Endocrinol.* **120**, 89–96.

Cooke, B. A. (1990). Is cyclic AMP an obligatory second messenger for luteinizing hormone? *Mol. Cell. Endocrinol.* **69**, C11–C15.

Cooke, B. A., Platts, E. A., Abayasekera, D. R. E., and Rose, M. P. (1989). Mechanisms of hormone-induced desensitization of adenylate cyclase. *Biochem. Soc. Trans.* **17**, 633–635.

Cooke, B. A., Dirami, G., Chaudry, L., Choi, M. S. K., Abayasekara, D. R. E., and Phipp, L. (1991). Release of arachidonic acid and the effects of corticosteroids on steroidogenesis in rat testis Leydig cells. *J. Steroid Biochem. Mol. Biol.* **40**, 465–471.

Costlow, M. E., and McGuire, W. L. (1977). Autoradiographic localization of the binding of ^{125}I-labelled prolactin to rat tissues in vitro. *J. Endocrinol.* **75**, 221–226.

Daehlin, L., Thore, J., Bergman, B., Damber, J. E., and Selstam, G. (1985). Direct inhibitory effects of natural and synthetic oestrogens on testosterone release from human testicular tissue in vitro. *Scand. J. Urol. Nephrol.* **19**, 7–12.

Davis, J. S., West, L. A., Weakland, L. L., and Farese, R. V. (1986a). Human chorionic gonadotropin activates the inositol 1,4,5-trisphosphate signalling system in bovine luteal cells. *FEBS Lett.* **208**, 287–291.

Davis, J. S., Weakland, L. L., West, L. A., and Farese, R. V. (1986b). Luteinizing hormone stimulates the formation of inositol trisphosphate and cyclic AMP in rat granulosa cells. Evidence for phospholipase C generated second messengers in the action of luteinizing hormone. *Biochem. J.* **238**, 597–604.

de Kretser, D. M., and Kerr, J. B. (1988). The cytology of the testis. In "The Physiology of Reproduction" (E. Knobil and J. Neill, eds.), pp. 837–932. Raven Press, New York.

Ding, Y.-Q., and Huhtaniemi, I. (1989). Human serum LH inhibitor(s): Behaviour and contribution to in vitro bioassay of LH using dispersed mouse Leydig cells. *Acta Endocrinol. (Copenh.)* **121**, 46–54.

Dix, C. J., Habberfield, A. D., Sullivan, M. H. F., and Cooke, B. A. (1984). Inhibition of steroid production in Leydig cells by non-steroidal and anti-inflammatory and related compounds: Evidence for involvement of lipoxygenase products in steroidogenesis. *Biochem. J.* **219**, 529–537.

Duchatelle, P., and Joffre, M. (1990). Potassium and chloride conductances in rat Leydig cells: Effects of gonadotrophins and cyclic adenosine monophosphate. *J. Physiol.* **428**, 15–37.

Dufau, M. L. (1988). Endocrine regulation and communicating functions of the Leydig cell. *Ann. Rev. Physiol.* **50**, 483–508.

Dufau, M. L., Pock, R., Neubauer, A., and Catt, K. J. (1976). In vitro bioassay of LH in human serum: The rat interstitial cell testosterone (RITC) assay. *J. Clin. Endocrinol. Metab.* **42**, 958–969.

Dufau, M. L., Tsuruhara, T., Horner, K. A., Podesta, E., and Catt, K. J. (1977). Intermediate role of cyclic adenosine-3′:5′-monophosphate and protein kinase during gonadotropin-induced steroidogenesis in testicular interstitial cells. *Proc. Natl. Acad. Sci. U.S.A.* **74**, 3419–3423.

Evain, D., Morera, A. M., and Saez, J. M. (1976). Glucocorticoid receptors in interstitial cells of the rat testis. *J. Steroid Biochem.* **7**, 1135–1139.

Foo, N.-C., Carter, D., Murphy, D., and Ivell, R. (1991). Vasopressin and oxytocin gene expression in rat testis. *Endocrinology* **128**, 2118–2128.

Forti, G., Santoro, S., Grisolia, G. A., Bassi, F., Boninsegni, R., Fiorelli, G., and Serio, M. (1981).

Spermatic and peripheral plasma concentrations of testosterone and androstenedione in prepubertal boys. *J. Clin. Endocrinol. Metab.* **53**, 883–886.

George, F. W., and Ojeda, S. R. (1987). Vasoactive intestinal peptide enhances aromatase activity in the neonatal rat ovary before development of primary follicles or responsiveness to follicle stimulating hormone. *Proc. Natl. Acad. Sci. U.S.A.* **84**, 5803–5807.

Gudermann, T., Birnbaumer, M., and Birnbaumer, L. (1992a). Evidence for dual coupling of the murine luteinizing hormone receptor to adenylyl cyclase and phosphoinositide breakdown and Ca^{2+} mobilization. *J. Biol. Chem.* **267**, 4479–4488.

Gudermann, T., Nichols, C., Levy, F. O., Birnbaumer, M., and Birnbaumer, L. (1992b). Ca^{2+} mobilization by the LH receptor expression in *Xenopus* oocytes independent of 3',5'-cyclic adenosine monophosphate formation: Evidence for parallel activation of two signaling pathways. *Mol. Endocrinol.* **6**, 272–278.

Guo, H., Chalkins, J. H., Sigel, M. M., and Lin, T. (1990). Interleukin-2 is a potent inhibitor of Leydig cell steroidogenesis. *Endocrinology* **127**, 1234–1239.

Hafiez, A. A., Lloyd, C. W., and Bartke, A. (1972). The role of prolactin in the regulation of testis function: the effects of prolactin and luteinizing hormone on the plasma levels of testosterone and androstenedione in hypophysectomized rats. *J. Endocrinol.* **52**, 327–332.

Hales, D. B., and Payne, A. H. (1989). Glucocorticoid-mediated repression of P450$_{scc}$ mRNA and *de novo* synthesis in cultured Leydig cells. *Endocrinology* **124**, 2099–2104.

Hall, P. F. (1988). Testicular steroid synthesis: organization and regulation. *In* "The Physiology of Reproduction" (E. Knobil, and J. Neill, eds.), pp. 975–998. Raven Press, New York.

Hall, P. F., Osawa, S., and Mrotek, J. J. (1981). The influence of calmodulin on steroid synthesis in Leydig cells from rat testis. *Endocrinology* **190**, 1677–1682.

Hardy, M. P., Kelce, W. R., Klinefelter, G. R., and Ewing, L. L. (1990). Differentiation of Leydig cell precursors in vitro: A role for androgen. *Endocrinology* **127**, 488–490.

Hipkin, R. W., and Moger, W. H. (1991). Interaction between cyclic nucleotide second messenger systems in murine Leydig cells. *Mol. Cell. Endocrinol.* **82**, 251–257.

Hsueh, A. J. W., and LaPolt, P. S. (1992). Molecular basis of gonadotropin receptor regulation. *Trends Endocrinol. Metab.* **3**, 164–170.

Huhtaniemi, I. T., and Pelliniemi, L. J. (1992). Fetal Leydig cells: Cellular origin, morphology, life span and special functional features. *Proc. Soc. Exp. Biol. Med.* **201**, 125–140.

Huhtaniemi, I. T., and Warren, D. W. (1990). Ontogeny of pituitary-gonadal interactions; recent advances and controversies. *Trends Endocrinol. Metab.* **1**, 356–362.

Huhtaniemi, I. T., Warren, D. W., and Catt, K. J. (1984). Functional maturation of rat testis Leydig cells. *Ann. N.Y. Acad. Sci.* **438**, 283–303.

Huhtaniemi, I. T., Eskola, V., Pakarinen, P., Matikainen, T., and Sprengel, R. (1992). The murine LH and FSH receptor genes: Transcription initiation sites, putative promoter sequences and promoter activity. *Mol. Cell. Endocrinol.* **88**, 55–66.

Inoue, Y., and Rebois, R. V. (1989). Protein kinase C can desensitize the gonadotropin-responsive adenylate cyclase in Leydig tumor cells. *J. Biol. Chem.* **264**, 8504–8908.

Janszen, F. H. A., Cooke, B. A., van Driel, M. J. A., and van der Molen, H. J. (1976). The effect of calcium ions on testosterone production in Leydig cells from rat testis. *Biochem. J.* **160**, 433–437.

Ji, I., and Ji, T. H. (1991). Exons 1–10 of the rat LH receptor encode a high affinity hormone binding site and exon 11 encodes G-protein modulation and a potential second hormone binding site. *Endocrinology* **128**, 2648–2650.

Jost, A. (1970). Hormonal factors in the sex differentiation of the mammalian foetus. *Philos. Trans. R. Soc. Lond. B.* **259**, 119–130.

Kasson, B. G., Adashi, E. Y., and Hsueh, A. J. W. (1986). Arginine vasopressin in the testis: An intragonadal peptide control system. *Endocr. Rev.* **7**, 156–168.

Kerr, J. B., and Sharpe, R. M. (1985). Follicle-stimulating hormone induction of Leydig cell maturation. *Endocrinology* **116**, 2592–2604.

Kimura, K., Katoh, N., Sakurada, K., and Kubo, S. (1984). Phospholipid-sensitive Ca²⁺-dependent protein kinase system in testis, localization and endogenous substrates. *Endocrinology* **115,** 2391–2399.

Koo, Y. B., Ji, I., Slaughter, R. G., and Ji, T. H. (1991). Structure of the luteinizing hormone receptor and multiple exons of the coding sequence. *Endocrinology* **128,** 2297–2308.

LaPolt, P. S., Jia, X.-C., Sincich, C., and Hsueh, A. J. W. (1991). Ligand-induced down-regulation of testicular and ovarian luteinizing hormone (LH) receptors is preceded by tissue-specific inhibition of alternatively processed LH receptor transcripts. *Mol. Endocrinol.* **5,** 397–403.

Lin, T., Haskell, J., Vinson, N., and Terracio, L. (1986). Characterization of insulin and insulin-like growth factor I receptors of purified Leydig cells and their role in steroidogenesis in primary culture: A comparative study. *Endocrinology* **119,** 1641–1647.

Lin, T., Blaisdell, J., and Haskell, J. F. (1988). Hormonal regulation of type I insulin-like growth factor receptors of Leydig cells in hypophysectomized rats. *Endocrinology* **123,** 2134–2140.

Lin, T., Chalkins, J. K., Morris, P. L., Vale, W., and Bardin, C. W. (1989). Regulation of Leydig cell function in primary culture by inhibin and activin. *Endocrinology* **125,** 2134–139.

Lobie, P. E., Breipohl, W., Garcia-Aragon, J., and Waters, M. J. (1990). Cellular localization of the growth hormone receptor/binding protein in the male and female reproductive systems. *Endocrinology* **126,** 2214–2221.

Loosfelt, H., Misrahi, M., Atger, M., Salesse, R., Thi, M. T. V. H.-L., Jolivet, A., Guiochon-Mantel, A., Sar, H., Jallal, B., Garnier, J., and Milgrom, E. (1989). Cloning and sequencing of porcine LH-hCG receptor cDNA: Variants lacking transmembrane domain. *Science* **245,** 525–528.

Lopez-Ruiz, M. P., Choi, M. S. K., Rose, M. P., West, A. P., and Cooke, B. A. (1992). Direct effects of arachidonic acid on protein kinase C and LH-stimulated steroidsogenesis in rat Leydig cells: Evidence for tonic inhibitory control of steroidogenesis by protein kinase C. *Endocrinology* **130,** 1122–1130.

McFarland, K. C., Sprengel, R., Phillips, H. S., Köhler, M., Rosemblit, N., Nikolics, K., Segaloff, D. L., and Seeburg, P. H. (1989). Lutropin-choriogonadotropin receptor: An unusual member of the G protein-coupled receptor family. *Science* **245,** 494–499.

Magoffin, D. A., and Erickson, G. F. (1982). Prolactin inhibition of luteinizing hormone-stimulated androgen synthesis in ovarian interstitial cells cultured in defined medium: Mechanism of action. *Endocrinology* **111,** 2001–2007.

Marrama, P., Zaidi, A. A., Montanini, V., Calani, M. F., Cioni, K., Carani, C., Morabito, F., Resentini, M., Bonati, B., and Baraghini, G. F. (1983). Age and sex related variations in biological activity and immunoreactive serum luteinizing hormone. *J. Endocrinol. Invest.* **6,** 427–433.

Mathews, L. S., Enberg, B., and Norstedt, G. (1989). Regulation of rat growth hormone receptor gene expression. *J. Biol. Chem.* **264,** 9905–9910.

Mayerhofer, A., Bartke, A., and Steger, R. W. (1989). Catecholamine effects on testicular testosterone production in the gonadally active and gonadally regressed adult golden hamster. *Biol. Reprod.* **40,** 752–761.

Meidan, R., and Hsueh, A. J. W. (1985). Identification and characterization of arginine vasopressin receptors in the rat testis. *Endocrinology* **116,** 416–423.

Meikle, A. W., Benson, S. J., Lui, X.-H., Boam, W. D., and Stingham, J. D. (1989). Nonesterified fatty acids modulate steroidogenesis in mouse Leydig cells. *Am. J. Physiol.* **257,** E937–E942.

Melsert, R., and Rommerts, F. F. G. (1987). Leydig cell steroid production in the presence of LH is further stimulated by rat testicular fluid, fetal calf serum and bovine follicular fluid, but not by rat and bovine serum. *J. Endocrinol.* **115,** R17–R20.

Metsikkö, M. K., Petäjä-Repo, U. E., Lakkakorpi, J. T., and Rajaniemi, H. J. (1990). Structural features of the LH/GC receptor. *Acta Endocrinol. (Copenh.)* **122**, 545–522.

Moger, W. H. (1991). Evidence for compartmentalization of adenosine 3′,5′-monophosphate (cAMP)-dependent protein kinases in rat Leydig cells using site-selective cAMP analogs. *Endocrinology* **128**, 1414–1418.

Morera, A. M., Cochet, C., Keramidas, M., Chauvin, M. A., De Peretti, E., and Benahmed, M. (1988). Direct regulating effects of transforming growth factor beta on the Leydig cell steroidogenesis in primary culture. *J. Steroid Biochem.* **30**, 443–447.

Mukhopadhyay, A. K., and Schumacher, M. (1985). Inhibition of hCG-stimulated adenylate cyclase in purified mouse Leydig cells by the phorbol ester PMA. *FEBS Lett.* **187**, 56–60.

Mukhopadhyay, A. K., Schumacher, M., and Leidenberger, F. A. (1986). Steroidogenic effect of atrial natriuretic factor in isolated mouse Leydig cells is mediated by cyclic GMP. *Biochem. J.* **239**, 463–467.

Mukhopadhyay, A. K., Helbing, J., and Leidenberger, F. A. (1989). The role of Ca^{2+} and calmodulin in the regulation of atrial nautriuretic peptide-stimulated guanosine 3′,5′-cyclic monophosphate accumulation by isolated mouse Leydig cells. *Endocrinology* **125**, 686–692.

Niemi, M., and Ikonen, M. (1961). Steroid 3β-ol-dehydrogenase activity in foetal Leydig's cells. *Nature (London)* **189**, 592–593.

Nikula, H., and Huhtaniemi, I. (1989). Effects of protein kinase C activation on cyclic AMP and testosterone production of rat Leydig cells in vitro. *Acta Endocrinol. (Copenh.)* **121**, 327–333.

Nikula, H., Naor, Z., Parvinen, M., and Huhtaniemi, I. (1987). Distribution and activation of protein kinase C in the rat testis tissue. *Mol. Cell. Endocrinol.* **49**, 39–49.

Nishizuka, Y. (1986). Studies and perspectives of protein kinase C. *Science* **233**, 305–312.

Nishizuka, Y. (1988). The molecular heterogeneity of protein kinase C and its implications for cellular regulation. *Nature (London)* **334**, 661–665.

Odell, W. D., and Swerdloff, R. S. (1976). Etiology of sexual maturation: A model system based on the sexually maturing rat. *Rec. Prog. Horm Res.* **32**, 245–288.

Ortlip, S. A., Li, S. A., and Li, J. J. (1981). Characterization of specific glucocorticoid receptor in the Syrian hamster testis. *Endocrinology* **109**, 1331–1338.

Pakarinen, P., Vihko, K., Voutilainen, R., and Huhtaniemi, I. (1990). Differential response of luteinizing hormone receptor and steroidogenic enzyme gene expression to human chorionic gonadotropin stimulation in the neonatal and adult rat testis. *Endocrinology* **127**, 2469–2447.

Pala, A., Coghi, I., Spampinato, G., Di Gregorio, R., Strom, R., and Carenza, L. (1988). Immunochemical and biological characteristics of a human autoantibody to human chorionic gonadotropin and luteinizing hormone. *J. Clin. Endocrinol. Metab.* **67**, 1317–1321.

Papadopoulos, V., Kamtchouing, P., Boujrad, N., Pisselet, C., Perreau, C., Locatelli, A., Drosdowsky, M. A., Hochereau de Reviers, M. T., and Carreau, S. (1990). Evidence for LH-inhibiting activity in ovine peripheral and testicular blood. *Acta Endocrinol. (Copenh.)* **123**, 345–352.

Pelosin, J.-M., Ricouart, A., Sergheraert, C., Benahmed, M., and Chambaz, E. M. (1991). Expression of protein kinase C in various steroidogenic cell types. *Mol. Cell. Endocrinol.* **75**, 149–155.

Pereira, M. E., Segaloff, D. L., Ascoli, M., and Eckstein, F. (1987). Inhibition of choriogonadotropin-activated steroidogenesis in cultured Leydig tumor cells by the Rp diastereoisomer of adenosine 3′:5′-cyclic phosphorothioate. *J. Biol. Chem.* **262**, 6093–6100.

Pereira, M. E., Segaloff, D. L., and Ascoli, M. (1988). Inhibition of gonadotropin-responsive adenylate cyclase in MA-10 Leydig tumor cells by epidermal growth factor. *J. Biol. Chem.* **263**, 9761–9766.

Phillips, D. M., Lakshmi, V., and Monder, C. (1989). Corticosteroid 11β-dehydrogenase in rat testis. *Endocrinology* **125**, 209–216.

Platts, E. A., Schulster, D., and Cooke, B. A. (1988). The inhibitory GTP-binding protein (G$_i$) occurs in rat Leydig cells and is differentially modified by lutropin and 12-O-tetradecan-oylphorbol 13-acetate. *Biochem. J.* **253**, 895–899.

Podesta, E. J., Solano, A. R., and Lemos, J. R. (1991). Stimulation of an individual cell with peptide hormone in a prescribed region of its membrane results in a compartmentalized cyclic AMP-dependent protein kinase response. *J. Mol. Endocrinol.* **6**, 269–279.

Posner, B. I., Kelly, P. A., Shiu, R. P. L., and Friesen, H. G. (1974). Studies of insulin, growth hormone, and prolactin binding: Tissue distribution, species variation and characterization. *Endocrinology* **95**, 521–531.

Rabinovici, J., and Jaffe, R. B. (1990). Development and regulation of growth and differentiated function in human and subhuman primate fetal gonads. *Endocr. Rev.* **11**, 532–557.

Rajalakshmi, M., Robertson, D. M., Choi, S. K., and Diczfalusy, E. (1979). Biologically active luteinizing hormone (LH) in plasma. III. Validation of the in vitro bioassay when applied to male plasma and the possible role of steroidal precursors. *Acta Endocrinol. (Copenh.)* **90**, 585–598.

Rajaniemi, H., and Vanha-Perttula, T. (1972). Evidence for LH and FSH binding protein(s) in human and rat serum. *Horm. Metab. Res.* **5**, 261–266.

Rebois, R. V., and Patel, J. (1985). Phorbol ester causes desensitization of gonadotropin-responsive adenylate cyclase in a murine Leydig tumor cell line. *J. Biol. Chem.* **260**, 8022–8031.

Rigaudière, N., Loubassou, S., Grizard, G., and Boucher, D. (1988). Characterization of insulin binding and comparative action of insulin and insulin-like growth factor I on purified Leydig cells from the adult rat. *Int. J. Androl.* **11**, 165–178.

Rivarola, M. A., Bergada, C., and Cullen, M. (1970). hCG stimulation test in prepubertal boys with cryptorchidism, in bilateral anorchia and in male pseudohermaphroditism. *J. Clin. Endocrinol. Metab.* **31**, 526–530.

Rodriguez, M. C., Xie, Y.-B., Wang, H., Collison, K., and Segaloff, D. L. (1992). Effects of truncations of the cytoplasmic tail of the luteinizing hormone/chorionic gonadotropin receptor on receptor-mediated hormone internalization. *Mol. Endocrinol.* **6**, 327–336.

Rommerts, F. F. G., and van der Molen, H. J. (1989). Testicular steroidogenesis. In "The Testis" (H. Burger and D. de Kretser, eds.), 2d Ed., pp. 303–328. Raven Press, New York.

Saez, J. M., Morera, A. M., Haour, F., and Evain, D. (1977). Effects of in vivo administration of dexamethasone, corticotropin and human chorionic gonadotropin on steroidogenesis and protein and DNA synthesis of testicular interstitial cells in prepubertal rats. *Endocrinology* **101**, 1256–1263.

Sairam, M. R. (1990). Complete dissociation of gonadotropin receptor binding and signal transduction in mouse Leydig tumour cells. *Biochem. J.* **265**, 667–674.

Samson, W. K. (1992). Natriuretic peptides. A family of hormones. *Trends Endocrinol. Metab.* **3**, 86–90.

Sanchez-Yagüe, J., Rodriguez, M. C., Segaloff, D. L., and Ascoli, M. (1992). Truncation of the cytoplasmic tail of the lutropin/choriogonadotropin receptor prevents agonist-induced uncoupling. *J. Biol. Chem.* **267**, 7217–7220.

Segaloff, D. L., Sprengel, R., Nikolics, K., and Ascoli, M. (1990). Structure of the lutropin/choriogonadotropin receptor. *Rec. Prog. Horm. Res.* **46**, 261–303.

Sharpe, R. M. (1984). Intratesticular factors controlling testicular function. *Biol. Reprod.* **30**, 29–49.

Sharpe, R. M. (1990). Intratesticular control of steroidogenesis. *Clin. Endocrinol. (Oxford)* **33**, 787–807.

Sharpe, R. M., and Cooper, I. (1987). Comparison of the effects on purified Leydig cells of four hormones (oxytocin, vasopressin, opiates and LHRH) with suggested paracrine roles in the testis. *J. Endocrinol.* **113**, 89–96.

Sokka, T., and Huhtaniemi, I. (1990). Ontogeny of gonadotrophin receptors and gonadotrophin stimulated cAMP production in the neonatal rat ovary. *J. Endocrinol.* **127**, 297–303.

Sokka, T., Hämäläinen, T., and Huhtaniemi, I. (1992). Functional LH receptor appears in the neonatal rat ovary after changes in the alternative splicing pattern of the LH receptor mRNA. *Endocrinology* **130**, 1738–1740.

Sprengel, R., Braun, T., Nikolics, K., Segaloff, D. L., and Seeburg, P. H. (1990). The testicular receptor for follicle stimulating hormone: Structure and functional expression of cloned cDNA. *Mol. Endocrinol.* **4**, 525–530.

Sullivan, M. H. F., and Cooke, B. A. (1986). The role of Ca^{2+} in steroidogenesis in Leydig cells. Stimulation of intracellular free Ca^{2+} by lutropin (LH), luliberin (LHRH) agonist and cyclic AMP. *Biochem. J.* **236**, 45–51.

Tahri-Joutei, A., and Pointis, G. (1988). Time-related effects of arginine vasopressin on steroidogenesis in cultured mouse Leydig cells. *J. Reprod. Fertil.* **82**, 247–254.

Tahri-Joutei, A., and Pointis, G. (1989). AVP receptors of mouse Leydig cells are regulated by LH and E_2 and influenced by experimental cryptorchidism. *FEBS Lett.* **254**, 189–193.

Teerds, K. J., De Rooij, D. G., Roomerts, F. F. G., and Wensing, C. J. G. (1988). The regulation of the proliferation and differentiation of rat Leydig cells after EDS administration or daily hCG treatment. *J. Androl.* **9**, 343–351.

Themmen, A. P. N., Hoogerbrugge, J. W., Rommerts, F. F. G., and van der Mole, H. J. (1985). Is cAMP the obligatory second messenger in the action of lutropin on Leydig cell steroidogenesis? *Biochem. Biophys. Res. Commun.* **128**, 1164–1172.

Themmen, A. P. N., Hoogerbrugge, J. W., Rommerts, F. F. G., and van der Mole, H. J. (1986). The possible role of protein kinase C and phospholipids in the regulation of steroid production in rat Leydig cells. *FEBS Lett.* **203**, 116–120.

Thomas, J. P., Dorflinger, L. J., and Behrman, H. R. (1978). Mechanism of the rapid antigonadotropic action of prostaglandins in cultured luteal cells. *Proc. Natl. Acad. Sci. U.S.A.* **75**, 1344–1338.

Thompson, E. B. (1992). Comment: Single receptors, dual second messengers. *Mol. Endocrinol.* **6**, 501.

Tiong, T. S., and Herington, A. C. (1991). Tissue distribution, characterization, and regulation of messenger ribonucleic acid for growth hormone receptor and serum binding protein in the rat. *Endocrinology* **129**, 1628–1634.

Tsai-Morris, C. H., Buczko, E., Wang, W., and Dufau, M. L. (1990). Intronic nature of the rat luteinizing hormone receptor gene defines a soluble receptor subspecies with hormone binding. *J. Biol. Chem.* **265**, 19385–19388.

Tsai-Morris, C. H., Buczko, E., Wang, W., Xie, X.-Z., and Dufau, M. L. (1991). Structural organization of the rat luteinizing hormone (LH) receptor gene. *J. Biol. Chem.* **266**, 11355–11359.

van Buul-Offers, S. (1983). Hormonal and other inherited growth disturbances in mice with special reference to the Snell dwarf mouse. A review. *Acta Endocrinol. (Copenh.) (Suppl. 258)* **103**, 1–47.

Vassart, G., Parmentier, M., Libert, F., and Dumont, J. (1991). Molecular genetics of the thyrotropin receptor. *Trends Endocrinol. Metab.* **2**, 151–156.

Vigier, B., Forest, M. G., Eychenne, B., Bezard, J., Garrigou, O., Robel, P., and Josso, N. (1989). Anti-müllerian hormone produced endocrine sex reversal of fetal ovaries. *Proc. Natl. Acad. Sci. U.S.A.* **86**, 3684–3688.

Vinggaard, A. M., and Hansen, H. S. (1991). Phorbol ester and vasopressin activate phospholipase D in Leydig cells. *Mol. Cell. Endocrinol.* **79**, 157–165.

Vollmar, A. M., Friedrich, A., and Schulz, R. (1990). Atrial natriuretic peptide precursors material in rat testis. *J. Androl.* **11**, 471–477.

Wahlström, T., Huhtaniemi, I., Hovatta, O., and Seppälä, M. (1983). Localization of LH, FSH, PRL and their receptors in human and rat testis using immunohistochemistry and radioreceptor assay. *J. Clin. Endocrinol. Metab.* **57**, 825–830.

Wang, H., Segaloff, D. L., and Ascoli, M. (1991a). Lutropin/choriogonadotropin down-regulates its receptor by both receptor-mediated enocytosis and a cAMP-dependent reduction in receptor mRNA. *J. Biol. Chem.* **266**, 780–785.

Wang, H., Segaloff, D. L., and Ascoli, M. (1991b). Epidermal growth factor and phorbol ester reduce the levels of the cognate mRNA for the LH/hCG receptor. *Endocrinology* **128**, 2651–2653.

Wang, H., Nelson, S., Ascoli, M., and Segaloff, D. L. (1992). The 5′-flanking region of the rat luteinizing hormone chorionic gonadotropin receptor gene confers Leydig cell expression and negative regulation of gene transcription by 3′,5′-cyclic adenosine monophosphate. *Mol. Endocrinol.* **6**, 320–326.

Warren, D. W. (1989). Development of the inhibitory guanine nucleotide-binding regulatory protein in the rat testis. *Biol. Reprod.* **40**, 1208–1214.

Warren, D. W., Huhtaniemi, I. T., Tapanainen, J., Dufau, M. L., and Catt, K. J. (1984). Ontogeny of gonadotropin receptors in the fetal and neonatal rat testis. *Endocrinology* **114**, 470–476.

Wartenberg, H. (1989). Differentiation and development of the testis. In "The Testis" (H. Burger and D. de Kretser, eds.), 2d Ed., pp. 67–118. Raven Press, New York.

Welsh, T. H., Jr., and Hsueh, A. J. W. (1982). Mechanism of the inhibitory action of epidermal growth factor on testicular androgen biosynthesis in vitro. *Endocrinology* **110**, 1498–1506.

Welsh, T. H., Jr., Kasson, B. G., and Hsueh, A. J. W. (1986). Direct biphasic modulation of gonadotropin-stimulated testicular androgen biosynthesis by prolactin. *Biol. Reprod.* **34**, 796–804.

West, A. P., and Cooke, B. A. (1992). Secretion of LH binding proteins from MA10 cells. *Miniposters of the 7th European woskshop on Molecular and Cellular Endocrinology of the Testis,* Castle Elmau, Germany.

Wilson, J. D., George, F. W., and Griffin, J. E. (1981). The hormonal control of sexual development. *Science* **211**, 1278–1284.

Xie, Y.-B., Wang, H., and Segaloff, D. L. (1990). Extracellular domain of lutropin/choriogonadotropin receptor expressed in transfected cells binds choriogonadotropin with high affinity. *J. Biol. Chem.* **265**, 21411–21414.

Zipf, W. B., Payne, A. H., and Kelch, R. P. (1978). Prolactin, growth hormone, and luteinizing hormone in the maintenance of testicular luteinizing hormone receptors. *Endocrinology* **103**, 595–600.

12

Growth Factors in the Control of Testicular Function

DAVID M. ROBERTSON, GAIL P. RISBRIDGER,
MARK HEDGER, & ROBERT I. McLACHLAN

I. Introduction

The pituitary follicle stimulating (FSH) and luteinizing (LH) hormones are well recognized to be the key endocrine factors governing testicular development and function. LH, through Leydig cell stimulation of testosterone, maintains extratesticular testosterone levels required for tissue androgenization as well as high intratesticular levels required for spermatogenesis. FSH appears to be involved primarily in Sertoli cell division and maturation, and possibly in Leydig cell maturation and spermatogenesis. The localization of LH and FSH receptors on the Leydig cell and Sertoli cell, respectively, and of the testosterone receptor on the Sertoli and Leydig cell and on surrounding peritibular myoid cells suggests that these hormones exert direct effects on these tissues.

Based on findings in other tissues, local factors are now known to be essential in the regulation of complex differentiation processes. A wide range of known growth factors identified in other systems has been tested in the testis as a means of understanding intratesticular control mechanisms. As a result, changes in a number of testicular processes have been observed. The difficulty with this approach is that, although numerous factors exerting testicular actions have been identified, their physiological importance is unclear since the underlying processes associated with testicular function are so poorly understood. In this chapter we highlight examples of testicular function in which local factors are of likely, if unproven, importance.

These functions are:

1. Leydig cell proliferation during development and following testicular damage
2. immunoregulation of the testis
3. Sertoli and germ cell interactions during spermatogenesis
4. spermatogenesis

In this chapter we use the term "growth factor" loosely to mean any factor other than FSH, LH, and testosterone that is believed to act by a receptor-mediated mechanism to regulate cellular function, irrespective of its origin. Generally, this term refers to polypeptides likely to be of testicular origin.

II. Role of Growth Factors in the Development and Function of Leydig Cells

The dominant role of Leydig cells in the testicular interstitium is producing steroid hormones, particularly testosterone. Most of the work on the interaction between Leydig cells and growth factors in the testis has been considered in this context, that is, as the ability of growth factors to regulate Leydig cell steroidogenesis. However, Leydig cells make many other proteins and factors that can be used to monitor Leydig cell activity, for example, β-endorphin, corticotrophin releasing factor (CRF), and immunoactive inhibin, but little is known about their regulation by key growth factors. Further, several different types of Leydig cells are studied that can represent Leydig cells isolated from distinct species (e.g., human, rodent, porcine) or from particular stages of development (fetal or adult). Different types of Leydig cells may have specific responses to growth factors. This section of this chapter focuses on the role of growth factors in the development and function of Leydig cell populations.

The appearance of at least two generations of Leydig cells has been recorded in the mammalian testis. Leydig cells present during fetal life are referred to as fetal Leydig cells to distinguish them from the adult Leydig cells present in the mature testis. The fetal Leydig cells generally are believed to develop from gonadal mesenchyme during gestation and decline after birth; the adult Leydig cells emerge as a second generation of Leydig cells during the prepubertal period of testicular development and persist through adulthood. In most species (e.g., human and rat), only two generations of Leydig cells have been identified, but the pig testis is a notable exception in which an additional population of Leydig cells is present in the perinatal period following the fetal but prior to adult Leydig cell generation (Whitehead, 1905). Surprisingly, the roles of growth factors in Leydig cell development have not been investigated more thoroughly. This lack of study may be the result of the fact that very little was known about the development of Leydig cells until recently. Hence the regulation of these processes is poorly understood.

A. Development of Fetal Leydig Cells

During gestation, the elongated mesenchymal precursor cells that lie between the developing seminiferous cords give rise to the fetal Leydig cells, which are characterized by their unique morphology and biochemistry. Although LH stimulates fetal Leydig cell testosterone production, its role in their development is equivocal. Leydig cell differentiation occurs in the anencephalic fetus, indicating that pituitary hormones per se may not be required. The numbers of fetal Leydig cells declines postnatally, although the precise fate of these cells in the adult testis remains unclear. Many of the studies of fetal Leydig cell function and growth factor action have been performed using rat tissues. In the rat, fetal Leydig cells have been isolated and shown to produce β-endorphin and CRF as well as testosterone (Fabbri et al., 1985), but the action of growth factors on fetal Leydig cells has been studied almost exclusively in terms of an action on androgen synthesis. Using postnatal rat testis Leydig cells, epidermal growth factor (EGF; Sordiollet et al., 1991) and activin inhibit androgen production whereas inhibin stimulates steroidogenesis (Meidan et al., 1985; Hsueh et al., 1987).

In porcine species, the postnatal testis is used routinely as a source of the second of the three generations of Leydig cells. The action of a number of growth factors has been determined in terms of their ability to modulate the steroidogenic pathway. These factors include insulin-like growth factor 1 (IGF-1; Bernier et al., 1986; Benhamed et al., 1987; Perrard-Sapori et al., 1987), transforming growth factor β (TGFβ), activin, which is structurally related to TGFβ (Avallet et al., 1987; Mauduit et al., 1992), tumor necrosis factor α (TNFα; Mauduit et al., 1991), and fibroblast growth factor (FGF; Sordoillet et al., 1988). Specific receptor binding sites on Leydig cells, through which these factors can act, have been demonstrated in only a few instances, for example, for FGF (Sordoillet et al., 1992). The individual actions of these hormones can be inhibitory or stimulatory, or have no overall effect on testosterone production. In the last instance, however, the absence of an effect may occur because the hormone, for example, activin, decreases one steroidogenic enzyme activity (cholesterol side-chain cleavage cytochrome P450) but stimulates another (3β-hydroxysteroid dehydrogenase/isomerase), resulting in no net effect (Mauduit et al., 1992). The local production of these growth factors has been postulated to allow them to act in the paracrine manner to regulate testicular steroidogenesis. Only one study (Bernier et al., 1986) has reported that another action of IGF-1 is to stimulate DNA synthesis by these cells; this result has not been investigated further, but may be relevant to the action of IGF-1 on immature Leydig cell mitosis (see subsequent discussion).

B. Development of Adult Leydig Cells

The adult Leydig cells in most species generally are believed to be derived from the mesenchymal cells, that is, undifferentiated connective tissue, that are located in the interstitium. In the rat, this process of development has been studied extensively using quantitative morphometric techniques (Hardy *et al.*, 1989). The final number of Leydig cells in the adult rat testis was shown to arise from an initial differentiation of mesenchymal precursor cells during puberty followed by a period of mitosis of histologically recognizable Leydig cells between days 28 and 52 of age. These immature–adult-type Leydig cells appear to differ from mature Leydig cells in the adult testis. For example, the major steroid products are reduced androgens rather than testosterone.

The action of growth factors on immature Leydig cells isolated from the prepubertal testis has been determined in several laboratories. EGF appears to stimulate steroidogenesis (Verhoeven and Cailleau, 1986), whereas FGF and platelet derived growth factor (PDGF) inhibit steroidogenesis (Fauser *et al.*, 1988; Murono and Washburn, 1990a,b). Binding sites for PDGF have been localized to postpubertal, that is, mature adult Leydig cells (Gnessi *et al.*, 1992). IGF-1 and its binding protein are produced in variable amounts by Sertoli cells, peritubular cells, and Leydig cells derived from the immature testis (Cailleau *et al.*, 1990), but most of the available evidence suggests that the Leydig cells themselves are a major target for Leydig cell-derived IGF-1, which stimulates steroidogenesis (Verhoeven and Cailleau, 1989). Whether or not IGF-1 binding protein increases or decreases the *in vivo* effects of IGF-1 remains to be determined.

With renewed interest in the development of Leydig cells, transforming growth factor α (TGFα) was reported to be localized to Leydig cells in the pubertal testis (Teerds *et al.*, 1990). This result led to a study to determine the effects of growth factors (known to be present in the testis) on DNA synthesis by immature Leydig cells (Kahn *et al.*, 1992a). The results showed that, in synergy with LH, the growth factors TGFα, TGFβ, and insulin/IGF-1 stimulated thymidine incorporation by immature Leydig cell cultures. Similar actions of an unidentified factor purified from human follicular fluid (termed SIP; steroid inducing protein) also were reported by Kahn *et al.* (1992b). These data demonstrate that growth factors may be involved in the differentiation of Leydig cells, but how this involvement occurs is still unresolved. The main difficulty with interpeting these data is determining the cellular site of action of these growth factors. Kahn *et al.* (1992a,b) have not used any purification procedures (such as elutriation on density gradient separation) to obtain Leydig cells, although the methods used are justified by stating that

90% of the cells stain for 3β-hydroxysteroid dehydrogenase (3β-HSD). Hardy *et al.* (1990) showed already that Leydig cell precursors also stain positively for 3β-HSD. Hence, the preparation used by Kahn and co-workers could contain a mixture of Leydig cell precursors and immature Leydig cells. Further, autoradiographs of the cultures performed by Kahn *et al.* (1992a) show that only a small proportion ($\sim 16\%$) of the cells is labeled under maximal stimulation with insulin and TGFα. The experimental protocols involving [^{3}H]thymidine uptake will not permit the detection of all the cells undergoing, or capable of, DNA synthesis. Therefore this percentage underestimates the proportion of cells capable of mitosis. However, the action of the growth factors cannot be assumed to be limited to histologically recognizable Leydig cells, since a proportion of the 3β-HSD positive cells in the interstitium will be Leydig cell precursors, and other cells in the interstitium as well as Leydig cells have been reported to undergo DNA synthesis (Moore *et al.*, 1992). As discussed earlier, the immature Leydig cells are known to undergo a round of mitosis, following differentiation from the mesenchymal precursors, to acquire the full numbers of Leydig cells present in the adult testis. The hypothesis that growth factors such as IGF-1 are involved in stimulating the mitotic division of these Leydig cells and their precursors remains to be tested.

In the adult testis, the actions of growth factors on mature Leydig cells has focused exclusively on their role as putative paracrine or autocrine regulators of steroidogenesis. The role of cytokines has been examined by Lin and co-workers (1987), who reported that the interleukins (IL)-1α, -1β, and -2 inhibit steroid biosynthesis *in vitro*. The inhibitory action of IL-1β is more complex and appears to involve the down-regulation of mRNA levels of IGF-1, which itself can act to enhance androgen production (Lin *et al.*, 1992). Although this group reports an inhibition of steroidogenesis by interleukins, others report that IL-1 can stimulate steroidogenesis (Warren *et al.*, 1990; Moore and Moger, 1991). Other growth factors such as TGFβ and activin have been shown to inhibit testosterone production by adult Leydig cells over 2 days of culture (Lin *et al.*, 1987,1989; Zwain *et al.*, 1991). The implications of these contradictory observations are unclear. Without the use of conditions that maintain LH responsiveness and stabilize steroidogenic activity at levels similar to those estimated to occur *in vivo* (see Risbridger and Hedger, 1992), determining whether or not the growth factors have actions on steroidogenesis per se or whether the changes are a result of exposure to less than optimal conditions in culture is difficult (Risbridger, 1992).

In summary, the role of growth factors in Leydig cell function has been determined in the context of the regulation of Leydig cell steroidogenesis. These studies have been performed using Leydig cells from different stages of development of the testis and from different species (mainly rat and pig). However, the development of Leydig cells from mesenchymal precursors and the mitosis of Leydig cells also have been examined. An understanding of these processes is emerging gradually. In these contexts, the local regulatory actions of growth factors (acting with LH) may be more important than an action on steroidogenesis per se.

III. Growth Factors and Interactions between the Testis and Immune System

Considerable experimental evidence suggests that cells of the immune system also should be considered in discussions of testicular cell interactions. In the testis, the seminiferous epithelium is separated from the vasculature by a loose connective interstitial tissue (Fawcett, 1973). In addition to connective tissue cells and the Leydig cells, the interstitium contains the testicular lymphatics and resident macrophages, which are ultrastructurally and functionally similar to macrophages found in other tissues (Miller *et al.*, 1983; Wei *et al.*, 1988). In the human and rat testis, macrophages are numerous, although in some species such as the sheep they are relatively rare (Pollanen and Maddocks, 1988). Lymphocytes have been identified in the interstitial tissue of both normal and infertile human testes, in addition to those of other species (El-Demiry *et al.*, 1987; Pöllänen and Niemi, 1987). Further, as in other tissues, testicular inflammation following infection or other stimuli involves invasion by peripheral monocytes, lymphocytes, and polymorphonuclear cells (PMNs) (Bergh *et al.*, 1987; Tung *et al.*, 1987). Mast cells, however, tend to be few in the normal rodent testis, and observed occasionally in human testicular interstitial tissue (Nistal *et al.*, 1984; Gaytan *et al.*, 1989). Immune cells usually are not observed within the seminiferous epithelium under normal conditions (El-Demiry *et al.*, 1987; Pöllänen and Niemi, 1987).

A. Intratesticular Immune Responses

From both clinical and experimental studies, the male germ cells and sperm are well known to be highly immunogenic (Alexander and Anderson, 1987; Tung *et al.*, 1987; Yule *et al.*, 1988), presumably because mature spermatogenic cells appear long after the development of tolerance to "self" during the perinatal/neonatal period. In the adult animal, cellular and humoral immune responses to most "non-self" antigens are initiated by antigen-presenting cells (APC) that include the macrophages (Weaver and Unanue, 1990). These cells process and present the antigen on their surfaces in conjunction with specific self-recognition molecules [major histocompatibility complex (MHC) class II antigens], to antigen-specific T helper lymphocytes (T_h cells). The subsequent immune response is promoted by leukotactic or growth and differentiation factors secreted by the APCs, which induce and control lymphocyte activation and proliferation and the inflammatory event. The best characterized of these factors are IL-1, IL-6, and TNFα (Muller *et al.*, 1989; Heinrich *et al.*, 1990). In response, the lymphocytes secrete their own immunoregulatory growth and differentiation factors, including IL-2 and interferon-gamma (IFN-γ) (McDonald and Nabholz, 1986). The active arm of the immune response involves macrophages, killer cells, and cytotoxic T lympohcytes (T$_c$ cells), also under the control of growth and differentiation factors including those already described (Zinkernagel and Doherty, 1974). All the cell types necessary for a normal immune response, therefore, have access to the testis, but under normal conditions no overt immune response to the testicular germ cell autoantigens occurs.

The majority of germ cells are protected by the blood–testis barrier comprising tight junctions between adjacent Sertoli cells in the seminiferous epithelium (Fawcett, 1973). However, germ cell autoantigens in the testis are not sequestered completely by this barrier and are demonstrably accessible to circulating autoantibodies and activated lymphocytes (Tung *et al.*, 1987; Yule *et al.*, 1988). In addition, the testis of rodent species, at least, is an immunoprivileged site for transplanted tissue (Head *et al.*, 1983; Head and Billingham, 1985). Allografts and xenografts survive considerably longer within the testicular interstitial tissue than in most other tissues, but are rejected rapidly when the host is immunized actively against the donor tissue (Head *et al.*, 1983). Collectively, these data indicate that the initial phases of immune response within the testis are suppressed chronically, indicating the existence of a general immunoprotective regulation of macrophage or lymphocyte function within the testicular environment (reviewed by Mahi-Brown *et al.*, 1988; Hedger, 1989).

In summary, the immune response is induced, promoted, and regulated by growth and differentiation factors secreted by the immune cells, which will have the potential to affect testicular cell functions as well. In turn, strong evidence suggests that the testis regulates local immune responses, presumably by production of factors that control immune cell growth and differentiation. Also evident is that the majority of these interactions are likely to be confined to the testicular interstitium.

B. Testis–Lymphocyte Interactions

1. Control of lymphocyte activation—testicular factors

Lymphosuppressive proteins have been observed in conditioned medium collected from cultures of seminiferous tubule cells (Wyatt *et al.*, 1988; Selawry *et al.*, 1991) and Leydig cells (Hedger *et al.*, 1990; Pöllänen *et al.*, 1990), in testis extracts (Emoto *et al.*, 1991; Sainio-Pöllänen *et al.*, 1991), and in testicular interstitial fluid (Pöllänen *et al.*, 1988; Emoto *et al.*, 1989). However, when several studies are compared, both inhibitory and stimulatory actions on lymphocyte functions *in vitro* have been reported for similar testicular extracts, testicular cell cultures, or testicular fluids (Khan *et al.*,

1987,1988; Gustaffson *et al.*, 1988; Syed *et al.*, 1988), indicating that the results observed may have been at least partially dependent on the experimental design employed. The eventual isolation or identification of the lymphoregulatory factors in these preparations and subsequent investigation of their mode of action and physiological regulation should be more informative.

Paradoxically, some studies have established that the Sertoli cell secretes an "IL-1-like" factor that has biochemical and immunological identity with the α isoform of IL-1 (IL-1α) (Khan *et al.*, 1987,1988; Syed *et al.*, 1988; Gerard *et al.*, 1991). IL-1, which appears as two bioactive isoforms (α and β) with only 30% sequence homology, is produced in numerous tissues and affects the growth and differentiation of several different cell types (Granholm and Soder, 1991). In the seminiferous tubules, IL-1 has been implicated in the control of spermatogonial mitosis (Pöllänen *et al.*, 1989; Parvinen *et al.*, 1991; Soder *et al.*, 1991). In addition to IL-1, the tubules appear capable of producing several other factors that stimulate lymphoproliferation *in vitro* (Pöllänen *et al.*, 1990; Granholm *et al.*, 1992), including a factor with IL-6 bioactivity (Syed *et al.*, 1992). Testicular interstitial fluid also is capable of stimulating lymphocyte proliferation *in vitro* (Gustaffson *et al.*, 1988; Hedger *et al.*, 1990). Although the activity has a heterogeneous molecular weight profile, part of the activity has similar properties to the seminiferous tubule IL-1α (Gustaffson *et al.*, 1988). Since testicular interstitial cells do not secrete IL-1 constitutively (Khan *et al.*, 1987) and nonstimulated macrophages usually do not produce IL-1, IL-1 from the tubules must gain access to the interstitium.

Our own studies have indicated that the gonadal glycoproteins, inhibin and activin, may be involved in controlling immune responses within the testis (Hedger *et al.*, 1989,1990). Both inhibin and activin display structural and sequence homology with the TGFβ family, all of which are potent regulators of immune cell function. Activin appears to have a distinct receptor family (Mathews and Vale, 1991; Attisano *et al.*, 1992). Inhibin is produced by both the Sertoli cell and the Leydig cell (Bicsak *et al.*, 1987; Risbridger *et al.*, 1989) and is present at relatively high levels in testicular interstitial fluid (Robertson *et al.*, 1988). The Leydig cell is implicated as the main potential source of activin in the adult testis (Lee *et al.*, 1989; Roberts *et al.*, 1989; Shaha *et al.*, 1989).

In our studies, bovine inhibin stimulated lectin-induced adult rat thymocyte proliferation *in vitro*, whereas bovine activin almost completely suppressed proliferation, with a relative activity of 0.5–1.0% that of porcine TGFβ (Hedger *et al.*, 1989). A monomeric form of the β subunit, isolated from bovine ovarian follicular fluid and human recombinant sources, also suppresses thymocyte proliferation, although this form is less potent (45%) than the activin dimer (Robertson *et al.*, 1992). Activin also inhibited IFN-γ secretion by peripheral blood lymphocytes (Petraglia *et al.*, 1991). Inhibin was chemotactic for monocytes *in vitro*, whereas this activity was antagonized by activin (Petraglia *et al.*, 1991). Since TGFβ regulates macrophage functions in addition to lymphocytes (Wahl *et al.*, 1989), a role for activin and, consequently, for inhibin in regulating macrophages in the testis is also possible.

Several other growth factors and regulatory peptides produced by the Sertoli, peritubular, or Leydig cells under various conditions, including IGF-I, clusterin, and substance P, also affect lymphocyte functions *in vitro* (reviewed by Pöllänen *et al.*, 1990). In particular, TGFβ has been reported to be produced by the Sertoli cells, at least in the immature rat testis (Skinner and Moses, 1989; Watrin *et al.*, 1991). The data, therefore, indicate that a large number of potentially lymphoproliferative and lymphosuppressive activities are present in the seminiferous tubules and interstitial fluid of the testis. The role of many of these growth factors is presumably in sperm development and cellular communication. These molecules may only coincidentally affect lymphocyte proliferation *in vitro*. The immunosuppressed state of the testis would be caused by a balance between these opposing growth-regulating activities.

2. Effects of Lymphocyte Products on Leydig Cell and Sertoli Cell Functions

Some data suggest that lymphocytes, in turn, influence testicular function. IL-2, which is required for T-lymphocyte proliferation, also inhibits LH-stimulated testosterone production by adult Leydig cells *in vitro* (Guo *et al.*, 1990), suggesting a direct effect of activated lymphocytes on testicular steroidogenesis. IL-2 also has a stage-dependent stimulatory effect on transferrin secretion by Sertoli cells *in vitro* (Boockfor and Schwarz, 1991). Moreover, the hypothalamic peptide, LH releasing hormone (LHRH), has been established to be produced by peripheral lymphocytes (Emanuele *et al.*, 1990; Azad *et al.*, 1991). Although LHRH has been known for some time to stimulate Leydig cell testosterone production, the levels of LHRH in the testis are extremely low and the source of testicular LHRH has not been identified (see review by Hedger *et al.*, 1987). IFN-α, a product of leukocytes exposed to viral infection, also inhibits LH-stimulated Leydig cell testosterone production *in vitro* (Orava *et al.*, 1985). However, given the small number of lymphocytes present in the noninflamed testis, the influence of these lymphocytic factors on testicular function is likely to be slight under normal conditions.

C. Testis–Macrophage Interactions

1. Functions of the testicular macrophage

Morphological studies have identified extensive cytoplasmic contacts with adjacent membrane specializations between the macrophages and the Leydig cells in the rat testis (Miller *et al.*, 1983; Geierhaas *et al.*, 1991). In the mouse, macrophages as well as lymphoid cells bind specifically to isolated Leydig cells *in vitro* (Born and Wekerle, 1981; Rivenson *et al.*, 1981). Moreover, morphological evidence exists for a functional interaction between the Leydig cell and the testicular macrophage in the rat, in experimental cryptor-chidism and after vasectomy (Bergh, 1987; Geierhaas *et al.*, 1991). Surpris-ingly, transient testicular inflammation with accumulation of interstitial fluid and spermatogenic disruption can be induced in the rat by stimulating the Leydig cells with LH, human chorionic gonadotropin (hCG) or LHRH (Set-chell and Sharpe, 1981; Bergh *et al.*, 1987), whereas concurrent phagocytic stimulation of the macrophages potentiates these responses (Kerr and Sharpe, 1989). These data suggest a close functional interaction between the testicular macrophage and the Leydig cell in particular.

Although testicular macrophages are known to express MHC II antigens (Head and Billingham, 1985; El-Demiry *et al.*, 1987; Hedger and Eddy, 1987), relatively little is known about the immunoregulatory functions of the testic-ular macrophage or about the possible control of these functions by the testis. However, some studies suggest that the inflammatory functions of the testic-ular macrophage may be reduced or inhibited (Bergh *et al.*, 1992; Guzzardi and Hedger, 1992; Maddocks *et al.*, 1992). In addition to lymphocyte-acti-vating factors and mediators of inflammation, macrophages can secrete im-munosuppressive activities, particularly TGFβ and an IL-1 receptor antago-nist (Assoian *et al.*, 1987; Dinarello and Thompson, 1991). Our own studies have demonstrated that testicular macrophages are a source of pro-opiome-lanocortin-derived peptides (Li *et al.*, 1991) that have regulatory effects on lymphocyte activity and IL-1 activity (Johnson *et al.*, 1982; Brown and van-Epps, 1986). The inflammatory functions of the testicular macrophages re-main an important area for future study.

2. Regulation of testicular cells by macrophage products

Although early studies reported a stimulatory effect of unstimulated testicular macrophages on Leydig cell steroidogenesis *in vitro* (Yee and Hut-son, 1985), this result was not consistent with our own data showing inhibi-tion of hCG-stimulated, but not basal, Leydig cell testosterone production by testicular macrophage-secreted factors (Sun *et al.*, 1993). Also, unstimulated testicular macrophages have been reported to cause a small inhibition of basal testosterone production in coculture with adult rat Leydig cells, whereas activation of the macrophages with lipopolysaccharide (LPS) re-sulted in a stimulation of testosterone production (Lombard-Vignon *et al.*, 1991). The reasons for these apparently contradictory results must be clari-fied but may be attributable to the differences in culture conditions employed in each study. The macrophage factors involved in any of these studies have not been identified, although IL-1 is implicated. Autoradiography with iodi-nated IL-1α has localized IL-1 receptors to the interstitial tissue in the mouse testis, with low levels of incorporation over the tubules (Takao *et al.*, 1990). Further, several studies have reported that IL-1β stimulates basal and inhibits LH/hCG-stimulated testosterone production by adult Leydig cells, although

IL-1α has little or no effect (Calkins *et al.*, 1988,1990a). However, we observed no effect at all of IL-1α or IL-1β on adult rat Leydig cells cultured under a range of incubation conditions (Sun et al., 1993). Other studies have reported that TNFα, either alone or in the presence of IL-1β (Calkins *et al.*, 1990b; Mauduit *et al.*, 1991), and TGFβ (Avallet *et al.*, 1987; Lin *et al.*, 1987) also inhibit LH-stimulated testosterone production by adult Leydig cells. Therefore, macrophage control of Leydig cell steroidogenesis appears to involve the secretion of several factors.

In addition to effects on Leydig cell steroidogenesis, TGFβ also stimulates peritubular cell activity in culture (Skinner and Moses, 1989; Tung and Fritz, 1991) and Sertoli cell lactate production (Esposito *et al.*, 1991). IL-6 and TNFα regulate stage-specific transferrin secretion by seminiferous tubules *in vitro* (Boockfor and Schwarz, 1991). However, the problem with interpreting these actions of macrophage products on tubular cells, in particular, is that cells other than the macrophage within the tubules are a potential source of these factors.

D. Physiological Significance

Until additional data concerning these interactions between the immune system and the testis are accumulated, the physiological importance of moment-to-moment testis–immune cell interactions mediated by locally produced growth factors in normal testicular processes must remain conjectural. However, these interactions, or failure of these interactions, clearly may have serious consequences for testicular function during and after testicular inflammation. Although complete infertility after mumps orchitis (the most common form of clinical testicular infection) is very rare, impaired fertility as a result of reduced sperm count, motility, and morphology occurs in 7–13% of adult mumps orchitis cases (Manson, 1990). Leydig cell function is depressed during the acute phase of acute mumps orchitis and, in some cases, can lead to long-term Leydig cell impairment and androgen deficiency (Adamopoulos *et al.*, 1978; Aiman *et al.*, 1980). However, the unexplained development of antisperm antibodies, which occurs in approximately 10% of infertile men, is a much more prevalent clinical condition (Alexander and Anderson, 1987). This condition is likely to be the most clinically significant problem associated with a breakdown in the normal testis–immune cell interaction mechanisms.

IV. Role of Growth Factors in Sertoli Cell Development and Function

Current perspectives on Sertoli cell function center on its role in supporting spermatogenesis, which is presumed to involve the interplay of regulatory factors originating from the interstitial space and those originating from the developing germ cells. The nature of the spermatogenic process implies close regulation of activities. The Sertoli cell appears to play a crucial role in the regulation of spermatogenesis, since much of the spermatogenic process is isolated from other testicular activities by the blood–testis barrier, with access only via the Sertoli cell. However, whether the Sertoli cell provides any additional function other than support for the spermatogenic process is unclear.

In the prepubertal testis, Sertoli cells proliferate under FSH control. *In vivo* and *in vitro* studies have shown that FSH stimulates thymidine uptake (Griswold *et al.*, 1976) whereas withdrawal of FSH by immunoneutralization of FSH (Chemes *et al.*, 1979; Orth, 1984) inhibits cell division. Immunoneutralization of endorphin in immature rats resulted in testicular hypertrophy, suggesting that opioids, probably of testicular origin (Tsong *et al.*, 1982), may antagonize this process (Orth, 1986). The data suggest that FSH is the primary Sertoli cell mitogen.

The cessation of Sertoli cell division occurs with the formation of Sertoli cell–Sertoli cell junctions corresponding to days 13–19 in the rat (Hagenas *et al.*, 1981) and with the differentiation of the Sertoli cell into its mature adult form. This latter process is dependent on FSH and LH (Chemes *et al.*, 1979), implicating testosterone or other Leydig cell factors.

Sertoli cells produce or express a number of known growth factors. Note that *in vitro* studies examining the effects of growth factors on Sertoli cell function have largely used cells from immature animals in culture with a limited number of biochemical, or other, end points. These studies, although giving some information on the regulation of immature Sertoli cell function, provide little information on the role of these factors on Sertoli cell function *in vivo* and less for the adult. Further, results of many of the earlier studies were confounded by the presence of contaminating peritubular cells that were responsive to many growth factors (Skinner, 1991). These cells, in contrast to Sertoli cells, readily proliferate in culture, particularly when cultured in serum, resulting in a high contamination in long-term cultures.

An EGF-like activity previously observed by Holmes *et al.* (1986) was identified in both Sertoli and peritubular cell culture medium. This activity is believed to be TGFα (Skinner *et al.*, 1989), based on the presence of TGFα mRNA and immunoactivity in these culture media, the absence of EGF expression in both cell types, and the presence of EGF receptors only on peritubular cells. TGFα had minimal effects on transferrin production by Sertoli cells, although a stimulatory effect was observed in cocultures of peritubular cells and Sertoli cells. Nonetheless, EGF stimulates the production of IGF-1 (Chatelian *et al.*, 1987), lactate (Mallea *et al.*, 1986), inhibin (Morris *et al.*, 1988), transferin, and androgen binding protein (ABP) (Perez-Infante *et al.*, 1986). TGFβ bioactivity and mRNA have been identified in immature Sertoli and peritubular cell cultures (Skinner and Moses, 1989), although TGFβ has no demonstrable effect on transferrin production by Sertoli cells cultured alone or in coculture with peritubular cells. TGFβ, however, had effects on peritubular cell protein synthesis.

Sertoli cells produce IGF-1 (Casella *et al.*, 1987) and an IGF-1-like molecule (Smith *et al.*, 1987) as well as exhibiting surface receptors for IGF-1 (Oonk and Grootegoed, 1988). IGF-1 stimulates glucose transport (Mita *et al.*, 1985) and lactate production (Oonk and Grootegoed, 1988). IGF-1 stimulates tritiated thymidine incorporation by immature Sertoli cells (Borland *et al.*, 1984). FGF (Ueno *et al.*, 1987) has been isolated from, and an FGF-like molecule (Smith *et al.*, 1989) identified in, testicular tissue.

In addition, some of these factors, for example, TGFβ and TGFα, cause marked morphological changes *in vitro* because they promote peritubular cell migration and colony formation and the formation of peritubular–Sertoli cell clusters, activities that may be involved in the formation of the exterior wall of the seminiferous tubule (Skinner and Moses, 1989; Skinner *et al.*, 1989).

Under appropriate conditions of high density, Sertoli cells can form junctional complexes that produce a barrier similar to the blood–testis barrier *in vivo* (Byers *et al.*, 1986; Hadley *et al.*, 1987). Using bicameral chambers, the effects of factors on the directional flow of a number of Sertoli cell markers have been examined. FSH, testosterone (Janecki and Steinberger, 1986; Ailenberg *et al.*, 1990), EGF, IGF-1 and 2 (Spaliviero and Handelsman, 1991), and conditioned medium from pachytene spermatocyte cultures (Djakiew and Dyn, 1988) modify the vectorial secretion of transferrin and specifically induced proteins, implicating these growth factors in the paracrine regulation of Sertoli cell vectorial secretion.

The action of testosterone on Sertoli cell function is poorly resolved. Testosterone receptors have been identified on Sertoli and peritubular cells but not on germ cells (Takeda *et al.*, 1990). Since testosterone alone can promote spermatogenesis qualitatively, the action of testosterone must be mediated by regulatory factors. Earlier studies in which an effect of testosterone on Sertoli cell cultures was observed are attributed in part to the action of testosterone on the contaminating peritubular cells (Skinner and Fritz, 1985). Studies by Skinner have identified a peritubular cell protein that is stimulated by testosterone and can stimulate Sertoli cell function markedly in terms of transferrin and ABP production. This protein, termed P-Mod-S consists of two forms (A and B) that appear to act on Sertoli cells through a cGMP-mediated response (Skinner *et al.*, 1988; Norton and Skinner, 1989). These studies strongly implicate the peritubular cells in regulating Sertoli cell activity through the production of intermediate regulatory proteins. The direct effects of testosterone on Sertoli cell function are less clearly understood. How these actions and those of P-Mod-S modulate spermatogenesis is unknown. Testosterone has been shown to stimulate the synthesis of specific proteins (Roberts and Griswold, 1989), but the actions or roles of these proteins are unclear.

Evidence of regulation of Sertoli cell function by germ cells at different stages of differentiation has been shown clearly by the production pattern of a number of Sertoli cell products at different stages of the seminiferous tubule cycle. Original studies reviewed by Parvinen (1982) following dissection of the seminiferous tubule into discrete stages and, more recently, by the use of model systems that permit synchronization of the spermatogenic process have shown cyclic activities of a large number of Sertoli cell proteins and their mRNAs. These proteins include ABP and transferrin (and their receptors), FSH, testosterone, retinoic acid receptor, plasminogen activator (Vihko *et al.*, 1989), clusterin, cathepsin, and inhibin [Morales *et al.*, 1987; Gonzales *et al.*, 1989; Linder *et al.*, 1991). The pattern of these proteins appears as either of two modes with peaks of activities at Stages VII–IX or XIII–III, suggesting that germ cells at different stages differentially regulate Sertoli cell function (Linder *et al.*, 1991).

Additional studies have shown that damage of the seminiferous epithelium by cryptorchidism (Jégou *et al.*, 1983), heat treatment (Jégou *et al.*, 1984), vitamin D deficiency (Morales and Griswold, 1987; Bartlett *et al.*, 1989; van Beek and Meistrich, 1991), specific cytotoxins (McKinnell and Sharpe, 1992), or irradiation (Pinon-Lataillade *et al.*, 1988) resulted in an inhibition of Sertoli cell function in association with a specific germ cell loss. Inhibin and ABP levels are lowest at the time of late spermatid loss, whereas they remain unchanged at the time pachytene spermatocytes and early spermatids are depleted (Jégou, 1991). However, studies (McKinnell and Sharpe, 1992) that monitored the production of seven androgen-sensitive proteins in seminiferous tubules at various times after selective depletion of pachytene spermatocytes with the cytotoxin methoxyacetic acid suggested that their production was modified differentially according to the composition of the germ cell complement present. These studies suggest that each germ cell type may modulate Sertoli cell function in a specific way, presumably via appropriate factors.

This hypothesis has been tested by investigating the effects of coculture of isolated germ cells, their conditioned media, or seminiferous tubule segments on Sertoli cell function. Sertoli cell aromatase activity was inhibited by conditioned medium from tubules at spermatogenic Stages VII and VIII (Boitini and Ritzen, 1981), although the inhibitor probably is of peritubular cell origin (Verhoeven and Cailleau, 1988). The presence of enriched preparations of either pachytene spermatocytes or early spermatids inhibited RNA and DNA synthesis (Rivarola *et al.*, 1986); enhanced the FSH-stimulated production of transferrin, inhibin, ABP, and specifically induced peptides (Djakiew and Dym, 1988; Jégou, 1991); stimulated phosphorylation of specific proteins (Ireland and Welsh, 1987); stimulated adenylate cyclase (Welsh *et al.*, 1985); and modulated preproenkephalin expression (Fujisawa *et al.*, 1992). However, the conclusions drawn from *in vivo* studies, for example, about the effects of germ cell depletion on inhibin and ABP production, were not mirrored by parallel *in vitro* studies (Jégou, 1991), a difference that is not yet understood. The molecular size of the germ cell factor responsible for stimulating expression of preproenkephalin gene is > 30 kDa (Fujisawa *et al.*, 1992) whereas the factor stimulating ABP and transferrin synthesis is 10–30 kDa (Jégou, 1991). Nerve growth factor (NGF) may be one of these agents. NGF is expressed in pachytene spermatocytes and early spermatids (Ayer-LeLievre *et al.*, 1988) and NGF receptors have been identified on Sertoli cells (Persson *et al.*, 1990). The biological role of NGF in these processes is unclear.

Other factors that stimulate Sertoli cell function or proliferation have been identified, although their overall role is unclear. Opioid peptides and their mRNAs have been identified in the testis; receptors are found on spermatogenic and Sertoli cells but not on Leydig cells. β-Endorphin inhibits Sertoli cell proliferation, basal and FSH-stimulated ABP production, and FSH-induced inhibin and adenylyl cyclase activity (Fabbri *et al.*, 1985; Morris *et al.*, 1987) whereas passive immunization of prepubertal rats with endorphin antiserum (Orth, 1986) or opioid antagonists results in increased thymidine incorporation, testis weight, and ABP secretion (Gerendal *et al.*, 1986). Sertoli cells produce a 1450-nucleotide form of proenkephalin mRNA that is stimulated by FSH and dibutyl cyclic AMP (dbcAMP) but not by testosterone.

A mitogen that specifically stimulates a human epidermoid carcinoma cell line has been purified partially from Sertoli cell culture medium (Lamb *et al.*, 1991). This protein is acid and heat stable with a molecular mass of 14 kDa. Bellvé and colleagues (Bellvé and Feig, 1984; Braunhut *et al.*, 1990) have extensively purified a 3T3 fibroblast mitogen from calf testis that has many of the attributes of heparin-binding growth factor, although its structure has not been established. This protein, termed seminiferous growth factor (SGF), induces proliferation of Sertoli cells in culture (Bellvé and Feig, 1984) and of a cell line of mouse Sertoli cell origin (Braunhut *et al.*, 1990). In contrast, addition of FSH, LH, and bovine aFGF and bFGF was without effect on this cell line. Kancheva *et al.*, (1990) identified a factor (molecular mass > 8 kDa; heat and trypsin sensitive) produced by immature Sertoli cells that stimulates proliferation of 3T3 fibroblasts and prespermatogonia in testicular cord cultures from 3.5-day-old rats. The biological role of these mitogens is unknown. Future research awaits their isolation and characterization.

V. Growth Factors in the Control of Spermatogenesis

The key difficulty in assessing potential regulators of spermatogenesis has been the lack of an *in vitro* model system that permits normal germ cell development and considers the complex cellular architecture of the normal testes. Isolated germ cells survive poorly *in vitro* whereas their viability is enhanced greatly by coculture with Sertoli cells (Tres and Kierszenbaum,

1983). Nonetheless, these cells do undergo some limited differentiation in this setting. Pachytene spermatocytes have been reported to differentiate into early spermatids, that is, to pass through the meiotic step in such a coculture system (Le Magueresse-Battistoni *et al.*, 1991). Previously this event had been observed *in vitro* only using short-term cultures of isolated tubule segments (Parvinen *et al.*, 1983; Toppari and Parvinen, 1985). In these systems, meiotic division occurred in chemically defined medium in the absence of added hormones. Thus, although Sertoli cells clearly enhanced germ cell differentiation *in vitro*, the mechanism of their beneficial effect is obscure. As a whole, key elements involved in normal germ cell differentiation, particularly as related to spermiogenesis, are missing. The "conventional" hormones such as FSH, testosterone, retinol, and insulin are generally without effect in these *in vitro* systems. Accordingly, several additional growth factors have been examined for their potential contribution to spermatogenesis.

Numerous growth factors have been shown to be present in or secreted by Sertoli and germ cells and therefore are candidates for controlling spermatogenesis (see Skinner, 1991). Unfortunately, direct proof of their physiological role or of their involvement in disorders of fertility is lacking. Several growth factors for which evidence suggests a potentially important role in spermatogenesis are described here.

A. Activin

Activin, the homodimer of the inhibin β_A (activin A) or β_B (activin B) subunits or the $\beta_A\beta_B$ subunit heterodimer (activin AB), may play an important local role in spermatogenesis (Mather *et al.*, 1990). Inhibin β_A immunoreactivity was reported in the cytoplasm of both spermatocytes and spermatids, whereas its presence on the plasma membrane of spermatids suggests its secretion by that cell type (Ogawa *et al.*, 1992). Activin immunoreactivity was not present on spermatogonia or spermatozoa, nor has inhibin α subunit immunoreactivity been reported in germ cells.

The presence of receptors for both hormones seems likely on germ cells. Stage-specific binding of fluorescein isothiocyanate-conjugated labeled inhibin and activin to germ cells has been demonstrated. Activin binds primarily to spermatogonia whereas inhibin binds to all germ cell types (Woodruff *et al.*, 1992). The differential pattern of binding supports the existence of independent receptors for these two hormones. Evidence that the observed binding on testicular cells is caused by specific receptors is weak. To date, only the activin receptor has been characterized as a transmembrane serine kinase (Mathews and Vale, 1991; Attisano *et al.*, 1992). Studies by de Winter *et al.* (1992), employing a cDNA probe directed to the extracellular domain of the activin receptor, provided evidence of two activin receptor mRNAs (4 kb and 6 kb) of which the 4-kb form was localized to pachytene spermatocytes and round spermatids but was absent from elongating spermatids. Both mRNAs were identified in Sertoli cells, with very low expression in Leydig cells. These findings suggest that activin may have multiple actions in the control of testicular function. The cloning of a new member of the TGFβ/activin receptor family from a rat Sertoli cell cDNA library (Baarends *et al.*, 1992) raises the possibility that other activin-related growth factors play a role in germ cell regulation.

The most convincing evidence that activin may have a physiological role in controlling spermatogenesis comes from a study using the coculture of germ and Sertoli cells derived from immature rats (Mather *et al.*, 1990). Activin A increased thymidine incorporation by spermatogonia 2- to 4-fold and, in addition, stimulated the reaggregation of cultured cells into tubule-like structures. This type of reaggregation generally is thought to require the presence of peritubular myoid cells. Therefore, the action of activin in their absence suggests an unusual and unique effect of activin as a mitogenic and a differentiating factor for the seminiferous tubule. Inhibin had no effect on the cultured cells, nor did it block activin action as it does in many other biological systems. Nonetheless, highly purified inhibin has been described to reduce spermatogonia number *in vivo* in hamsters (van Dissel-Emiliani *et al.*, 1989). The discrepancy between these two reports has not been explained, but may relate to the purity of the hormone preparations used or to the absence of cell types involved in the mediation of inhibin action from the immature coculture system.

B. Interleukin 1

Whole testes and isolated tubules produce large amounts of IL-1-like activity (Khan *et al.*, 1987). IL-1 synthesis begins at the same time as active spermatogenesis in the pubertal rat (Syed *et al.*, 1988; Pöllänen *et al.*, 1990). Subsequently, the Sertoli cell has been shown to be the site of production of IL-1 bioactivity. An IL-1α antiserum can neutralize this bioactivity (Gerard *et al.*, 1991). Binding of radiolabeled recombinant human IL-1 to mouse testes membranes has been demonstrated (Takao *et al.*, 1990). Autoradiography showed its primary localization to the interstitium, where it is known to modulate Leydig cell androgen secretion. Since IL-1 has diverse roles in the immune and other systems, these data suggest that Il-1 may be an important modulator of testicular cell–cell communication.

In a series of publications, Parvinen and colleagues have shown that Il-1 production is stage specific (Söder *et al.*, 1991). High levels were found in conditioned media from cultured tubules from most stages of the cycle but very low levels were apparent in Stages VII–VIII, which coincided with low DNA synthesizing activity by spermatogonia, leading to speculation that IL-1 was involved in spermatogonial proliferation. Intratesticular IL-1 has been reported to increase thymidine incorporation, probably by spermatogonia, in hypophysectomized rats (Pöllänen *et al.*, 1989). In short-term tubule cultures, recombinant human IL-1 stimulated thymidine incorporation by type A spermatogonia of Stage I, in addition to enhancing meiotic DNA synthesis by preleptotene spermatocytes that developed in Stage VIIa cultures (Parvinen *et al.*, 1991). IL-1α was concluded to act as a regulator of DNA synthesis in both meiotic phases of spermatogenesis. Overall, a good case can be made for IL-1 as an important potential regulator of spermatogenesis.

C. Insulin-like Growth Factor 1

Sertoli cells produce IGF-1, but its receptors are present on Sertoli cells as well as on spermatocytes and spermatids. Testicular IGF-1 mRNA levels are increased in immature hypophysectomized animals by LH, FSH, and growth hormone (Closset *et al.*, 1989). IGF-1 levels also are increased in the rat testes by retinol (Bartlett *et al.*, 1990) and IGF-1 receptors have been reported on both rat (Tres *et al.*, 1986) and human (Vannelli *et al.*, 1986) germ cells. Although actions of IGF-1 on Sertoli cells have been described (Borland *et al.*, 1984), currently no evidence supports a direct action on germ cells. Further studies are required.

D. Other Growth Factors

Several other growth factors are present in the seminiferous epithelium, although their importance in spermatogenesis has not been established. These factors include TGFα and TGFβ, FGF, and SGF.

References

Adamopoulos, D. A., Lawrence, D. M., Vassilopoulos, P., Contoyiannis, P. A., and Swyer, G. I. M. (1978). Pituitary-testicular interrelationships in mumps orchitis and other viral infections. *Br. Med. J.* **1**, 1177–1180.

Ailenberg, M., McCabe, D., and Fritz, I. B. (1990). Androgens inhibit plasminogen activator activity by Sertoli cells in culture in a two-chambered assembly. *Endocrinology* **126**, 1561–1568.

Aiman, J., Brenner, P. F., and MacDonald, P. C. (1980). Androgen and estrogen production in elderly men with gynecomastia and testicular atrophy after mumps orchitis. *J. Clin. Endocrinol. Metab.* **50**, 380–386.

Alexander, N. J., and Anderson, D. J. (1987). Immunology of semen. *Fertil. Steril.* **47**, 192–205.

Assoian, R. K., Fleurdelys, B. E., Stevenson, H. C., Miller, P. J., Madtes, D. K., Raines, E. W., Ross, R., and Sporn, M. B. (1987). Expression and secretion of type β transforming growth factor by activated human macrophages. *Proc. Natl. Acad. Sci. U.S.A.* **84**, 6020–6024.

Attisano, L., Wrana, J. L., Cheifetz, S., and Massague, J. (1992). Novel activin receptors: Distinct genes and alternative mRNA splicing generate a repertoire of serine/threonine kinase receptors. *Cell* **68**, 97–108.

Avallet, O., Vigier, M., Perrard-Sapori, M. H., and Saez, J. M. (1987). Transforming growth factor β inhibits Leydig cell functions. *Biochem. Biophys. Res. Commun.* **146**, 575–581.

Ayer-LeLievre, C., Olson, L., Ebendal, T., Hallbook, F., and Persson, H. (1988). Nerve growth factor mRNA and protein the testis and epididymis of mouse and rat. *Proc. Natl. Acad. Sci. U.S.A.* **86**, 2628–2632.

Azad, N., Emanuele, N. V., Halloran, M. M., Tentler, J., and Kelley, M. R. (1991). Presence of luteinizing hormone-releasing hormone (LHRH) mRNA in rat spleen lymphocytes. *Endocrinology* **128**, 1679–1681.

Baarends, W. M., Marjolen, J. L., van Helmond, M. J. L., Post, M., Hoogerbrugge, J. W., Themmen, A. P. N., and Grootegoed, J. A. (1992). Cloning of a new member of the TGF-β/activin receptor family, mainly expressed in the gonads. American Endocrine Society Meeting, Abstract 215.

Bartlett, J. M. S., Weinbauer, G. F., and Neischlag, E. (1989). Stability of spermatogenic synchronization achieved by depletion and restoration of vitamin A deficiency. *Biol. Reprod.* **42**, 603–612.

Bartlett, J. M. S., Spiteri-Grech, J., and Nieschlag, E. (1990). Regulation of insulin-like growth factor I and stage-specific levels of epidermal growth factor in stage synchronized rat testes. *Endocrinology* **127**, 747–758.

Bellve, A. R., and Feig, L. A. (1984). Cell proliferation in the mammalian testis: biology of the seminiferous growth factor (SGF). *Rec. Prog. Horm. Res.* **40**, 531–567.

Benhamed, M., Morera, A. M., Chauvin, M. C., and de Peretti, E. (1987). Somatomedin C/isulin/like factor 1 as a possible intratesticular regulator of Leydig cell activity. *Mol. Cell. Endocrinol.* **50**, 69–77.

Bergh, A. (1987). Treatment with hCG increases the size of Leydig cells and testicular macrophages in unilaterally cryptorchid rats. *Int. J. Androl.* **10**, 765–772.

Bergh, A. (1985). Effect of cryptorchidism on the morphology of testicular macrophages: Evidence for a Leydig cell-macrophage interaction in the rat testis. *Int. J. Androl.* **8**, 86–96.

Bergh, A., Rooth, P., Widmark, A., and Damber, J.-E. (1987). Treatment of rats with hCG induces inflammation-like changes in the testicular microcirculation. *J. Reprod. Fertil.* **79**, 135–143.

Bergh, A., Damber, J.-E., and van Rooijen, N. (1992). The hCG-induced inflammatory response is enhanced in macrophage-depleted testes. *Proceedings of the 7th European Testis Workshop on Molecular and Cellular Endocrinology of the Testis*, Abstract 103.

Bernier, M., Chatelain, P., Mather, J. P., and Saez, J. M. (1986). Regulation of gonadotropin receptors, gonadotropin responsiveness, and cell multiplication by somatomedin-C and insulin in cultured pig Leydig cells. *J. Cell Physiol.* **129**, 257–263.

Bicsak, T. A., Vale, W., Vaughan, J., Tucker, E. M., Cappel, S., and Hsueh, A. J. W. (1987). Hormonal regulation of inhibin production by cultured Sertoli cells. *Mol. Cell. Endocrinol.* **49**, 211–217.

Boitini, C., and Ritzen, E. M. (1981). Inhibition of rat Sertoli cell aromatase by factor(s) secreted specifically at spermatogenic stages VII and VIII. *Mol. Cell. Endocrinol.* **23**, 11–22.

Boockfor, F. R., and Schwarz, L. K. (1991). Effects of inteleukin-6, interleukin-2, and tumor necrosis factor α on transferrin release from Sertoli cells in culture. *Endocrinology* **129**, 256–262.

Borland, K., Mita, M., Oppenheimer, C. L., Blinoerman, L. A., Massague, J., Hall, P. F., and Czech, M. P. (1984). The actions of insulin-like growth factors I and II on cultured Sertoli cells. *Endocrinology* **114**, 240–246.

Born, W., and Wekerle, H. (1981). Selective, immunologically nonspecific adherence of lymphoid and myeloid cells to Leydig cells. *Eur. J. Cell Biol.* **25**, 76–81.

Braunhut S. J., Rufo, G. A., Ernisee, B. J., Zheng, W., and Bellve, A. R. (1990). The seminiferous growth factor induces proliferation of TM4 cells in serum-free medium. *Biol. Reprod.* **42**, 639–648.

Brown, S. L., and van Epps, D. E. (1986). Suppression of T lymphocyte chemotactic factor production by the opioid peptides β-endorphin and met-enkephalin. *J. Immunol.* **134**, 3384–3390.

Byers, S. W., Hadley, M. A., Djakiew, D., and Dym, M. (1986). Growth and characterisation of polarised monolayers of epididymal epithelial cells and Sertoli cells in dual environment culture chambers. *J. Androl.* **7**, 59–68.

Cailleau, J., Vermeire, S., and Verhoeven, G. (1990). Independent control of the production of insulin-like growth factor I and its binding protein by cultured testicular cells. *Mol. Cell. Endocrinol.* **69**, 79–89.

Calkins, J. H., Sigel, M. M., Nankin, H. R., and Lin, T. (1988). Interleukin-1 inhibits Leydig cell steroidogenesis in primary culture. *Endocrinology* **123**, 1605–1610.

Calkins, J. H., Guo, H., Sigel, M. M., and Lin, T. (1990a). Differential effects of recombinant interleukin-1α and β on Leydig cell function. *Biochem. Biophys. Res. Commun.* **167**, 548–553.

Calkins, J. H., Guo, H., Sigel, M. M., and Lin, T. (1990b). Tumor necrosis factor-α enhances inhibitory effects of interleukin-1β on Leydig cell steroidogenesis. *Biochem. Biophys. Res. Commun.* **166**, 1313–1318.

Casella, S. J., Smith, E. P., van Wyk, J. J., Joseph, D. R., Hynes, M. A., Hoyt, E. C., and Lund, P. K. (1987). Isolation of rat testis cDNAs encoding an insulin-like growth factor I precursor. *DNA* **6**, 325–330.

Chatelian, P. G., Naville, D., and Saez, J. M. (1987). Somatomedin C/insulin-like growth factor 1-like material secreted by porcine Sertoli cells in vivo: Characterization and regulation. *Biochem. Biophys. Res. Commun.* **146**, 1009–1017.

Chemes, H. E., Dym, M., and Raj, H. G. M. (1979). Hormonal regulation of Sertoli cell differentiation. *Biol. Reprod.* **21**, 251–262.

Closset, J., Gothot, A., Sente, B., Scippo, M. L., Igout, A., Vandenbroeck, M., Dombrowicz, D., and Hennen, G. (1989). Pituitary hormones dependent expression of insulin-like growth factors I and II in the immature hypophysectomized rat testis. *Mol. Endocrinol.* **3**, 1125–1131.

De Winter, J. P., Themmen, A. P. N., Hoogerbrugge, J. W., Klaij, I. A., Grootegoed, J. A., and de Jong, F. H. (1992). Activin receptor mRNA expression in rat testicular cell types. *Mol. Cell. Endocrinol.* **83**, R1–R8.

Dinarello, C. A., and Thompson, R. C. (1991). Blocking IL-1: Interleukin 1 receptor antagonist in vivo and in vitro. *Immunol. Today* **12**, 404–410.

Djakiew, D., and Dym, M. (1988). Pachytene spermatocyte proteins influence Sertoli cell function. *Biol. Reprod.* **39**, 1193–1205.

El-Demiry, M. I., Hargreave, T. B., Busuttil, A., Elton, R., James, K., and Chisholm, G. D. (1987). Immunocompetent cells in human testis in health and disease. *Fertil. Steril.* **48**, 470–479.

Emanuele, N. V., Emanuele, M. A., Tentler, J., Kirsteins, L., Azad, N., and Lawrence, A. M. (1990). Rat spleen lymphocytes contain an immunoactive and bioactive luteinizing hormone-releasing hormone. *Endocrinology* **126**, 2482–2486.

Emoto, M., Yagyu, Y., Nishikawa, F., Katsui, N., Kita, E., and Kashiba, S. (1989). Effects of mouse testicular extract on immunocompetant cells. *Am. J. Reprod. Immunol.* **21**, 61–66.

Emoto, M., Nishikawa, F., Oku, D., Hamuro, A., Kita, E., and Kashiba, S. (1991). Suppressive effect of a mouse testicular extract on lymphocyte activation. *Int. J. Androl.* **14**, 291–302.

Esposito, G., Keramidas, M., Mauduit, C., Feige, J. J., Morera, A. M., and Benahmed, M. (1991). Direct regulating effects of transforming growth factor-β 1 on lactate production in cultured porcine Sertoli cells. *Endocrinology* **128**, 1441–1449.

Fabbri, A., Tsai-Morris, C. H., Luna, S., Fraioli, F., and Dufau, M. L. (1985). Opiate receptors are present in the rat testis. Identification and localisation in Sertoli cells. *Endocrinology* **117**, 2544–2546.

Fauser, B. C. J. M., Baird, A., and Hseuh, A. J. W. (1988). Fibroblast growth factor inhibits luteinizing hormone-stimulated androgen production by cultured rat testicular cells. *Endocrinology* **123**, 2935–2941.

Fawcett, D. W. (1973). Observations on the organization of the interstitial tissue of the testis and on the occluding cell junctions in the seminiferous epithelium. *Adv. Biosci.* **10**, 83–99.

Fujisawa, M., Bardin, C. W., and Morris, P. L. (1992). A germ cell factor(s) modulates preproenkephalin gene expression in rat Sertoli cells. *Mol. Cell. Endocrinol.* **84**, 79–88.

Gaytan, F., Carrera, G., Pinilla, L., Aguilar, R., and Bellido, C. (1989). Mast cells in the testis, epididymis and accessory glands of the rat. Effects of neonatal steroid treatment. *J. Androl.* **10**, 351–358.

Geierhaas, B., Bornstein, S. R., Jarry, H., Scherbaum, W. A., Herrmann, M., and Pfeiffer, E. F. (1991). Morphological and hormonal changes following vasectomy in rats, suggesting a functional role for Leydig-cell associated macrophages. *Horm. Metab. Res.* **23**, 373–378.

Gérard, N., Syed, V., Bardin, W., Genetet, N., and Jégou, B. (1991). Sertoli cells are the site of inerleukin-1α synthesis in rat testis. *Mol. Cell. Endocrinol.* **82**, R13–R16.

Gerendal, I., Shaha, C., Gunsalus, G. L., and Bardin, C. W. (1986). The effects of opioid receptor antagonists suggest that testicular opiates regulate Sertoli and Leydig cell function in the neonatal rat. *Endocrinology* **118**, 2039–2044.

Gnessi, L., Emidi, A., Farini, D., Scarpa, S., Modesti Am Ciampani, T., Silvestroni, L., and Spera, G. (1992). Rat Leydig cells bind platelet-derived growth factor through specific receptors and produce platelet-derived growth factor-like molecules. *Endocrinology* **130**, 2219–2224.

Gonzales, G. F., Risbridger, G. P., Ishida, H., Hodgson, Y. M., Pollanen, P., and de Kretser, D. M. (1989). Stage-specific inhibin secretion by rat seminiferous tubules. *Reprod. Dev. Fertil.* **1**, 275–279.

Granholm, T., and Soder, O. (1991). Constitutive production of lymphocyte activating factors by normal tissues in the adult rat. *J. Cell. Biochem.* **46**, 143–151.

Granholm, T., Creasy, D. M., Pollanen, P., and Soder, O. (1992). Di-*n*-pentyl phthalate-induced inflammatory changes in the rat testis are accompanied by local production of a noval lymphocyte activating factor. *J. Reprod. Immunol.* **21**, 1–14.

Griswold, M. D., Mabley, E. R., and Fritz, I. B. (1976), FSH stimulation of DNA synthesis in Sertoli cells in culture. *Mol. Cell. Endocrinol.* **4**, 139–149.

Guo, H., Calkins, J. H., Sigel, M. M., and Lin, T. (1990). Interleukin-2 is a potent inhibitor of Leydig cell steroidogenesis. *Endocrinology* **127**, 1234–1239.

Gustafsson, K., Soder, O., Pollanen, P., and Ritzen, E. M. (1988). Isolation and partial characterization of an interleukin-1-like factor from rat testis interstitial fluid. *J. Reprod. Immunol.* **14**, 139–150.

Guzzardi, V., and Hedger, M. P. (1992). Comparison of bioactive interleukin-1 secretion and MHC II expression by testicular and peritoneal macrophages. *Proceedings of the 7th European Testis Workshop on Molecular and Cellular Endocrinology of the Testis*, Abstract 109.

Hadley, M. A., Djakiew, D., Byers, S. W., and Dym, M. (1987). Polarised secretion of androgen-binding protein and transferrin by Sertoli cells grown in a bicameral culture system. *Endocrinology* **20**, 1097–1103.

Hagenas, L., Ploen, L., Ekwall, H., Osman, D. I., and Ritzen E. M. (1981). Differentiation of the rat seminiferous tubules between 13 and 19 days of age. *Int. J. Androl.* **4**, 257–264.

Hardy, M. P., Zirkin, B. R., and Ewing, L. R. (1989). Kinetic studies on the development of the adult population of Leydig cells in tests of the pubertal rat. *Endocrinology* **124**, 762–770.

Hardy, M. P., Kelce, W. R., Klinefelter, G. R., and Ewing, L. L. (1990). Differentiation of Leydig cell precursors *in vitro*: A role for androgen. *Endocrinology* **127**, 488–490.

Head, J. R., and Billingham, R. E. (1985). Immune privilege in the testis. II. Evaluation of potential local factors. *Transplantation* **40**, 269–275.

Head, J. R., Neaves, W. B., and Billingham, R. E. (1983). Immune privilege in the testis, I. Basic parameters of allograft survival. *Transplantation* **36**, 423–431.

Hedger, M. P. (1989). The testis: An 'immunologically suppressed' tissue? *Reprod. Fertil. Dev.* **1,** 75–79.

Hedger, M. P., and Eddy, E. M. (1987). The heterogeneity of isolated rat Leydig cells separated on Percoll density gradients: An immunological, cytochemical and functional analysis. *Endocrinology* **121,** 1824–1838.

Hedger, M. P., Robertson, D. M., and de Kretser, D. M. (1987). LHRH and "LHRH-like" factors in the male reproductive tract. In "LHRH and Its Analogs: Contraceptive and Therapeutic Applications" (B. Vickery and J. Nestor, eds.), Pt. 2, pp. 141–160. MTP Press, Lancaster.

Hedger, M. P., Drummon, A. E., Robertson, D. M., Risbridger, G. P., and de Kretser, D. M. (1989). Inhibin and activin regulate [^{3}H]-thymidine uptake by rat thymocytes and 3T3 cells in vitro. *Mol. Cell. Endocrinol* **61,** 133–138.

Hedger, M. P., Quin, J.-X., Robertson, D. M., and de Kretser, D. M. (1990). Intragonadal regulation of immune system functions. *Reprod. Fertil. Dev.* **2,** 263–280.

Heinrich, P. C., Castell, J. V., and Andus, T. (1990). Interleukin-6 and the acute phase response. *Biochem. J.* **265,** 621–636.

Holmes, S. D., Spotts, G., and Smith, R. G. (1986). Rat Sertoli cells secrete a growth factor that blocks epidermal growth factor (EGF) binding to its receptor. *J. Biol. Chem.* **261,** 4076–4080.

Hseuh, A. J., Dahl, K. D., Vaughen, J., Tucker, E., Rivier, J., Bardin, C. W., and Vale, W. (1987). Heterodimers and homodimers of inhibin subunits have different paracrine action in the modulation of luteinizing hormone stimulated-androgen biosynthesis. *Proc. Natl. Acad. Sci. U.S.A.* **84,** 5682–5286.

Ireland, M. E., and Welsh, M. J. (1987). Germ cell stimulation of Sertoli cell protein phosphorylation. *Endocrinology* **120,** 1317–1326.

Janecki, A., and Steinberger, A. (1986). Polarized Sertoli cell functions in a new-two compartment culture system. *J. Androl.* **7,** 69–71.

Jeégou, B. (1991). Spermatids as regulators of Sertoli cell function. *Ann. N.Y. Acad. Sci.* **637,** 340–353.

Jégou, B., Risbridger, G. P., and de Kretser, D. M. (1983). Effects of experimental cryptorchidism on testicular function in adult rats. *J. Androl.* **4,** 88–94.

Jégou, B., Laws, A. O., and de Kretser, D. M. (1984). Changes in testicular function induced by short term exposure of the rat testis to heat. Further evidence for interaction of germ cells, Sertoli cells and Leydig cells. *Int. J. Androl.* **7,** 244–257.

Johnson, H. M., Smith, E. M., Torres, B. A., and Blalock, J. E. (1982). Regulation of the in vitro antibody response by neuroendocrine hormones. *Proc. Natl. Acad. Sci. U.S.A.* **79,** 4171–4174.

Kancheva, L. S., Martinova, Y. S., and Giorgiev, V. D. (1990). Prepubertal rat Sertoli cells secrete a mitogenic factor(s) that stimulates germ and somatic cell proliferation. *Mol. Cell. Endocrinol.* **69,** 121–127.

Kerr, J. B., and Sharpe, R. M. (1989). Macrophage activation enhances the human chorionic gonadotrophin-induced disruption of spermatogenesis in the rat. *J. Endocrinol.* **121,** 285–292.

Khan, S. A., Soder, O., Syed, V., Gustafsson, K., Lindh, M., and Ritzen, E. M. (1987). The rat testis produces large amounts of an interleukin-1-like factor. *Int. J. Androl.* **10,** 495–503.

Khan, S. A., Schmidt, K., Hallin, P., di Pauli, R., de Geyter, C., and Nieschlag, E. (1988). Human testis cytosol and ovarian follicular fluid contain high amounts of interleukin-1-like factor(s). *Mol. Cell. Endocrinol.* **58,** 221–230.

Khan, S., Teerds, K., and Dorrington, J. (1992a). Growth factor requirements for DNA synthesis by Leydig cells from the immature rat. *Biol. Reprod.* **46,** 335–341.

Khan, S. A., Teerds, K., and Dorrington, J. (1992b). Steroidogenesis-inducting protein promotes deoxyribonucleic acid synthesis in Leydig cells from immature rats. *Endocrinology* **130,** 599–605.

Lamb, D. L., Spotts, G. S., Shubhada, S., and Baker, K. R. (1991). Partial characterisation of a unique mitogenic activity secreted by rat Sertoli cells. *Mol. Cell. Endocrinol.* **79**, 1–12.

Lee, W., Mason, A. J., Schwall, R., Szonyi, E., and Mather, J. P. (1989). Secretion of activin by interstitial cells in the testis. *Science* **243**, 396–398.

Le Magueresse-Battistoni, B., Gérard, N., and Jégou B. (1991). Pachytene spermatocytes can achieve meiotic process in vitro. *Biochem. Biophys. Res. Commun.* **2**, 1115–1121.

Li, L., Hedger, M. P., Clements, J. A., and Risbridger, G. P. (1991). Localization of immunoreactive β-endorphin and adrenocorticotropic hormone, and pro-opiomelanocortin mRNA to testicular interstitial tissue macrophages. *Biol. Reprod.* **45**, 282–289.

Lin, T., Blaisdell, J., and Haskell, J. F. (1987). Transforming growth factor-β inhibits Leydig cell steroidogenesis in primary culture. *Biochem. Biophys. Res. Commun.* **146**, 387–394.

Lin, T., Calkins, J. H., Morris, P. L., Vale, W., and Bardin, C. W. (1989). Regulation of Leydig cell function in primary culture by inhibin and activin. *Endocrinology* **125**, 2134–2140.

Lin, T., Wang, D., Nagpal, M. L., Chang, W., and Calkins, J. (1992). Down regulation of Leydig cell insulin-like growth factor-I gene expression by interleukin-1. *Endocrinology* **130**, 1217–1224.

Linder, C. C., Hecker, L. L., Roberts, K. P., Kim, K. H., and Griswold, M. D. (1991). Expression of receptors during the cycle of the seminiferous epithelium. *Ann. N.Y. Acad. Sci.* **637**, 313–321.

Lombard-Vignon, N., Grizard, G., and Boucher, D. (1991). Influence of rat testicular macrophages on Leydig cell function in vitro. *Int. J. Androl.* **15**, 144–159.

McDonald, H. R., and Nabholz, M. (1986). T-cell activation. *Ann. Rev. Cell Biol.* **2**, 231–253.

McKinnell, C., and Sharpe, R. M. (1992). The role of specific germ cell types in modulation of the secretion of androgen-regulated proteins (ARPs) by stage VI-VIII seminiferous tubules from the adult rat. *Mol. Cell. Endocrinol.* **83**, 219–231.

Maddocks, S., Sowerbutts, S., van Rooijen, N., and Kerr, J. B. (1992). Macrophage depletion in the rat testis and its effect on the hCG response. *Proceedings of the 7th European Testis Workshop on Molecular and Cellular Endocrinology of the Testis*, Abstract 104.

Mahi-Brown, C. A., Yule, T. D., and Tung, K. S. K. (1988). Evidence for active immunological regulation in prevention of testicular autoimmune disease independent of the blood-testis barrier. *Am. J. Reprod. Immunol. Microbiol.* **16**, 165–170.

Mallea, L. E., Machado, A. J., Navaroli, F., and Rommerts, F. F. G. (1986). Epidermal growth factor stimulates lactate production and inhibits aromatisation in cultured Sertoli cells from immature rats. *Int. J. Androl.* **9**, 201–208.

Manson, A. L. (1990). Mumps orchitis. *Urology* **36**, 355–358.

Mather, J. P., Attie, K. M., Woodruff, T. K., Rice, G. C., and Phillips, D. M. (1990). Activin stimulates spermatogonial proliferation in germ-Sertoli cell cocultures from immature rat testis. *Endocrinology* **127**, 3206–3214.

Mathews, L. S., and Vale, W. W. (1991). Expression cloning of an activin receptor, a predicted transmembrane serine kinase. *Cell* **65**, 973–982.

Mauduit, C., Hartmann, D. J., Chauvin, M. A., Revol, A., Morera, A. M., and Benhamed, M. (1991). Tumor necrosis factor inhibits gonadotropin action in cultured porcine Leydig cells: Site(s) of action. *Endocrinology* **129**, 2933–2940.

Mauduit, C., Chauvin, M. A., de Peretti, E., Morera, A., and Benhamed, M. (1991). Effect of activin A in dehydroepiandrosterone and testosterone secretion by primary immature porcine Leydig cells. *Biol. Reprod.* **45**, 101–109.

Meidan, R., McAllister, J. M., and Hseuh, A. J. W. (1985). Hormonal regulation of androgen biosynthesis by primary cultures of testis cells from neonatal rats. *Endocrinology* **116**, 2473–2482.

Miller, S. C., Bowman, B. M., and Rowland, H. G. (1983). Structure, cytochemistry, endocytic activity, and immunoglobulin (Fc) receptors of rat testicular interstitial-tissue macrophages. *Am. J. Anat.* **168**, 1–13.

Mita, M., Price, J. M., and Hall, P. F. (1985). Metabolism of round spermatids from rats with lactate as the preferred substrate. *Biol. Reprod.* **26**, 445–455.

Moore, A., Findlay, K., and Morris, I. D. (1992). In vitro DNA synthesis in Leydig cells and other interstitial cells in the rat testis. *J. Endocrinol.* **134**, 247–255.

Moore, C., and Moger, W. H. (1991). Interleukin-1α-induced changes in androgen and cyclic andosine 3′,5′ monophosphate release in adult rat Leydig cells in culture. *J. Endocrinol.* **129**, 381–390.

Morales, C. R., and Griswold, M. D. (1987). Retinol induced stage syncyhronisation in seminiferous tubules of the rat. *Endocrinology* **121**, 432–434.

Morales, C. R., Alcivar, A. A., Hecht, N. B., and Griswold, M. D. (1987). Specific mRNAs in Sertoli and germ cells of testis from stage specific rats. *Mol. Endocrinol.* **3**, 725–733.

Morris, P. L., Vale, W. W., and Bardin, C. W. (1987). Beta-endorphin regulation of FSH-stimulated inhibin production is a component of a short loop system in testis. *Biochem. Biophys. Res. Commun.* **148**, 1513–1519.

Morris, P. L., Vale, W. W., Cappel, S., and Bardin, C. W. (1988). Inhibin production by primary Sertoli cell enriched cultures: regulation by follicle-stimulating hormone, androgens, and epidermal growth factor. *Endocrinology* **122**, 717–725.

Mueller, D. L., Jenkins, M. K., and Schwartz, R. H. (1989). Clonal expansion versus functional clonal inactivation: a costimulatory signalling pathway determines the outcome of T cell antigen receptor occupancy. *Ann. Rev. Immunol.* **7**, 445–480.

Murono, E. P., and Washburn, A. L. (1990a). Fibroblast growth factor inhibits 5α-reductase activity in cultured immature Leydig cells. *Mol. Cell. Endocrinol.* **68**, R19–R23.

Murono, G. P., and Washburn, A. (1990b). Platelet derived growth factor inhibits 5α-reductase and 5α,3β-hydroxysteroid dehydrogenase activities in cultured immature Leydig cells. *Biochem. Biophys. Res. Commun.* **169**, 1224–1229.

Nistal, M., Santamaria, L., and Paniagua, R. (1984). Mast cells in the human testis and epididymis from birth to adulthood. *Acta Anatomica* **119**, 155–160.

Norton, J. N., and Skinner, M. K. (1989). Regulation of Sertoli cell function and differentiation through the actions of a testicular paracrine factor, P-Mod-S. *Endocrinology* **124**, 2711–2719.

Ogawa, K., Kurohmaru, M., Shiota, K., Takahashi, M., and Hayashi, Y. (1992). Immunoreactive βA subunit of inhibin/activin is present in cytoplasm of rat spermatogenic cells. *J. Reprod. Dev.* **38**, 5–9.

Oonk, R. B., and Grootengoed, J. A. (1988). Insulin-like growth factor (IGF-1) receptors on Sertoli cells from immature rats and age dependent testicular binding of IGF-1 and insulin. *Mol. Cell. Endocrinol.* **55**, 33–43.

Orava, M., Cantell, K., and Vihko, R. (1985). Human leukocyte interferon inhibits human chorionic gonadotropin stimulated testosterone production by porcine Leydig cells in culture. *Biochem. Biophys. Res. Commun.* **127**, 809–815.

Orth, J. M. (1984). The role of FSH in controlling Sertoli cell proliferation in testis of fetal rats. *Endocrinology* **115**, 1248–1255.

Orth, J. M. (1986). FSH-induced Sertoli cell proliferation in the developing rat is modified by β-endorphin produced in the testis. *Endocrinology* **119**, 1876–1879.

Parvinen, M. (1982). Regulation of the seminiferous epithelium. *Endocr. Rev.* **3**, 404–417.

Parvinen, M., Wright, W. W., Phillips, D. M., Mather, J. P., Musto, N. A., and Bardin, C. W. (1983). Spermatogenesis *in vitro*: Completion of meiosis and early spermiogenesis. *Endocrinology* **112**, 1150–1152.

Parvinen, M., Soder, O., Mali, P., Froysa, B., and Ritzen, E. M. (1991). *In vitro* stimulation of stage-specific deoxyribonucleic acid synthesis in rat seminiferous tubule segments by interleukin-1α *Endocrinology* **129**, 1614–1620.

Perez-Infante, V., Bardin, C. W., Gunsalus, G. L., Musto, N. A., Rich, K. A., and Mather, J. P.

(1986). Differential regulation of testicular transferrin and androgen-binding protein secretion in primary cultures of rat Sertoli cells. *Endocrinology* **118**, 383–392.

Perrard-Sapori, M. H., Chatelain, P. C., Rogemond, N., and Saex, J. M. (1987). Modulation of Leydig cell functions by culture with Sertoli cells or with Sertoli-cell conditioned medium: Effect of insulin, somatomedin-C and FSH. *Mol. Cell. Endocrinol.* **50**, 193–201.

Persson, H., Ayer-LiLievre, C., Soder, O., Villar, M. J., Metsis, M., Olson, L., Ritzen, M., and Hokfelt, T. (1990). Expression of beta-nerve growth factor receptor mRNA in Sertoli cells downregulated by testosterone. *Science* **247**, 704–707.

Petraglia, F., Sacerdote, P., Cossarizza, A., Angioni, S., Genazzani, A. D., Franceschi, C., Muscettola, M., and Grasso, G. (1991). Inhibin and activin modulate human monocyte chemotaxis and human lymphocyte interferon-α production. *J. Clin. Endocrinol. Metab.* **72**, 496–502.

Pinon-Lataillade, G., Velez de la Calle, J. F., Viguier-Martinez, Garnier, D. H., Folliot, R., Maas, J., and Jégou, B. (1988). Influence of germ cells upon Sertoli cells during continuous low dose gamma-irradiation of adult rats. *Mol. Cell. Endocrinol.* **58**, 51–63.

Pöllänen, P., and Maddocks, S. (1988). Macrophages, lymphocytes and MHC II antigen in the ram and rat testis. *J. Reprod. Fertil.* **82**, 437–445.

Pöllänen, P., and Niemi, M. (1987). Immunohistochemical identification of macrophages, lymphoid cells and HLA antigens in the human testis. *Int. J. Androl.* **10**, 37–42.

Pöllänen, P., Soder, O., and Uksila, J. (1988). Testicular immunosuppressive protein. *J. Reprod. Immunol.* **14**, 125–138.

Pöllänen, P., Soder, O., and Parvinen, M. (1989). Interleukin-1α stimulation of spermatogonial proliferation in vivo. *Reprod. Fertil. Dev.* **1**, 85–87.

Pöllänen, P., von Euler, M., and Soder, O. (1990). Testicular immunoregulatory factors. *J. Reprod. Immunol.* **18**, 51–76.

Risbridger, G. P. (1992). Local regulation of Leydig cell function by inhibitors of steroidogenic activity. *Cell Biol. Int. Report* **16**, 399–406.

Risbridger, G. P., and Hedger, M. P. (1992). Adult rat Leydig cell cultures: Minimum requirements for maintenance of luteinizing hormone responsiveness and testosterone production. *Mol. Cell. Endocrinol.* **83**, 125–132.

Risbridger, G. P., Clements, J., Robertson, D. M., Drummond, A. E., Muir, J., Burger, H. G., and de Kretser, D. M. (1989). Immuno- and bioactive inhibin and inhibin α subunit expression in rat Leydig cell cultures. *Mol. Cell. Endocrinol.* **66**, 119–122.

Rivarola, M. A., Sanchez, P., and Saez, J. M. (1986). Inhibition of RNA and DNA synthesis in Sertoli cells by co-culture with spermatogenic cells. *Int. J. Androl.* **9**, 424–434.

Rivenson, A., Ohmori, T., Hamazaki, M., and Madden, R. (1981). Cell surface recognition: spontaneous identification of mouse Leydig cells by lymphocytes, macrophages and eosinophiles. *Cell. Mol. Biol.* **27**, 49–56.

Roberts, K., and Griswold, M. D. (1989). Testosterone induction of cellular proteins in cultured Sertoli cells from hypophysectomised rats and rats of different ages. *Endocrinology* **125**, 1174–1179.

Roberts, V., Meunier, H., Sawchenko, P. E., and Vale, W. (1989). Differential production and regulation of inhibin subunits in rat testicular cell types. *Endocrinology* **125**, 2350–2359.

Robertson, D. M., Hayward, S., Irby, D., Jacobsen, J., Clarke, L., McLachlan, R. I., and de Kretser, D. M. (1988). Radioimmunoassay of rat serum inhibin: Changes after PMSG stimulation and gonadectomy. *Mol. Cell. Endocrinol.* **58**, 1–8.

Robertson, D. M., Foulds, L. M., Prisk, M., and Hedger, M. P. (1992). Inhibin/activin β-subunit monomer: Isolation and characterization. *Endocrinology* **130**, 1680–1687.

Sainio-Pollanen, S., Pollanen, P., and Setchell, B. P. (1991). Testicular immunosuppressive activity in experimental hypogonadism and cryptorchidism. *J. Reprod. Immunol.* **20**, 59–72.

Selawry, H. P., Kotb, M., Herrod, H. G., and Lu, Z. N. (1991). Production of a factor, or factors, suppressing IL-2 production and T cell proliferation by Sertoli cell-enriched preparations. A potential role for isolet transplantation in an immunologically privileged site. *Transplantation* **52**, 846–850.

Setchell, B. P., and Sharpe, R. M. (1981). The effect of human chorionic gonadotrophin on capillary permeability, extracellular fluid volume and flow of lymph and blood in the testis of rats. *J. Endocrinol.* **91**, 245–254.

Shaha, C., Morris, P. L., Chen, C. L. C., Vale, W., and Bardin, C. W. (1989). Immunostainable inhibin subunits are in multiple types of testicular cells. *Endocrinology* **125**, 1941–1950.

Skinner, M. K. (1991). Cell–cell interactions in the testis. *Endocr. Rev.* **12**, 45–77.

Skinner, M. K., and Fritz, I. B. (1985). Androgen stimulation of Sertoli cell function is enhanced by peritubular cells. *Mol. Cell. Endocrinol.* **40**, 115–122.

Skinner, M. K., and Moses, H. L. (1989). Transforming growth factor β gene expression and action in the seminiferous tubule: Peritubular cell–Sertoli cell interactions. *Mol. Endocrinology* **3**, 625–634.

Skinner, M. K., Fetterolf, P. M., and Anthony, C. T. (1988). Purification of a paracrine factor, P-Mod-S, produced by testicular peritubular cells that modulates Sertoli cell function. *J. Biol. Chem.* **263**, 2884–2890.

Skinner, M. K., Takacs, K., and Coffey, R. J. (1989). Transforming growth factor-αgene expression in the seminiferous tubule: Peritubular cell–Sertoli cell interactions. *Endocrinology* **124**, 845–854.

Smith, E. P., Svaboda, M. E., Van Wyk, J. J., Kierszenbaum, A. L., and Tres, L. L. (1987). Paritial characterisation of a somatomedin-like peptide from the medium of cultured rat Sertoli cells. *Endocrinology* **120**, 186–193.

Smith, E. P., Hall, S. H., Monaco, L., French, F. S., Wilson, E. M., and Conti, M. (1989). A rat Sertoli cell factor similar to basic fibroblast growth factor increases c-fos messenger RNA in cultured Sertoli cells. *Mol. Endocrinol.* **3**, 954–961.

Soder, O., Syed, V., Callard, G. V., Toppari, J., Pollanen, P., Parvinen, M., Froysa, B., and Ritzen, E. M. (1991). Production and secretion of an interleukin-1-like factor is stage-dependent and correlates with spermatogonial DNA synthesis in the rat seminiferous epithelium. *Int. J. Androl.* **14**, 223–231.

Sordoillet, C., Chauvin, M. A., Revol, A., Morera, A., and Benhamed M. (1988). Fibroblast growth factor is a regulator of testosterone secretion in cultured immature Leydig cells. *Mol. Cell. Endocrinol.* **58**, 283–286.

Sordoillet, C., Chauvin, M. A., de Peretti, E., Morera, A., and Benhamed, M. (1991). Epidermal growth factor directly stimulates steroidigenesis in primary culture of Leydig cells: Actions and sites of action. *Endocrinology* **128**, 2160–2168.

Sordoillet, C., Savona, C., Chauvin, M. A., de Peretti, E., Fiege, J. J., Morera, A. M., and Benahmed, M. (1992). Basic fibroblast growth factor enhances testosterone secretion in cultured porcine Leydig cells: Site(s) of action. *Mol. Cell. Endocrinol.*, **89**, 163–171.

Spaliviero, J. A., and Handelsman, D. J. (1991). Effect of epidermal and insulin-like growth factors on vestorial secretion of transferrin by rat Sertoli cells in vitro. *Mol. Cell. Endocrinol.* **81**, 95–104.

Sun, X.-R., Hedger, M. P., and Risbridger, G. P. (1993). The effect of testicular macrophages and interleukin-1 on testosterone production by purified adult rat Leydig cells cultured under in vitro maintenance conditions. *Endocrinology*, **132**, 186–192.

Syed, V., Soder, O., Arver, S., Lindh, M., Khan, S., and Ritzen, E. M. (1988). Ontogeny and cellular origin of an interleukin-1-like factor in the reproductive tract of the male rat. *Int. J. Androl.* **11**, 437–447.

Syed, V., Gerard, N., Kaipia, A., Bardin, C. W., Parvinen, M., and Jegou, B. (1992). Identification, ontogeny and regulation of an interleukin-6 like (IL-6) factor in the rat seminiferous tubule. *Endocrinology*, **132**, 293–299.

Takao, T., Mitchell, W. M., Tracey, D. E., and de Souza, E. B. (1990). Identification of interleukin-1 receptors in mouse testis. *Endocrinology* **127**, 251–258.

Takeda, H., Chodak, G., Mutchnik, S., Nakamoto, T., and Chang, C. (1990). Immunohistochemical localization of androgen receptors with mono- and polyclonal antibodies to androgen receptor. *J. Endocrinol.* **126**, 17–25.

Teerds, K. J., Rommerts, F. F. G., and Dorrington, J. H. (1990). Immunohistochemical detection of transforming growth factor α in Leydig cells during the development of the rat testis. *Mol. Cell. Endocrinol.* **69**, R1–R6.

Toppari, J., and Parvinen, M. (1985). *In vitro* differentiation of rat seminiferous tubular segments from defined stages of the epithelial cycle morphologic and immunolocalization analysis. *J. Androl.* **6**, 334–343.

Tres, L. L., and Kierszenbaum, A. L. (1983). Viability of rat spermatogenic cells *in vitro* is facilitated by their coculture with Sertoli cells in serum-free hormone-supplemented medium. *Proc. Natl. Acad. Sci. U.S.A.* **80**, 3377–3381.

Tres, L. L., Smith, E., van Wyk, J. J., and Kierszenbaum, A. L. (1986). Immunoreactive sites and accumulation of somatomedin C in rat Sertoli spermatogenic cell cultures. *Exp. Cell Res.* **162**, 33–50.

Tsong, S-D., Phillips, D. M., Halmi, N., Krieger, D., and Bardin, C. W. (1982). β-Endorphin is present in the male reproductive tract of five species. *Biol. Reprod.* **27**, 755–764.

Tung, K. S. K., Yule, T. D., Mahi-Brown, C. A., and Listrom, M. B. (1987). Distribution of histopathology and Ia positive cells in actively induced and passively transferred experimental autoimmune orchitis. *J. Immunol.* **138**, 752–759.

Tung, P. S., and Fritz, I. B. (1991). Transforming growth factor-β and platelet-derived growth factor synergistically stimulate contraction by testicular peritubular cells in culture in serum-free medium. *J. Cell. Physiol.* **146**, 386–393.

Ueno, N., Baird, A., Esch, F., Lin, N., and Guillemin, R. (1987). Isolation and partial characterisation of basic fibroblast growth factor from bovine testis. *Mol. Cell. Endocrinol.* **49**, 189–194.

van Beek, M. E., and Meistrich, M. L. (1991). Stage-synchronised seminiferous epithelium in rats after manipulation of retinol levels. *Biol. Reprod.* **45**, 235–244.

van Dissel-Emiliani, FM., Grootenhuis, A. J., de Jong, F. H., and de Rooij, D. G. (1989). Inhibin reduces spermatagonial numbers in testis of adult mice and Chinese hamsters. *Endocrinol.* **125**, 1899–1903.

Vanelli, B. G., Barni, T., Orlando, C., Natali, A., Serio, M., and Balhoni, G. C. (1986). Insulin-like growth factor-I (IGF-I) and IGF-I receptor in human testis: An immunohistological study. *Fertil. Steril.* **49**, 666–673.

Verhoeven, G., and Cailleau, J. (1986). Stimulatory effects of epidermal growth factor on steroidogenesis in Leydig cells. *Mol. Cell. Endocrinol.* **47**, 99–106.

Verhoeven, G., and Cailleau, J. (1988). Follicle-stimulating hormone and androgens increase increase the concentrations of the androgen receptor in Sertoli cells. *Endocrinology* **122**, 1541–1550.

Verhoeven, G., and Cailleau, J. (1989). Tubule-Leydig cell interactions. *In* "Perspectives in Andrology" (M. Serio, ed.), Vol. 53, pp. 227–234. Raven Press, New York.

Vihko, K. K., Penttla, T.-L. Parvinen, M., and Belin, D. (1989). Regulation of urokinase- and tissue-type plasminogen activator gene expression in the rat seminiferous epithelium. *Mol. Endocrinol.* **3**, 52–59.

Wahl, S. M., McCartney-Francis, N., and Mergenhagen, S. E. (1989). Inflammatory and immunomodulatory roles of TGF-β. *Immunol. Today* **10**, 258–261.

Warren, D. W., Pasupuleti, V., Lu, Y., Platler, B. W., and Horton, R. (1990). Tumor necrosis factor and interleukin-1 stimulate testosterone secretion in adult male rat Leydig cells *in vitro*. *J. Androl.* **11**, 353–360.

Watrin, F., Scotto, L., Assoian, R. K., and Wolgemuth, D. J. (1991). Cell lineage specificity of expression of the murine transforming growth factor-β3 and transforming growth factor-β1 genes. *Cell. Growth Diff.* **2**, 77–83.

Weaver, C. T., and Unanue, E. R. (1990). The costimulatory function of antigen-presenting cells. *Immunol. Today* **11**, 49–55.

Wei, R. Q., Yee, J. B., and Straus, D. C., and Hutson, J. C. (1988). Bactericidal activity of testicular macrophages. *Biol. Reprod.* **38**, 830–835.

Welsh, M. J., Ireland, M. E., and Treisman, G. J. (1985). Stimulation of rat Sertoli cell adenylate cyclse by germ cells in vitro. *Biol. Reprod.* **33**, 1006–1056.

Whitehead, R. H. (1905). Studies on the interstitial cells of Leydig. *Am. J. Anat.* **3**, 167.

Woodruff, T. K., Borree, J., Attie, K. M., Cox, E. T., Rice, G. C., and Mather, J. P. (1992). Stage-specific binding of inhibin and activin to subpopulations of rat germ cells. *Endocrinology* **130**, 871–881.

Wyatt, C. R., Law, L., Magnuson, J. A., Griswold, M. D., and Magnuson, N. S. (1988). Suppression of lymphocyte proliferation by proteins secreted by cultured Sertoli cells. *J. Reprod. Immunol.* **14**, 27–40.

Yee, J. B., and Hutson, J. C. (1985). Effects of testicular macrophage-conditioned medium on Leydig cells in culture. *Endocrinology* **116**, 2682–2684.

Yule, T. D., Montoya, G. D., Russell, L. D., Williams, T. M., and Tung, K. S. K. (1988). Autoantigenic germ cells exist outside the blood testis barrier. *J. Immunol.* **141**, 1161–1167.

Zinkernagel, R. M., and Doherty, P. C. (1974). Restriction of *in vitro* T cell mediated cytotoxicity in lymphocytic choriomeningitis within a syngeneic or semi-allogenic system. *Nature (London)* **248**, 701–702.

Zwain, I. H., Morris, P., and Cheng, C. Y. (1991). Identification of an inhibitory factor from a Sertoli clonal cell line (TM4) that modulates Leydig cell steroidogenesis. *Mol. Cell. Endocrinol.* **80**, 115–126.

13

Vascular Controls in Testicular Physiology

ANDERS BERGH & JAN-ERIK DAMBER

I. Introduction

The vascular system is of obvious importance to the function of all tissues and organs, including the testis. The entering blood delivers oxygen, nutrients, regulatory hormones, and, when necessary, immunoglobulins and immunocompetent cells. Waste and secretory products leave the organ by the venous blood or by the lymph. In addition to these general functions, the testicular blood vessels show several unique adaptations in structure and function. Some of these peculiarities are obviously necessary to secure physiological processes unique to the testis. For example, spermatogenesis cannot take place at normal body temperature, so the vasculature is adapted to enable temperature control. The rationale for other local adaptations is more obscure, but they may be related to other more or less unique features of testicular function such as the blood–testis barrier, the locally immunosuppressed environment, and others that are still unknown.

In this chapter, we review the current knowledge of the role of the testicular vasculature in the overall function of the mammalian testis. For details on anatomy, physiology, and species variations, the reader is referred to several previous reviews (Setchell, 1970, 1990; Gunn and Gould, 1975; Free, 1977; Setchell and Brooks, 1988; Veijola, 1992).

II. Functional Anatomy of the Testicular Vasculature

A. Arteries

The organization of the vascular system is related to the position of the testis (Setchell, 1970). The testicular artery originates from the upper part of the abdominal aorta. In animals with scrotal testes, this artery follows a long and tortuous course through the abdomen; in the lower part, it is enveloped by the venous return channels, the pampiniform plexus (Fig. 1). After giving off branches to the caput epididymis, the testicular artery serves as the principal end artery supplying the testis. The corpus and cauda epididymis and the testicular capsule are supplied by branches from the deferential artery.

After penetrating the capsule, the artery encircles the testis before entering the parenchyma, then runs toward the rete and mediastinum (Setchell, 1970; Free, 1977). Additional branching then occurs and a dual capillary network is formed (Kormano, 1967a; Murakami *et al.*, 1989; Fig. 2): the intertubular (running parallel to tubules in the triangular interstitial columns between three adjoining tubules) and the peritubular (running between different interstitial columns and forming a dense network lying close to individual tubules). The peritubular network is, however, less developed in some species, including humans (Suzuki and Nagano, 1986).

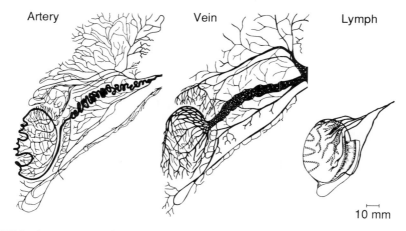

FIGURE 1 Arrangement of arteries, veins, and lymph vessels supplying the rat testis and epididymis. The testicular artery is coiled in the spermatic cord and is an end artery that supplies the entire testicular parenchyma. Two branches to the epididymis originate from the testicular artery proximal to the major testicular coiling. The venous drainage forms the testicular plexus and the pampiniform plexus surrounding the testicular artery. Most of the lymph vessels originate from an area close to the rete testis. Modified with permission from Desjardins (1989) and Perez-Clavier and Harrison (1977).

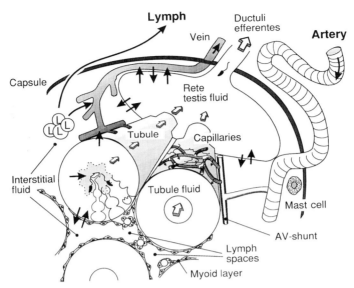

FIGURE 2 Interior of the testes showing the relationship between the rete testis and the testicular vasculature. Groups of Leydig cells (L) are surrounded by the interstitial fluid, which leaves the testis by the lymph vessels. Arrows indicate possible sites of fluid exchange between venous blood flow and the rete testis, the interstitial compartment, and the interior of the tubules and the rete testis fluid. Note also the close relationship between the inflowing artery and the rete testis. Mast cells are located under the testicular capsule, also near the rete testis. (For details and further explanation, see text.)

B. Capillaries and Local Variations

The capillaries, both peritubular and intertubular, are covered by a continuous nonfenestrated endothelium with few vesicles (Fawcett *et al.*, 1970; Mayerhofer and Bartke, 1990). The size of the interendothelial cell clefts has not been studied systematically. Fawcett *et al.* (1970) described 20-nm wide clefts, but their actual size is probably considerably smaller since they do not allow passage of horseradish peroxidase (5 nm; Weihe *et al.*, 1979). Testicular capillaries are impermeable to dyes that rapidly penetrate other vascular beds. The blood–testis barrier originally was suggested to be located in the vessel walls (Kormano, 1967a). For some unknown reason, the endothelial cells have a high number of microvillous processes on their surface (Gabbiani and Majno, 1969).

The anatomical organization of the testis, in which the two principal functions of hormone and sperm production occur in different compartments each of which, at least in part, has its own capillary network (Weerasooriya and Yamamoto, 1985; Murakami *et al.*, 1989) in which flow apparently is controlled independently (Kormano, 1970), clearly suggests local variations within the organ. The intratubular network has been suggested to supply and drain Leydig cells preferentially and the peritubular network has been suggested to vascularize the tubules (Setchell, 1978; Murakami et al., 1989). The presence of 10- to 20-μm thick arteriovenous anastomoses in the interstitial columns (Murakami et al., 1989; see subsequent text) indicates that such shunting (bypassing tubules) takes place and explains why the interstitial oxygen tension is only 63% of that in the subcapsular veins (Free, 1977). The functional role and the regulation of these shunts is unknown, but local control within the organ could be of functional importance. Spermatogenesis is stage dependent. One way to secure local transport demands could be modulating local perfusion and permeability. Indeed, capillary density and filling has been suggested to vary along individual tubules (Ichev and Barkardjiev, 1971), but this possibility has not been studied further.

The resorption of fluid and substances from the rete testis and the efferent ducts region may require local adaptations in the surrounding vasculature. Capillaries in this area are more permeable than in the rest of the testis (Kormano, 1967a). The microvasculature in the capsule is also more permeable than in the rest of the testis (Pöllänen and Setchell, 1989), but the functional role of this difference is unknown. The concept that the microvessels in the capsule and around the rete may be different also is supported by the observation that lymphocyte accumulation starts here in cases of autoimmune orchitis (Itoh et al., 1991).

C. Veins

Numerous postcapillary venules are formed in the peritubular and intratubular regions. They enlarge and run either directly to veins on the testicular surface (Weerasooriya and Yamamoto 1985; Murakami et al., 1989) or, in some species, to a central vein near the mediastinum (Setchell, 1970). The numerous subcapsular veins run toward the upper pole of the organ. In some species (with a superficially located rete), these veins form a plexus over the rete (Fig. 1). The superficial and central veins drain into the pampiniform plexus at the cranial pole of the organ. Note that the incoming and the outflowing blood may pass closely to the rete testis (Free, 1977; Suzuki and Nagano, 1986), and testosterone is known to diffuse rapidly from the overlying venous plexus to the rete testis fluid (Free and Jaffe, 1982). Secretory products from Sertoli cells (e.g., inhibin) and rete testis fluid are resorbed by the venous blood at this site (Maddocks and Sharpe, 1989a). The full functional significance of the close relationship between blood vessels and the rete and the bidirectional exchange of substances remains to be explored (Fig. 2).

A portion of the venous drainage from the testis may reach the ipsilateral part of the prostate gland via veins that follow the ductus deferens. Factors that influence flow and content in the venous blood consequently may influence the accessory sex organs locally (Pierrepoint *et al.*, 1975).

D. Pampiniform Plexus

During evolution, the pampiniform plexus and the scrotum are developed simultaneously (Setchell, 1970). The plexus is assumed to be developed to allow countercurrent exchange of heat between the arterial and venous blood and, in this way, reduce testicular temperature (Setchell, 1970; Waites, 1970). On the other hand, note that the pampiniform plexus is not very effective in protecting the testis from changes in temperature during heat exposure. Its capacity is decreased markedly by moderate reductions in blood flow (Waites, 1970; Sealfon and Zorgniotti, 1991).

Another suggested function for the curious vascular arrangement within the spermatic cord is the transport of different substances from the arteries to the veins. Although this exchange has been shown to take place for a number of marker substances, respiratory gases, vasoactive substances (see subsequent text), and testosterone (Free, 1977), little is known about the physiological rationale for the exchange of such factors (Setchell and Brooks, 1988).

Arteriovenous shunts also may be present in the spermatic cord. At least in some animals, such as sheep and pigs, the existence of such shunts has been shown conclusively (Setchell and Brooks, 1988), but in the rat the issue is more controversial (Damber and Janson, 1978a; Weerasooriya and Yamamoto, 1985; Maddocks and Sharpe, 1989b). Arteriovenous transfer of different substances may be of limited importance to testicular function if the amount of substances transferred is constant. Why would shunts exist if countercurrent change is important for testicular function? On the other hand, if this transfer is variable and is regulated (by nerves or hormones?) according to different physiological demands, it could have major impact on testicular function. However, whether this is the case has not been studied.

E. Lymph Vessels

The organization of testicular lymph vessels is related closely to the general morphology of the interstitial tissue, which is quite variable among species (Fawcett *et al.*, 1973). The lymph vessels are either sinusoids (rodents) or discrete lymph capillaries (ram, boar, man). The lymphatic endothelium contains gaps (Holstein *et al.*, 1979), allowing passage of macromolecules from the interstitial fluid. The lymph is drained into a subalbugineal plexus which is drained via lymph trunks along the seminiferous cord. Lymph from the lower part of the testis is drained toward the epididymis (Perez-Clavier and Harrison, 1977; Fig. 1). The lymph flows through the renal or iliac lymph nodes, but some lymph flows directly into the cisterna chyli (Kazeem, 1991).

When radioactive albumin is injected into the testis, it is mainly transported from the testis via the lymph and not via the venous blood. Radioactivity accumulates in the ipsilateral epididymis and prostate (Setchell, 1990). Therefore, changes in lymph flow and composition may influence the accessory sex organs locally.

Ligation of the lymph trunks in the spermatic cord results in interstitial edema, marked degeneration of the seminiferous epithelium, and a reduced testosterone content in the testicular vein, but blood flow apparently is unaffected (Kotani *et al.*, 1974; Perez-Clavier and Harrison, 1978). The mechanisms behind these changes are unknown.

F. Vascular Innervation

In most vascular beds, one important part of the control system is innervation, but the role of nerves in testicular vascular control is largely unknown. Adrenergic nerves can be found in the spermatic cord and along interstitial blood vessels, particularly in and near the capsule (Hodson, 1970; Setchell and Brooks, 1988) and around the rete testis (Bell and McLean, 1973). The vascular tone in the testis is under nervous control; sympathetic nerve stimulation induces testicular vasoconstriction (Setchell, 1970; Free, 1977). On the other hand, blocking of α receptors has little effect on flow, suggesting a low resting sympathetic tone in testicular vessels (Setchell, 1970). The testis also receives cholinergic fibers (Bell and McLean, 1973; Langford and Silver, 1974) particularly in the capsular region. Peptidergic nerves containing neuropeptide Y (NPY), substance P, and vasoactive intestitinal peptide (VIP) supply the capsule and, to some extent, also supply the intratesticular blood vessels (Larson, 1977; Alm *et al.*, 1978,1980; Allen *et al.*, 1989). The testis also is innervated by serotonergic nerves (Campos *et al.*, 1990). The regional variation within the testis, with more dense innervation in the capsular region and around the rete, suggests possible local regulatory roles.

III. Testicular Blood Flow

Testicular blood flow is the main pathway for the transport of nutrients, regulatory hormones, and secretory products to and from the testis. As in all other organs, control of blood flow is therefore important, and may be particularly critical for the testis since the concentration of oxygen in the seminiferous tubules is very low. The tubules are actually on the brink of hypoxia (Setchell, 1970,1978) and Leydig cell function is critically dependent on oxygen (Payne *et al.*, 1985). The most effective ways to modulate the total amount of blood flowing to the testis are changing the diameter of the artery outside the testis or influencing arteriovenous shunting in the spermatic cord (as discussed earlier). The observations of local reactive hyperemia (Free, 1977), the apparent autoregulation of capillary pressure in the testis (Sweeney *et al.*, 1991), and the local control of vasomotion (see subsequent text) suggest local controls inside the testis. On the other hand, blood flow to the testis does not vary with temperature (Setchell, 1970; Damber and Janson, 1978b) unless the heating is unphysiological. Testosterone secretion is correlated positively to blood pressure (Damber and Janson, 1978a), suggesting a lack of autoregulation in the testicular vascular bed.

Is it possible to modulate testicular function directly by changing blood flow? Two principal arguments suggest that this may be the case.

1. In testes with damage to the seminiferous tubules, for example, abdominally located or irradiated testes (Damber *et al.*, 1985; Setchell, 1986,1990), testosterone secretion is reduced. The reduced testosterone secretion appears to be related primarily to a subnormal blood flow and not to Leydig cell dysfunction. The reason flow is reduced is not known (see subsequent text). However, disturbances in blood flow are important for the understanding of testicular endocrine pathophysiology.

2. A strong positive correlation exists between testicular blood flow and the output of testosterone into the spermatic vein (Damber and Janson, 1978a). Consequently, testosterone secretion may be influenced rapidly by changes in blood flow (provided that arterial blood contains a constant concentration of luteinizing hormone (LH) and other factors that directly influence steroidogenesis in the Leydig cells). Several factors are known to influence total testicular blood flow. Those that are considered important are discussed here and listed in Table 1.

A. Catecholamines and Acetylcholine

Adrenaline and noradrenaline, in large doses, cause a rapid and pronounced testicular vasoconstriction, but this finding is inconsistent when lower doses are used (Free, 1977; Damber and Janson, 1978c). A low dose of adrenaline has been shown to induce vasodilation (Free, 1977). The fact that extensive treatment with catecholamines is necessary to induce vasoconstriction makes the relationship to testicular function uncertain, although microcirculatory events other than total testicular blood flow, for example, vasomotion (see subsequent text), may be influenced by catecholamines. A vasodilatory drug such as acetylcholine is without marked effect on testicular blood vessels (Setchell, 1970; Free, 1977; Noordhuizen-Stassen *et al.*, 1983).

TABLE 1   Factors Known to Influence Total Testicular Blood Flow

Effects within minutes	Effects within hours
Factors increasing flow LH? adenosine kallikrein	Factors increasing flow LH/hCG LHRH testosterone AVP
Factors decreasing flow serotonin histamine epinephrine norepinephrine prostaglandins AVP	Factors decreasing flow LH/hCG LHRH

B. Serotonin, Histamine, Prostaglandins, Kallikrein, and Adenosine

Large doses of serotonin, histamine, and substance 48/80 may reduce testicular blood flow and cause seminiferous tubule damage (Free, 1977; Nemetallah and Ellis, 1985a). However we found that histamine does not influence testicular blood flow (unpublished observations). Injection of a low dose of serotonin into a testicular vein reduces arterial blood flow (Free, 1977). Serotonin produced by testicular mast cells (Lombard-des Couttes *et al.*, 1974) may leave the testis via the venous blood and, by countercurrent exchange in the plexus, may be involved in the physiological control of testicular blood flow (Free, 1977). Interestingly, the plasma concentration of serotonin is reduced acutely by testosterone, human chorionic gonadotropin (hCG), and LH releasing hormone (LHRH) treatment (Serova *et al.*, 1987).

Local treatment of the rat testes with different prostaglandins reduces blood flow. The decrease in plasma testosterone brought about by prostaglandins (Saksena *et al.*, 1973) probably is mediated by a parallel reduction in testicular blood flow (Free, 1977). Prostaglandins are produced by testicular tissue, particularly in the capsule but also in Leydig cells (Haour *et al.*, 1979; Nemetallah and Ellis, 1985b). These prostaglandins may reach the testicular artery by vein–arterial transfer within the pampiniform plexus (Free, 1977) and thus modulate blood flow. The physiological significance of this event, however, is unknown.

Kallikrein (Saito and Kumamoto, 1988) and adenosine (Fleet *et al.*, 1982) increase testicular blood flow but the functional significance of this observation is unknown.

C. Luteinizing Hormone/Human Chorionic Gonadotrophin

Treatment of adult rats with hCG initially results in a decrease in testicular blood flow that occurs 2–6 hr after treatment (Wang *et al.*, 1984; van Vliet *et al.*, 1988), but thereafter is followed by an increase at 16–24 hr after treatment (Damber *et al.*, 1981; Setchell and Sharpe, 1981). A similar pattern is observed after LH treatment (Fig. 3), although the time–response curve is somewhat different (Widmark *et al.*, 1989). Hypophysectomy also has been shown to decrease testicular blood flow (Daehlin *et al.*, 1985; Setchell 1990). Administration of hCG can restore blood flow in such animals (Daehlin *et al.*, 1985). LH also may have acute effects on the testicular vasculature, since a slight decrease of testicular vascular resistance has been shown to occur after a 20-min infusion of LH (Damber and Janson, 1978c), but no change in testicular blood flow follows endogenous LH pulses (Setchell 1986). A high dose of hCG results in focal necrosis of the seminiferous tubules in some rats, possibly because of the initial decrease in blood flow (van Vliet *et al.*, 1988).

One problem in interpreting LH/hCG-induced blood flow changes is generated by the relatively high doses needed to induce a response. The relationship to physiology is, therefore, somewhat obscure. Until blood flow

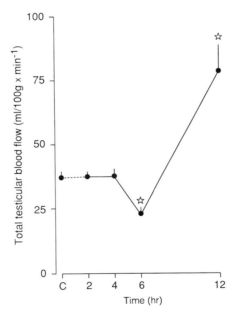

FIGURE 3   Total blood flow of rat testis as measured with radioactive microspheres at different times after subcutaneous injection of 25 μg LH. A significant decrease of blood flow occurred 6 hrs after injection followed by a significant increase of blood flow 12 hrs after treatment. Reprinted with permission from Widmark *et al.* (1989).

changes are shown to be involved in a close time relationship with the metabolic changes in the Leydig cells that are induced by LH stimulation, claiming that the LH/hCG-induced blood flow changes (in contrast to permeability changes; see subsequent discussion) are of any major importance to the regulation of the endocrine response of the testis to LH will be difficult.

D. Luteinizing Hormone Releasing Hormone

LHRH agonists (LHRH-a) constitute another group of substances known to influence testicular blood flow. Treatment of hypophysectomized rats with LHRH-a increases blood flow within 2–4 hr (Damber et al., 1984). Further, intratesticular LHRH-a treatment in a low dose that does not affect endogenous LH decreases flow, whereas treatment with larger doses that increase LH induce an increased flow in the injected testis but not in the contralateral control testis (Widmark et al., 1986a). This result indicates that LH and LHRH-a may act synergistically at the testicular level to stimulate the blood flow of the testis.

A large single dose of LHRH to intact rats may result in focal necrosis of the seminiferous tubules (Habenicht and Müller, 1988). Long-term treatment with LHRH results in induction of fenestrae and open interendothelial cell gaps in the testicular microvessels (Mayerhofer and Dube, 1989). This response is similar to the one seen after hCG treatment in a large dose. The tubular necrosis, but not the vascular effect, can be prevented by simultaneous treatment with indomethacin (Habenicht and Müller, 1988).

E. Arginine Vasopressin

Arginine vasopressin (AVP) is produced in the testis and influences testicular steroidogenesis (Kasson et al., 1986). Local injection of AVP in low doses reduces testicular blood flow within minutes; this decrease can be blocked completely by an AVP antagonist (Widmark et al., 1991). This early vascular effect of AVP is likely to be mediated directly via receptors on testicular vessels because vasoconstriction is a well-known effect of AVP in several other vascular beds (Altura and Altura, 1984). However, AVP does not seem crucial to the maintenance of testicular vascular tone, since treatment of rats with an AVP antagonist did not induce any effect on testicular circulation (Widmark et al., 1991).

IV. Testicular Microcirculation

A. Testicular Microvessels — General Aspects of Structure and Function

The testicular microvasculature shows several unique adaptations, both in structure and in function. The vascular permeability to macromolecules is very high (Setchell, 1990). The pressure in testicular capillaries is the lowest among organs studied and only slightly higher than central venous pressure (Sweeney *et al.*, 1991; Fig. 4). The reason for the high precapillary resistance is unknown, but is probably related to the scrotal position of the testis and the long, coiled, and rather narrow artery (Free, 1977). Despite the low pressure gradient, fluid is filtered from the vasculature and the rate of lymph flow is high. How is this possible?

The principal explanation is probably the high vascular permeability, which results in an interstitial protein concentration that is almost similar to that in serum (Setchell, 1978, 1990; Sweeney *et al.*, 1991; Fig. 5). Consequently, nearly no plasma onconic pressure gradient opposes filtration of fluid from the arterial side of the microvessels. A low capillary pressure thus may be sufficient for filtration. Hypothetically, the high vascular permeability to macromolecules could be a necessary adaptation to secure filtration in an organ in which the capillary pressure (because of the peculiar anatomy) is low. Factors that decrease permeability therefore could be deleterious to the organ, but such substances have not been identified.

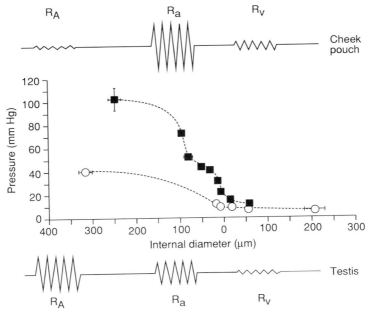

FIGURE 4 Hydrostatic pressure distribution in hamster testis (O) compared to that in hamster cheek pouch (■) and the estimated distribution of hemodynamic resistance in the two vascular beds. Height of each resistance symbol is proportional to present contribution of that segment to total vascular resistance for that tissue. R_A, Arterial resistance; R_a, arteriolar resistance; R_v, venular resistance. Reprinted with permission from Sweeney *et al.* (1991).

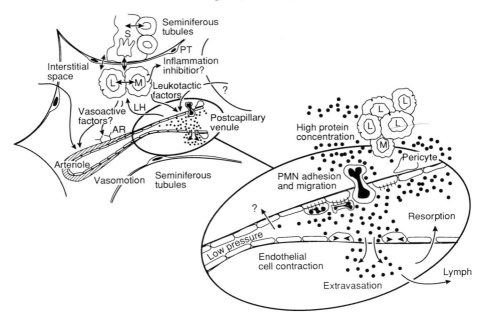

FIGURE 5 Suggested sites for endocrine, paracrine, and autocrine regulation of the testicular microvessels. LH that reaches the testes through the blood induces increased metabolism
of the Leydig cells (L). The Leydig cells — by interacting with the macrophages or Sertoli
cells or by producing vasoactive factors and testosterone — interact with the testicular
microvessels at the precapillary, capillary, and postcapillary levels. At the precapillary
level, control of hydrostatic pressure and vasomotion may occur by interaction with the
vascular androgen receptors (AR). At the capillary and postcapillary levels, permeability
changes take place, possibly involving attraction and penetration of polymorphonuclear
leukocytes. This change in permeability controls the extravasation and resorption of fluid
and macromolecules. (For details, see discussion in the text.)

Alternatively, a high basal vascular permeability could be obligatory for
some particular testicular function. Macromolecules such as albumin serve as
carrier molecules for steroids (Sakiyama *et al.*, 1985). Unrestricted passage
may facilitate testicular secretions. A high permeability for macromolecules
such as lipoproteins and immunoglobulins may serve some unknown role in
testicular function. If the high vascular permeability is the primary factor
then the high precapillary resistance could be a necessary secondary adaptation. Under high permeability conditions, factors that decrease precapillary
resistance probably cause a major transport of fluid from the testicular
vasculature. The factors that control capillary pressure are, however, unknown but this value apparently remains constant over a wide variation in
systemic systolic pressure (Sweeney *et al.*, 1991). Local injection of fluid into
the testis, increasing interstitial pressure, raises the testicular venous pressure

by unknown mechanisms, suggesting a regulatory system (as found in the brain) that aims to stabilize the perfusion through the capillary bed (Free, 1977). Increased venous and, consequently, capillary pressure, as in varicocele (see subsequent text) probably does result in an abnormal fluid filtration out into the interstitial space (Sweeney et al., 1991).

In the near absence of a colloid osmotic pressure gradient between the vasculature and the interstitium, that testicular lymph flow is high is not surprising (Free, 1977; Setchell, 1990). Lymph flow can be increased further by raising venous pressure (Setchell, 1990) or by increasing vascular permeability with LH/hCG (see subsequent text). Understanding why fluid obviously is resorbed to a large extent at the venous side of the microcirculation is more difficult (Fig. 5). Hypothetically, local variations in interstitial pressure caused by contraction in the testicular capsule, the peritubular myoid cells, or the cremaster muscle could be involved (Free, 1977). Indeed slight variations in interstitial pressure in the testis disappear when the capsule is opened (Free, 1977). Another important factor favoring fluid resorption is testicular vasomotion (see subsequent discussion).

B. Precapillary Control of Microcirculation — Vasomotion

When studying testicular microcirculation in the rat with laser Doppler flowmetry (Damber et al., 1982,1983), the blood flow signal is characterized by large rhythmical variations with 7–11 peaks per min. These variations are the result of local variations in erythrocyte velocity (Damber et al., 1986). In a group of capillaries supplied by the same arteriole, blood flow is synchronized; periods of high erythrocyte velocity alternate with periods of very slow or no flow (Fig. 6). These variations in blood flow apparently are related to rhythmical changes in the precapillary sphincters, called vasomotion (Damber et al., 1986). Vasomotion is a general phenomenon and is present in most vascular beds. Vascular resistance is related inversely to vasomotion amplitude. Vasomotion also is involved in the formation of interstitial fluid. During periods with high flow a net filtration of fluid from the vasculature occurs, but during periods with slow blood flow the flow of fluid from the vasculature is reversed (Intaglietta, 1988). This relative change also occurs in the testis, since the relative hematocrit in single microvessels fluctuates in phase with vasomotion (Damber et al., 1986; Fig. 6). Vasomotion is caused by spontaneous pacemaker activity in smooth muscle cells in the walls of arterioles, and probably can be modulated by humoral and neuronal factors (Intaglietta, 1988). However, the mechanisms that control arteriolar vasomotion in general are largely unknown.

In the testis, vasomotion is influenced by hormones and other factors (Table 2). Vasomotion is not present in immature rats but can be induced by hCG treatment (Damber et al., 1990). Vasomotion disappears after hypophysectomy and after Leydig cell depletion, but can be restored by hCG or

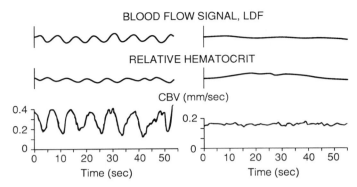

FIGURE 6 Testicular microcirculation as recorded by laser Doppler flowmetry (LDF) and videophotometric capillaroscopy (CBV). Changes in relative hematocrit were registered simulateneously. *(Left)* Control. *(Right)* Blood flow in a rat treated with 200 IU hCG 16–20 hr prior to experiments. Simultaneous analyses of CBV, LDF, and relative hematocrit showed rhythmical oscillations of the three parameters with the same frequency. These rhythmical oscillations in blood flow (CBV, LDF) or relative hematocrit dissappear completely after hCG treatment. Reprinted with permission from Damber *et al.* (1986).

testosterone treatment, respectively (Widmark, 1987; Damber *et al.*, 1992). Some vasoconstrictors such as catecholamines inhibit vasomotion (Damber *et al.*, 1982), but others such as AVP do not influence vasomotion (Widmark *et al.*, 1991).

LH, hCG, or LHRH-a treatment inhibits vasomotion within 2–6 hr of treatment (Damber *et al.*, 1986,1987; Widmark *et al.*, 1986a). The volume of interstitial fluid is increased simultaneously. Inhibition of vasomotion apparently restricts fluid resorption to the vasculature, but the increase in intersti-

TABLE 2   Factors Known to Influence Testicular Vasomotion

Situations in which vasomotion is inhibited
     hypoxia
     hypovolemia
     hCG/LH
     LHRH
     epinephrine
     norepinephrine
     cryptorchidism
     varicocele
Situations in which vasomotion is induced
     sexual maturation
     testosterone
     hCG

tial fluid volume also is caused by simultaneously occurring changes in venular permeability (see subsequent discussion).

No direct relationship exists between the microcirculatory blood flow pattern and total testicular blood flow. Blood flow may be low, normal, or high with and without vasomotion (Widmark, 1987), probably because the overall vascular resistance of the testis is related to a complicated balance between events occurring both at the precapillary (changes in the mean diameters of resistance vessels and changes in vasomotion amplitude) and at the postcapillary side of the vascular bed. The physiological role and control of testicular vasomotion is still largely unknown, but can be postulated to be particularly large in a microcirculation in which filtration and resorption apparently are dependent to such a large extent on changes in microvascular pressure (Setchell, 1990; Sweeney *et al.*, 1991).

C. Postcapillary Control of Vascular Permeability

Endogenous LH pulses or exogenously administered LH doses within minutes result in increased testosterone secretion, indicating an apparently unrestricted vascular passage for molecules of the size of LH (28 kDa; Desjardins, 1989). Injected labeled albumin leaks rapidly out into the interstitial space within minutes (Sweeney *et al.*, 1991). The concentration of albumin and other macromolecules, except for the largest immunoglobulins, is almost identical in plasma and testicular lymph (Beh *et al.*, 1974; Setchell, 1990). All these findings, suggesting a high vascular permeability, are in sharp contradiction with the morphological appearance of the capillary endothelium (see previous text) since endothelia of this type only allow slow passage of macromolecules (Schnitzer, 1992). The explanation for this apparent paradox is likely to be that the size of the interendothelial cell spaces in postcapillary venules is variable and that these spaces may, in response to hormone stimulation, open to allow rapid macromolecular passage. Specific transport mechanisms (Setchell, 1986,1990) also may facilitate transendothelial passage.

LH/hCG and LHRH-a stimulation is able, in a dose-dependent way, to induce endothelial cell contraction and opening of the junctions between endothelial cells in postcapillary venules (Bergh *et al.*, 1987,1990; Veijola, 1992; Fig. 7). To some extent, this process involves the secretion of leukotactic factors that cause polymorphonuclear leukocyte accumulation and migration. In the unstimulated testes, the endothelial junctions apparently are more or less closed, but their exact dimensions have not been studied (Bergh *et al.*, 1990). Some junctions are opened after physiological LH pulses and allow an inflammation-like leakage of both macromolecules and polymorphonuclear leukocytes into the interstitial space (Bergh *et al.*, 1990). However, whether this behavior is the only explanation for the high protein

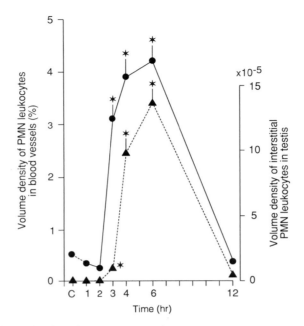

FIGURE 7 Volume density of polymorphonuclear (PMN) leukocytes in blood vessels expressed as percentage of blood vessel volume (●) and volume density of interstitial PMN leukocytes in the interstitial space relative to total testis volume (▲) in control rats and in rats 1–12 hr after subcutaneous treatment with 25 μg LH. * indicates significant change in comparison to control rats. Reprinted with permission from Widmark *et al.* (1989).

content of testicular interstitial fluid in both stimulated and unstimulated testes is unknown. For example, the contribution of the basally more leaky vessels (see previous discussion) in the capsule and rete region is unknown.

A similar hormonal regulation of vascular permeability is observed in the ovary, in which the preovulatory LH peak is able to induce polymorphonuclear leukocyte accumulation and inflammation-like increases in vascular permeability with kinetics and morphological appearance very similar to those in the testis (see Gerdes *et al.*, 1992, for details). The physiological reason that the inflammatory system apparently is used to regulate vascular permeability in the gonads is unknown.

Interendothelial cell junctions also may be opened by mechanisms other than LH or LHRH stimulation. Treatment with cadmium causes a major increase in permeability and eventually a hemorrhagic necrosis of the organ (Gunn and Gould, 1975), but this effect is probably a direct effect on the endothelium (Bergh, 1990). Unilateral torsion of the human testis results in vascular stagnation, platelet accumulation, disappearance of microvilli, and opening of interendothelial cell gaps in the contralateral testis (Chakraborty *et al.*, 1985). The mechanism that mediates these changes is not known.

D. Development and Seasonal Variations

The structure and function of the testicular vasculature changes during development and seasonal changes. The peritubular capillary network is not developed fully until 25–35 days of age in the rat (Kormano, 1967b) and puberty in humans (Free, 1977). Vascular permeability, as measured by the albumin space, shows significant changes during the sexual maturation in rats (Setchell *et al.*, 1988). A marked increase in permeability of the vasculature to IgG occurs at puberty (Pöllänen and Setchell, 1989). Differences exist in the hCG-induced increase in vascular permeability between immature and adult rats, and vasomotion is not established until at 24 days of age (Damber *et al.*, 1990).

The factors that regulate the development of the testicular vasculature during sexual maturation are not known. The increased gonadotropic stimulation of the testis, the maturation of the Leydig cells, and the maturation of the seminiferous tubules — which could signal the development of the peritubular capillary network — are probably all important. Mayerhofer and Bartke (1990) suggested that angiogenic factors might be involved, since the development of the vasculature in the postnatal testis of the golden hamster, and during photoperiod-related seasonal transition from reproductive quiescence to reproductive activity in the adult golden hamster (Mayerhofer *et al.*, 1989), share typical morphological features with the angiogenic process in other organs. Also other species exhibit similar indications of a close relationship between maturational and seasonal changes in the reproductive capacity of the testis and the blood flow (Joffre, 1977). However, whether these changes in the testicular vasculature are primary events and a prerequisite for testicular development or whether the changes only occur as a consequence of varying metabolic demands is not known.

E. Microcirculation and the Local Immune Environment

Testicular germ cells are antigenic. Therefore local immune responses are suppressed to avoid autoimmune orchitis. How this suppression is accomplished has not been established, but immunosuppressive substances are produced locally (Pöllänen *et al.*, 1990). Testicular microvessels probably are involved in establishing this local milieu. Endothelial cells in general, by expressing different adhesion molecules, regulate the traffic of immunocompetent cells into organs and, by contraction, regulate vascular permeability and immunoglobulin passage into the interstitial space (Pober and Cotran, 1990). The observation that vascular permeability is not increased in the testis by common inflammation mediators such as histamine, bradykinin, serotonin, and leukotriene B4 (Bergh and Söder, 1990) suggests local adaptations of endothelial cells that could be necessary to protect the local immune environment.

V. Local Cellular Control of the Testicular Vasculature

Testicular function is, to a large extent, dependent on complicated and largely unexplored paracrine and autocrine regulatory interactions between different testicular cells (Sharpe, 1984,1990; Sharpe *et al.*, 1990). The composition of interstitial fluid, which is a medium for such communication, obviously is influenced by changes in blood flow and permeability. Factors that influence the vasculature locally thus may regulate the uptake of circulating substances and the local paracrine milieu within the organ. The cellular mechanisms that control flow and permeability within the testis are discussed in this section (Fig. 5).

A. Leydig Cells

Several lines of evidence suggest that Leydig cells are involved in the local control of the testicular vasculature. Testosterone is apparently the most important factor. Other Leydig cell products could be involved, but the experimental evidence for such involvement is less convincing.

1. Testosterone and Other Steroids

Total testicular blood flow is reduced markedly in Leydig cell-depleted rats. Vasomotion is inhibited, but flow and vasomotion are normalized within 8–24 hr of treatment with testosterone (Damber *et al.*, 1992). Testosterone, but no other Leydig cell product, apparently is necessary for normal flow. The muscular layer of intratesticular arteries expresses androgen receptors (Bergh and Damber, 1992). Very low levels of testosterone, considerably lower than those needed to normalize spermatogenesis, are needed to normalize vasomotion in Leydig cell-depleted rats (Collin *et al.*, 1993), suggesting that testosterone could have direct vascular effects. Obviously the testosterone effect also could be mediated by the seminiferous tubules (see subsequent text).

Stimulation of Leydig cells with LH/hCG or LHRH alters total testicular blood flow, inhibits vasomotion, and induces an inflammation-like response. These responses are not seen in Leydig cell-depleted animals (Bergh *et al.*, 1988). Whether the precapillary effects of Leydig cell stimulation on flow and vasomotion are related to increased testosterone is unknown. Androgen receptors are not present on testicular capillaries or on postcapillary venules

(Bergh and Damber, 1992). The hCG-induced inflammation-like increase in vascular permeability is not inhibited in testes in which steroidogenesis is blocked (Veijola and Raijaniemi, 1985; Sowerbutts *et al.*, 1986; Widmark *et al.*, 1986b). These observations suggest that the hCG-induced inflammatory response is mediated by Leydig cell products other than steroids. Whether these substances influence vessels directly or indirectly via the tubules, however, is unknown.

2. CRF, ACTH, β-Endorphin, and α-MSH

Leydig cells and testicular macrophages apparently synthesize β-endorphin and adrenal corticotropic hormone (ACTH) (Li *et al.*, 1991). The secretion of β-endorphin is increased by corticotropin releasing factor (CRF) and LH (Fabbri *et al.*, 1990). Interestingly, the hCG-induced increase in vascular permeability is enhanced by naloxone treatment, suggesting that β-endorphin may inhibit this response (Valenca and Negro-Vilar, 1985). This suggestion is in keeping with the general edema-inhibiting effects of β-endorphin (Denko and Gabriel, 1985).

ACTH, α-melanocyte stimulating hormone (α-MSH), and CRF are produced locally in the testis by Leydig cells (Fabbri *et al.*, 1990). In other tissues, locally produced CRF is a potent stimulator of inflammatory responses (Karalis *et al.*, 1991) and both ACTH and α-MSH are inhibitors of inflammation (Mason and van Epps, 1989), but whether these substances have similar effects in the testis is unknown.

3. Testicular Renin–Angiotensin System

Leydig cells probably secrete renin and angiotensin II under acute stimulation by LH/hCG (Okuyama *et al.*, 1988). Angiotensin II is a potent vasoconstrictor in other vascular beds, but its effect in the testis has not been studied.

4. Oxytocin

Oxytocin is synthesized in Leydig cells and probably is involved in the modulation of seminiferous tubule contractility (Nicholson *et al.*, 1991). Oxytocin is a potent vasoconstrictor in most vascular beds (Altura and Altura, 1984). The effect of oxytocin on testicular blood flow apparently has not been studied.

5. Substance P

Leydig cells probably synthesize substance P (Chiwakata *et al.*, 1989). Interestingly, substance P is able to increase vascular permeability and cause polymorphonuclear leukocyte accumulation in several organs, partly by causing mast cell degranulation (see subsequent discussion), but whether substance P influences testicular microcirculation has not been studied.

6. Prostaglandins and Leukotrienes

Leydig cells synthesize both prostaglandins and leukotrienes. Their synthesis apparently is increased by LH/hCG stimulation (Haour et al., 1979; Sullivan and Cooke, 1986). Prostaglandins are known to modulate blood flow in the testis (see previous text). Leukotrienes modulate flow and permeability in various tissues but their effects on testicular blood flow have not been studied. The hCG-induced increase in vascular permeability does, however, occur after inhibition of prostaglandin synthesis (Veijola and Rajaniemi, 1985; Sowerbutts et al., 1986). No increase in permeability in the testis is seen after leukotriene B4 treatment (Stenson et al., 1986), suggesting that the inflammatory response is not caused by these substances.

B. Seminiferous Tubules

Up to 90% of the testicular volume is composed of seminiferous tubules. Therefore, that the tubules could not control blood flow and vascular permeability would be very surprising. Indirect evidence also suggests that such control exists. Under most conditions, but not all, a close correlation is found between testis weight (tubule weight) and testicular blood flow (Setchell, 1990). Selective depletion of elongated spermatids from the testis results in an increase in interstitial fluid volume in the testis. Therefore spermatids were suggested to influence Sertoli cell function and Sertoli cells were suggested to secrete factors that influence the blood vessels (Sharpe et al., 1991). The mechanism by which the tubules influence the vasculature is unknown. Effects caused by testosterone-dependent metabolites (perhaps adenosine; see previous discussion) or secretion of more specific vasoactive factors are both a possibility. Follicle-stimulating hormone (FSH) treatment does not influence testicular blood flow (Bergh et al., 1992a).

1. Testicular LHRH

An LHRH-like peptide probably is produced in Sertoli cells. Via action on LHRH receptors on Leydig cells, this molecule is a possible paracrine regulator of Leydig cell function (Sharpe, 1984). Interestingly, LHRH treatment—by effects not related to increased LH—results in a response, both at the pre- and at the postcapillary parts of the vasculature, similar to that to LH/hCG treatment (see previous discussion). The local synthesis of testicular LHRH is increased by hCG (Sharpe, 1984). Therefore, some of the vascular effects seen after LH/hCG treatment may be caused by local secretion of LHRH from the tubules. The tubules, via LHRH secretion, may decrease or increase testicular blood flow in a dose-related response (see previous text).

Note that LHRH agonist treatment, within minutes, results in vasoconstriction and increased permeability in the hamster cheek pouch (Habenicht and Müller, 1988). Some LHRH antagonists bind to mast cells and cause histamine release (Sundaram *et al.*, 1988). Thus, some of the LHRH effects are not necessarily mediated via changes in Leydig cell function.

2. Arginine Vasopressin

AVP is synthesized locally in the testis, but the regulation and cellular site are largely unknown. Synthesis in the seminiferous tubules and in Leydig cells has been suggested (Pontis and Tahri-Joutei, 1990). AVP secretion from the seminiferous tubules could be one way for the tubules to reduce testicular blood flow (see preceding discussion).

3. Testicular Interleukin 1

Large amounts of an interleukin 1 (IL-1) α-like substance are produced constitutively by Sertoli cells (Söder *et al.*, 1989). Testicular macrophages and Leydig cells also have been suggested to secrete IL-1 (Wang *et al.*, 1991; Guzzardi and Hedger, 1992). In most vascular beds, IL-1 is able to influence blood flow and promote polymorphonuclear leukocyte adhesion, and thus increase vascular permeability in postcapillary venules. However, no continuous inflammatory reaction occurs in the testis under basal conditions, suggesting that the locally produced IL-1 could be blocked by some inhibitory factor. In the skin and the ovary, where large amounts of IL-1 also are produced constitutively, IL-1 activity is blocked by a locally produced IL-1 receptor antagonist (Bigler *et al.*, 1992; Hurwitz *et al.*, 1992).

The role of IL-1 in the control of testicular vasculature is largely unknown. Local injection of IL-1β into the testis (in contrast to several other inflammation mediators) mimics the hCG-induced testicular inflammatory response (Bergh and Söder, 1990), but the effects of IL-1 on blood flow and vasomotion have never been studied. We previously suggested that the testis was insensitive to local injection of human IL-1α (Bergh and Söder, 1990), but this suggestion is an oversimplification since a very large dose is, indeed, able to induce a response (A. Bergh, unpublished results). Most interestingly, the sensitivity to IL-1α and β (but not to other mediators) is increased considerably by hCG pretreatment (A. Bergh, unpublished results).

Collectively, these data suggest that an IL-1 antagonist may be present in the testis and hypothetically is down-regulated by hCG treatment. Further studies are needed to elucidate the role of the testicular IL-1 system in the control of the testicular vasculature.

4. Endothelin 1

Endothelin 1, a highly potent vasoconstrictor, is probably secreted by Sertoli cells (see Conti *et al.*, 1993). The functional significance is, however, unknown.

C. Testicular Mast Cells

Mast cells are present in the pampiniform plexus in some species (Rerkamnuaychoke *et al.*, 1989) and also are observed to be closely related to the subcapsular parts of the testicular artery (Sowerbutts *et al.*, 1986). The significance of mast cells in testicular vascular control is not known, but their anatomical position clearly suggests a role. Moreover, hCG treatment results in mast cell degranulation in the rat testis (Morales and Heyn, 1992). Mast cell products are known to decrease testicular blood flow (see previous discussion). Histamine and serotonin, however, do not modulate vascular permeability in the testis (see previous discussion), but treatment with a serotonin antagonist (Sowerbutts *et al.*, 1986) partly blocks the hCG-induced increase in albumin space and interstitial fluid volume (by inhibiting changes in flow?). Evidently, further studies should be undertaken to elucidate the role of mast cells in testicular vascular control.

D. Testicular Macrophages

The testicular interstitium in rats contains numerous resident macrophages (Miller *et al.*, 1983), the cell size of which is increased by treatment with hCG (Bergh, 1987). Since macrophages in general secrete a number of vasoactive and proinflammatory substances, they have been suggested to be involved in the control of the testicular vasculature, particularly the hCG-induced inflammatory response (Bergh and Söder, 1990). This hypothesis has now been tested. Unilateral macrophage depletion (by local injection of liposome-entrapped dichloromethylene diphosphonate) does not influence spermatogenesis but results in a gradual decrease in Leydig cell testosterone secretion. Vasomotion is normal and the hCG-induced inhibition of vasomotion occurs, suggesting that macrophages are not important for vasomotion (Bergh *et al.*, 1993a). In contrast, the hCG-induced inflammatory response is enhanced in macrophage-depleted testes (Bergh *et al.*, 1993b), suggesting that one role of testicular macrophages could be secreting a factor that inhibits hCG-induced inflammation. The nature of this suggested macrophage product is unknown, but it could be an IL-1 antagonist (see previous discussion).

VI. Testicular Pathophysiology and Microcirculation

Changes in testicular blood flow and microcirculation play an important role in testicular pathophysiology as well. A few examples are discussed in this section.

Varicocele is a common vascular abnormality consisting of dilation of the internal spermatic vein and pampiniform plexus, resulting in increased testicular temperature that is associated with subfertility (Lewis and Harrison, 1982). Total testicular blood flow is decreased in a monkey varicocele model (Harrison *et al.*, 1983). Vasomotion is inhibited in a rat model (Nagler *et al.*, 1987). This inhibition and the likely increase in microvascular pressure (Sweeney *et al.*, 1991) probably cause increased filtration and lymph flow. Although the exact mechanism behind the varicocele-induced impairment of spermatogenesis is unknown, these studies suggest that alteration in testicular microcirculation play a significant role.

Cryptorchidism results in morphological and functional changes in all the major cell types in the testis (Bergh, 1989). In human cryptorchid testes, the blood vessel walls have an increased thickness (Francavilla *et al.*, 1979); similar changes are observed in a rat model (Hjertkvist and Bergh, 1993). Testicular blood flow, vasomotion, and vascular permeability (Damber *et al.*, 1989; Hjertkvist and Bergh, 1993) also are influenced by experimental cryptorchidism. HCG treatment induces a tremendous increase in testicular vascular permeability, leading to a glaucoma-like increase of intratesticular pressure (Hjertkvist *et al.*, 1988). Similar changes are noted also in hCG-treated cryptorchid boys (Hjertkvist *et al.*, 1993). These data suggest that the ability of LH/hCG to increase vascular permeability (and to cause tubule damage; see previous discussion) may be of importance to testicular pathophysiology as well.

Arteriosclerosis often is associated with focal hyalinization and tubular fibrosis (Sasano and Ichijo, 1969; Suoranta, 1971). These changes may be involved in the age-related decline in testicular function.

VII. Concluding Remarks

The uptake of substances and testicular secretions and, consequently, overall testicular function can be influenced markedly by changes in blood flow and permeability. However, the cells and factors involved in the physiological regulation of the vasculature, and how vascular control is integrated into the overall hormonal and paracrine network that regulates testicular physiology, still remain largely unexplored.

Acknowledgments

This work was supported by grants from the Swedish Medical Research Council (Project 5935) and the Maud and Birger Gustavsson Foundation.

462 Anders Bergh & Jan-Erik Damber

References

Allen, L. G., Wilson, F. J., and MacDonald, G. J. (1989). Neuropeptide Y-containing nerves in rat gonads: Sex difference and development. *Biol. Reprod.* 40, 371–378.

Alm, P., Alumets, J., Brodin, E., Håkanson, R., Nilson, G., Sjöberg, N-O., and Sundler, F. (1978). Peptidergic (substance P) nerves in the genito-urinary tract. *Neurosci.* 3, 419–425.

Alm, P., Alumets, J., Håkanson, R., Owman, Ch., Sjöberg, N-O., Sundler, F., and Walles, B. (1980). Origin and distribution of VIP-nerves in the genito-urinary tract. *Cell Tissue Res.* 205, 337–347.

Altura, B. M., and Altura, B. T. (1984). Actions of vasopressin, oxytocin and synthetic analogs on vascular smooth muscle. *Fed. Proc.* 43, 80–86.

Beh, K. J., Watson, D. L., and Lascelles, A. K. (1974). Concentration of immunoglobulins and albumin in lymph collected from various regions of the body of the sheep. *Austr. J. Biol. Exp. Sci.* 52, 81–86.

Bell, C., and McLean, J. R. (1973). The autonomic innervation of the rat testicular capsule. *J. Reprod. Fertil.* 32, 253–258.

Bergh, A. (1987). Treatment with hCG increases the size of Leydig cells and testicular macrophages in unilaterally cryptorchid rats. *Int. J. Androl.* 10, 765–772.

Bergh, A. (1989). Experimental models of cryptorchidism. In "The Cryptorchid Testis" (B. Keel and T. Abney, eds.), pp. 15–34. CRC Press, Boca Raton, Florida.

Bergh, A. (1990). The acute vascular effects of cadmium in the testis do not require the presence of Leydig cells. *Toxicology* 63, 183–186.

Bergh, A., and Damber, J.-E. (1992). Immunohistochemical demonstration of androgen receptors on testicular blood vessels. *Int. J. Androl.* 15, 425–434.

Bergh, A., and Söder, O. (1990). Interleukin-1β but not interleukin 1α induces acute inflammation-like changes in the testicular microcirculation of adult rats. *J. Reprod. Immunol.* 17, 155–165.

Bergh, A., Rooth, P., Damber, J., and Widmark, A. (1987). Treatment of rats with hCG induces inflammation-like changes in the testicular microcirculation. *J. Reprod. Fertil.* 79, 135–143.

Bergh, A., Damber, J. E., and Widmark, A. (1988). Hormonal control of testicular blood flow, microcirculation and vascular permeability. In "Molecular and Cellular Endocrinology of the Testis" (B. A. Cooke and R. M. Sharpe, eds.), pp. 123–133. Raven Press, New York.

Bergh, A., Damber, J.-E., and Widmark, A. (1990). A physiological increase in LH may influence vascular permeability in the rat testis. *J. Reprod. Fertil.* 89, 23–31.

Bergh, A., Damber, J.-E., Lieu, L., and Widmark, A. (1992a). Does follicle-stimulating hormone or pregnant mare serum gonadotrophin influence testicular blood flow in rats? *Int. J. Androl.* 15, 365–371.

Bergh, A., Damber, J.-E., and van Rooijen, N. (1993a). Liposome mediated depletion of macrophages—An experimental approach to study the role of testicular macrophages. *J. Endocrinol.* 136, 407–413.

Bergh, A., Damber, J.-E., and van Rooijen, N. (1993b). The hCG-induced inflammation-like response is enhanced in macrophage-depleted rat testes. *J. Endocrinol.* 136, 415–420.

Bigler, C. F., Norris, D. A., Weston, W. L., and Arend, W. P. (1992). Interleukin-1 receptor antagonist production by human keratinocytes. *J. Invest. Dermatol.* 98, 38–44.

Campos, M. B., Vitale, M. L., Calandra, R. S., and Chicchio, S. R. (1990). Serotoninergic innervation of the rat testis. *J. Reprod. Fertil.* 88, 475–479.

Chakraborty, J., Sinhia Hikim, A. P., and Jhunjhunwala, J. S. (1985). Stagnation of blood in the microvasculature of the affected and contralateral testes in men with short-term torsion of the spermatic cord. *J. Androl.* 6, 291–299.

Chiwakata, C., Brackmann, B., Hunt, N., Davidoff, M., Schulze, W., and Ivell, R. (1989).

Tachykinin (substance-P) gene expression in Leydig cells of the human and mouse testis. *Endocrinology* **128**, 2441–2448.

Collin, O., Bergh, A., Damber, J.-E., and Widmark, A. (1993). Control of testicular vasomotion by testosterone and tubular factors. *J. Reprod. Fertil.* **97**, 115–121.

Conte, D., Questino, P., Filio, S., Norido, M., Isodori, A., Romanelli, F. (1993). Endothelin stimulates testosterone secretion by rat Leydig cells. *J. Endocrinol.* **136**, R1–R4.

Daehlin, L., Damber, J.-E., Selstam, G., and Bergman, B. (1985). Effects of human chorionic gonadotrophin, oestradiol and estromustine on testicular blood flow in hypophysectomized rats. *Int. J. Androl.* **8**, 58–68.

Damber, J.-E., and Janson, P.O. (1978a). Testicular blood flow and testosterone concentrations in spermatic venous blood of anaesthetized rats. *J. Reprod. Fertil.* **52**, 265–269.

Damber, J.-E., and Jansson, P. O. (1978b). The influence of scrotal warming on testicular blood flow and indocrine function in the rat. *Acta Physiol. Scand.* **104**, 61–67.

Damber, J.-E., and Janson, P. (1978c). The effects of LH, adrenaline and noradrenaline on testicular blood flow and plasma testosterone concentrations in anaesthetized rats. *Acta Endocr. (Copenh.)* **88**, 390–396.

Damber, J.-E., Selstam, G., and Wang, J. (1981). Inhibitory effect on estradiol-17β on human chorionic gonadotrophin-induced increment of testicular blood flow and plasma testosterone concentration in rats. *Biol. Reprod.* **25**, 555–559.

Damber, J.-E., Lindahl, O., Selstam, G., and Tenland, T. (1982). Testicular blood flow measured with a laser Doppler flowmeter: Acute effects of catecholamines. *Acta Physiol. Scand.* **115**, 209–215.

Damber, J.-E., Lindahl, O., Selstam, G., and Tenland, T. (1983). Rhythmical oscillations in rat testicular microcirculation as recorded by laser Doppler flowmetry. *Acta Physiol. Scand.* **118**, 117–123.

Damber, J.-E., Bergh, A., and Daehlin, L. (1984). Stimulatory effect on an LHRH-agonist on testicular blood flow in hypophysectomized rats. *Int. J. Androl.* **7**, 236–243.

Damber, J.-E., Bergh, A., and Daehlin, L. (1985). Testicular blood flow, vascular permeability, and testosterone production after stimulation of unilaterally cryptorchid adult rats with human chorionic gonadotropin. *Endocrinology* **117**, 1906–1913.

Damber, J.-E., Bergh, A., Fagrell, B., Lindahl, O., and Rooth, P. (1986). Testicular microcirculation in the rat studied by videophotometric capillaroscopy, fluorescence microscopy and laser Doppler flowmetry. *Acta Physiol. Scand.* **126**, 371–376.

Damber, J.-E., Bergh, A., and Widmark, A. (1987). Effect of an LHRH-agonist on testicular microcirculation in hypophysectomized rats. *Int. J. Androl.* **10**, 785–791.

Damber, J.-E., Bergh, A., and Widmark, A. (1989). Effects of hormones on testicular microvasculature. *In* "Perspectives in Andrology" (M. Serio, ed.), Vol. 53, pp. 97–109. Raven Press, New York.

Damber, J.-E., Bergh, A., and Widmark, A. (1990). Age-related differences in testicular microcirculation. *Int. J. Androl.* **13**, 197–206.

Damber, J.-E., Maddocks, S., Widmark, A., and Bergh, A. (1992). Testicular blood flow and vasomotion can be maintained by testosterone in Leydig cell-depleted rats. *Int. J. Androl.* **15**, 385–393.

Denko, C. W., and Gabriel, P. (1985). Effects of peptide hormones in ureate crystal inflammation. *J. Rheumatol.* **12**, 971–975.

Desjardins, C. (1989). The microcirculation of the testis. *Ann. N.Y. Acad. Sci.* **564**, 243–249.

Fabbri, A., Tinajero, J. C., and Dufau, M. L. (1990). Corticotrophin-releasing factor is produced by rat Leydig cells and has a major local antireproductive role in the testis. *Endocrinology* **127**, 1541–1543.

Fawcett, D. W., Leak, L. V., and Heidger, P. M. (1970). Electron microscope observations on the structural components of the blood-testis barrier. *J. Reprod. Fertil. (Suppl.)* **10**, 105–122.

Fawcett, D. W., Neawes, W. B., and Flores, M. M. (1973). Comparative observations on the intertubular lymphatics and the organization of the interstitial tissue of the mammalian testis. *Biol. Reprod.* **9**, 500–532.

Fleet, I. R., Laurie, M. S., Noordhuizen-Stassen, E. N., Setchell, B. P., and Wensing, C. J. G. (1982). The flow of blood from the artery to vein in the spermatic cord of the ram, with some observations on reactive hyperaemia in the testis and the effects of adenosine and noradrenaline. *J. Physiol.* **332**, 44P–45P.

Francavilla, S., Santiemma, V., Francavilla, F., De Martino, C., Santucci, R., and Fabbrini, A. (1979). Ultrastructural changes in the seminiferous tubule wall and intertubular blood vessels in human cryptorchdism. *Arch. Androl* **2**, 21–30.

Free, M. J. (1977). Blood supply to the testis and its role in local exchange and transport of hormones. *In* "The Testis" (A. D. Johnson and W. R. Gomes, eds.), Vol. IV, pp. 39–90. Academic Press, New York.

Free, M. J., and Jaffe, R. A. (1982). Evidence for direct transfer of 1,2-3H-testosterone from testicular veins to the rete testis in rats. *J. Androl.* **3**, 101–107.

Gabbiani, G., and Majno, G. (1969). Endothelial microvilli in the vessels of the rat Gasserian ganglion and testis. *Z. Zellforsch.* **97**, 11–17.

Gerdes, U., Gåfvels, M., Bergh, A., and Cajander, S. (1992). Localized increases in ovarian vascular permeability and leukocyte accumulation after induced ovulation in the rabbit. *J. Reprod. Fertil.* **95**, 539–550.

Gunn, S. A., and Gould, T. C. (1975). Vasculature of the testes and adnexa. *In.* "Handbook of Physiology" (D. W. Hamilton and R. O. Greep, eds.), pp. 219–224. Williams & Wilkins, Baltimore.

Guzzardi, V., and Hedger, M. P. (1992). Comparison of bioactive interleukin-1 secretion and MHC II expression by testicular and peritoneal macrophages. *Proceedings of the 7th European Workshop on Molecular and Cellular Endocrinology of the Testis,* Abstract 109.

Habenicht, U.-F., and Müller, B. (1988). Disturbance of peripheral microcirculation by LHRH-agonists. II. *Andrologia* **20**, 23–32.

Haour, F., Mather, J., Saez, J. M., Kousnetzova, B., and Dray, F. (1979). Role of prostaglandins in Leydig cell stimulation by hCG. *INSERM* **79**, 75–88.

Harrison, R. M., Lewis, R. W., and Roberts, J. A. (1983). Testicular blood flow and fluid dynamics in monkeys with surgically induced varicoceles. *J. Androl.* **4**, 256–265.

Hjertkvist, M., and Bergh, A. (1993). The time-response and magnitude of hCG-induced vascular changes are different in scrotal and abdominal testes. *Int. J. Androl.* **16**, 62–70.

Hjertkvist, M., Bergh, A., and Damber J.-E. (1988). HCG treatment increases intratesticular pressure in the abdominal testis of unilaterally cryptorchid rats. *J. Androl.* **9**, 116–120.

Hjertkvist, M., Läckgren, G., Plöen, L., and Bergh, A. (1993). Does hCG treatment induce inflammation-like changes in undescended testes in cryptorchid boys? *J. Pediatr. Surg.* **28**, 254–258.

Hodson, N. (1970). The nerves of the testis, epididymis, and scrotum. *In* "The Testis" (A. D. Johnson, W. R. Gomes, and N. L. Vandemark, eds.), Vol. I, pp. 47–99. Academic Press, New York.

Holstein, A. F., Orlandini, G. E., and Möller, R. (1979). Distribution and fine structure of the lymphatic system in the human testis. *Cell Tissue Res.* **200**, 15–27.

Hurwitz, A., Loukides, J., Ricciarelli, E., Botero, L., Katz, E., McAllister, J. M., Garcia, J. E., Rohan, R., Adashi, E. Y., and Hernandez, E. R. (1992). Human intraovarian interleukin-1 (IL-1) system: Highly compartmentalized and hormonally dependent regulation of the genes encoding IL-1, its receptor, and its receptor antagonist. *J. Clin. Invest.* **89**, 1746–1754.

Ichev, K., and Barkardjiev, A. (1971). The blood supply of the male sexual gland. *Nauchni Tr. Vissh. Med. Inst. Sofia* **50**, 9–14.

Intaglietta, M. (1988). Arteriolar vasomotion: normal physiological activity or defence mechanism? *Diab. Metab.,* **14,,** 489–494.

Itoh, M., Hiramine, C., and Hojo, K. (1991). A new murine model of autoimmune orchitis induced by immunization with viable syngenic testicular germ cels alone. I. Immunological and histological studies. *Clin. Exp. Immunol.* **83**, 137–142.

Joffre, M. (1977). Relationship between testicular blood flow, testosterone secretion and spermatogenic activity in young and adult wild red foxes *(Vulpes vulpes)*. *J. Reprod. Fertil.* **51**, 35–40.

Karalis, K., Sano, H., Redwine, J., Listwak, S., Wilder, R. L., and Chrousos, G. P. (1991). Autocrine and paracrine inflammatory actions of corticotrophin-releasing hormone in vivo. *Science* **254**, 421–423.

Kasson, B. G., Adashi, E. Y., and Hsueh, A. J. W. (1986). Arginine vasopressin in the testis: An intragonadal peptide control system. *Endocr. Rev.* **7**, 156–168.

Kazeem, A. A. (1991). Species variation in the extrinsic lymphatic drainage of the rodent testis: Its role within the context of an immunologically privileged site. *Lymphology* **24**, 140–144.

Kormano, M. (1967a). An angiographic study of the testicular vasculature in the postnatal rat. *Z. Anat. Entwickl.* **126**, 138–153.

Kormano, M. (1967b). Dye permeability and alkaline phosphatase activity of testicular capillaries in the postnatal rat. *Histochemie* **9**, 327–338.

Kormano, M. (1970). An experimental technique of in vivo high resolution microangiography. *Br. J. Radiol.* **43**, 180–183.

Kotani, M., Seiki, K., and Hattori, M. (1974). Retardation of spermatogeneis and testosterone secretion after ligature of the lymphatic draining of the testes of rabbits. *Endocr. Jap.* **21**, 1–8.

Langford, G. A., and Silver, A. (1974). Histochemical localization of acetylcholinesterase-containing nerve fibres in the testis. *J. Physiol.* **242**, 9P–10P.

Larson, L-I. (1977). Occurrence of nerves containing vasoactive intestinal polypeptide immunoreactivity in the male genital tract. *Life Sci.* **21**, 503–508.

Lewis, R. W., and Harrison, R. M. (1982). Diagnosis and treatment of varicocele. *Clin. Obstet. Gynecol.* **25**, 501–523.

Li, H., Hedger, M. P., Clements, J. A., and Risbridger, G. P. (1991). Localization of immunoreactive β-endrophin and adrenocorticotrophic hormone and pro-opiomelanocortin mRNA to rat testicular interstitial tissue macrophages. *Biol. Reprod.* **45**, 282–289.

Lombard-des Couttes, M. N., Falk, B., Owman, C. H., Rosengren, E., Sjöberg, N. O., and Walles, B. (1974). On the question of content and distribution of amines in the rat testis during development. *Endocrinology* **95**, 1746–1749.

Maddocks, S., and Sharpe, R. M. (1989a). The route of secretion of inhibin from the rat testis. *J. Endocrinol.* **120**, R5–R8.

Maddocks, S., and Sharpe, R. M. (1989b). Dynamics of testosterone secretion by the rat testis: Implications for measurements of intratesticular levels of testosterone. *J. Endocrinol.* **120**, 323–329.

Mason, M. J., and van Epps, D. (1989). Modulation of IL-1, tumor necrosis factor, and C5a-mediated murine neutrophil migration by α-melanocyte-stimulating hormone. *J. Immunol.* **142**, 1646–1651.

Mayerhofer, A., and Bartke, A. (1990). Developing testicular microvasculature in the golden hamster, *Mesocricetus auratus:* A model for angiogenesis under physiological conditions. *Acta Anat.* **139**, 78–85.

Mayerhofer, A., and Dube, D. (1989). Chronic administration of a gonadotrophin-releasing hormone (GnRH) agonist affects testicular microvasculature. *Acta Endocr. (Copenh.)* **120**, 75–80.

Mayerhofer, A., Hikim, A. P. S., Bartke, A., and Russel, L. D. (1989). Changes in the testicular microvasculature during photoperiod-related seasonal transition from reproductive quiescence to reproductive activity in the adult golden hamster. *Anat. Rec.* **224**, 495–507.

Miller, S. C., Bowman, B. M., and Rowland, H. G. (1983). Structure, cytochemistry, endocytic activity and immunoglobulin (Fc) receptors of rat testicular interstitial-tissue macrophages. *Am. J. Anat.* **168**, 1–13.

Morales, B. A., and Heyn, R. (1992). HCG induces degranulation in testicular mast cells. *9th Workshop on Development and Function of the Reproductive Organs.* Ares Serono Symposia, Abstract 40.

Murakami, T., Uno, Y., Ohtsuka, A., and Taguchi, T. (1989). The blood vascular architecture of the rat testis: A scanning electron microscopic study of corrosion casts followed by light microscopy of tissue sections. *Arch. Histol. Cytol.* **52**, 151–172.

Nagler, H. M., Lizza, E. F., House, S. D., Tomachefsky, P., and Lipowsky, H. H. (1987). Testicular hemodynamic changes after surgical creation of a varicocele in the rat: Intravital microscopic observations. *J. Androl.* **8**, 292–298.

Nemetallah, B. R., and Ellis, C. L. (1985a). Ablation of the blood-testis barrier in rats and guinea pigs by 48 80, a histamine releaser and cadmium cloride. *Arch. Androl.* **15**, 41–48.

Nemetallah, B. R., and Ellis, L. C. (1985b). Prostaglandin dehydrogenase activity of rat and rabbit testicular tissues and accessory gland before and after castration. *J. Androl.* **6**, 97–101.

Nicholson, H. D., Guldenar, S. E. F., Boer, G. J., and Pickering, B. T. (1991). Testicular oxytocin: effects of intratesticular oxytocin in the rat. *J. Endocrinol.* **130**, 231–238.

Noordhuizen-Stassen, E. N., Beijer, H. J. M., Charbon, G. A., and Wensing, C. J. G. (1983). The effect of norepinephrine, isoprenaline and acetylcholine on the testicular and epididymal circulation in the pig. *Int. J. Androl.* **6**, 44–56.

Okuyama, A., Nonomura, N., Koh, E., Kondoh, N., Takeyama, M., Nakamura, M., Namiki, M., Fujioka, H., Matsumoto, K., and Matsuda, M. (1988). Induction of renin-angiotensin system in human testis in vivo. *Arch. Androl.* **21**, 29–35.

Payne, A. H., Quinn, P. Q., and Sheela-Rani, C. S. (1985). Regulation of cytochrome p-450 and testosterone production in Leydig cells. *Rec. Progr. Horm. Res.* **41**, 153–172.

Perez-Clavier, R., and Harrison, R. G. (1977). The pattern of lymphatic drainiage of the rat testis. *J. Anat.* **127**, 93–100.

Perez-Clavier, R., and Harrison, R. G. (1978). The effect of interruption of lymphatic drainage from the rat testis. *J. Pathol.* **124**, 219–225.

Pierrepoint, C. G., Davies, P., Millington, D., and John, B. (1975). Evidence that the deferential veins acts as local transport system for androgen in the rat and the dog. *J. Reprod. Fertil.* **43**, 293–303.

Pober, J. S., and Cotran, R. S. (1990). The role of endothelial cells in inflammation. *Transplantation* **50**, 537–544.

Pöllänen, P., and Setchell, B. P. (1989). Microvascular permeability to IgG in the rat testis at puberty. *Int. J. Androl.* **12**, 206–218.

Pöllänen, P., von Euler, M., and Söder, O. (1990). Testicular immunoregulatory factors. *J. Reprod. Immunol.* **18**, 51–76.

Pontis, G., and Tahri-Joutei, A. (1990). Controle intragonadique de la fonction testiculaire par des peptides de type neurohypophysiare. *Ann. Endocr.* **5**, 209–217.

Rerkamnuaychoke, W., Nishida, T., Kurohmaru, M., and Hayashi, Y. (1989). Vascular morphology of the golden hamster spermatic cord. *Arch. Histol. Cytol.* **52**, 183–190.

Saito, S., and Kumamoto, Y. (1988). Effect of kallikrein on testicular blood circulation. *Arch. Androl.* **20**, 51–65.

Sakiyama, R., Pardridge, W. M., and Musto, N. A. (1985). Influx of testosterone-binding blobulin and TeBG-bound sex steroid hormones into rat testis and prostate. *J. Clin. Endocr. Metab.* **67**, 98–103.

Saksena, S. K., Safoury, S. E., and Bartke, A. (1973). Prostaglandins E2 and F2α decrease plasma testosterone levels in the male rat. *Prostaglandins* **4**, 235–242.

Sasano, N., and Ichijo, S. (1969). Vascular pattern of the human testis with special reference to its senile changes. *Tohoku J. Exp. Med.* **99**, 269–280.

Schnitzer, J. E. (1992). gp60 is an albumin-binding glycoprotein expressed by continuous endothelium involved in albumin transcytosis. *Am. J. Physiol.* **262**, H246–H254.

Sealfon, A. I., and Zorgniotti, A. W. (1991). A theoretical model for testis thermoregulation. *In* "Temperature and Environmental Effects on the Testis" (A. W. Zorgniotti, ed.), pp. 123–135. Plenum Press, New York.

Serova, L. I., Kudryavtseva, N. N., Popova, N. K., Naumenko, E. V., and Parvez, S. H. (1987). Hormones of the hypothalamic-pituitary-testicular complex in the peripheral control of serotonin. *Biogen. Amines* **4**, 145–151.

Setchell, B. P. (1970). Testicular blood supply, lymphatic drainage, and secretion of fluid. *In* "The Testis" (A. D. Johnson, W. R. Gomes, and N. L. Vandemark, eds.), Vol. I, pp. 101–239. Academic Press, New York.

Setchell, B. P. (1978). "The Mammalian Testis." Elek Books, London.

Setchell, B. P. (1986). Movement of fluids and substances in the testis. *Aust. J. Biol. Sci.* **339**, 193–207.

Setchell, B. P. (1990). Local controls of testicular fluids. *Reprod. Fertil. Dev.* **2**, 291–309.

Setchell, B. P., and Brooks, D. E. (1988). Anatomy, vasculature, innervation, and fluids of the male reproductive tract. *In* "The Physiology of Reproduction" (E. Knobil, J. Neill, L. L. Ewing, G. S. Greenwald, C. L. Market, and D. W. Pfaff, eds.), pp. 753–836. Raven Press, New York.

Setchell, B. P., and Sharpe, R. M. (1981). The effect of human chorionic gonadotrophin on capillary permeability, extracellular fluid volume and flow of lymph and blood in the testis of rats. *J. Endocrinol.* **91**, 245–254.

Setchell, B. P., Pöllänen, P., and Zupp, J. L. (1988). Development of the blood-testis barrier and changes in vascular permeability at puberty in rats. *Int. J. Androl.* **11**, 225–233.

Sharpe, R. M. (1984). Intratesticular factors controlling testicular function. *Biol. Reprod.* **30**, 29–49.

Sharpe, R. M. (1990). Intratesticular control of steroidogenesis. *Clin. Endocinol.* **33**, 787–807.

Sharpe, R. M., Maddocks, S., and Kerr, J. B. (1990). Cell-cell interactions in the control of spermatogenesis as studied using Leydig cell destruction and testosterone replacement. *Am. J. Anat.* **188**, 3–20.

Sharpe, R. M., Bartlett, J. M. S., and Allenby, G. (1991). Evidence for the control of testicular interstitial fluid volume in the rat by specific germ cell types. *J. Endocrinol.* **128**, 359–367.

Söder, O., Pöllänen, P., Syed, V., Holst, M., Granholm, T., Arver, S., von Euler, M., Gustafsson, K., Fröysa, B., Parvinen, M., and Ritzen, E. M. (1989). Mitogenic factors in the testis. *In* "Perspectives in Andrology" (M. Serio, ed.), pp. 215–226. Raven Press, New York.

Sowerbutts, S. F., Jarvis, L. G., and Setchell, B. P. (1986). The increase in testicular vascular permeability induced by human chorionic gonadotrophin involves 5-hydroxytryptamine and possibly oestrogens but not testosterone, prostaglandins histamine or bradykinin. *Austr. J. Exp. Biol. Med. Sci.* **64**, 137–147.

Sundaram, K., Didolkar, A., Thau, R., Chadhuri, M., and Schmidt, F. (1988). Antagonists of luteinizing hormone releasing hormone bind to rat mast cells and induce histamine release. *Agents Actions* **25**, 307–313.

Suoranta, H. (1971). Changes in the small blood vessels of the adult human testis in relation to age and to some pathological conditions. *Virchows Arch. (A)* **352**, 165–171.

Stenson, W. F., Chang, K., and Williamson, J. R. (1986). Tissue differences in vascular permeability induced by leukotriene B_4 and prostaglandin E_2 in the rat. *Prostaglandins* **32**, 5–17.

Sullivan, M. H., and Cooke, B. A. (1986). The role of Ca2+ in steroidogenesis in Leydig cells. Stimulation of intracellular free Ca^{2+} by lutropin (LH), luliberin (LHRH) agonist and cyclic AMP. *Biochem. J.* **236**, 45–51.

Suzuki, F., and Nagano, T. (1986). Microvasculature of the human testis and excurrent duct system. *Cell Tissue Res.* **243**, 79–89.

Sweeney, T., Rozum, J. S., Desjardins, C., and Gore, R. W. (1991). Microvascular pressure distribution in the hamster testis. *Am. J. Physiol.* **260**, H1581–1589.

Valenca, M. M., and Negro-Vilar, A. (1985). Proopiomelanocortin-derived peptides in testicular interstitial fluid: Characterization and changes in secretion after human chorionic gonadotrophin or luteinizing hormone-releasing hormone analog treatment. *Endocrinology* **118**, 32–37.

van Vliet, J., Rommerts, F. F. G., de Rooij, D. G., Buwalda, G., and Wensing, C. J. G. (1988). Reduction of testicular blood flow and focal degeneration of tissue in the rat after administration of human chorionic gonadotrophin. *J. Endocrinol.* **117**, 51–57.

Veijola, M. (1992). Regulation of vascular permeability in the rat testis by gonadotrophins. *Acta Univ. Ouluensis Ser. D Med.* **241**, 15–62.

Veijola, M., and Raijaniemi, H. (1985). The hCG-induced increase in hormone uptake and interstitial fluid volume in the rat testis is not mediated by steroids, prostaglandins or protein synthesis. *Int. J. Androl.* **8**, 69–79.

Wang, D., Nagpal, M. L., Calkins, J. H., Chang, W., Sigel, M. M., and Lin, T. (1991). Interleukin-1β induces interleukin-1α messenger ribonucleic acid expression in primary cultures of Leydig cells. *Endocrinology* **129**, 2862–2866.

Wang, J. M., Gu, C. H., Qian, Z. M., and Jing, E. W. (1984). Effect of gossypol on testicular blood flow and testosterone production in rats. *J. Reprod. Fertil.* **71**, 127–133.

Waites, G. M. H. (1970). Temperature regulation and the testis. *In* "The Testis" (A. D. Johnson, W. R. Gomes, and N. L. Vandemark, eds.), Vol. I, pp. 241–279. Academic Press, New York.

Weerasooriya, T. R., and Yamamoto, T. (1985). Three-dimensional organization of the vasculature of the rat spermatic cord and testis. A scanning electron-microscopic study of vascular corrosion casts. *Cell Tissue Res.* **241**, 317–323.

Weihe, E., Nimmrich, H., Metz, J., and Forssman, W. G. (1979). Horseradish peroxidase as a marker for capillary permeability studies. *J. Histochem. Cytochem.* **27**, 1357–1359.

Widmark, A. (1987). Testicular microcirculation: An experimental study in the rat. Ph. D. Thesis. University of Umeå, Sweden.

Widmark, A., Damber, J.-E., and Bergh, A. (1986a). Testicular vascular resistance in the rat after intratesticular injection of an LRH-agonist. *Int. J. Androl.* **9**, 416–423.

Widmark, A., Damber, J.-E., and Bergh, A. (1986b). The relationship between human chorionic gonadotropin-induced changes in testicular microcirculation and the formation of testicular interstitial fluid. *J. Endocrinol.* **109**, 419–425.

Widmark, A., Damber, J.-E., and Bergh, A. (1989). High and low doses of luteinizing hormone induce different changes in testicular microcirculation. *Acta Endocrinol.* **121**, 621–627.

Widmark, A., Damber, J.-E., and Bergh, A. (1991). Arginine–vasopressin induced changes in testicular blood flow. *Int. J. Androl.* **14**, 58–65.

Index